The Blue Marble—Land Surfaces

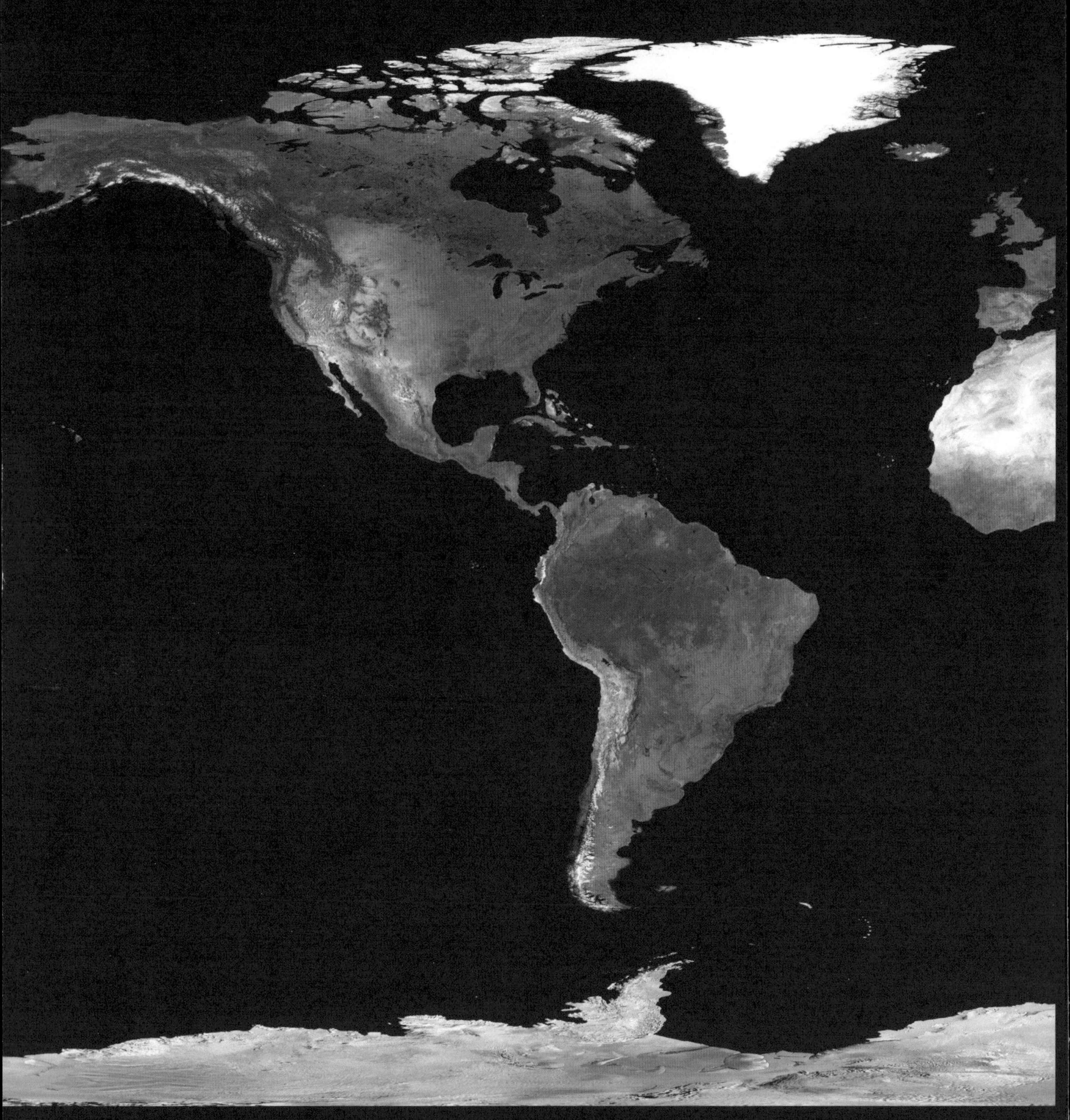

The "blue marble" image is the most detailed true-color image of Earth to date. Using a collection of satellite-based observations, scientists and visual designers stitched together months of observations of the land surface, oceans, sea ice, and

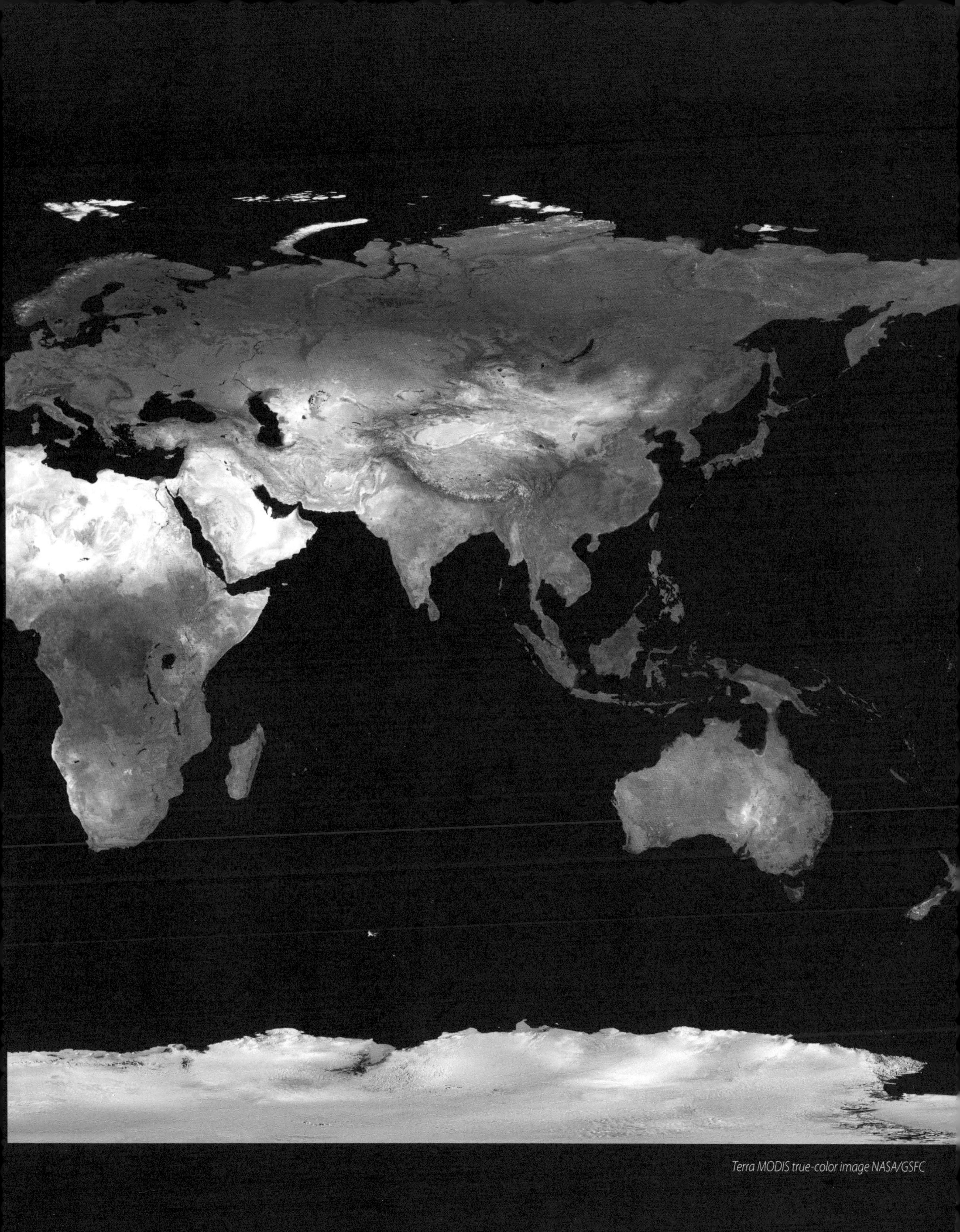

Terra MODIS true-color image NASA/GSFC

Introduction to Contemporary Geography

James M. Rubenstein
MIAMI UNIVERSITY, OXFORD, OHIO

William H. Renwick
MIAMI UNIVERSITY, OXFORD, OHIO

Carl T. Dahlman
MIAMI UNIVERSITY, OXFORD, OHIO

PEARSON

Boston Columbus Indianapolis New York San Francisco Upper Saddle River Amsterdam Cape Town Dubai London Madrid Milan Munich Paris Montreal Toronto Delhi Mexico City Sao Paulo Sydney Hong Kong Seoul Singapore Taipei Tokyo

Pearson
Geography Editor: Christian Botting
Marketing Manager: Maureen McLaughlin
Project Editor: Anton Yakovlev
VP/Executive Director, Development: Carol Trueheart
Development Editors: Jonathan Cheney and Melissa Parkin
Media Producer: Ziki Dekel
Assistant Editor: Kristen Sanchez
Editorial Assistant: Bethany Sexton
Marketing Assistant: Nicola Houston
Managing Editor, Geosciences and Chemistry: Gina M. Cheselka
Project Manager, Production: Maureen Pancza
Project Manager, Full Service: Cindy Miller
Compositor: Element Thomson North America
Senior Technical Art Specialist: Connie Long
Image Lead: Maya Melenchuk
Photo Researcher: Stefanie Ramsay, Bill Smith Group
Operations Specialist: Michael Penne
Cover Photos (front and back): Niterói Museum of Contemporary Art, Rio de Janeiro

Dorling Kindersley
Design Development, Page Design, and Layouts: Stuart Jackman
Page Layout: Anthony Limerick
Cover Design: Stuart Jackman

Credits and acknowledgments borrowed from other sources and reproduced, with permission, in this textbook appear on the appropriate page within the text.

CIP data available upon request.

KV 10.08.2018 0803

ISBN-10: 0-321-80319-1; ISBN-13: 978-0-321-80319-1 [Student Edition]
ISBN-10: 0-321-81984-5; ISBN-13: 978-0-321-81984-0 [Instructor's Review Copy]

Brief Contents

Contents

1 THINKING GEOGRAPHICALLY

2 WEATHER, CLIMATE, AND CLIMATE CHANGE

3 **LANDFORMS**

4 **BIOSPHERE**

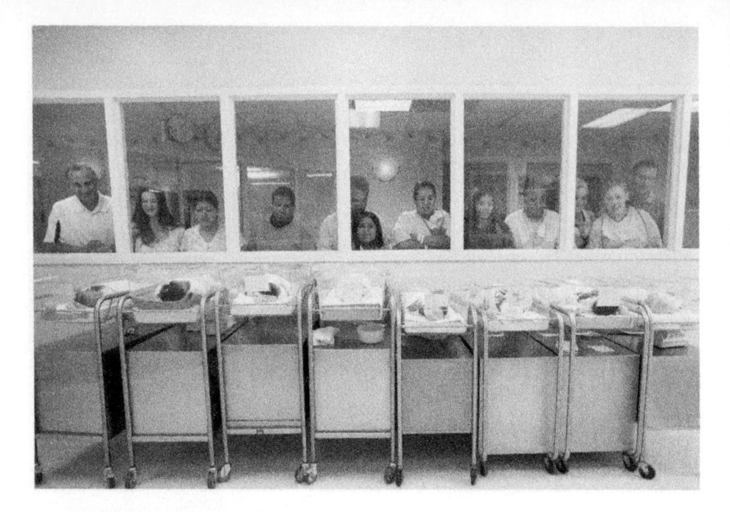

5 **POPULATION**

6 **MIGRATION**

7 LANGUAGES AND RELIGIONS

8 POLITICAL GEOGRAPHY

9 **DEVELOPMENT**

How does development vary among regions?

How can countries promote development?

What are future challenges for development?

10 **FOOD AND AGRICULTURE**

What do people eat?

How is agriculture distributed?

What challenges does agriculture face?

11 **INDUSTRY**

12 **SERVICES AND SETTLEMENTS**

13 URBAN PATTERNS

14 RESOURCES

Preface

Welcome to a new kind of geography textbook! We live in a visual age, and geography is a highly visual discipline, so Pearson—the world's leading publisher of geography textbooks—invites you to study geography as a visual subject.

GEOGRAPHY IS CONTEMPORARY

The main purpose of this book is to introduce you to the study of geography as a science by emphasizing the relevance of geographic concepts to human problems. It is intended for use in college-level introduction to geography courses. The book is written for students who have not previously taken a college-level geography course.

A central theme in this book is a tension between two important realities of the twenty-first-century world—globalization and cultural diversity. In many respects we are living in a more unified world economically, culturally, and environmentally. The actions of a particular corporation or country affect people around the world. This book argues that, after a period when globalization of the economy, culture, and environment has been a paramount concern in geographic analysis, local diversity now demands equal time. Every place on Earth displays a unique combination of climate, landforms, and vegetation. People are taking deliberate steps to retain distinctive cultural identities. They are preserving little-used languages, fighting fiercely to protect their religions, and carving out distinctive economic roles.

Recent world events lend a sense of urgency to geographic inquiry. More than a decade into the twenty-first century, we continue to face wars in unfamiliar places and experience economic struggles unprecedented in the lifetimes of students or teachers. Geography's spatial perspectives help to relate economic change to the distributions of cultural features such as languages and religions, demographic patterns such as population growth and migration, and natural resources such as energy, water quality, and food supply.

For example, geographers examine the prospects for water crises by relating the distributions of water resources and consumption. Geographers find that the major water demand is located in places with different social, economic, and political institutions than the regions where precipitation and runoff are ample. Geographers seek first to describe the distribution of water uses such as irrigated agriculture, and then to explain the relationships between these distributions and other human and physical phenomena.

CHAPTER ORGANIZATION AND FEATURES

Each chapter is organized into between 11 and 16 two-page modular "spreads" that follow a consistent pattern:

- **Introductory module.** The first spread includes a short introduction to the chapter, as well as an outline of between 9 and 13 issues that will be addressed in the chapter. The key issues are grouped into several overarching Key Questions for that chapter.
- **Quick Response (QR) Codes.** Each introductory module includes a QR code, enabling students on the go to link smartphones from the book to various websites relevant to each chapter, providing easy and immediate access to up-to-date information, data, and statistics from sources such as the United Nations and the U.S. Geological Survey.
- **Topic modules.** Between 9 and 13 modules cover the principal topics of the chapter. Each of these two-page spreads is self-contained and organized around Key Questions and Main Points, making it easier for an instructor to shuffle the order of presentation. A numbering

system also facilitates finding material on a particular spread. Module features include:

- **Key Questions and Main Points.** Each chapter module is framed by conceptual "Key Questions," which ask students to take a bigger picture approach to the concept, and by two "Main Points," which students should understand after studying the module. These Key Questions and Main Points serve as an outline for the topics covered in the chapter, and are revisited and summarized in the end-of-chapter Review.
- **Integration of Photos and Text.** The best possible images have been carefully chosen to complement and expand upon content and concepts.
- **The latest science, statistics, and associated imagery.** Data sources include the 2010 U.S. Census and the 2011 Population Reference Bureau World Population Data. Material includes coverage of recent political conflicts, economic difficulties, and the Japan earthquake and tsunami.
- **MapMaster™ Interactive Maps.** These layered thematic interactive maps act as a mini-GIS, allowing students to layer different data at global and regional scales to examine the resulting spatial patterns and practice critical thinking. The interactive maps, with associated tasks and questions, are integrated into select chapter modules and into all end-of-chapter review modules, encouraging students to login to MasteringGeography™ to access these exciting geospatial media to practice visual analysis and critical thinking.
- **Google Earth™ Explorations.** Images integrated into select chapter modules and at the end of the chapter pose questions to be answered through Google Earth, the recognized leader in desktop geospatial imagery.

- **Chapter Review module.** Following the topic modules are concluding spreads that review the chapter's main concepts and key terms while providing students with opportunities to interact with media and engage in critical thinking. The Chapter Review module features include:
  - **Key Questions.** The Key Questions presented on the introductory spread are repeated, along with an outline summary of Main Points made in the chapter that address the questions.
  - **On the Internet.** URLs are listed for several useful Internet sites related to the themes of the chapter.
  - **Thinking Geographically.** These critical-thinking questions give students a chance to practice higher-order thinking.
  - **Interactive Mapping.** Using Pearson's MapMaster interactive mapping media, available in MasteringGeography, students create custom layered maps and answer questions about spatial relationships of different data. Teachers have the option of assigning these questions through MasteringGeography.
  - **Explore.** Using Google Earth students inspect imagery from places around the world and answer questions based on their observations. Teachers, have the option of assigning these questions in MasteringGeography.
  - **Key Terms.** The key terms in each chapter are indicated in bold type when they are introduced. These terms are defined both at the end of the chapter and at the end of the book.
  - **Looking Ahead.** Each chapter concludes with a brief preview of the next chapter and highlights connections between chapters.
- The new **MasteringGeography™** platform is linked to the Key Questions and Main Points and contains a wide range of assignable and self study resources and activities designed to reinforce basic concepts in geography, including MapMaster™ interactive maps, Geoscience Animations, Google Earth™ activities, geography videos, and more.

HOW TO USE THIS BOOK'S MEDIA

Introduction to Contemporary Geography features an innovative integration of media and connections to the MasteringGeography platform, giving students *and* instructors flexible self-study and assessment options to extend the book with current data, interactive mapping, process visualization, and exciting geospatial tools.

- **Quick Response Codes.** Traditional books are challenged to provide students with quick and easy access to original sources and up-to-date data. *Quick Response codes* integrated into the beginning of each chapter help solve this problem, enabling students to use their mobile devices to easily and instantly access websites with current data and information related to chapter topics.
- **MapMaster™ Interactive Maps.** Maps are an important part of the geographer's toolset, but traditional print maps are limited in their ability to allow students to dynamically isolate or compare different spatial data. Available in MasteringGeography both for student self-study and for teachers as assignable and automatically-gradable assessment activities, *MapMaster Interactive Maps* act as mini-GIS tools, allowing students to overlay, isolate, and examine different thematic data at regional and global scales.

  Select chapter modules and all chapter review modules from the book present MapMaster maps, along with activities and questions, encouraging students to login to the MasteringGeography Study Area on their own to explore additional map data layers to complete the activities and extend their learning beyond the book's maps.

  Teachers have the option of assigning these short answer questions for credit in MasteringGeography. Teachers also have access to a separate large suite of MapMaster activities for each chapter, including hundreds of multiple-choice questions that can be customized, assigned, and automatically graded by the MasteringGeography system, for a wide range of interactive mapping assessment activity options.
- **Google Earth™.** Geobrowser technology provides unparalleled opportunity for students to get a sense of place and explore Earth's physical and cultural landscapes with mashups of various data and digital media.

  Select chapter modules and all chapter review modules present *Google Earth* imagery and activities, encouraging students to connect the print book to this exciting tool to browse the globe and explore different data, perform visual and spatial analysis tasks, and extend their learning beyond the book's photos and figures.

  Teachers have the option of assigning these short answer questions for credit, and also have access to a separate large suite of Google Earth *Encounter* activities for each chapter, including hundreds of associated multiple-choice questions that can be customized, assigned, and automatically graded by the MasteringGeography system.

  For classes that do not use MasteringGeography, the Google Earth *Encounter* activities are also available via a set of standalone workbooks and websites (see the Teaching and Learning Package section of this Preface for more information).
- **Geoscience Animations.** Static 2-D print figures do not always present a convenient way to visualize complicated physical processes that occur over vast expanses of space and time. Available in MasteringGeography both for student self study and as assignable and automatically-gradable assessment activities, *Geoscience Animations* provide students with dynamic visualizations of the most complex physical processes, with voiceover narrative and text transcripts to help guide them through the animations. Icons for the animations are integrated into select chapter modules, encouraging students to login to the Study Area of MasteringGeography to access the media on their own, while teachers have the option of assigning the animations with automatically graded questions.

OUTLINE OF TOPICS

This book discusses the following main topics:

What basic concepts do geographers use?

Geographers employ several concepts to describe the distribution of people and activities across Earth, to explain reasons underlying the observed distribution, and to understand the significance of the arrangements. Chapter 1 provides an introduction to ways that geographers think about the world.

What are Earth's key physical processes?

Chapters 2 through 4 offer an overview of Earth's physical environment. The discussion emphasizes processes operating in the landscape, such as atmospheric circulation, landform change, and ecosystem dynamics.

Where are people located in the world?

Why do some places on Earth contain large numbers of people or attract newcomers whereas other places are sparsely inhabited? Chapters 5 and 6 examine the distribution and growth of the world's population, as well as the movement of people from one place to another.

How are different cultural groups distributed?

Chapters 7 and 8 analyze the distribution of different cultural traits and beliefs and the political challenges that result from those spatial patterns. Chapter 7 examines two key elements of cultural identity: language and religion. Chapter 8 looks at political problems that arise from cultural diversity. Geographers look for similarities and differences in the cultural features at different places, the reasons for their distribution, and the importance of these differences for world peace.

How do people earn a living in different parts of the world?

Human survival depends on acquiring an adequate food supply. One of the most significant distinctions among people globally is whether they produce their food directly from the land or buy it with money earned by performing other types of work. Chapters 9 through 12 look at the three main ways of earning a living: agriculture, manufacturing, and services. Chapter 13 discusses cities, the centers for economic and cultural activities.

What issues result from using Earth's resources?

The final chapter is devoted to a study of issues related to the use of Earth's natural resources, a topic especially well suited to integrating elements of physical geography and human geography. Geographers recognize that cultural problems result from the depletion, destruction, and inefficient use of the world's natural resources.

THE TEAM

At this point in the preface, an author usually goes through the motions of perfunctorily thanking many people who performed jobs that resulted in the book's production. In this case, collaborative partnership is a better way to describe the process. Let's face it, some textbooks have been slow to adapt to our visual age. This is because the steps involved in producing most textbooks haven't changed much. The book passes from one to another like a baton in a relay race; those responsible for producing a book's graphics typically start their work only after the author's words have been written, reviewed, and approved.

In contrast, this book started as a genuine partnership among the key editorial and production teams. Each two-page module was assembled in the reverse order of traditional textbooks. Instead of beginning with an author's complete manuscript, this book started with a sketch of a visual concept for each two-page module in the book. What would be the most important geographic idea presented on the spread, and what would be the most effective visual way to portray that idea? The maps and images were placed on the page first, and then the text was written around the graphics.

The traditional separation of editorial and production personnel did not occur, and in fact the lines between the two were deliberately blurred. Key members of the team included Stuart Jackman, Christian Botting, Anton Yakovlev, Melissa Parkin, Jonathan Cheney, and Cindy Miller.

Stuart Jackman, Design Director at DK Education, is the creative genius responsible for the spectacular graphics. Stuart and the DK team deserve the lion's share of the credit for giving this book the best graphics in geography.

Christian Botting, Geography Editor at Pearson Education, led the team with both the big picture and the reality checks. Christian mastered the many challenges posed by our untraditional workflow.

Anton Yakovlev, Geography Project Manager at Pearson Education, was the ringmaster. Anton kept track of what was where and who was doing what, and joined in the many discussions on design elements.

Senior Development Editor Melissa Parkin and Executive Development Editor Jonathan Cheney contributed substantially to this complex and creative project.

Cindy Miller, Vice President of Higher Education at the Element division of Thomson Digital, led the unusually complex task of managing the flow of copyediting and other production tasks for this project.

Many others have contributed to the success of this project. At DK, Anthony Limerick, Senior Designer, helped create the book's distinctive layouts. Sophie Mitchell, Publisher at DK Education, provided the strategic vision for the design team. At Spatial Graphics, Kevin Lear and Andy Green developed the maps and line art throughout the book.

At Pearson Education, Managing Editor Gina Cheselka and Production Project Manager Maureen Pancza were a major organizing force in the nonstandard production workflow. Editorial Assistant Bethany Sexton organized the substantial reviewing process for the project. Marketing Manager Maureen McLaughlin expertly created the marketing package for this unique book. Media Producer Ziki Dekel managed the production of the MasteringGeography program and senior Media Producer Angela Bernhardt managed the production of MapMaster interactive maps.

REVIEWERS

We would like to extend a special thanks to our colleagues who served as reviewers on this new textbook and the overlapping material from *Contemporary Human Geography*:

Roger Balm, *Rutgers University*
Joby Bass, *University of Southern Mississippi*
Steve Bass, *Mesa Community College*
David C. Burton, *Southmoore High School*
Michelle Calvarese, *California State University, Fresno*
Craig S. Campbell, *Youngstown State University*
Edward Carr, *University of South Carolina*
Carolyn Coulter, *Atlantic Cape Community College*
Stephen Davis, *University of Illinois*, Chicago
Owen Dwyer, *Indiana University-Purdue University*, Indianapolis
Leslie Edwards, *Georgia State University*
Caitie Finlayson, *University of Florida*
Barbara E. Fredrich, *San Diego State University*
Piper Gaubatz, *University of Massachusetts*, Amherst
Daniel Hammel, *University of Toledo*
James Harris, *Metropolitan State College of Denver*
Leila Harris, *University of Wisconsin*
Susan Hartley, *Lake Superior College*
Marc Healy, *Elgin Community College*
Scot Hoiland, *Butte College*
Wilbur Hugli, *University of West Florida*
Anthony Ijomah, *Harrisburg Area Community College*
Karen Johnson-Webb, *Bowling Green State University*
Oren Katz, *California State University*, Los Angeles
Marti Klein, *Saddleback College*
Olaf Kuhlke, *University of Minnesota*, Duluth
Dean P. Lambert, *San Antonio College*
Peter Landreth, *Westmont High School*
Jose López-Jiménez, *Minnesota State University*, Mankato
Claudia Lowe, *Fullerton College*
Ken Lowrey, *Wright State University*
Jerry Mitchell, *University of South Carolina*
Eric C. Neubauer, *Columbus State Community College*
Ray Oman, *University of the District of Columbia*
Lynn Patterson, *Kennesaw State University*
Tim Scharks, *Green River Community College*
Debra Sharkey, *Cosumnes River College*
Wendy Shaw, *Southern Illinois University*, Edwardsville
Laurel Smith, *University of Oklahoma*
Richard Smith, *Harford Community College*
James Tyner, *Kent State University*
Richard Tyre, *Florida State University*
Daniel Vara, *College Board Advanced Placement Human Geography Consultant*
Anne Will, *Skagit Valley College*
Lei Xu, *California State University*, Fullerton
Daisaku Yamamoto, *Central Michigan University*
Robert C. Ziegenfus, *Kutztown University of Pennsylvania*

About the Authors

James M. Rubenstein received his Ph.D. from Johns Hopkins University in 1975. He is Professor of Geography at Miami University, Oxford, Ohio, where he teaches urban and human geography. Dr. Rubenstein is the author of *The Cultural Landscape*, the bestselling textbook for college and high school human geography, as well as *Contemporary Human Geography*. Dr. Rubenstein also conducts research in the automotive industry and has published three books on the subject—*The Changing U.S. Auto Industry: A Geographical Analysis* (Routledge); *Making and Selling Cars: Innovation and Change in the U.S. Auto Industry* (The Johns Hopkins University Press); and *Who Really Made Your Car? Restructuring and Geographic Change in the Auto Industry* (W.E. Upjohn Institute, with Thomas Klier).

William H. Renwick earned a B.A. from Rhode Island College in 1973 and a Ph.D. in geography from Clark University in 1979. He has taught at the University of California, Los Angeles, and Rutgers University, and is Professor of Geography at Miami University, Oxford, Ohio. Dr. Renwick is co-author of *Introduction to Geography: People, Places & Environment.* A physical geographer with interests in geomorphology and environmental issues, his research focuses on impacts of human acivities on rivers and lakes, particularly in agricultural landscapes in the Midwest. When time permits, he studies these environments from the seat of a wooden boat.

Carl T. Dahlman earned degrees in sociology, music, and urban affairs before receiving his Ph.D. in geography from the University of Kentucky in 2001. He is Associate Professor of Geography at Miami University, Oxford, Ohio, where his teaching focuses on political geography, migration and mobility, and globalization. Dr. Dahlman is co-author of *Introduction to Geography: People, Places & Environment.* His current research includes the role of European integration in the geopolitics of Southeastern Europe, and he has published a book on the subject *Bosnia Remade: Ethnic Cleansing and Its Reversal* (Oxford University Press, with Gearóid Ó Tuathail). He enjoys photography and traveling with his son.

About Our Sustainability Initiatives

Pearson recognizes the environmental challenges facing this planet, as well as acknowledges our responsibility in making a difference. This book is carefully crafted to minimize environmental impact. The binding, cover, and paper come from facilities that minimize waste, energy consumption, and the use of harmful chemicals. Pearson closes the loop by recycling every out-of-date text returned to our warehouse.

Along with developing and exploring digital solutions to our market's needs, Pearson has a strong commitment to achieving carbon-neutrality. As of 2009, Pearson became the first carbon- and climate-neutral publishing company. Since then, Pearson remains strongly committed to measuring, reducing, and offsetting our carbon footprint.

The future holds great promise for reducing our impact on Earth's environment, and Pearson is proud to be leading the way. We strive to publish the best books with the most up-to-date and accurate content, and to do so in ways that minimize our impact on Earth. To learn more about our initiatives, please visit www.pearson.com/responsibility.

The Teaching and Learning Package

This edition provides a complete introductory geography program for students and teachers.

FOR STUDENTS AND TEACHERS:

MasteringGeography™ with Pearson eText www.masteringgeography.com

The Mastering platform is the most effective and widely used tutorial, homework and assessment system for the sciences. It helps instructors maximize class time with customizable, easy-to-assign, and automatically graded assessments that motivate students to learn outside of class and arrive prepared for lecture. These assessments can easily be customized and personalized for an instructor's individual teaching style. The powerful gradebook and diagn ostic features provide unique insight into student and class performance even before the first test. As a result, instructors can spend class time where students need it most. The Mastering system empowers students to take charge of their learning through activities aimed at different learning styles, and engages them in learning science through practice and step-by-step guidance–at their convenience, 24/7. MasteringGeography offers:

- Assignable activities that include MapMaster™ interactive maps, Geoscience Animations, geography videos, *Encounter* Google Earth Explorations, Thinking Spatially and Data Analysis activities, end-of-chapter questions, reading quizzes, *Test Bank* questions, and more.
- Student study area with MapMaster™ interactive maps, Geoscience Animations, geography videos, glossary flashcards, "In the News" RSS feeds, reference maps, an optional Pearson eText, and more.

MasteringGeography icons are integrated within the chapters of the text to highlight various online self-study media.

MasteringGeography with Pearson eText gives students access to the text whenever and wherever they can access the Internet. The eText pages look exactly like the printed text, and include powerful interactive and annotation functions, including links to the multimedia.

Teaching College Geography: A Practical Guide for Graduate Students and Early Career Faculty (0136054471)

This two-part resource provides a starting point for becoming an effective geography teacher from the very first day of class. Divided in two parts, Part One addresses "nuts-and-bolts" teaching issues. Part Two explores being an effective teacher in the field, supporting critical thinking with GIS and mapping technologies, engaging learners in large geography classes, and promoting awareness of international perspectives and geographic issues.

Aspiring Academics: A Resource Book for Graduate Students and Early Career Faculty (0136048919)

Drawing on several years of research, this set of essays is designed to help graduate students and early career faculty start their careers in geography and related social and environmental sciences. *Aspiring Academics* stresses the interdependence of teaching, research, and service–and the importance of achieving a healthy balance of professional and personal life–while doing faculty work. Each chapter provides accessible, forward-looking advice on topics that often cause the most stress in the first years of a college or university appointment.

Practicing Geography: Careers for Enhancing Society and the Environment (0321811151)

This book examines career opportunities for geographers and geospatial professionals in business, government, non-profit, and educational sectors. A diverse group of academic and industry professionals share insights on career planning, networking, transitioning between employment sectors, and balancing work and home life. The book illustrates the value of geographic expertise and technologies through engaging profiles and case studies of geographers at work.

AAG Community Portal

This web site is intended to support community-based professional development in geography and related disciplines. Here you will find activities providing extended treatment of the topics covered in both books. The activities can be used in workshops, graduate seminars, brown bags, and mentoring programs offered on campus or within an academic department. You can also use the discussion boards and contributions tool to share advice and materials with others. www.personhighered.com/aag/

Geoscience Animation Library 5th edition DVD-ROM (0321716841)

This resource offers over 100 animations covering the most difficult-to-visualize topics in physical geology, physical geography, oceanography, meteorology, and earth science. The animations are provided as Flash files and pre-loaded into PowerPoint® slides for both Windows and Mac. This library was created through a unique collaboration among Pearson's leading geoscience authors—including Robert Christopherson, Darrel Hess, Frederick Lutgens, Aurora Pun, Gary Smith, Edward Tarbuck, and Alan Trujillo.

Television for the Environment *Earth Report* Geography Videos on DVD (0321662989)

This three- DVD set is designed to help students visualize how human decisions and behavior have affected the environment and how individuals are taking steps toward recovery. With topics ranging from the poor land management promoting the devastation of river systems in Central America to the struggles for electricity in China and Africa, these 13 videos from Television for the Environment's global *Earth Report* series recognize the efforts of individuals around the world to unite and protect the planet.

Television for the Environment *Life* Human Geography Videos on DVD (0132416565)

This three-DVD set is designed to enhance any human geography course. It contains 14 complete video programs (average length 25 minutes) covering a wide array of issues affecting people and places in the contemporary world, including international immigration, urbanization, global trade, poverty, and environmental destruction. The videos included on these DVDs are offered at the highest quality to allow for full-screen viewing on a computer and projection in large lecture classrooms.

Television for the Environment *Life* World Regional Geography Videos on DVD (013159348X)

From the Television for the Environment's global *Life* series this two-DVD set brings globalization and the developing world to the attention of any world regional geography course. These 10 full-length video programs highlight matters such as the growing number of homeless children in Russia, the lives of immigrants living in the United States trying to aid family still living in their native countries, and the European conflict between commercial interests and environmental concerns.

FOR STUDENTS:

Goode's World Atlas' 22nd edition by Rand McNally (0321652002)

Goode's World Atlas has been the world's premier educational atlas since 1923, and for good reason. It features over 250 pages of maps, from definitive physical and political maps to important thematic maps that illustrate the spatial aspects of many important topics. The *22nd Edition* includes 160 pages of new, digitally produced reference maps, as well as new thematic maps on global climate change, sea level rise, CO_2 emissions, polar ice fluctuations, deforestation, extreme weather events, infectious diseases, water resources, and energy production.

***Encounter Human Geography* Workbook and Website** by Jess C. Porter (0321682203)

Encounter Human Geography provides rich, interactive explorations of human geography concepts through Google Earth™. Students explore the globe through themes such as population, sexuality and gender, political geography, ethnicity, urban geography, migration, human health, and language. All chapter explorations are available in print format as well as online quizzes, accommodating different classroom needs. All worksheets are accompanied with cooresponding Google Earth™ media files, available for download for those who do not use MasteringGeography, from www.mygeoscienceplace.com.

***Encounter World Regional Geography* Workbook and Website** by Jess C. Porter (0321681754)

Encounter World Regional Geography provides rich, interactive explorations of world regional geography concepts through Google Earth™ Students explore the globe through the themes of environment, population, culture, Geopolitics, and economy and development, answering multiple choice and short-answer questions. All chapter explorations are available in print format as well as online quizzes, accommodating different classroom needs. All worksheets are accompanied with cooresponding Google Earth™ media files, available for download for those who do not use MasteringGeography, from www.mygeoscienceplace.com.

Encounter Geosystems Workbook and Website by Charlie Thomsen (0321636996)

Encounter Geosystems is a print workbook and website that provides rich, interactive explorations of physical geography concepts through Google Earth™. All chapter explorations are available in print format as well as online quizzes, accomodating different classroom needs. All worksheets are accompanied with cooresponding Google Earth media files, available for download available for download for those who do not use MasteringGeography, from www.mygeoscienceplace.com.

Dire Predictions: Understanding Global Warming by Michael Mann, Lee R. Kump (0136044352)

For any science or social science course in need of a basic understanding of IPCC reports. Periodic reports from the Intergovernmental Panel on Climate Change (IPCC) evaluate the risk of climate change brought on by humans. But the sheer volume of scientific data remains inscrutable to the general public, particularly to those who may still question the validity of climate change. In just over 200 pages, this practical text presents and expands upon the essential findings in a visually stunning and undeniably powerful way to the lay reader. Scientific findings that provide validity to the implications of climate change are presented in clear-cut graphic elements, striking images, and understandable analogies.

FOR TEACHERS:

Instructor Resource DVD for Introduction to Contemporary Geography (0321812832)

The Instructor Resource DVD provides high-quality electronic versions of photos and illustrations form the book in JPEG, pdf, and PowerPoint formats, as well as customizable PowerPoint™ lecture presentations, Classroom Response System questions in PowerPoint, and the *Instructor Resource Manual* and *Test Bank* in Microsoft Word and TestGen formats. For easy reference and identification, all resources are organized by chapter.

Instructor Resource Manual (download only) *for Introduction to Contemporary Geography*

by Jose López-Jiménez (0321822919)

The *Instructor Resource Manual* is intended as a resource for both new and experienced instructors. It includes lecture outlines, additional source materials, teaching tips, advice about how to integrate visual supplements (including the Web-based resources), and various other ideas for the classroom. http://www.pearsonhighered.com/irc

TestGen® Computerized Test Bank (download only) *for Introduction to Contemporary Geography* by Iddi Adam (0321815300)

TestGen® is a computerized test generator that lets instructors view and edit *Test Bank* questions, transfer questions to tests, and print the test in a variety of customized formats. This *Test Bank* includes over 1,000 multiple choice, true/false, and short answer/essay questions. Questions are correlated to the revised U.S. National Geography Standards and Bloom's Taxonomy to help instructors better map the assessments against both broad and specific teaching and learning objectives. The *Test Bank* questions are also tagged to chapter specific learning outcomes. The *Test Bank* is available in Microsoft Word®, and is importable into Blackboard. http://www.pearsonhighered.com/irc

Modular Springboard

With an innovative integration of visuals, text, active learning tools, and online media, ***Introduction to Contemporary Geography*** is a modular and highly graphical springboard to introductory geography—ideal for contemporary students and learning styles.

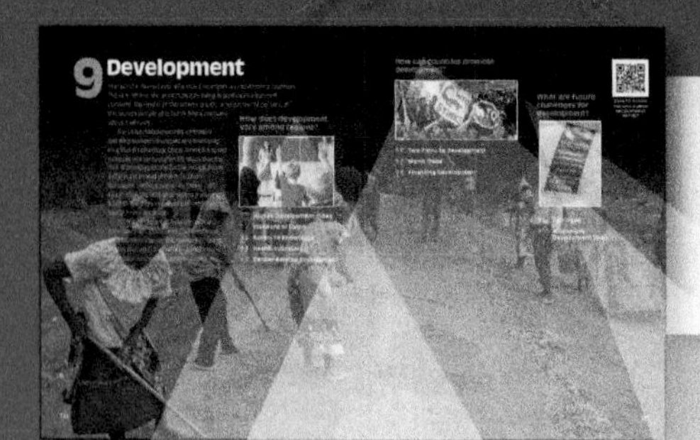

Every chapter is organized around conceptual **"Key Questions,"** which ask students to take a bigger picture approach to the concept, with each module framed by two unique **"Main Points,"** which students should understand after reading the module.

How is agriculture distributed?

What do people eat?

What challenges does agriculture face?

The book's modular organization consists of chapters made up of self-contained two-page spreads, a reliable presentation that offers the instructor flexibility. Each module uses integrated visuals, text, active learning tools, and online media, to effectively convey the concept at hand.

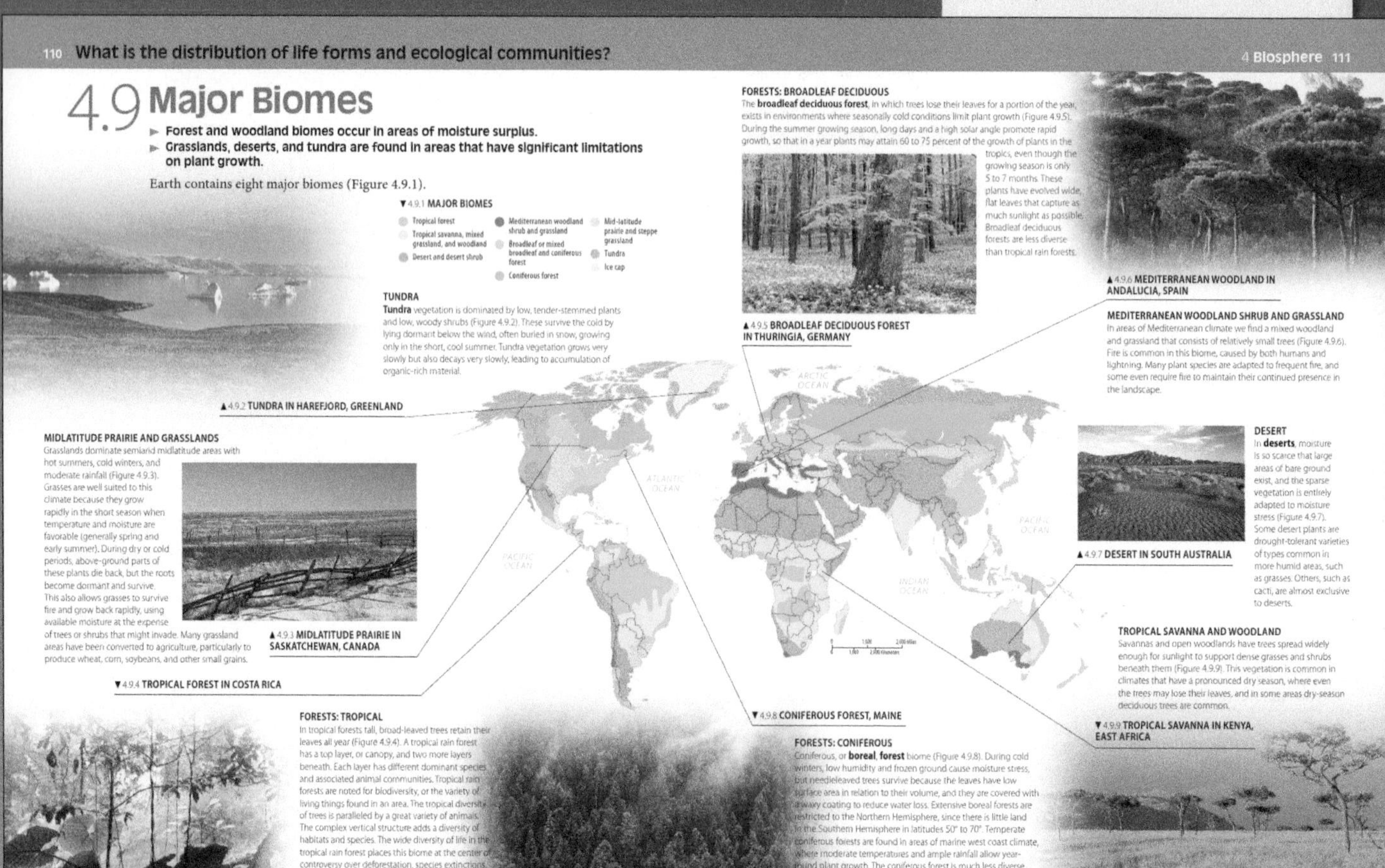

110 **What is the distribution of life forms and ecological communities?** | 4 **Biosphere** 111

4.9 Major Biomes

- **Forest and woodland biomes occur in areas of moisture surplus.**
- **Grasslands, deserts, and tundra are found in areas that have significant limitations on plant growth.**

Earth contains eight major biomes (Figure 4.9.1).

▼ 4.9.1 **MAJOR BIOMES**

TUNDRA

Tundra vegetation is dominated by low, tender-stemmed plants and low, woody shrubs (Figure 4.9.2). These survive the cold by lying dormant below the wind, often buried in snow, growing only in the short, cool summer. Tundra vegetation grows very slowly but also decays very slowly, leading to accumulation of organic-rich material.

▲ 4.9.2 **TUNDRA IN HAREFJORD, GREENLAND**

MIDLATITUDE PRAIRIE AND GRASSLANDS

Grasslands dominate semiarid midlatitude areas with hot summers, cold winters, and moderate rainfall (Figure 4.9.3). Grasses are well suited to this climate because they grow rapidly in the short season when temperature and moisture are favorable (generally spring and early summer). During dry or cold periods, above-ground parts of these plants die back, but the roots become dormant and survive. This also allows grasses to survive fire and grow back rapidly, using available moisture at the expense of trees or shrubs that might invade. Many grassland areas have been converted to agriculture, particularly to produce wheat, corn, soybeans, and other small grains.

▲ 4.9.3 **MIDLATITUDE PRAIRIE IN SASKATCHEWAN, CANADA**

▼ 4.9.4 **TROPICAL FOREST IN COSTA RICA**

FORESTS: TROPICAL

In tropical forests tall, broad-leaved trees retain their leaves all year (Figure 4.9.4). A tropical rain forest has a top layer, or canopy, and two more layers beneath. Each layer has different dominant species and associated animal communities. Tropical rain forests are noted for biodiversity, or the variety of living things found in an area. The tropical diversity of trees is paralleled by a great variety of animals. The complex vertical structure adds a diversity of habitats and species. The wide diversity of life in the tropical rain forest places this biome at the center of controversy over deforestation, species extinctions, and biodiversity.

FORESTS: BROADLEAF DECIDUOUS

The **broadleaf deciduous forest**, in which trees lose their leaves for a portion of the year, exists in environments where seasonally cold conditions limit plant growth (Figure 4.9.5). During the summer growing season, long days and a high solar angle promote rapid growth, so that in a year plants may attain 60 to 75 percent of the growth of plants in the tropics, even though the growing season is only 5 to 7 months. These plants have evolved wide, flat leaves that capture as much sunlight as possible. Broadleaf deciduous forests are less diverse than tropical rain forests.

▲ 4.9.5 **BROADLEAF DECIDUOUS FOREST IN THURINGIA, GERMANY**

▲ 4.9.6 **MEDITERRANEAN WOODLAND IN ANDALUCIA, SPAIN**

MEDITERRANEAN WOODLAND SHRUB AND GRASSLAND

In areas of Mediterranean climate we find a mixed woodland and grassland that consists of relatively small trees (Figure 4.9.6). Fire is common in this biome, caused by both humans and lightning. Many plant species are adapted to frequent fire, and some even require fire to maintain their continued presence in the landscape.

DESERT

In **deserts**, moisture is so scarce that large areas of bare ground exist, and the sparse vegetation is entirely adapted to moisture stress (Figure 4.9.7). Some desert plants are drought-tolerant varieties of types common in more humid areas, such as grasses. Others, such as cacti, are almost exclusive to deserts.

▲ 4.9.7 **DESERT IN SOUTH AUSTRALIA**

TROPICAL SAVANNA AND WOODLAND

Savannas and open woodlands have trees spread widely enough for sunlight to support dense grasses and shrubs beneath them (Figure 4.9.9). This vegetation is common in climates that have a pronounced dry season, where even the trees may lose their leaves, and in some areas dry-season deciduous trees are common.

▼ 4.9.8 **CONIFEROUS FOREST, MAINE**

FORESTS: CONIFEROUS

Coniferous, or **boreal**, **forest** biome (Figure 4.9.8). During cold winters, low humidity and frozen ground cause moisture stress, but needleleaved trees survive because the leaves have low surface area in relation to their volume, and they are covered with a waxy coating to reduce water loss. Extensive boreal forests are restricted to the Northern Hemisphere, since there is little land in the Southern Hemisphere in latitudes 50° to 70°. Temperate coniferous forests are found in areas of marine west coast climate, where moderate temperatures and ample rainfall allow year-round plant growth. The coniferous forest is much less diverse than the tropical rain forest.

▼ 4.9.9 **TROPICAL SAVANNA IN KENYA, EAST AFRICA**

Innovative End-of-Chapter Tools

Extend student learning with a rich suite of critical-thinking and media-rich activities.

Review of the Key Questions and the associated Main Points from all modules in the chapter

Thinking Geographically critical-thinking questions

MapMaster™ Interactive Mapping activities

Key Terms definitions

This chapter has introduced ways in which geographers think about the world, as well as key concepts in understanding geography.

Key Questions

Where is the world's population distributed?

- Global population is highly concentrated; two-thirds of the world's people live in four clusters (Europe, East Asia, Southeast Asia, and South Asia).
- Population density varies around the world partly in response to resources.

Why does population growth vary among countries?

- A population increases because of fertility and decreases because of mortality.
- The demographic transition is a process of change in a country's population from a condition of high birth and death rates, with little population growth, to a condition of low birth and death rates, with low population growth.
- More than 200 years ago, Thomas Malthus argued that population was increasing more rapidly than the food supply; some contemporary analysts believe that Malthus' prediction is accurate in some regions.

How might population change in the future?

- Most countries in Europe and North America face slow or even declining population in the future.
- World population growth is slowing in part because birth rates are declining.
- Meanwhile, death rates are increasing in some countries because of chronic disorders associated with aging and in some developing countries because of infectious diseases.

▼5.CR.1 **VERY HIGH ARITHMETIC DENSITY: MARKET, DARAW, EGYPT**
What is the evidence that human behavior is affected by a high population density?

Thinking Geographically

The U.S. Census Bureau is allowed to utilize statistical sampling to determine much of the information about the people of the United States, such as age and gender. However, for determining the total population of each state and congressional district, the Census Bureau is required to count only the people for whom a census form was completed.

1. **What are the advantages of using each of the two approaches to counting the population?**

Some humans live at very high density (Figure 5.CR.1). Scientists disagree about the effects of high density on human behavior. Some laboratory tests have shown that rats display evidence of increased aggressiveness, competition, and violence when very large numbers of them are placed in a box.

2. **Is there any evidence that high density might cause humans to behave especially violently or less aggressively?**

Members of the baby-boom generation – people born between 1946 and 1964 – constitute nearly one-third of the U.S. population.

3. **As they grow older, what impact will baby boomers have on the entire American population in the years ahead?**

On the Internet

The Population Reference Bureau (PRB) provides authoritative demographic information for every country and world region at its website **www.prb.org.**

The Population Division of the United Nations Department of Economic and Social Affairs provides tables on population, births, and deaths for every country, at **http://esa.un.org/unpd/wpp/unpp/panel_population.htm**, or scan the QR code at the beginning of the chapter.

Interactive Mapping:

POPULATION DISTRIBUTION IN SOUTHWEST ASIA AND NORTH AFRICA

Population is highly clustered within Southwest Asia and Northern Africa.

Open: MapMaster Southeast Asia and North Africa in MasteringGEOGRAPHY

Select: *Population Density* from the *Population* menu, adjust opacity to 60%, then select *Physical Features* from the *Physical Environment* menu.

Most people live near what type of physical feature?

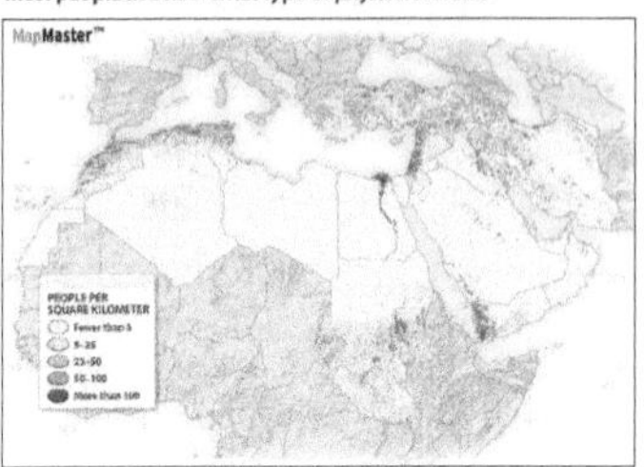

Explore

MAHĀMĪD, EGYPT

Use Google Earth to explore Mahāmīd, a town of 45,000 near the banks of the Nile River.

Fly to: *Mahāmīd, Luxor, Egypt.* Zoom in.

1. **What color is most of the land immediately in and around the town? Does this indicate that the land is used for agriculture or is it desert?**

Zoom out until you see the entire band of green surrounded by tan.

2. **How wide is the green strip? What does the tan color represent? What feature is in the middle of the green strip?**

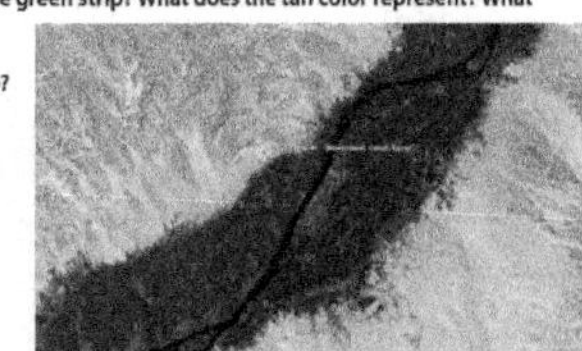

Key Terms

Agricultural density
The ratio of the number of farmers to the total amount of land suitable for agriculture.

Arable land
Land suited for agriculture.

Arithmetic density
The total number of people divided by the total land area.

Crude birth rate (CBR)
The total number of live births in a year for every 1,000 people alive in the society.

Crude death rate (CDR)
The total number of deaths in a year for every 1,000 people alive in the society.

Demographic transition
The process of change in a society's population from a condition of high crude birth and death rates and low rate of natural increase to a condition of low crude birth and death rates, low rate of natural increase, and a higher total population.

Dependency ratio
The number of people who are considered too young or too old to work (under age 15 or over age 64), compared to the number of people in their productive years.

Doubling time
The number of years needed to double a population, assuming a constant rate of natural increase.

Elderly support ratio
The number of working-age people (ages 15–64) divided by the number of persons 65 or older.

Epidemiologic transition
Distinctive causes of death in each stage of the demographic transition.

Epidemiology
Branch of medical science concerned with the incidence, distribution, and control of diseases that affect large numbers of people.

Infant mortality rate (IMR)
The total number of deaths in a year among infants under 1 year old for every 1,000 live births in a society.

Life expectancy
The average number of years an individual can be expected to live, given current social, economic, and medical conditions. Life expectancy at birth is the average number of years a newborn infant can expect to live.

Natural increase rate (NIR)
The percentage growth of a population in a year, computed as the crude birth rate minus the crude death rate.

Overpopulation
The number of people in an area exceeds the capacity of the environment to support life at a decent standard of living.

Pandemic
Disease that occurs over a wide geographic area and affects a very high proportion of the population.

Physiological density
The number of people per unit of area of arable land, which is land suitable for agriculture.

Population pyramid
A bar graph that displays the percentage of a place's population for each age and gender.

Total fertility rate (TFR)
The average number of children a woman will have throughout her childbearing years.

▶ LOOKING AHEAD

Population increases because of births and decreases because of deaths. The population of a place also increases when people move in and decreases when people move out. This element of population change—migration—is discussed in the next chapter.

On the Internet web links

Google Earth™ Explore image analysis questions

Looking Ahead chapter preview section

Current Data and Applications

The latest science, statistics, and imagery are used for the most contemporary introduction to geography.

5.6 Declining Birth Rates

- Some developing countries have lowered birth rates through improved education and health care.
- In other developing countries, distribution of contraceptives has reduced birth rates.

Population has been increasing at a much slower rate since the mid-twentieth century. After hitting a peak around 1970, the world NIR has been declining steadily through the late twentieth century and the early twenty-first. The NIR declined beginning in the 1960s in developed countries and beginning in the 1970s in developing ones (Figure 5.6.1).

▲ 5.6.1 NIR 1950–2010

In most countries, the decline in the NIR has occurred because of lower birth rates (Figure 5.6.2). Between 1980 and 2010, the CBR declined in every country except for three in Northern Europe—Denmark, Norway, and Sweden—where the CBR increased by only 1.

Two strategies have been successful in reducing birth rates. One alternative emphasizes reliance on education and health care, the other on distribution of contraceptives. Because of varied economic and cultural conditions, the most effective method varies among countries.

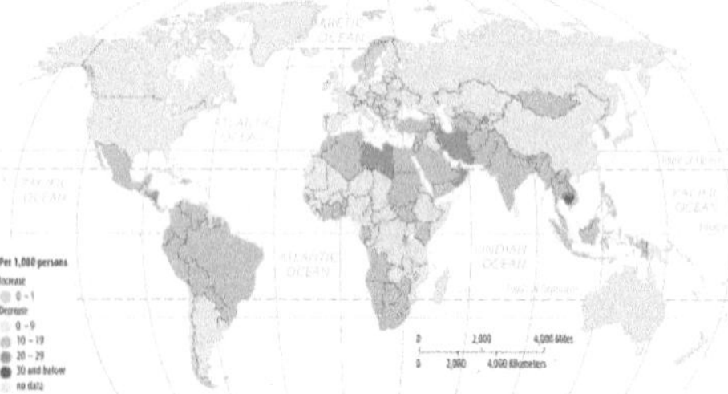

► 5.6.2 CRUDE BIRTH RATE CHANGE 1980–2010

LOWERING BIRTH RATES THROUGH EDUCATION AND HEALTH CARE

One approach to lowering birth rates emphasizes the importance of improving local economic conditions. A wealthier community has more money to spend on education and health-care programs that would promote lower birth rates.

According to this approach:

- With more women able to attend school and to remain in school longer, they would be more likely to learn employment skills and gain more economic control over their lives.
- With better education, women would better understand their reproductive rights, make more informed reproductive choices, and select more effective methods of contraception.
- With improved health-care programs, IMRs would decline through such programs as improved prenatal care, counseling about sexually transmitted diseases, and child immunization.
- With the survival of more infants ensured, women would be more likely to choose to make more effective use of contraceptives to limit the number of children.

Current data incorporates the latest science, statistics, and imagery, including data from the 2010 U.S. Census, the 2011 World Population Reference Bureau World Population Data, as well as coverage of events like the 2011 Japan earthquake and tsunami and much more. These figures illustrate declines in Natural Increase Rates and Birth Rates.

11 Industry

Manufacturing jobs are viewed as a special asset by communities around the world. They are seen as the "engine" of economic growth and prosperity. Different communities possess distinctive assets in attracting particular types of industries as well as challenges in retaining them.

A generation ago, industry was highly clustered in a handful of communities within developed countries, but industry has diffused to more communities, including some in developing countries. Meanwhile, a loss of manufacturing jobs has caused economic problems for communities in developed countries like the United States traditionally dependent on them.

Where is industry clustered?

What situation factors influence industrial location?

What site factors influence industrial location?

SCAN TO ACCESS U.S. LABOR STATISTICS

11.1 The Industrial Revolution
11.2 Distribution of Industry
11.3 Situation Factors in Locating Industry
11.4 Changing Steel Production
11.5 Changing Auto Production
11.6 Ship by Boat, Rail, Truck, or Air?
11.7 Site Factors in Industry
11.8 Textile and Apparel Production
11.9 Emerging Industrial Regions

STEEL WORKS LIAONING, CHINA

Quick Response (QR) Codes on the chapter opening pages enable students to link smartphones from the book to various open source geography websites relevant to each chapter, providing easy and immediate access to original sources and up-to-date data.

Students simply scan the QR codes in the book with their mobile smartphones, after following a few easy steps:

1. Download a QR reader from an app store (there are many free apps available) or use the built-in code reader if the device has one.
2. Open the QR code reader app on the phone and scan the code (as shown on the left).
3. The student's device will be automatically redirected to the website. *(Note: data usage charges may apply.)*

Interactive Activities

MapMaster™ interactive mapping and Google Earth™ activities and questions encourage students to use exciting and engaging geospatial media to explore concepts, practice critical image and data anlaysis, and extend chapter learning.

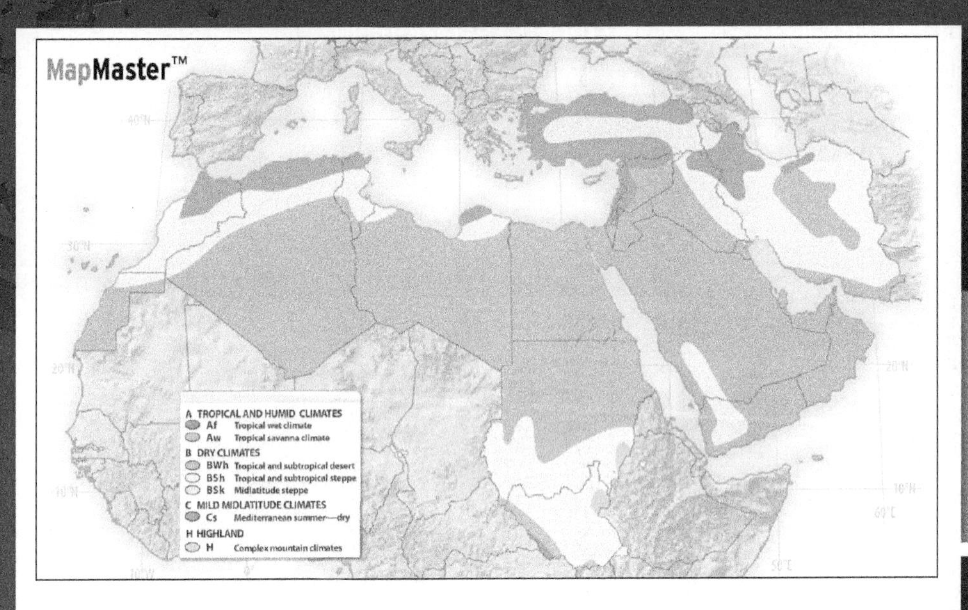

▲ 10.5.4 **CLIMATE REGIONS**
Climate influences the crop that is grown, or whether animals are raised instead of growing any crop.

Launch MapMaster Southwest Asia and North Africa in MasteringGEOGRAPHY

Select: *Climate* from the *Physical Environment* menu, then *Agricultural Regions* from the *Economic* menu.

What climate region is correlated with pastoral nomadism?

MapMaster™ Layered Thematic Interactive Mapping activities are provided within selected modules and at the end of each chapter in the text, linking students to the Student Study Area of MasteringGeography.™

▲ 11.3.1 **BULK-REDUCING INDUSTRY: COPPER**
Use Google Earth to explore Mount Isa, Australia's leading copper production center, which includes mining and smelting.

Fly to: *Parkside, Mount Isa, Australia.*

What is the large crater-like feature in the image?

Locate the plume of smoke just west of Parkside and drag the street view icon to the main road west of the Parkside label.

What type of structure is producing the smoke? Why is this structure located close to the other feature?

Google Earth Explore image analysis questions are provided within selected modules and at the end of each chapter, linking students from the print page to the dynamic geobrowsing technology of Google Earth.

MasteringGeography™

This online homework and tutoring system delivers self-paced activities that provide individualized coaching, focus on your course objectives, and are responsive to each student's progress. The Mastering system helps instructors maximize class time with customizable, easy-to-assign, and automatically graded assessments that motivate students to learn outside of class and arrive prepared for lecture.

www.masteringgeography.com

Proven Results

The Mastering platform is the only online homework system with research showing that it improves student learning. A wide variety of published papers based on NSF-sponsored research and tests illustrate the benefits of the Mastering program. Results documented in scientifically valid efficacy papers are available at www.masteringgeography.com/site/results.

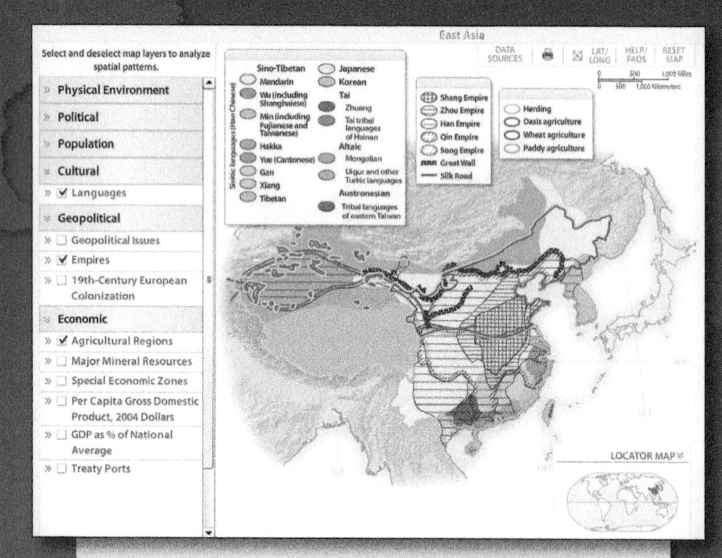

MapMaster™ Layered Thematic Interactive Map Activities

These act as a mini-GIS tool, allowing students to layer various thematic maps to analyze spatial patterns and data at regional and global scales. Multiple-choice and short-answer questions are organized around themes of Physical Environment, Population, Culture, Geopolitics, and Economy, giving instructors flexible, modular options for assessing student learning.

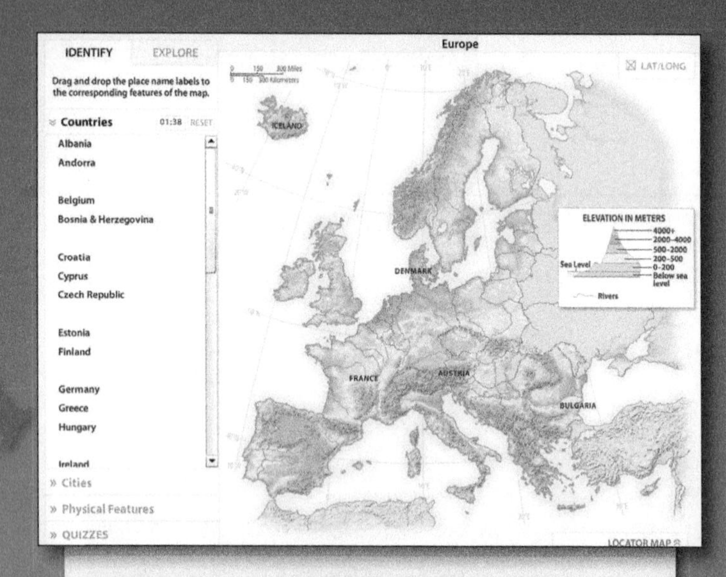

MapMaster™ Place Name Interactive Map Activities

These have students identify place names of political and physical features at regional and global scales, and explore select country data from the CIA World Factbook. Multiple-choice questions for the place name labeling activities and country data sets offer instructors flexible opportunities for summative assessment and pre- and post-testing.

Engaging Experiences

MasteringGeography provides a personalized, dynamic, and engaging experience for each student that strengthens active learning. Survey data show that the immediate feedback and tutorial assistance in MasteringGeography motivate students to do more homework. The result is that students learn more and improve their test scores.

Encounter Google Earth Explore Activities

Pearson's Encounter Activities provide rich, interactive explorations of geography concepts through Google Earth.™ All explorations include corresponding Google Earth KMZ media files, and questions include hints and specific wrong-answer feedback to help coach students towards mastery of the concepts.

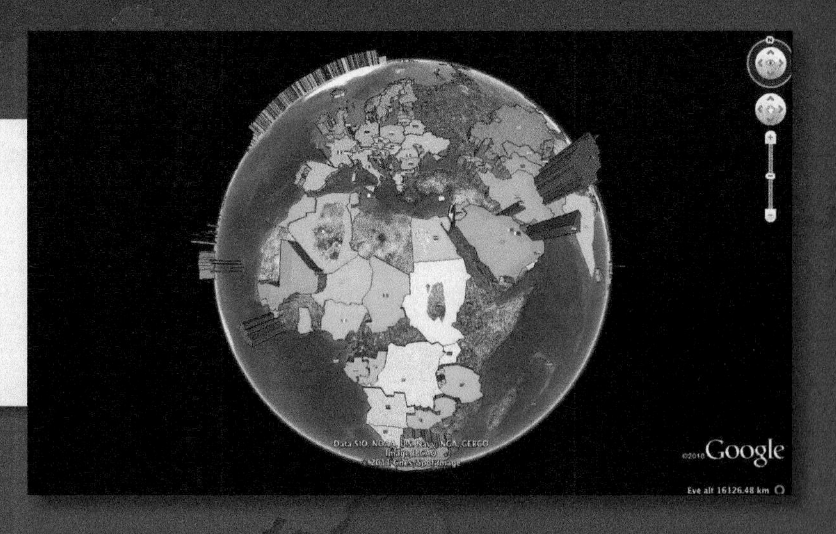

Geography Videos

A variety of short geography video activities provides students with a sense of place, helps them visualize concepts, and allows them means to explore a range of topics and places. Covering issues of economy, development, globalization, climate and climate change, culture, and more, there are 10–20 multiple choice and short answer questions for each of the 34 episodes. Students can access videos from the Study Area in MasteringGeography,™ and professors can assign video questions. These video activities allow instructors to test students' understanding and application of concepts, and offer hints and specific wrong-answer feedback.

Geoscience Animations

Geoscience Animation activities illuminate the most difficult-to-visualize topics from across the geosciences, such as the hydrologic cycle, plate tectonics, glacial advance and retreat, global warming, etc. Animations include audio narration, a text transcript, and assignable multiple-choice quizzes with specific wrong-answer feedback. Icons integrated throughout the physical geography chapters indicate to students when they can login to the Study Area of MasteringGeography to view the animations.

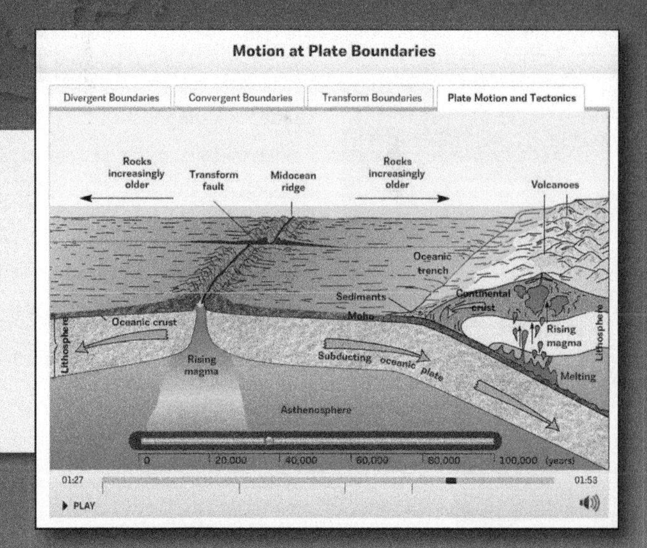

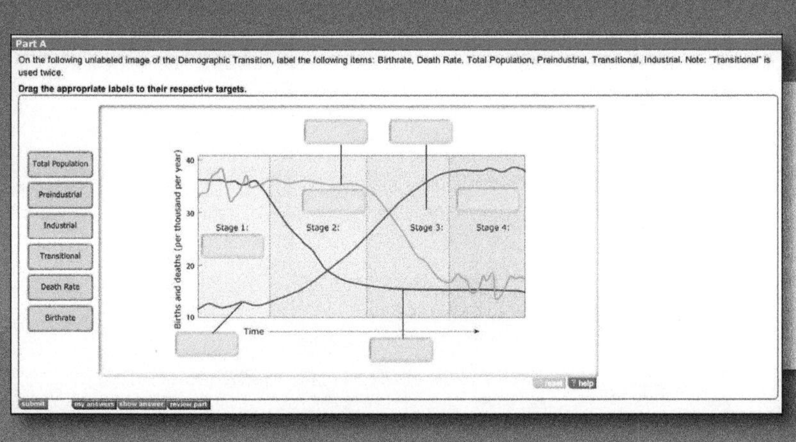

Thinking Spatially and Data Analysis

These activities help students develop spatial reasoning and critical thinking skills by identifying and labeling features from maps, illustrations, photos, graphs, and charts. Students then examine related data sets, answering multiple-choice and increasingly higher-order conceptual short-answer questions.

MasteringGeography™

www.masteringgeography.com

A Trusted Partner

The Mastering platform was developed by scientists for science students and instructors, and has a proven history with over 10 years of student use. Mastering currently has more than 1.5 million active registrations with active users in 50 states and in 41 countries.

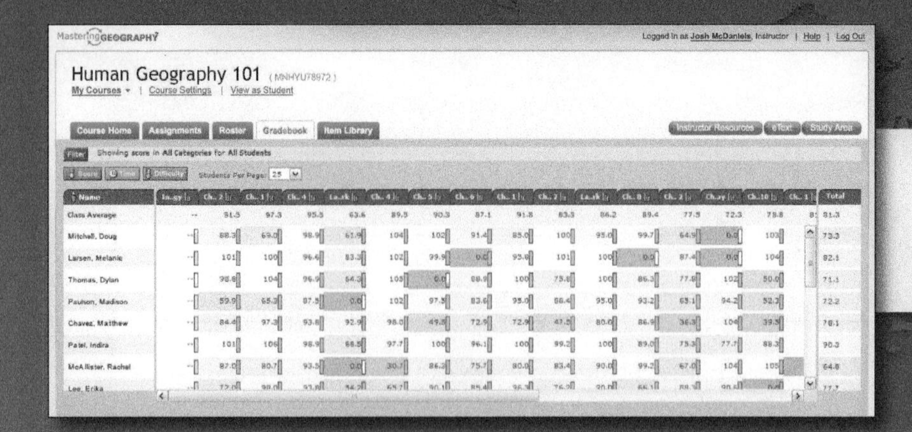

Gradebook

Every assignment is automatically graded. At a glance, shades of red highlight struggling students and challenging assignments.

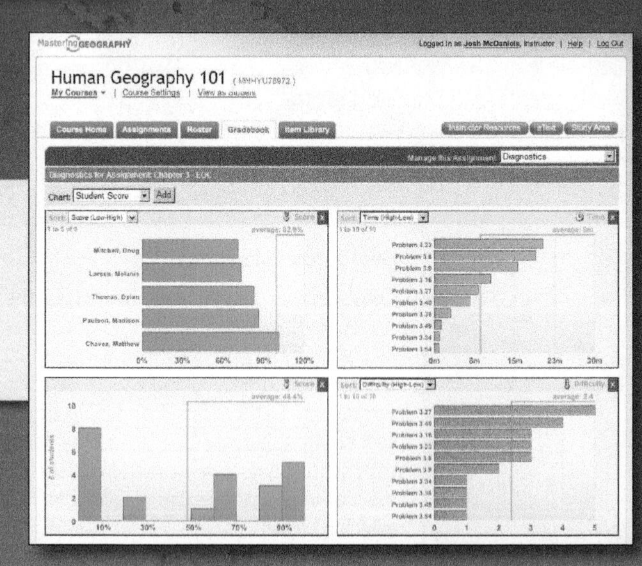

Gradebook Diagnostics

Gradebook Diagnostics provide unique insight into class and student performance. With a single click, charts summarize the most difficult problems, struggling students, grade distribution, and even score improvement over the duration of the course.

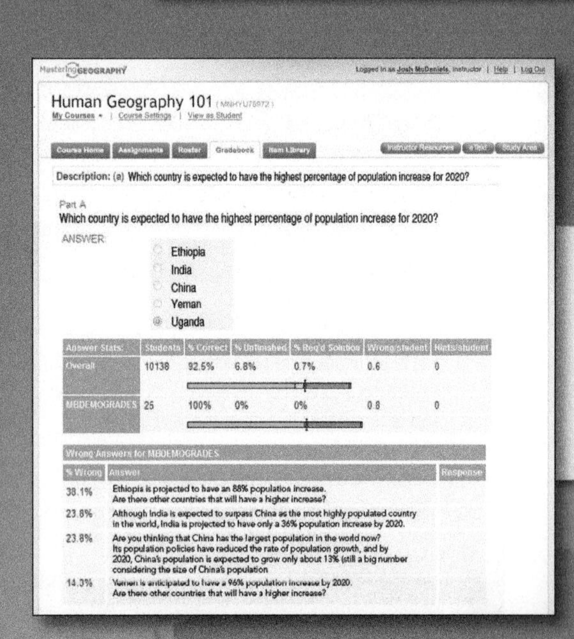

Gradebook Diagnostics

MasteringGeography provides at-a-glance statistics on your class as well as national results. Wrong-answer summaries give unique insight into your students' misconceptions and facilitate just-in-time teaching adjustments.

Learning Outcomes

MasteringGeography tracks student performance against each instructor's learning outcomes. Instructors can:

- Add their own or use the publisher-provided learning outcomes to track student performance and report it to their administrations.
- View class performance against the specified learning outcomes.
- Export results to a spreadsheet that can be customized further and/or shared with the chair, dean, administrator, or accreditation board.

Mastering offers a data-supported measure to quantify students' learning gains and to share those results quickly and easily.

Pearson eText

Pearson eText gives students access to the text whenever and wherever they can access the Internet. The eText pages look exactly like the printed text, and include powerful interactive and customization functions. Users can create notes, highlight text in different colors, create bookmarks, zoom, click hyperlinked words and phrases to view definitions, and view as a single page or as two pages. Pearson eText also links students to associated media files, enabling them to view an animation as they read the text, and offers a full text search and the ability to save and export notes. The Pearson eText for ***Introduction to Contemporary Geography*** also includes embedded URLs in the chapter text with active links to the Internet.

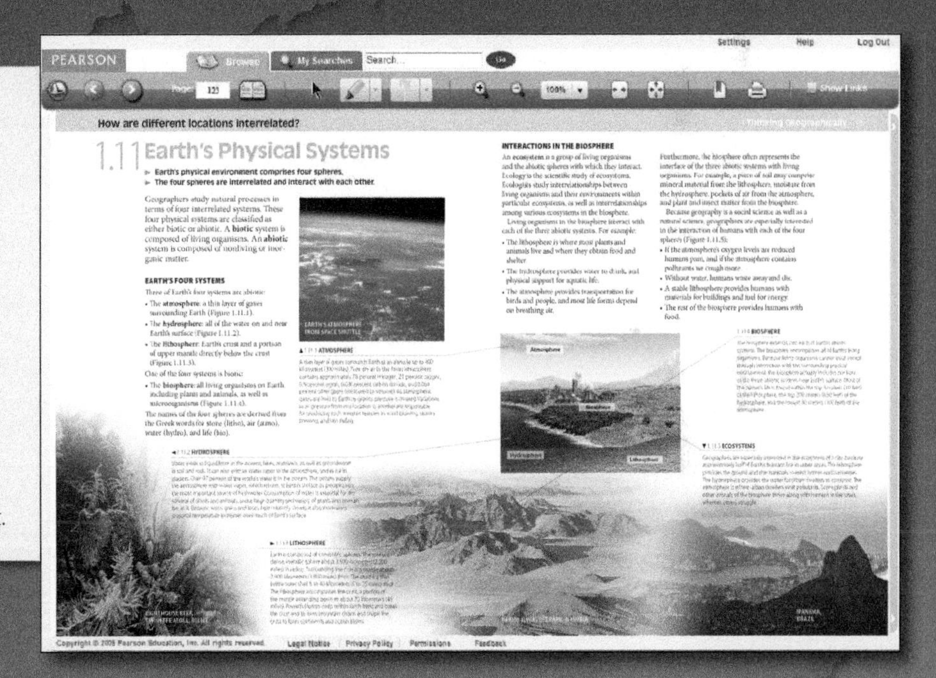

Student Resources in MasteringGeography

- MapMaster™ interactive maps
- Practice chapter quizzes
- Geography videos
- Geoscience Animations
- "In the News" RSS feeds
- Glossary flashcards
- Optional Pearson eText and more.

Instructor Resources in MasteringGeography

Assignable activities include:

- MapMaster™ interactive maps
- Geography Videos
- Encounter Google Earth™ Explorations
- Geoscience Animations
- Reading quizzes
- Test Bank questions
- Thinking Spatially and Data Analysis activities
- End-of-chapter questions and more

This book is dedicated to the memory of
Marye Stone Dahlman (1969–2011).

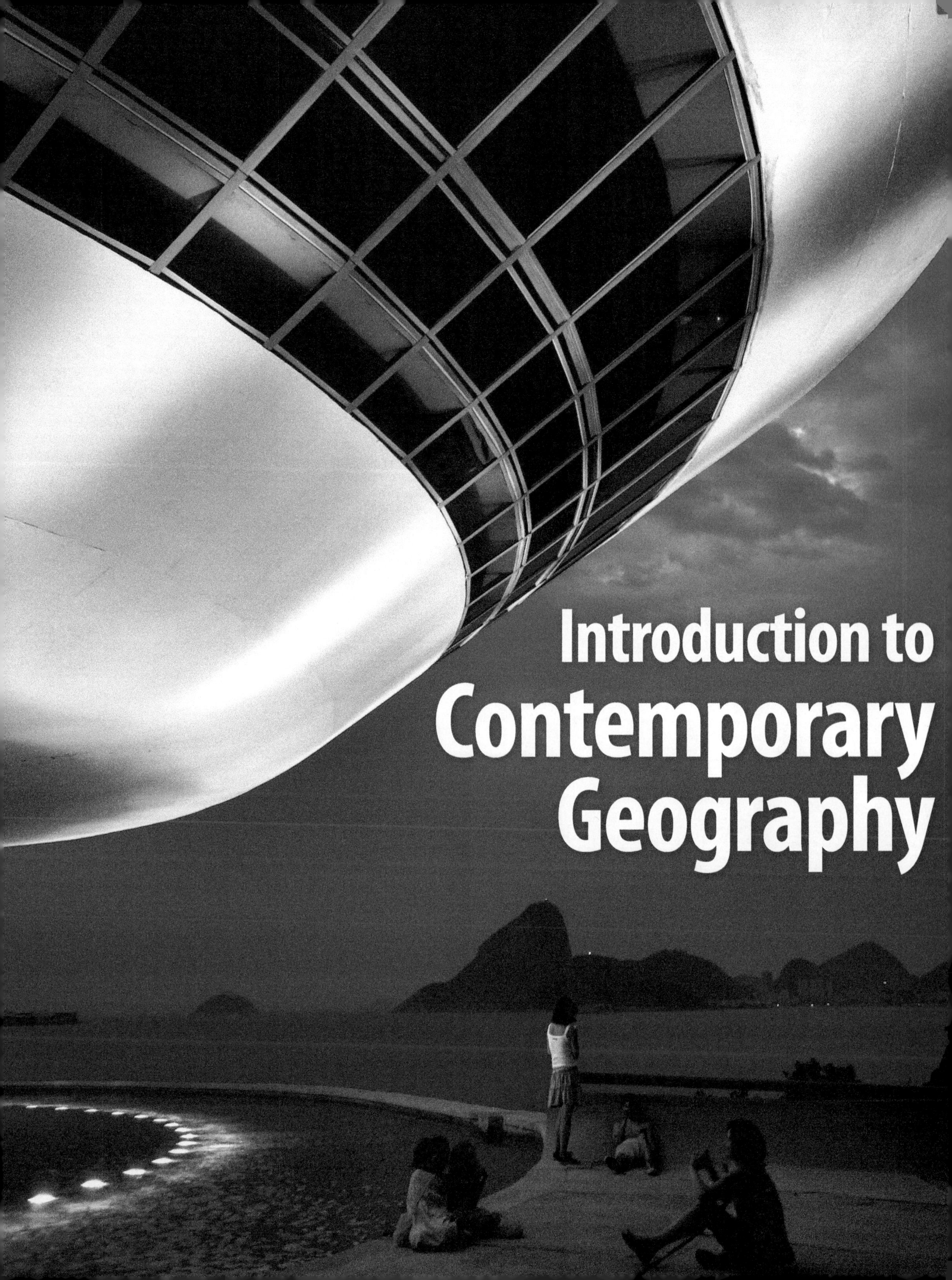
Introduction to
Contemporary
Geography

1 Thinking Geographically

Thinking geographically is one of the oldest human activities. Perhaps the first geographer was a prehistoric human who crossed a river or climbed a hill, observed what was on the other side, returned home to tell about it, and scratched a map of the route in the dirt. Perhaps the second geographer was a friend or relative who followed the dirt map to reach the other side.

Today, geographers are still trying to understand more about the world in which we live. Geography is the study of where natural environments and human activities are found on Earth's surface and the reasons for their location. This chapter introduces basic concepts that geographers use to study Earth's people and natural environment.

How do geographers describe where things are?

Why is each point on Earth unique?

BEACH AT IPANEMA, BRAZIL

How are different locations interrelated?

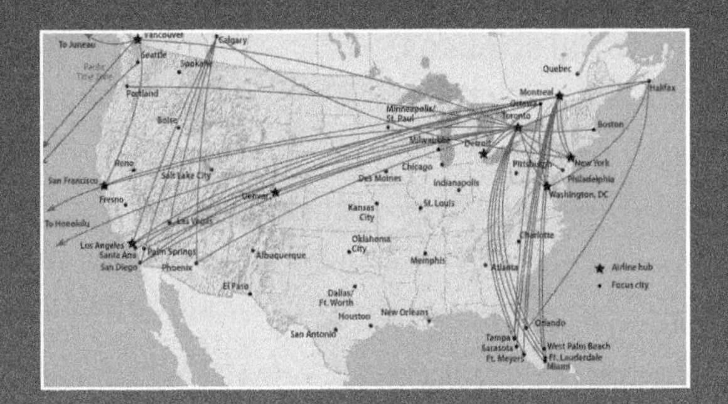

1.8 **Scale: From Global to Local**

1.9 **Space: Distribution of Features**

1.10 **Connection: Interaction Between Places**

SCAN TO ENTER THE WORLD OF GEOGRAPHY

How do people relate to their environment?

1.11 **Earth's Physical Systems**

1.12 **Human–Environment Interaction**

1.1 Welcome to Geography

- Geography combines natural and social sciences.
- Geographers explain *where* things are, *why* they are there, and why their locations have *significance*.

The word *geography* was invented by the ancient Greek scholar Eratosthenes (ca. 276–ca. 194 B.C.); it is based on two Greek words: *geo* meaning "Earth," and *graphy*, meaning "to write." Contemporary geography is more than only writing about Earth; it is a science.

GEOGRAPHY COMBINES NATURAL AND SOCIAL SCIENCE

The science of geography is divided broadly into two categories—physical geography and human geography. The beach at Ipanema, Brazil, displays features of interest to both physical and human geographers.

- Human geographers study cultural features, such as economic activities and cities (Figure 1.1.1).
- Physical geographers concentrate on the distribution of natural features, such as landforms and vegetation (Figure 1.1.2).

One of the distinctive features of geography is its use of natural science concepts to help understand human behavior, and conversely the use of social science concepts to help understand physical processes.

GEOGRAPHERS EXPLAIN *WHERE* AND *WHY*

To explain *where* things are, one of geography's most important tools is a map. Ancient and medieval geographers created maps to describe what they knew about Earth. Today, accurate maps are generated from electronic data.

Geographers study *why* every place on Earth is in some ways unique and in other ways related to other locations.

To explain why every place is unique, geographers have two basic concepts:

- A **place** is a specific point on Earth distinguished by a particular characteristic. Every place occupies a unique location, or position, on Earth's surface (Figure 1.1.3).
- A **region** is an area of Earth distinguished by a distinctive combination of cultural and physical features. (Figure 1.1.4).

To explain why different places are interrelated, geographers have three basic concepts:

- **Scale** is the relationship between the portion of Earth being studied and Earth as a whole. Geographers study every scale from the individual to the entire Earth, though increasingly they are concerned with global-scale patterns and processes (Figure 1.1.5).
- **Space** refers to the physical gap or interval between two objects. Geographers observe that many objects are distributed across space in a regular manner, for discernable reasons (Figure 1.1.6).
- **Connection** refers to relationships among people and objects across the barrier of space. Geographers are concerned with the various means by which connections occur (Figure 1.1.7).

To explain the underlying significance of observed spatial patterns, geographers examine the interrelationships between the natural environment and human behavior. Humans are influenced by nature and in turn alter natural processes.

▲ 1.1.1 **HUMAN GEOGRAPHY AT IPANEMA**
Ipanema is located in one of world's largest urban areas—Rio de Janeiro—and has an economy heavily dependent on tourism.

▲ 1.1.2 **PHYSICAL GEOGRAPHY AT IPANEMA BEACH**
Ipanema's physical landscape includes a distinctive landform, the twin peaks in the background known as Two Brothers.

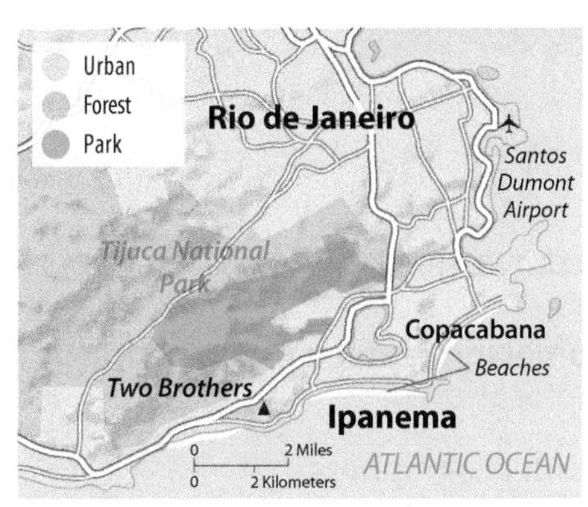

▲ 1.1.3 **PLACE**
Ipanema is located along the Atlantic Ocean within the city of Rio de Janeiro, Brazil.

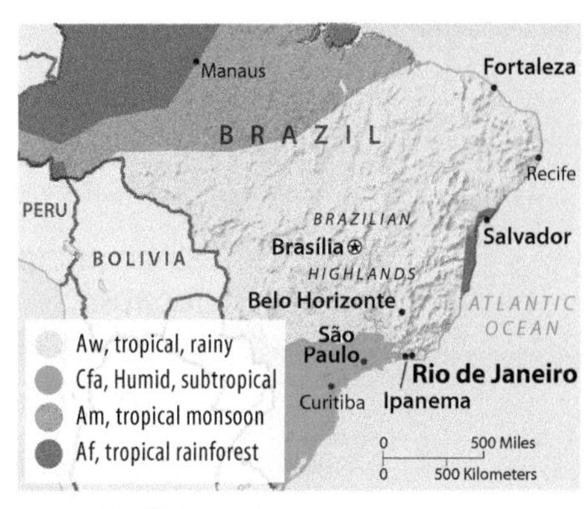

▲ 1.1.4 **REGION**
The climate region known by the notation Aw that includes Ipanema is warm all year round but nearly all of the year's rain falls between May and October. To be virtually guaranteed great beach weather, visit Ipanema in December. Other nearby climate regions have other patterns of temperature and rainfall.

▲ 1.1.5 **SCALE**
Geographers study trends of globalization and local diversity. For example, people in diverse locations play on sand, wearing very little clothing. At the same time, many cultures frown on public display of near-naked bodies.

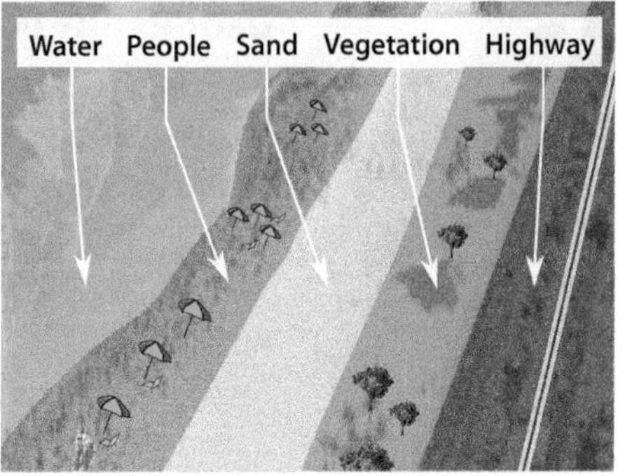

▲ 1.1.6 **SPACE**
People are not distributed randomly along Ipanema. The distribution of people follows a linear pattern parallel to the water's edge and away from the highway.

▲ 1.1.7 **CONNECTION**
Few humans live in isolation. Ipanema beach is connected with the rest of the city of Rio de Janeiro by a major highway. An airplane, such as the one in the photo, can connect Ipanema beach with the rest of the world.

1.2 Ancient and Medieval Geography

- **Since ancient times, maps have helped geographers to explain where things are located.**
- **Accurate mapmaking revived in the Middle Ages, and developed further during the Age of Exploration**

In the ancient world, geographers in Greece and neighboring Eastern Mediterranean lands, as well as in China, described and mapped Earth with increasing accuracy. A revival of geography and mapmaking occurred in Europe and Asia during the Middle Ages.

GEOGRAPHY IN THE ANCIENT WORLD

The earliest surviving maps were drawn in the Eastern Mediterranean in the seventh or sixth century B.C. (Figure 1.2.1). Major contributors to geographic thought in the Ancient Eastern Mediterranean included:

- Thales of Miletus (ca. 624–ca. 546 B.C.), who applied principles of geometry to measuring land area.
- Anaximander (610–ca. 546 B.C.), a student of Thales, who made a world map based on information from sailors, though he argued that world was shaped like a cylinder.
- Hecataeus (ca. 550–ca. 476 B.C.), who may have produced the first geography book, called *Ges Periodos,* ("Travels Around the Earth").
- Aristotle (384–322 B.C.), who was the first to demonstrate that Earth was spherical.

▼ 1.2.1 **THE OLDEST KNOWN MAP**
A map of a plan for the town of Çatalhöyük, located in present-day Turkey, dates from approximately 6200 B.C. Archaeologists found the map on the wall of a house that was excavated in the 1960s. The map is now in the Konya Archaeology Museum. (below right) A color version of the Çatalhöyük map. A volcano rises above the buildings of the city. (below left) A 3D reconstruction of the Çatalhöyük map.

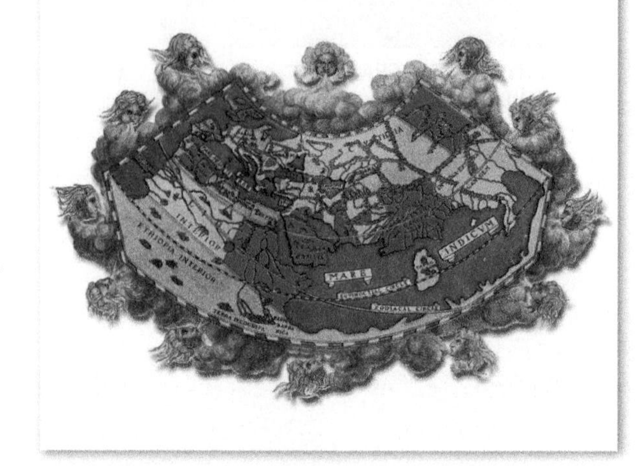

▲ 1.2.2 **WORLD MAP BY PTOLEMY, CA. 150 BC**
The map shows the known world at the height of the Roman Empire, surrounding the Mediterranean Sea and Indian Ocean.

- Eratosthenes (ca. 276–ca. 195 B.C.), the inventor of the word *geography*, who accepted that Earth was round (as few did in his day), calculated its circumference within 0.5 percent accuracy, accurately divided Earth into five climatic regions, and described the known world in one of the first geography books.
- Strabo (ca. 63 B.C–ca. A.D. 24), who described the known world in a 17-volume work *Geography.*
- Ptolemy (A.D. ca. 100–ca. 170), who wrote an eight-volume *Guide to Geography*, codified basic principles of mapmaking, and prepared numerous maps that were not improved upon for more than a thousand years (Figure 1.2.2).

Ancient Chinese geographic contributions included:

- *Yu Gong* ("Tribute of Yu"), a chapter of the book *Shu Jing* ("Classic of History"), which was the earliest surviving Chinese geographical writing, by an unknown author from the fifth century B.C., described the economic resources of the country's different provinces.
- Pei Xiu (A.D. 224–271), the "father of Chinese cartography," who produced an elaborate map of the country in A.D. 267.

GEOGRAPHY IN THE MIDDLE AGES

During the first millennium A.D., maps became less mathematical and more fanciful, showing Earth as a flat disk surrounded by fierce animals and monsters. Scientific mapmaking resumed during the Middle Ages, first in Asia and then in Europe. Leading medieval contributors to geography included:

- Muhammad al-Idrisi (1100–ca. 1165), a Muslim geographer who prepared a world map and geography text in 1154, building on Ptolemy's long-neglected work (Figure 1.2.3).
- Abu Abdullah Muhammad Ibn-Battuta (1304–ca. 1368), a Moroccan scholar, who wrote *Rihla* ("Travels") based on three decades of journeys covering more than 120,000 kilometers (75,000 miles) through the Muslim world of northern Africa, southern Europe, and much of Asia.

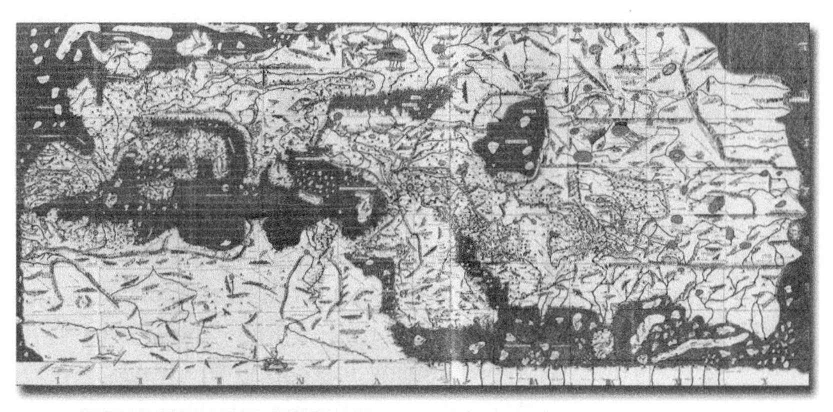

▲ 1.2.3 **WORLD MAP BY AL-IDRISI, 1154**
Al-Idrisi built on Ptolemy's map, which had been neglected for nearly a millennium.

GEOGRAPHY IN THE AGE OF EXPLORATION

With the discovery of the New World and the development of printing, scientific mapmaking in Europe made rapid progress.

- Martin Waldseemuller (ca. 1470–ca. 1521), a German cartographer who was credited with producing the first map to use the label "America;" he wrote on the map (translated from Latin) "from Amerigo the discoverer . . . as if it were the land of Americus, thus 'America'" (Figure 1.2.4).
- Abraham Ortelius (1527–1598), a Flemish cartographer, created the first modern atlas, and was the first to hypothesize that the continents were once joined together before drifting apart (Figure 1.2.5).

▲ 1.2.4 **WORLD MAP BY WALDSEEMULLER, 1507**
This was one of the first maps to depict the Western Hemisphere separated from Europe and Africa by the Atlantic Ocean, and the first to label the Western Hemisphere "America."

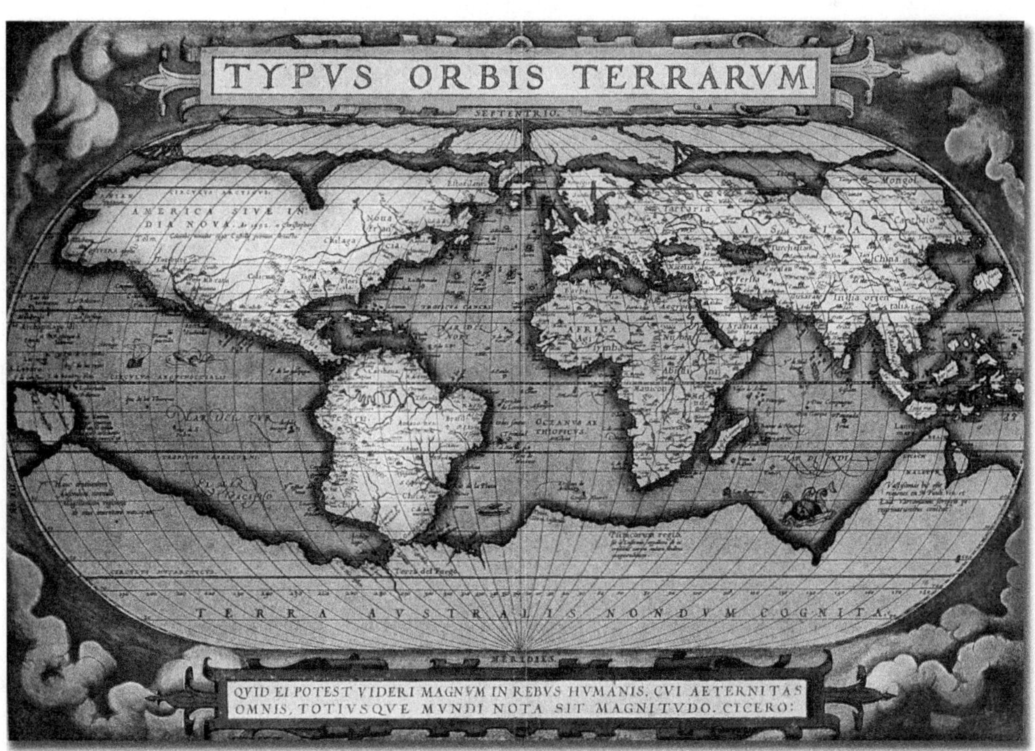

► 1.2.5 **WORLD MAP BY ORTELIUS, 1571**
This was one of the first maps to show the considerable extent of the Western Hemisphere, as well as the Antarctic land mass.

1.3 Reading Maps

- **A map is a scale model of all or a portion of Earth.**
- **A map is a flat depiction of a nearly round Earth.**

For centuries, geographers have worked to perfect the science of mapmaking, called **cartography**. A **map** is a scale model of the real world, made small enough to work with on a desk or computer. Maps serve two purposes:

- A reference tool. A map helps us to find the shortest route between two places and to avoid getting lost along the way.
- A communications tool. A map is often the best means for depicting the distribution of human activities or physical features, as well as for thinking about reasons underlying a distribution.

To make a map, a cartographer must make two decisions:

- How much of Earth's surface to depict on the map (map scale).
- How to transfer a spherical Earth to a flat map (projection).

MAP SCALE

Should the map show the entire globe, or just one continent, or a country, or a city? To make a map of the entire world, many details must be omitted because there simply is not enough space. Conversely, if a map shows only a small portion of Earth's surface, such as a street map of a city, it can provide a wealth of detail about a particular place.

The level of detail and the amount of area covered depend on the **map scale,** which is the relationship of a feature's size on a map to its actual size on Earth. Map scale is presented in three ways (Figure 1.3.1):

- A ratio or fraction.
- A written scale.
- A graphic scale.

◄ 1.3.1. **MAP SCALE**

The four images show Washington State, the Seattle metropolitan area, downtown Seattle, and Pike Place Market. These four images show map scale in three ways:

- *A ratio or fraction* shows the numerical ratio between distances on the map and Earth's surface. The Washington map has a ratio of 1:10,000,000, meaning that one unit (inch, centimeter, foot, finger length) on the map represents 10,000,000 of the same unit (inch, centimeter, foot, finger length) on the ground. The unit chosen for distance can be anything, as long as the units of measure on both the map and the ground are the same. The 1 on the left side of the ratio always refers to a unit of distance *on the map*, and the number on the right always refers to the *same unit* of distance *on Earth's surface.*
- *A written scale* describes this relation between map and Earth distances in words. For example, the statement "1 centimeter equals 10 kilometers" means that one centimeter on the map represents 10 kilometers on Earth's surface. Again, the first number always refers to map distance, and the second to distance on Earth's surface.
- *A graphic scale* usually consists of a bar line marked to show distance on Earth's surface. To use a bar line, first determine with a ruler the distance on the map in inches or centimeters. Then hold the ruler against the bar line and read the number on the bar line opposite the map distance on the ruler. The number on the bar line is the equivalent distance on Earth's surface.

1 centimeter on the map equals 100 kilometers on Earth; 1:10,000,000.

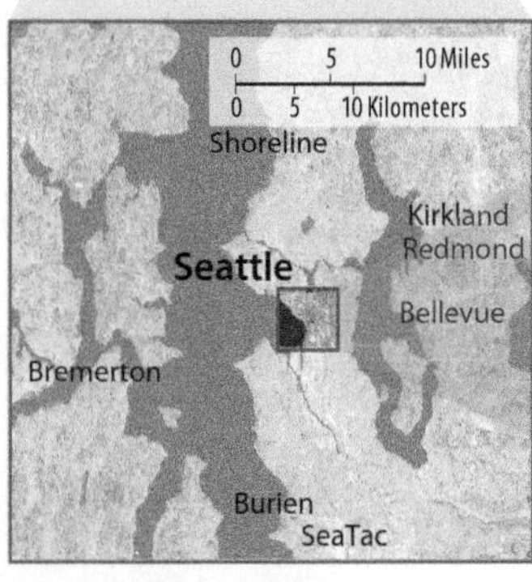

1 centimeter on the map equals 10 kilometers on Earth; 1:1,000,000.

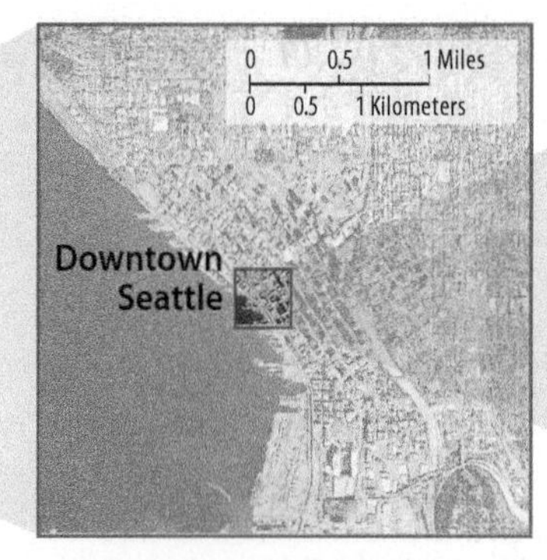

1 centimeter on the map equals 1 kilometer on Earth; 1:100,000.

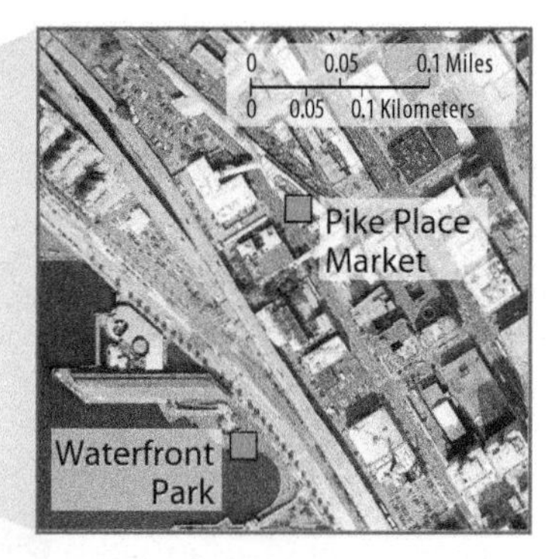

1 centimeter on the map equals 100 meters on Earth; 1:10,000.

PROJECTION

Earth is very nearly a sphere and therefore accurately represented in the form of a globe. However, a globe is an extremely limited tool with which to communicate information about Earth's surface. A small globe does not have enough space to display detailed information, whereas a large globe is too bulky and cumbersome to use. And a globe is difficult to write on, photocopy, display on a computer screen, or refer to in a car.

Earth's spherical shape poses a challenge for cartographers because drawing Earth on a flat piece of paper unavoidably produces some distortion. The scientific method of transferring locations on Earth's surface to a flat map is called **projection**. Several types of projections are shown in Figure 1.3.2.

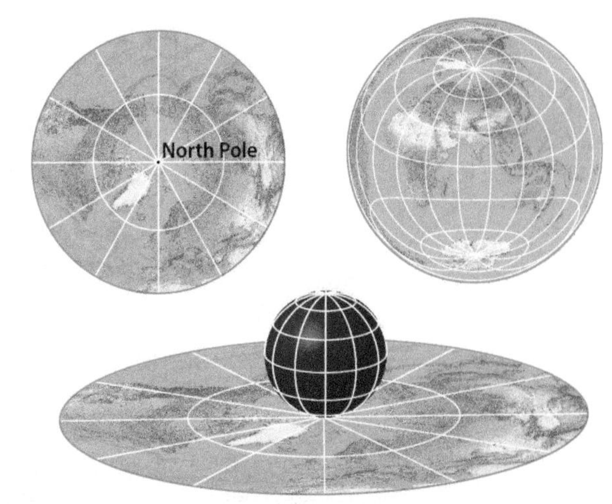

▲ 1.3.2.a **AZIMUTHAL PROJECTIONS**
Azimuthal projections are well-suited for larger areas and are used for most of the world maps.

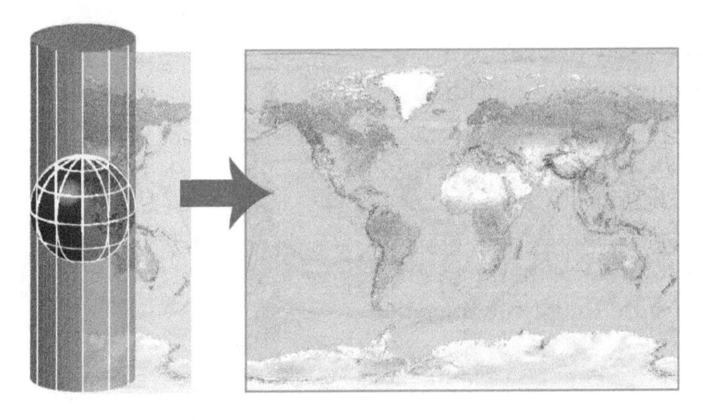

▲ 1.3.2.b **CYLINDRICAL PROJECTIONS**
Cylindrical projections are used for specialized maps. The most widely used cylindrical world map, created by Gerardus Mercator in 1569, was widely used by mariners.

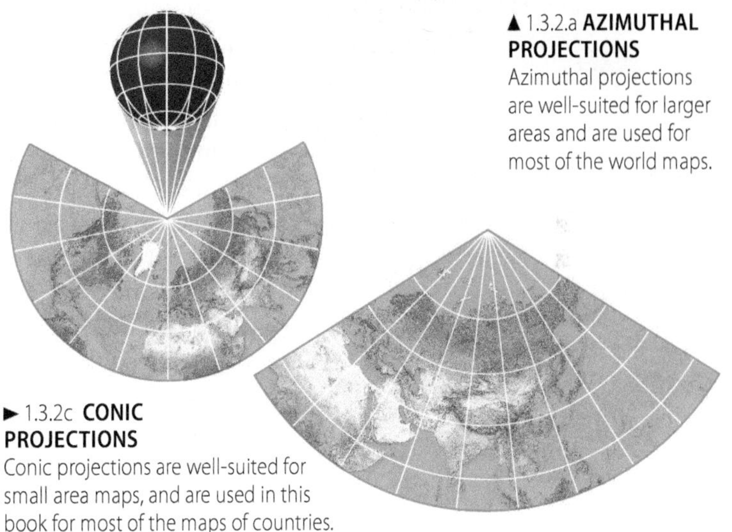

► 1.3.2c **CONIC PROJECTIONS**
Conic projections are well-suited for small area maps, and are used in this book for most of the maps of countries.

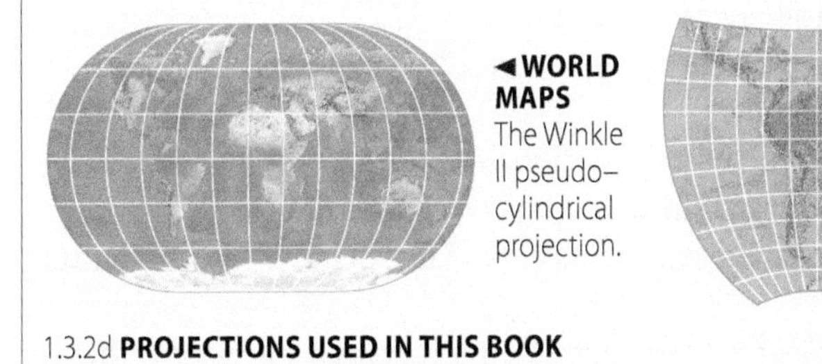

◄ **WORLD MAPS**
The Winkle II pseudo-cylindrical projection.

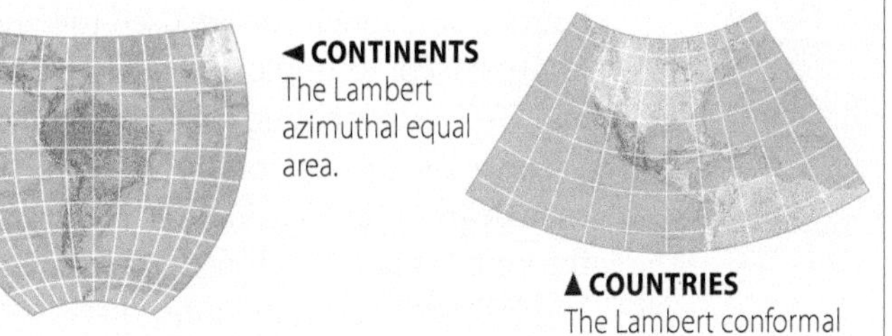

◄ **CONTINENTS**
The Lambert azimuthal equal area.

▲ **COUNTRIES**
The Lambert conformal conic projection.

1.3.2d **PROJECTIONS USED IN THIS BOOK**

MasteringGEOGRAPHY
Animations
Map Projections

A projection can result in four types of distortion (Figure 1.3.3):

1. The ***shape*** of an area can be distorted, so that it appears more elongated or squat than in reality.
2. The ***distance*** between two points may become increased or decreased.
3. The ***relative size*** of different areas may be altered, so that one area may appear larger than another on a map but is in reality smaller.
4. The ***direction*** from one place to another can be distorted.

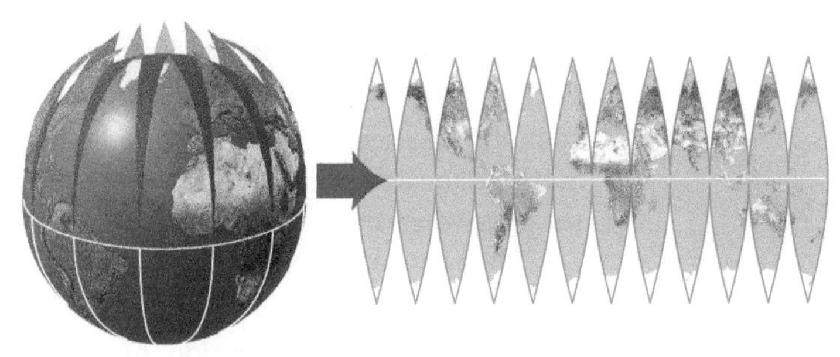

▲ 1.3.3 **DISTORTION** To make a map, the spherical Earth is divided into segments. Flattening these segments onto flat paper results in distortion.

1.4 The Geographic Grid

- The geographic grid divides Earth's surface into latitudes and longitudes.
- The geographic grid is the basis for time zones.

The **geographic grid** is a system of imaginary arcs drawn in a grid pattern on Earth's surface. The location of any place on Earth's surface can be described by these human-created arcs, known as meridians and parallels. The geographic grid plays an important role in telling time.

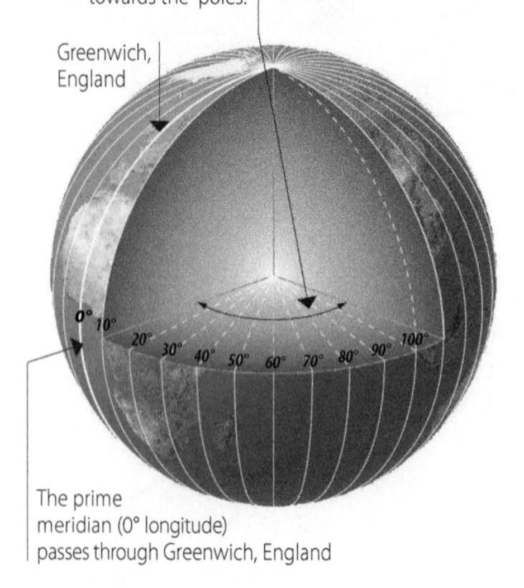

▲ 1.4.1 **LONGITUDE**
Meridians have numbers between 0° and 180° east or west longitude, depending if they are either east or west of the prime meridian.

LATITUDE AND LONGITUDE

Cartographers identify meridians and parallels through numbering systems:

- A **meridian** is an arc drawn between the North and South poles. The location of each meridian is identified on Earth's surface according to a numbering system known as **longitude**. The meridian that passes through the Royal Observatory at Greenwich, England, is 0° longitude, also called the **prime meridian**. The meridian on the opposite side of the globe from the prime meridian is 180° longitude (Figure 1.4.1).
- A **parallel** is a circle drawn around the globe parallel to the equator and at right angles to the meridians. The numbering system to indicate the location of a parallel is called **latitude.** The equator is 0° latitude, the North Pole 90° north latitude, and the South Pole 90° south latitude (Figure 1.4.2).

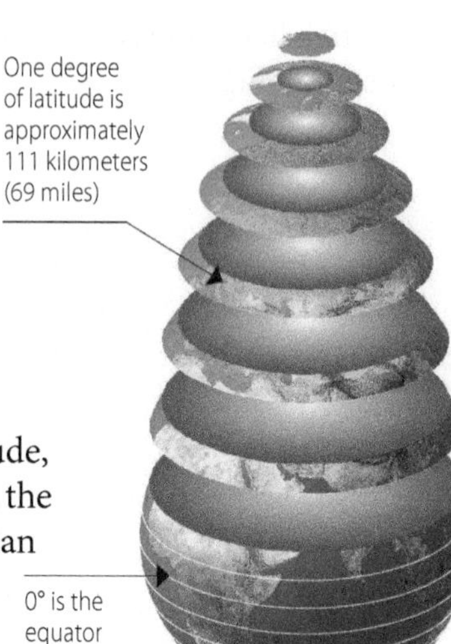

▲ 1.4.2 **LATITUDE**
Parallels range between 0° and 90° north or south latitude, depending if they are either north or south of the equator.

Latitude and longitude are used together to identify locations. For example, Philadelphia, Pennsylvania, is located at 40° north latitude and 75° west longitude (Figure 1.4.3).

The mathematical location of a place can be designated more precisely by dividing each degree into 60 minutes (′) and each minute into 60 seconds (″). For example, the official mathematical location of Philadelphia's City Hall is 39°57′8″ north latitude and 75°9′49″ west longitude.

Global Positioning Systems typically divide degrees into decimal fractions rather than minutes and seconds. The Bally Ribbon Mills factory in suburban Philadelphia, for example, is located at 40.400780° north latitude and 75.587439° west longitude.

Measuring latitude and longitude is a good example of how geography is partly a natural science and partly a study of human behavior.

- Latitudes are scientifically derived by Earth's shape and its rotation around the Sun. The equator (0° latitude) is the parallel with the largest circumference and is the place where every day has 12 hours of daylight. Even in ancient times, latitude could be accurately measured by the length of daylight and the position of the Sun and stars.
- Longitudes are a human creation. Any meridian could have been selected as 0° longitude because all have the same length and all run between the poles. The 0° longitude runs through Greenwich because England was the world's most powerful country when longitude was first accurately measured and the international agreement was made.

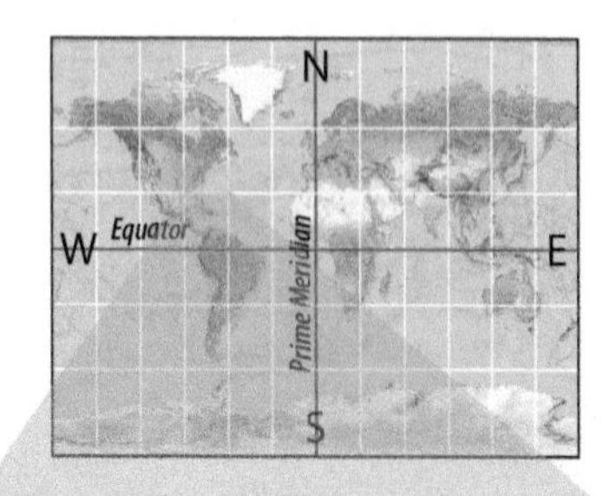

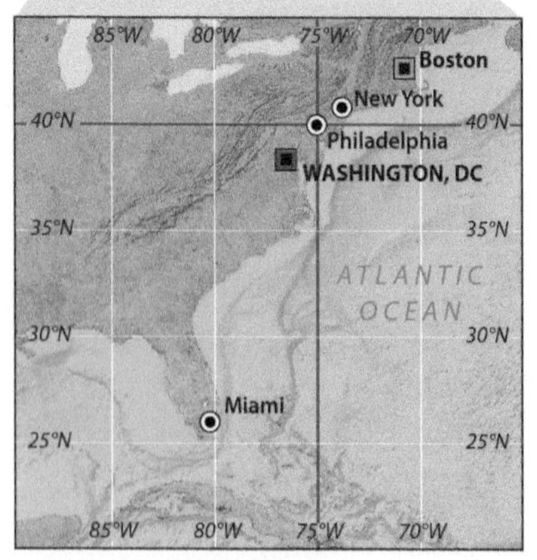

▲ 1.4.3 **HOW LATITUDE AND LONGITUDE WORK**
Philadelphia, Pennsylvania, is located near 40° north latitude and 75° west longitude.

TIME ZONES

Longitude plays an important role in calculating time. Earth makes a complete rotation every 24 hours and as a sphere is divided into 360° of longitude. Therefore, traveling 15° east or west is the equivalent of traveling to a place that is 1 hour earlier or later than the starting point (360° divided by 24 hours equals 15°).

By international agreement, **Greenwich Mean Time (GMT)** or Universal Time (UT), which is the time at the prime meridian (0° longitude), is the master reference time for all points on Earth.

As Earth rotates eastward, any place to the east of you always passes "under" the Sun earlier. Thus as you travel eastward from the prime meridian, you are "catching up" with the Sun, so you must turn your clock ahead from GMT by 1 hour for each 15°. If you travel westward from the prime meridian, you are "falling behind" the Sun, so you turn your clock back from GMT by 1 hour for each 15°.

Each 15° band of longitude is assigned to a standard time zone (Figure 1.4.4). The United States and Canada share four standard time zones:

- Eastern, near 75° west, is 5 hours earlier than GMT.
- Central, near 90° west, is 6 hours earlier than GMT.
- Mountain, near 105° west, is 7 hours earlier than GMT.
- Pacific, near 120° west, is 8 hours earlier than GMT.

The United States has two additional standard time zones:

- Alaska, near 135° west, is 9 hours earlier than GMT.
- Hawaii-Aleutian, near 150° west, is 10 hours earlier than GMT.

Canada has two additional standard time zones:

- Atlantic, near 60° west, is 4 hours earlier than GMT.
- Newfoundland is 3½ hours earlier than GMT; the residents of Newfoundland assert that their island, which lies between 53° and 59° west longitude, would face dark winter afternoons if it were in the Atlantic Time Zone, and dark winter mornings if it were 3 hours earlier than GMT.

The **International Date Line** for the most part follows 180° longitude. When you cross it heading eastward toward America you move the clock back 24 hours, or one entire day. You turn the clock ahead 24 hours if you are heading westward toward Asia.

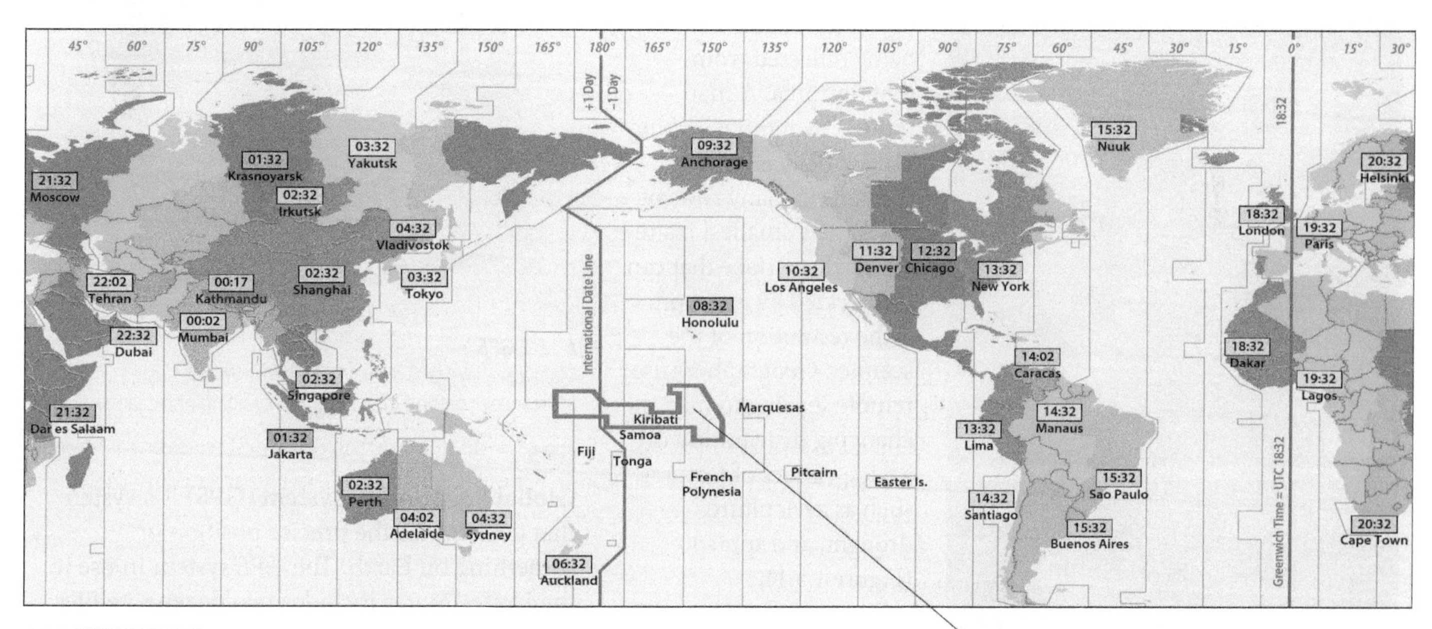

▲ 1.4.4 **TIME ZONES**

The eastern United States, which is near 75° west longitude, is therefore 5 hours earlier than GMT (the 75° difference between the prime meridian and 75° west longitude, divided by 15° per hour, equals 5 hours). Thus when the time in New York City in the winter is 1:32 PM, (or 13:32 hours, using a 24-hour clock), it is 6:32 PM (or 18:32 hours) GMT. During the summer, many places in the world, including most of North America, move the clocks ahead one hour; so in the summer when it is 6:32 PM GMT, the time in New York City is 2:32 PM.

If it is 1:32 PM, (or 13:32 hours) Sunday in New York, it is 6:32 PM Sunday in London, 7:32 PM (19:32) Sunday in Paris, 8:32 PM (20:32) Sunday in Helsinki, 9:32 PM (21:32) Sunday in Moscow, 2:32 AM Monday in Singapore, and 4:32 AM Monday In Sydney. Continuing farther east, it is 6:32 AM *Monday* in Auckland --but when you get to Honolulu, it is 8:32 AM *Sunday*, because the International Date Line lies between Auckland and Honolulu.

The International Date Line for the most part follows 180° longitude. However, in 1997, Kiribati, a collection of small islands in the Pacific Ocean, moved the International Date Line 3,000 kilometers (2,000 miles) to its eastern border near 150° west longitude. As a result, Kiribati is the first country to see each day's sunrise. Kiribati hoped that this feature would attract tourists to celebrate the start of the new millennium on January 1, 2000 (or January 1, 2001, when sticklers pointed out the new millennium really began). But it did not.

1.5 Geography's Contemporary Analytic Tools

- **Geographic study is aided by information from satellite imagery.**
- **Complex geographic data can be overlaid and analyzed through geographic information systems.**

Having largely completed the great task of accurately mapping Earth's surface, geographers have turned to **Geographic Information Science** (GIScience), which is the development and analysis of data about Earth acquired through satellite and other electronic information technologies. GIScience helps geographers to create more accurate and complex maps and to measure changes over time in the characteristics of places.

REMOTE SENSING

The acquisition of data about Earth's surface from a satellite orbiting Earth or from other long-distance methods is known as **remote sensing**. Remote-sensing satellites scan Earth's surface and transmit images in digital form to a receiving station on Earth's surface.

At any moment a satellite sensor records the image of a tiny area called a picture element or pixel. Scanners detect the radiation being reflected from that tiny area. A map created by remote sensing is essentially a grid containing many rows of pixels. The smallest feature on Earth's surface that can be detected by a sensor is the resolution of the scanner. Geographers use remote sensing to map the changing distribution of a wide variety of features, such as agriculture, drought, and sprawl (Figure 1.5.1).

◄ **1.5.1 REMOTE SENSING**
(top) Satellite image of South Dakota's Oahe Reservoir in 1999. Deep water is shown in dark blue and shallow water in light blue. (bottom) Satellite image of Oahe Reservoir in 2004. After several years of drought, the reservoir has less deep water, as shown in the reduction of the area of dark blue.

▲ 1.5.2 **GPS**
Many cars have GPS display on the instrument cluster, offering assistance in getting directions and avoiding traffic jams.

GPS

Global Positioning System (GPS) is a system that determines the precise position of something on Earth. The GPS system in use in the United States includes two dozen satellites placed in predetermined orbits, a series of tracking stations to monitor and control the satellites, and receivers that compute position, velocity, and time from the satellite signals.

GPS is commonly used in the navigation of aircraft and ships, and increasingly to provide directions for drivers of motor vehicles (Figure 1.5.2). Cell phones equipped with GPS allow individuals to share their whereabouts with friends and relatives.

GIS

A **geographic information system** (GIS) is a computer system that captures, stores, queries, analyzes, and displays geographic data. GIS can be used to produce maps (including those in this book) that are more accurate and attractive than those drawn by hand.

The position of any object on Earth can be measured and recorded with mathematical precision and then stored in a computer. A map can be created by asking the computer to retrieve a number of stored objects and combine them to form an image. Each type of information can be stored in a layer (Figure 1.5.3).

GIS enables geographers to calculate whether relationships between objects on a map are significant or merely coincidental. Layers can be compared to show relationships among different kinds of information. To protect hillsides from development, for example, a geographer may wish to compare a layer of recently built houses with a layer of steep slopes.

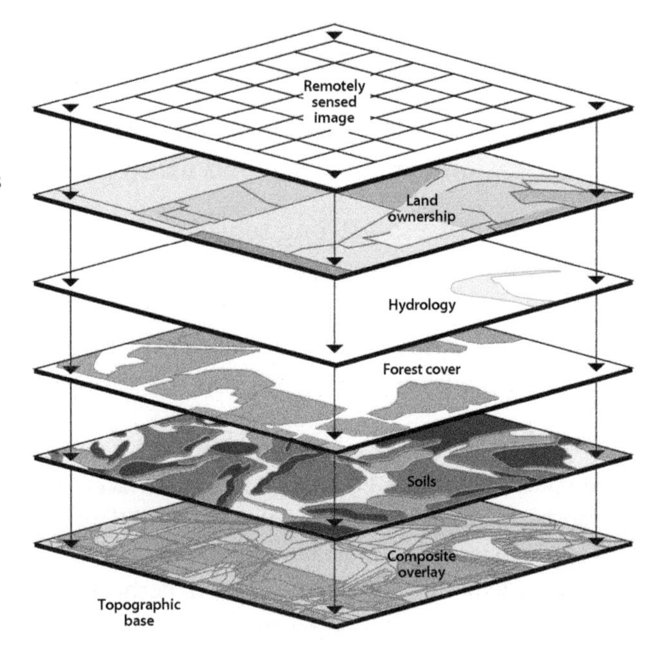

▲ 1.5.3 **GIS**
GIS involves storing information about a location in layers. Each layer represents a different piece of human or environmental information. The layers can be viewed individually or in combination.

◄ 1.5.4 **MASH-UP**
Use Google Earth to explore a mash-up of Central London and the subway ("Underground") stations near the London Eye (the Ferris wheel on the south bank of the River Thames).

Fly to: *London Eye, London, England.*

Use Google Earth's navigation tools and layer options to explore features of this area.

What is the distance from the Eye to the nearest Underground station?

MASH-UPS

Mapping services, such as Google Maps and Google Earth, give computer programmers access to the application programming interface (API), which is the language that links a database such as an address list with software such as mapping. The API for mapping software, available at such sites as www.google.com/apis/maps, enables a computer programmer to create a mash-up that places data on a map. The term mash-up refers to the practice of overlaying data from one source on top of one of the mapping services and comes from the hip-hop practice of mixing two or more songs.

Mash-ups assist in finding apartments, bars, hotels, sports facilities, and transit stops (Figure 1.5.4). Mapping software can show the precise location of commercial airplanes currently in the air, the gas stations with the cheapest prices, and current traffic tie-ups on highways and bridges.

1.6 Place: A Unique Location

- **Location is the position that something occupies on Earth.**
- **Location can be described using place names, situation, and site.**

Humans possess a strong sense of place—that is, a feeling for the features that contribute to the distinctiveness of a particular spot on Earth. Geographers think about where particular places are located and the combination of features that make each place on Earth distinct. Geographers describe a feature's place on Earth by identifying its **location**, which is the position that something occupies on Earth's surface.

PLACE NAMES

Because all inhabited places on Earth's surface—and many uninhabited places—have been named, the most straightforward way to describe a particular location is often by referring to its place name. A **toponym** is the name given to a place on Earth.

A place may be named for a person, perhaps its founder or a famous person with no connection to the community, such as George Washington. Some settlers selected place names associated with religion, such as St. Louis and St. Paul, whereas other names derive from ancient history, such as Athens, Attica, and Rome, or from earlier occupants of the place (Figure 1.6.1).

The Board of Geographical Names, operated by the U.S. Geological Survey, was established in the late nineteenth century to be the final arbiter of names on U.S. maps. In recent years the board has been especially concerned with removing offensive place names, such as those with racial or ethnic connotations.

► 1.6.1 **LONGEST U.S. PLACE NAME**
The longest place name in the United States may be Lake Chargoggagoggmanchauggagoggchaubunagungamaugg, Massachusetts. The name is thought to be a combination of three phrases: "you fish on your side" (Chargoggagogg), "I fish on my side" (Manchauggagogg), and "nobody fish in the middle" (Chaubunagungamaugg). The name is said to have been given by agreement of several Native American tribes living at opposite ends of the lake.

▲ 1.6.2 **SITUATION OF SINGAPORE**
The country of Singapore is situated near the Strait of Malacca, which is the major passageway for ships traveling between the South China Sea and the Indian Ocean. Some 50,000 vessels, one-fourth of the world's maritime trade, pass through the strait each year. The downtown area of the City of Singapore is situated near where the Singapore River flows into the Singapore Strait.

SITUATION

Situation is the location of a place relative to other places. Situation is a valuable way to indicate location, for two reasons:

- Situation helps us find an unfamiliar place by comparing its location with a familiar one. We give directions to people by referring to the situation of a place: "It's down past the courthouse, beside the large elm tree."
- Situation helps us understand the importance of a location. Many places are important because they are accessible to other places. For example, because of its situation, Singapore has become a center for the trading and distribution of goods for much of Southeast Asia (Figure 1.6.2).

SITE

Geographers can describe the location of a place by **site**, which is the physical character of a place. Important site characteristics include climate, water sources, topography, soil, vegetation, latitude, and elevation. The combination of physical features gives each place a distinctive character (Figure 1.6.3).

Humans have the ability to modify the characteristics of a site. The southern portion of New York City's Manhattan Island is twice as large today as it was in 1626, when Peter Minuit bought the island from its native inhabitants for the equivalent of $23.75 worth of Dutch gold and silver coins (Figure 1.6.4).

▲ 1.6.3 **SITE OF SINGAPORE**
The site of the country of Singapore is approximately 60 islands, the largest and most populous of which is named Pulau Ujong. The site of the City of Singapore is the southern portion of Pulau Ujong.

◀ 1.6.4 **CHANGING SITE OF NEW YORK CITY**
(top) The site of Manhattan Island has been altered by building on landfill in the Hudson and East rivers. Extending the island into the rivers provided more land for offices, homes, parks, warehouses, and docks. (left) During the late 1960s and early 1970s, the World Trade Center was built partially on landfill in the Hudson River from the colonial era. Battery Park City was built on landfill removed from the World Trade Center construction site.

Fly to: *World Trade Center, New York, NY, USA*

Click the World Trade Center site near the top of the image.

What are the plans for the site?

1.7 Region: A Unique Area

- **A region is an area of Earth with a unique combination of features.**
- **Three types of regions are functional, formal, and vernacular.**

The "sense of place" that humans possess may apply to a larger area of Earth rather than to a specific point. An area of Earth defined by one or more distinctive characteristics is a region. A region derives its unified character through the **cultural landscape**—a combination of cultural features such as language and religion, economic features such as agriculture and industry, and physical features such as climate and vegetation.

FUNCTIONAL REGION

A **functional region**, also called a nodal region, is an area organized around a node or focal point. The characteristic chosen to define a functional region dominates at a central focus or node and diminishes in importance outward. The region is tied to the central point by transportation or communications systems, or by economic or functional associations.

Geographers often use functional regions to display information about economic areas. The region's node may be a shop or service, and the boundaries of the region mark the limits of the trading area of the activity. People and activities may be attracted to the node, and information may flow from the node to the surrounding area.

Examples of functional regions include the reception area of a television station, the circulation area of a newspaper, and the trading area of a department store (Figure 1.7.1). A television station's signal is strongest at the center of its service area, becomes weaker at the edge, and eventually disappears. A department store attracts fewer customers and a newspaper fewer readers from the edge of a trading area.

New technology is breaking down traditional functional regions. Newspapers such as *USA Today, The Wall Street Journal*, and *The New York Times* are composed in one place, transmitted by satellite to printing machines in other places, and delivered by airplane and truck to yet other places. Television stations are broadcast to distant places by cable, satellite, or Internet. Customers can shop at distant stores by mail or Internet.

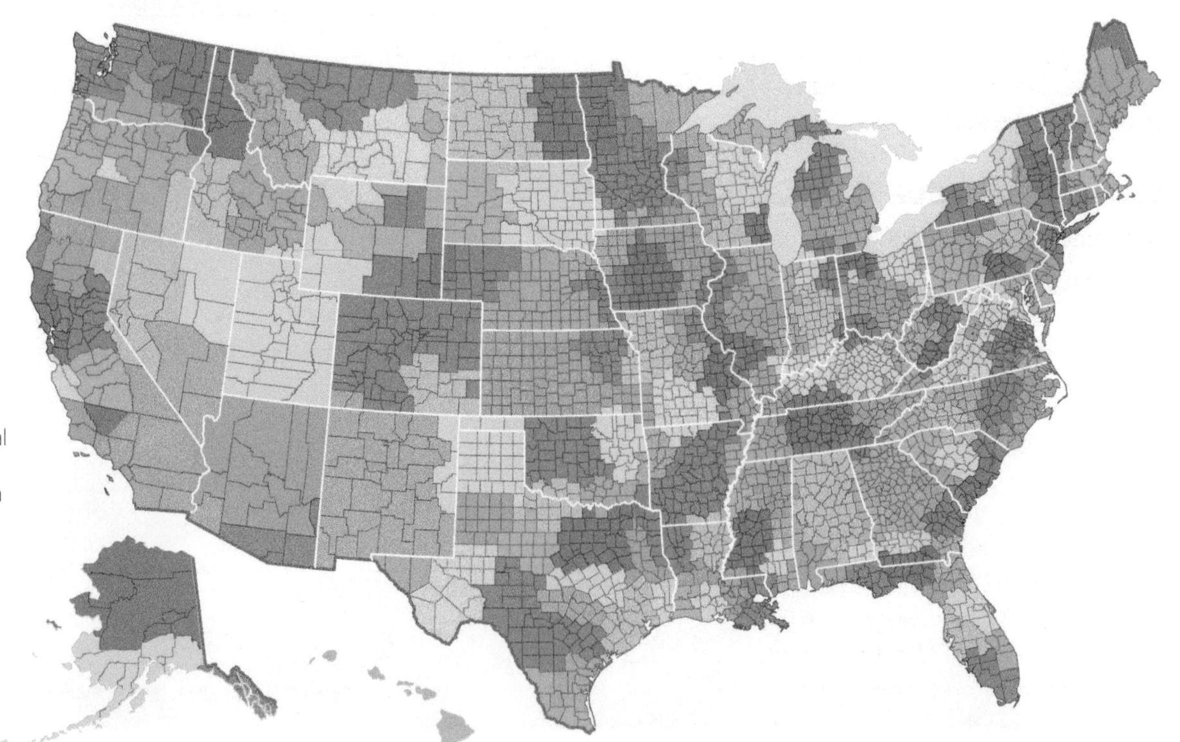

► 1.7.1 **FUNCTIONAL REGIONS**
The United States can be divided into functional regions based on television markets, which are groups of counties served by a collection of TV stations. Each functional region is known as a designated market areas (DMA), a term trademarked by Nielsen Media Research.

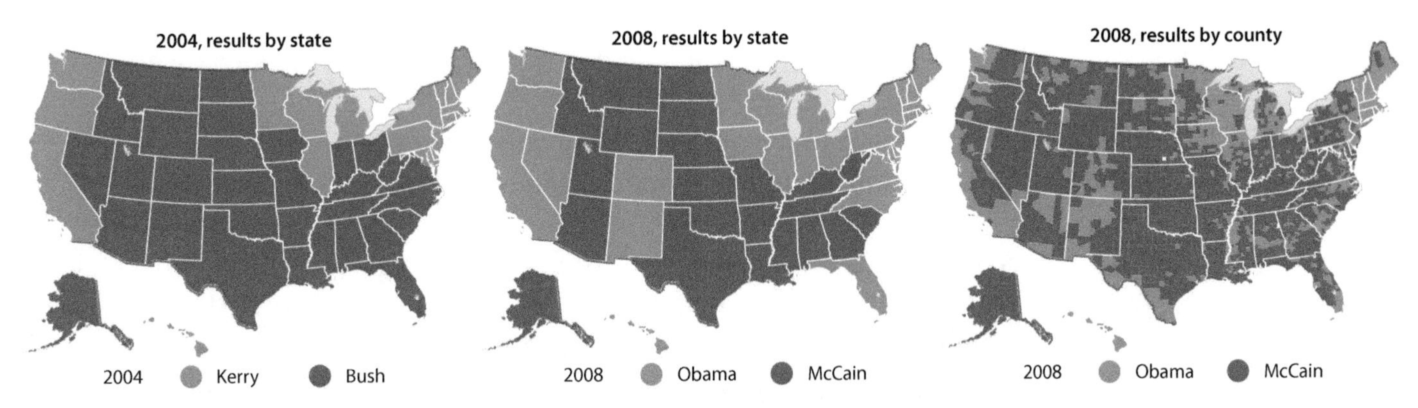

▲ 1.7.2 **FORMAL REGIONS**

(left) Presidential election results by state, 2004. Democrat John Kerry won the states in the Northeast, Upper Midwest, and Pacific Coast regions of the United States, while Republican George W. Bush won the remaining regions. (center) Presidential election results by state, 2008. Democrat Barack Obama won the election by capturing some states in regions that had been won entirely by the Republican four years earlier. (right) Presidential election results by county, 2008. Republican John McCain carried most of the land area of the United States, but Obama won the most votes by winning the most populous counties.

FORMAL REGION

A **formal region**, also called a uniform region or a homogeneous region, is an area within which everyone shares in common one or more distinctive characteristics. The shared feature could be a cultural value such as a common language, an economic activity such as production of a particular crop, or an environmental property such as climate. In a formal region the selected characteristic is present throughout.

Geographers typically identify formal regions to help explain broad global or national patterns, such as variations in religions and levels of economic development. The characteristic selected to distinguish a formal region often illustrates a general concept rather than a precise mathematical distribution.

Some formal regions are easy to identify, such as countries or local government units. Montana is an example of a formal region, characterized by a government that passes laws, collects taxes, and issues license plates with equal intensity throughout the state.

In other kinds of formal regions a characteristic may be predominant rather than universal. For example, we can distinguish formal regions within the United States characterized by a predominant voting for Republican candidates, although Republicans do not get 100 percent of the votes in these regions—nor in fact do they always win (Figure 1.7.2).

A cautionary step in identifying formal regions is the need to recognize the diversity of cultural, economic, and environmental factors, even while making a generalization. A minority of people in a region may speak a language, practice a religion, or possess resources different from those of the majority. People in a region may play distinctive roles in the economy and hold different positions in society based on their gender or ethnicity.

VERNACULAR REGION

A **vernacular region**, or perceptual region, is a place that people believe exists as part of their cultural identity. Such regions emerge from people's informal sense of place rather than from scientific models developed through geographic thought.

As an example of a vernacular region, Americans frequently refer to the South as a place with environmental, cultural, and economic features perceived to be quite distinct from the rest of the United States. Many of these features can be measured (Figure 1.7.3).

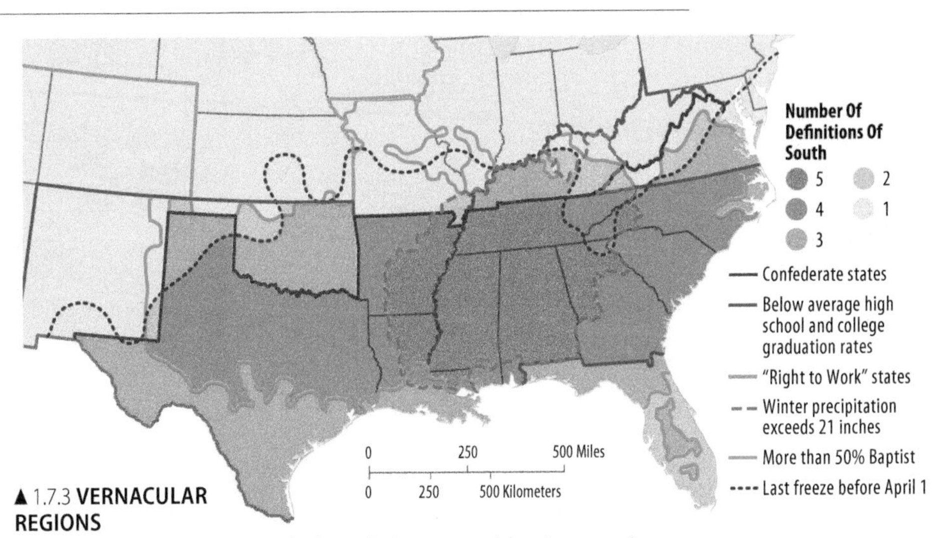

▲ 1.7.3 **VERNACULAR REGIONS**

The South is popularly distinguished as a distinct vernacular region according to a number of factors.

1.8 Scale: From Global to Local

- People are connected to a global economy and culture.
- Despite globalization, people play specialized economic roles and preserve cultural diversity.

Geographers think about scale at many levels, from local to global. At the global scale, encompassing the entire world, geographers tend to see broad patterns. At a local scale, such as a neighborhood within a city, geographers tend to see unique features. Geography matters in the contemporary world because it can explain human actions at all scales, from local to global.

Scale is an increasingly important concept in geography because of **globalization**, which is a force or process that involves the entire world and results in making something worldwide in scope. Globalization means that the scale of the world is shrinking—not literally in size, of course, but in the ability of a person, object, or idea to interact with a person, object, or idea in another place. People are plugged into a global economy and culture, producing a world that is more uniform, integrated, and interdependent (Figure 1.8.1).

▲ 1.8.1 **CHAOYANG CENTRAL BUSINESS DISTRICT, BEIJING, CHINA**

GLOBALIZATION OF THE ECONOMY

A few people living in very remote regions of the world may be able to provide all of their daily necessities. But most economic activities undertaken in one region are influenced by interaction with decision makers located elsewhere. The choice of crop is influenced by demand and prices set in markets elsewhere. The factory is located to facilitate bringing in raw materials and shipping out products to the markets.

Globalization of the economy has been led primarily by transnational corporations, sometimes called multinational corporations. A **transnational corporation** conducts research, operates factories, and sells products in many countries, not just where its headquarters and principal shareholders are located (Figure 1.8.2).

Every place in the world is part of the global economy, but globalization has led to more specialization at the local level. Each place plays a distinctive role, based on its local assets. A place may be near valuable minerals, or it may be inhabited by especially well-educated workers. Transnational corporations assess the particular economic assets of each place.

Modern technology provides the means to easily move money, materials, products, technology, and other economic assets around the world. Thanks to the electronic superhighway, companies can now organize economic activities at a global scale.

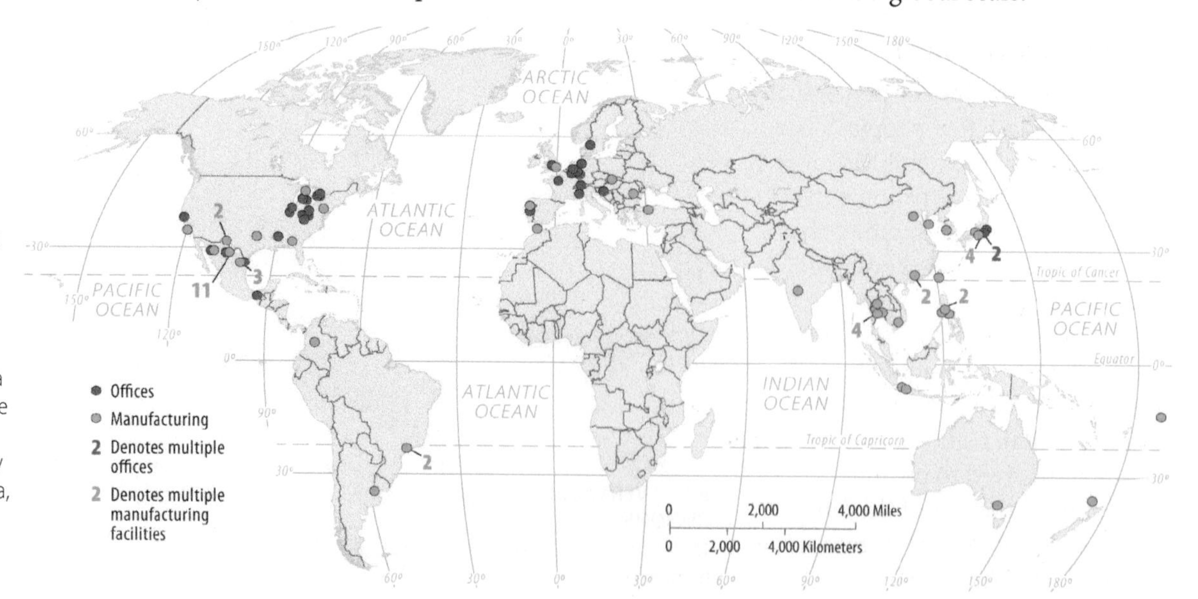

►1.8.2 **GLOBALIZATION OF THE ECONOMY** Yazaki, a transnational corporation that makes parts for cars, has factories primarily in Asia and Latin America, where labor costs are relatively low, and offices primarily in Europe, North America, and Japan, where most of the customers (carmakers) are located.

GLOBALIZATION OF CULTURE

Geographers observe that increasingly uniform cultural preferences produce uniform "global" landscapes of material artifacts and of cultural values. Fast-food restaurants, service stations, and retail chains deliberately create a visual appearance that varies among locations as little as possible so that customers know what to expect regardless of where in the world they happen to be (Figure 1.8.3).

CHINA

DUBAI

RUSSIA

THAILAND

JORDAN

JAPAN

▲ 1.8.3 **GLOBALIZATION OF CULTURE** "McDonald's" has more than 32,000 restaurants in 117 countries. To promote global uniformity of its restaurants, the company erects signs around the world that include two golden arches.

The survival of a local culture's distinctive beliefs, forms, and traits is threatened by interaction with such social customs as wearing jeans and Nike shoes, consuming Coca-Cola and McDonald's hamburgers, and displaying other preferences in food, clothing, shelter, and leisure activities.

Underlying the uniform cultural landscape is globalization of cultural beliefs and forms, especially religion and language. Africans, in particular, have moved away from traditional religions and have adopted Christianity or Islam, religions shared with hundreds of millions of people throughout the world. Globalization requires a form of common communication, and the English language is increasingly playing that role.

LOCAL DIVERSITY

As more people become aware of elements of global culture and aspire to possess them, local cultural beliefs, forms, and traits are threatened with extinction. Yet despite globalization, cultural differences among places not only persist but actually flourish in many places.

Global standardization of products does not mean that everyone wants the same cultural products. The communications revolution that promotes globalization of culture also permits preservation of cultural diversity.

Television, for example, is no longer restricted to a handful of channels displaying one set of cultural values. With the distribution of programming through cable and satellite systems, people may choose from hundreds of programs. With the globalization of communications, people in two distant places can watch the same television program. At the same time, with the fragmentation of the broadcasting market, two people in the same house can watch different programs.

Although consumers in different places express increasingly similar cultural preferences, they do not share the same access to them. And the desire of some people to retain their traditional culture in the face of increased globalization of cultural preferences, has led to political conflict and market fragmentation in some regions.

Globalization has not destroyed the uniqueness of an individual place's culture and economy. Human geographers understand that many contemporary social problems result from a tension between forces promoting global culture and economy on the one hand and preservation of local economic autonomy and cultural traditions on the other hand.

1.9 Space: Distribution of Features

- Three properties of distribution are density, concentration, and pattern.
- Gender and ethnicity are important examples of how patterns in space can vary.

Chess and computer games, require thinking about space. Pieces are arranged on the game board or screen in order to outmaneuver an opponent or form a geometric pattern. To excel at these games, a player needs spatial skills, the ability to perceive the future arrangement of pieces.

Similarly, spatial thinking is the most fundamental skill that geographers possess to understand the arrangement of objects across surfaces considerably larger than a game board. Geographers think about the arrangement of people and activities found in space and try to understand why those people and activities are distributed across space as they are.

Each human and natural object occupies space on Earth, which is the physical gap or interval between it and other objects. Geographers explain how features are arranged in space across Earth. On Earth as a whole, or within an area of Earth, features may be numerous or scarce, close together or far apart. The arrangement of a feature in space is known as its **distribution**.

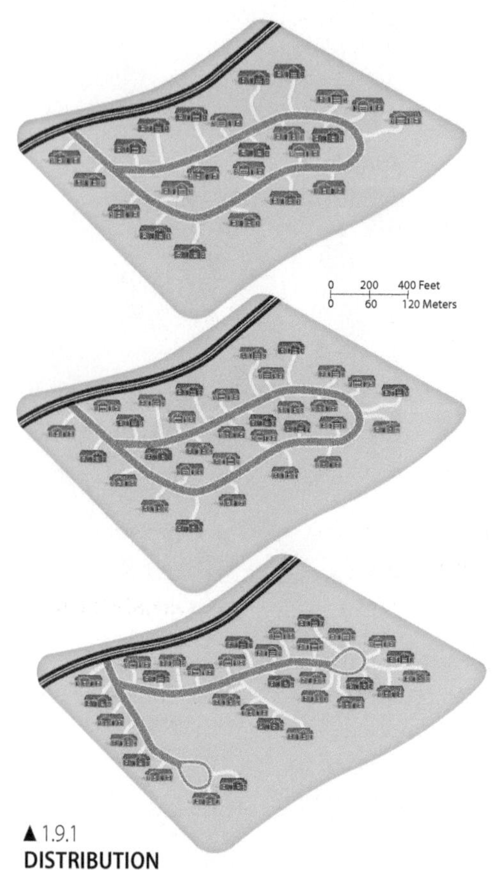

▲ 1.9.1
DISTRIBUTION
The top plan for a residential area has a lower density than the middle plan (24 houses compared to 32 houses on the same 30-acre piece of land), but both have dispersed concentrations. The middle and lower plans have the same density (32 houses on 30 acres), but the distribution of houses is more clustered in the lower plan. The lower plan has shared open space, whereas the middle plan provides a larger, private yard surrounding each house.

PROPERTIES OF DISTRIBUTION

Geographers identify three main properties of distribution across space:

- **Density**. The frequency with which something occurs in space is its density. The feature being measured could be people, houses, cars, volcanoes, or anything. The area could be measured in square kilometers, square miles, hectares, acres, or any other unit of area. A large population does not necessarily lead to a high density. Russia has a much larger population than the Netherlands, but the Netherlands has a much higher density because its land area is much smaller.
- **Concentration.** The extent of a feature's spread over space is its concentration. If the objects in an area are close together, they are clustered; if relatively far apart, they are dispersed. To compare the level of concentration most clearly, two areas need to have the same number of objects and the same size area.
- **Pattern.** The arrangement of objects in space is its pattern. Some features are organized in a geometric pattern, whereas others are distributed irregularly.

Concentration is not the same as density. Two neighborhoods could have the same density of housing but different concentrations. In a dispersed neighborhood each house has a large private yard, whereas in a clustered neighborhood the houses are close together and the open space is shared as a community park (Figure 1.9.1).

We can illustrate the difference between density and concentration at larger scales than a neighborhood. The changing distribution of baseball teams in the United States during the second half of the twentieth century resulted in a higher density of teams and a lower concentration (Figure 1.9.2).

Objects are frequently arranged in a square or rectangular pattern. Many American cities contain a regular pattern of streets, known as a grid pattern, which intersect at right angles at uniform intervals to form square or rectangular blocks. Other objects form a linear pattern, such as the arrangement of people and vegetation along the beach (refer to Figure 1.1.6).

◄ 1.9.2 **DISTRIBUTION OF MAJOR LEAGUE BASEBALL TEAMS**
The changing distribution of North American baseball teams illustrates the difference between density and concentration.

These six teams moved to other cities during the 1950s and 1960s:

Braves—Boston to Milwaukee in 1953, then to Atlanta in 1966
Browns—St. Louis to Baltimore (Orioles) in 1954
Athletics—Philadelphia to Kansas City in 1955, then to Oakland in 1968
Dodgers—Brooklyn to Los Angeles in 1958
Giants—New York to San Francisco in 1958
Senators—Washington to Minneapolis (Minnesota Twins) in 1961

These 14 teams were added between the 1960s and 1990s:

Angels—Los Angeles in 1961, then to Anaheim (California) in 1965
Senators—Washington in 1961, then to Arlington (Texas) (Texas Rangers) in 1971
Mets—New York in 1962
Astros—Houston (originally Colt .45s) in 1962
Royals—Kansas City in 1969
Padres—San Diego in 1969
Expos—Montreal in 1969, then to Washington (Nationals) in 2005
Pilots—Seattle in 1969, then to Milwaukee (Brewers) in 1970
Blue Jays—Toronto in 1977
Mariners—Seattle in 1977
Marlins—Miami (originally Florida) in 1993
Rockies—Denver (Colorado) in 1993
Rays—Tampa Bay (originally Devil Rays) in 1998
Diamondbacks—Phoenix (Arizona) in 1998

As a result of these relocations and additions, the density of teams increased, and the distribution became more dispersed.

GENDER AND ETHNIC DIVERSITY IN SPACE

Patterns in space vary according to gender and ethnicity. Consider the daily patterns of an "all-American" family of mother, father, son, and daughter. Leave aside for the moment that this type of family constitutes less than one-fourth of American households.

In the morning Dad drives from home to work, where he parks the car and spends the day; then, in the late afternoon, he drives home. Meanwhile, Mom takes the children to school, drives to the supermarket, visits Grandmother, walks the dog, collects the youngsters at school, takes them to sports practice or ballet lessons, and in between organizes the several thousand square feet of space that the family calls home. Most American women are now employed at work outside the home, adding a substantial complication to an already complex pattern of moving across urban space.

The importance of gender in space is learned as a child. Which child—the boy or girl—went to play ball, and which went to ballet lessons? To which activity is substantially more land allocated in a city—ballfields or dance studios (Figure 1.9.3)?

If the above family were parented by a same-sex couple, its connections with space would change. Similarly, effects of race and ethnicity on spatial interaction can be seen across America.

▲▼1.9.3 **GENDER AND SPACE**
Ballfields, which are more likely to be used by boys, take up more space in a community than ballet studios, which are more likely to be used by girls.

1.10 Connection: Interaction Between Places

- ▶ **A characteristic spreads from one place to another through diffusion.**
- ▶ **Connections between places result in spatial interaction.**

Geographers increasingly think about connections among places and regions. In the past, most forms of interaction among cultural groups resulted from the slow movement of settlers, explorers, and plunderers from one location to another. Today travel by motor vehicle or airplane is much quicker. More rapid connections have reduced the distance across space between places, not literally in miles, of course, but in time. Geographers apply the term **space-time compression** to describe the reduction in the time it takes for something to reach another place. Distant places seem less remote and more accessible to us (Figure 1.10.1). We know more about what is happening elsewhere in the world, and we know sooner.

But we do not even need to travel to know about another place. We can communicate instantly with people in distant places through computers and telecommunications, and we can instantly see people in distant places on television. These and other forms of communication have made it possible for people in different places to be aware of the same cultural beliefs, forms, and traits.

▲ 1.10.1 **SPACE-TIME COMPRESSION**

Transportation improvements have shrunk the world:

- In 1492, Christopher Columbus took 37 days (nearly 900 hours) to sail across the Atlantic Ocean from the Canary Islands to San Salvador Island.
- In 1912, the Titanic was scheduled to sail from Queenstown (now Cobh), Ireland, to New York in about 5 days, although two-thirds of the way across, after 80 hours at sea, it hit an iceberg and sank.
- In 1927, Charles Lindbergh was the first person to fly nonstop across the Atlantic, taking 33.5 hours to go from New York to Paris.
- In 1962, John Glenn, the first American to orbit in space, crossed over the Atlantic in about a quarter-hour and circled the globe three times in 5 hours.

RELOCATION DIFFUSION

The process by which a characteristic spreads across space from one place to another over time is **diffusion**. The place from which an innovation originates is called a **hearth**. Something originates at a hearth or node and diffuses from there to other places. Geographers document the location of nodes and the processes by which diffusion carries things elsewhere over time.

Diffusion occurs through cultural interaction involving persons, objects, or ideas. Geographers observe two basic types of diffusion: relocation diffusion and expansion diffusion. **Relocation diffusion** is the spread of an idea through physical movement of people from one place to another. When people move, they carry with them their culture, including language, religion, and ethnicity (Figure 1.10.2).

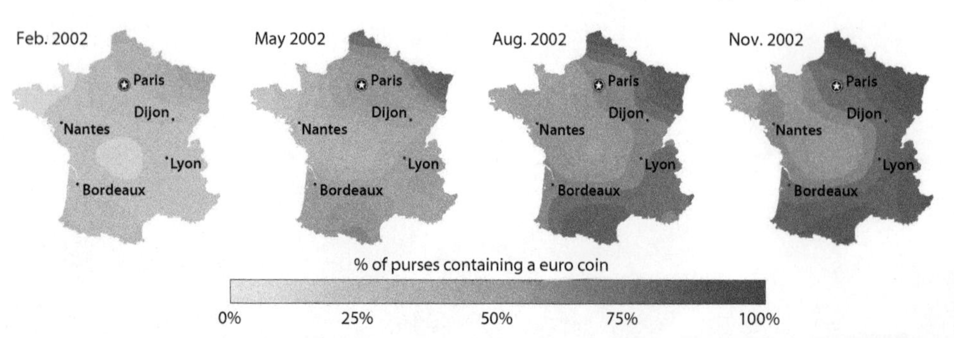

▼ 1.10.2. **RELOCATION DIFFUSION**

Introduction of a common currency, the euro, in 12 European countries on January 1, 2002, gave scientists an unusual opportunity to measure relocation diffusion. A single set of paper money was issued, but each of the 12 countries minted its own coins in proportion to its share of the region's economy. A country's coins were initially distributed only inside its borders, although the coins could also be used in the other 11 countries. French researchers took month-to-month samples to monitor the proportion of coins from the other 11 countries. The percentage of purses containing "foreign" euro coins is a measure of the level of relocation diffusion into France. Not surprisingly, diffusion occurred earlier and in higher percentages near the borders with other countries.

EXPANSION DIFFUSION

Expansion diffusion is the spread of a feature from one place to another in an additive process. This expansion may result from one of three processes:

Hierarchical diffusion is the spread of an idea from persons or nodes of authority or power to other persons or places. Hierarchical diffusion may result from the spread of ideas from political leaders, socially elite people, or other important persons to others in the community.

Contagious diffusion is the rapid, widespread diffusion of a characteristic throughout the population. As the term implies, this form of diffusion is analogous to the spread of a contagious disease, such as influenza. Ideas placed on the World Wide Web spread through contagious diffusion, because Web surfers throughout the world have access to the same material simultaneously—and quickly.

Stimulus diffusion is the spread of an underlying principle. For example, innovative features of Apple's iPhone and iPad operating systems have been adopted by competitors.

All three types of expansion diffusion occur much more rapidly in the contemporary world than in the past, because of widespread access to modern communications systems. Ideas are able to diffuse from one place to another, even if people are not actually relocating.

SPATIAL INTERACTION

When places are connected to each other through a network, geographers say there is **spatial interaction** between them. Typically, the farther away one group is from another, the less likely the two groups are to interact. Contact diminishes with increasing distance and eventually disappears. This trailing-off phenomenon is called **distance decay.**

Transportation systems form networks that facilitate relocation diffusion. Most airlines, for example, adopt a distinctive network called "hub-and-spokes" (Figure 1.10.3). Airlines fly planes from a large number of places into one hub airport within a short period of time and soon thereafter send the planes to another set of places (Figure 1.10.4).

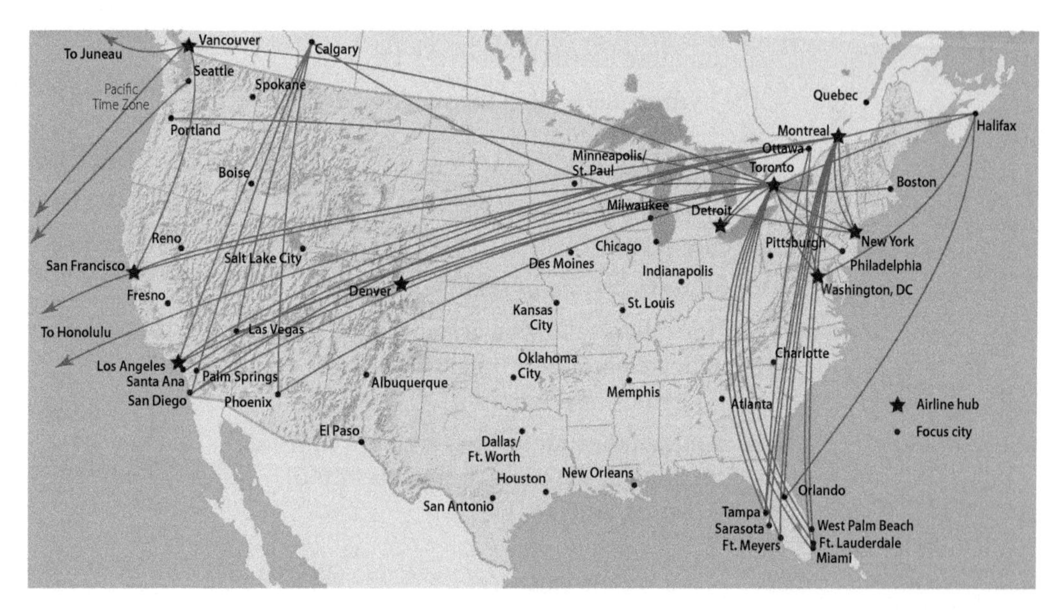

▲ 1.10.3 **SPATIAL INTERACTION: AIRLINE NETWORK**
Air Canada, like other major airlines, has configured its route network in a system known as "hub and spokes." A large percentage of Air Canada's flights originate or end at its principal hub in Toronto.

Electronic communications have removed barriers to interaction between people who are far from each other. The birth of these electronic communications was initially viewed as the "death" of geography, because they made it cheap and easy to stay in touch with people and events on the other side of the planet. In reality, geography matters even more than before. Internet access depends upon availability of electricity to either power a computer directly, or at least to recharge the battery in the computer or in a smart phone.

The Internet has also magnified the importance of geography, because when an individual is online the specific place in the world where the individual is located is known. This knowledge is valuable information for businesses that target advertisements and products to specific tastes and preferences of particular places.

▼ 1.10.4 **AIRLINE HUB**
Toronto Pearson International Airport, Air Canada's principal hub.

1.11 Earth's Physical Systems

- Earth's physical environment comprises four spheres.
- The four spheres are interrelated and interact with each other.

Geographers study natural processes in terms of four interrelated systems. These four physical systems are classified as either biotic or abiotic. A **biotic** system is composed of living organisms. An **abiotic** system is composed of nonliving or inorganic matter.

EARTH'S ATMOSPHERE FROM SPACE SHUTTLE

▲ 1.11.1 **ATMOSPHERE**

A thin layer of gases surrounds Earth at an altitude up to 480 kilometers (300 miles). Pure dry air in the lower atmosphere contains approximately 78 percent nitrogen, 21 percent oxygen, 0.9 percent argon, 0.036 percent carbon dioxide, and 0.064 percent other gases (measured by volume). As atmospheric gases are held to Earth by gravity, pressure is created. Variations in air pressure from one location to another are responsible for producing such weather features as wind blowing, storms brewing, and rain falling.

EARTH'S FOUR SYSTEMS

Three of Earth's four systems are abiotic:

- The **atmosphere**: a thin layer of gases surrounding Earth (Figure 1.11.1).
- The **hydrosphere**: all of the water on and near Earth's surface (Figure 1.11.2).
- The **lithosphere**: Earth's crust and a portion of upper mantle directly below the crust (Figure 1.11.3).

One of the four systems is biotic:

- The **biosphere**: all living organisms on Earth, including plants and animals, as well as microorganisms (Figure 1.11.4).

The names of the four spheres are derived from the Greek words for stone (litho), air (atmo), water (hydro), and life (bio).

◄ 1.11.2 **HYDROSPHERE**

Water exists in liquid form in the oceans, lakes, and rivers, as well as groundwater in soil and rock. It can also exist as water vapor in the atmosphere, and as ice in glaciers. Over 97 percent of the world's water is in the oceans. The oceans supply the atmosphere with water vapor, which returns to Earth's surface as precipitation, the most important source of freshwater. Consumption of water is essential for the survival of plants and animals, and a large quantity and variety of plants and animals live in it. Because water gains and loses heat relatively slowly, it also moderates seasonal temperature extremes over much of Earth's surface.

► 1.11.3 **LITHOSPHERE**

Earth is composed of concentric spheres. The core is a dense, metallic sphere about 3,500 kilometers (2,200 miles) in radius. Surrounding the core is a mantle about 2,900 kilometers (1,800 miles) thick. The crust is a thin, brittle outer shell 8 to 40 kilometers (5 to 25 miles) thick. The lithosphere encompasses the crust, a portion of the mantle extending down to about 70 kilometers (45 miles). Powerful forces deep within Earth bend and break the crust to form mountain chains and shape the crust to form continents and ocean basins.

LIGHTHOUSE REEF, TURNEFFE ATOLL, BELIZE

INTERACTIONS IN THE BIOSPHERE

An **ecosystem** is a group of living organisms and the abiotic spheres with which they interact. **Ecology** is the scientific study of ecosystems. Ecologists study interrelationships between living organisms and their environments within particular ecosystems, as well as interrelationships among various ecosystems in the biosphere.

Living organisms in the biosphere interact with each of the three abiotic systems. For example:

- The lithosphere is where most plants and animals live and where they obtain food and shelter.
- The hydrosphere provides water to drink, and physical support for aquatic life.
- The atmosphere provides the air for animals to breathe and protects them from the sun's rays.

Furthermore, the biosphere often represents the interface of the three abiotic systems with living organisms. For example, a piece of soil may comprise mineral material from the lithosphere, moisture from the hydrosphere, pockets of air from the atmosphere, and plant and insect matter from the biosphere.

Because geography is a social science as well as a natural science, geographers are especially interested in the interaction of humans with each of the four spheres (Figure 1.11.5):

- If the atmosphere's oxygen levels are reduced, or if the atmosphere contains pollutants, humans have trouble breathing.
- Without water, humans waste away and die.
- A stable lithosphere provides humans with materials for buildings and fuel for energy.
- The rest of the biosphere provides humans with food.

◄ 1.11.4 **BIOSPHERE**

The biosphere extends into each of Earth's abiotic systems. The biosphere encompasses all of Earth's living organisms. Because living organisms cannot exist except through interaction with the surrounding physical environment, the biosphere actually includes portions of the three abiotic systems near Earth's surface. Most of the planet's life is found within the top 3 meters (10 feet) of the lithosphere, the top 200 meters (650 feet) of the hydrosphere, and the lowest 30 meters (100 feet) of the atmosphere.

▼ 1.11.5 **ECOSYSTEMS**

Geographers are especially interested in the ecosystem of a city, because approximately half of Earth's humans live in urban areas. The lithosphere provides the ground and the materials to erect homes and businesses. The hydrosphere provides the water for urban dwellers to consume. The atmosphere is where urban dwellers emit pollutants. Some plants and other animals of the biosphere thrive along with humans in the cities, whereas others struggle.

NAMIB-NAUKLUFT PARK, NAMIBIA

IPANEMA, BRAZIL

1.12 Human–Environment Interaction

- **The environment can limit human actions, but people adjust to the environment.**
- **Humans are able to modify the environment, not always sensitively.**

Distinctive to geography is the importance given to relationships between human behavior and the natural environment. The geographic study of human–environment relationships is known as **cultural ecology.** Geographers are interested in two main types of human–environment interaction: how people adjust to their environment, and how they modify it.

POSSIBILISM: ADJUSTING TO THE ENVIRONMENT

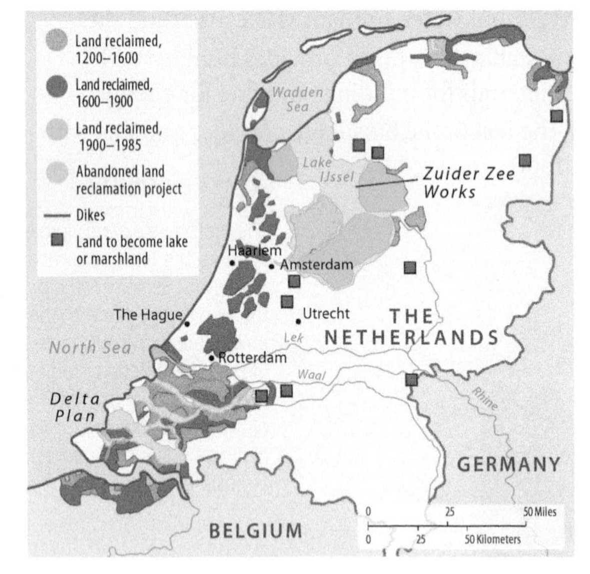

▼► 1.12.1 **MODIFYING THE ENVIRONMENT IN THE NETHERLANDS**
The Zuider Zee, an arm of the North Sea, once threatened the heart of the Netherlands with flooding. A dike completed in 1932 caused the Zuider Zee to be converted from a saltwater sea to a freshwater lake. The newly created body of water was named the IJsselmeer, or Lake IJssel, because the IJssel River now flows into it. Some of the lake has been drained to create several polders, encompassing an area of 1,600 square kilometers (620 square miles). The Dutch government has reserved most of the polders for agriculture to reduce the country's dependence on imported food.

A second ambitious dike project is the Delta Plan. Flowing through the Netherlands are several important rivers, including the Rhine (Europe's busiest river), the Maas (known as the Meuse in France), and the Scheldt (known as the Schelde in Belgium). As these rivers flow into the North Sea, they split into many branches and form a low-lying delta that is vulnerable to flooding. After a devastating flood in January 1953 killed nearly 2,000 people, the Delta Plan called for the construction of several dams to close off the waterways from the North Sea. The project took 30 years to build and was completed in the mid-1980s. (below) Polder near Loosdrecht, the Netherlands

A century ago, geographers argued that the physical environment *caused* social development, an approach called **environmental determinism.** For example, according to environmental determinism the temperate climate of northwestern Europe produced greater human efficiency as measured by better health conditions, lower death rates, and higher standards of living.

To explain relationships between human activities and the physical environment, modern geographers reject environmental determinism in favor of **possibilism**. According to possibilism the physical environment may limit some human actions, but people have the ability to adjust to their environment. People can choose a course of action from many alternatives in the physical environment. For example, people learn that different crops thrive in different climates; wheat is more likely than rice to be grown successfully in colder climates. Thus, under geography's possibilism approach, people choose the crops they grow in part by considering their environment.

MODIFYING THE ENVIRONMENT

Modern technology has altered the historic relationship between people and the environment. Humans now can modify a region's physical environment to a greater extent than in the past.

Few regions have been as thoroughly modified by humans as the Netherlands and southern Louisiana. With approximately 8,000 square kilometers (3,000 square miles) of land below sea level in both regions, extensive areas would be under water today were it not for massive projects to modify the environment by holding back the sea. The modifications undertaken by the Dutch have generally been more sensitive to environmental processes.

- **Modifying the Netherlands.** The Dutch have modified their environment with two distinctive types of construction projects—polders and dikes. A **polder** is a piece of land that is created by draining water from an area. All together, the Netherlands has 6,500 square kilometers (2,600 square miles) of polders, comprising 16 percent of the country's land area (Figure 1.12.1).

  Massive dikes have been built in two major locations—the Zuider Zee project in the north and the Delta Plan project in the southwest. These dikes prevent the North Sea, an arm of the Atlantic Ocean, from flooding much of the country.

- **Modifying southern Louisiana**. In an effort to protect New Orleans and other low-lying land from flooding, government agencies have constructed a complex system of levees, dikes, seawalls, canals, and pumps. Hurricane Katrina in 2005 showed that the efforts to control and tame all of the forces of nature in southern Louisiana and Gulf Coast regions of neighboring states were not successful.

  Hurricanes such as Katrina form in the Atlantic Ocean during the late summer and autumn and gather strength over the warm waters of the Gulf of Mexico. When it passes over land, a hurricane can generate a powerful storm surge that floods low-lying areas. Human geographers were especially concerned with the uneven impact of destruction and the incompetent response to the hurricane.

  Hurricane Katrina's victims were primarily poor, African American, and older individuals living in low-lying areas (Figure 1.12.2). The slow and incompetent response to the destruction by local, state, and federal emergency teams was blamed by many analysts on the victims' lack of a voice in the politics, economy, and social life of New Orleans and other impacted communities.

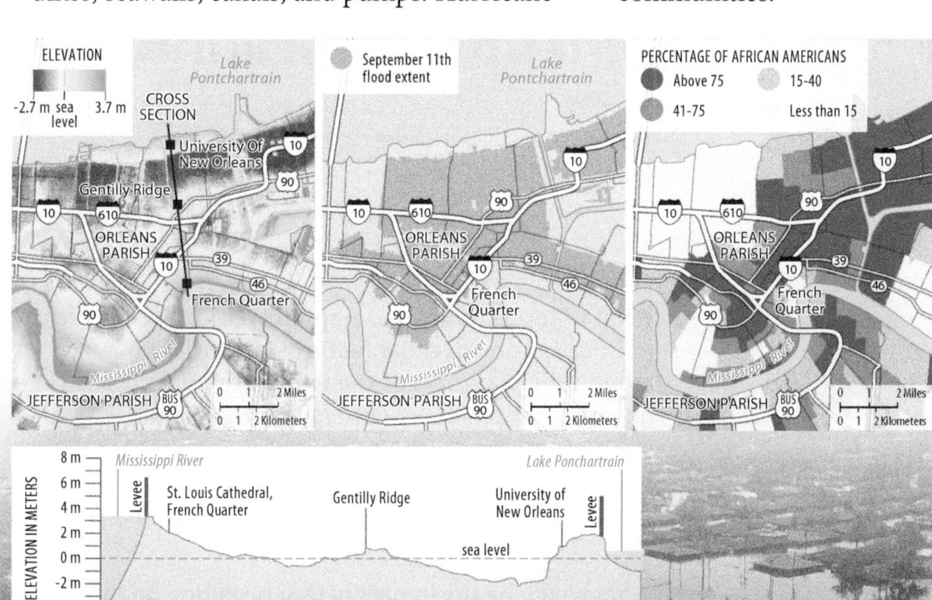

▼◄ 1.12.2 **MODIFYING THE ENVIRONMENT IN SOUTHERN LOUISIANA**
(left) Eighty percent of New Orleans is below sea level. (center and below) The day after Hurricane Katrina hit southern Louisiana, many of the levees in New Orleans broke, causing widespread flooding. (right) Neighborhoods inhabited primarily by African Americans were especially hard hit by flooding.

This chapter has introduced ways in which geographers think about the world, as well as key concepts in understanding geography.

Key Questions

How do geographers describe where things are?

- Geography describes the distribution of both natural and human phenomena.
- Geography began in ancient times as a descriptive aid for navigation and discovery.
- One of geography's most important tools since ancient times for describing the location of things on Earth has been the map.
- Satellite imagery and geographic information systems have enhanced geography's traditional ways of describing where things are.

Why is each point on Earth unique?

- Each place on Earth has distinct features of site and situation.
- Regions are areas of the world distinguished by a unique combination of features.

▼ 1.CR.1 **SATELLITE VIEW OF THE PERSIAN GULF**

"Persia" is an ancient name for present-day Iran. Most countries bordering the Gulf are Arab and prefer the name Arabian Gulf.

How are different locations interrelated?

- People are increasingly plugged into a global culture and economy.
- Many human and environmental features are distributed according to a regular arrangement.
- People and physical features in one place are connected to those elsewhere through processes of diffusion.

How do people relate to their environment?

- The biosphere, which comprises humans and other living things, is tied to processes in the atmosphere, hydrosphere, and lithosphere.
- Human actions are influenced by the environment, and in turn humans increasingly modify the environment.

Thinking Geographically

Using geographic tools such as maps and GIS is not simply a mechanical exercise. Nor are decisions confined to scale, projection, and layers. For example, many countries believe that the Persian Gulf should be labeled the Arabian Gulf (Figure 1.CR.1). Other countries believe that Kosovo should not be shown as a country.

1. **What criteria should geographers use to decide such politically controversial features? Majority vote among the countries of the world? The U.S. government position? Other ways to decide?**

Imagine that a transportation device (perhaps the one in *Star Trek* or *Harry Potter*) would enable all humans to travel instantaneously to any location on Earth.

2. **What would be the impact of that invention on the distribution of people, activities, and physical features across Earth?**

When earthquakes, hurricanes, and other environmental disasters strike, some humans tend to "blame" nature and see themselves as the innocent victims of Earth's sometimes harsh and cruel physical environment.

3. **To what extent do environmental hazards stem from physical systems and to what extent do they originate from human actions? Should victims blame nature, other humans, or themselves for the disaster? Why?**

Interactive Mapping

ELEVATION IN NORTH AMERICA

Launch MapMaster North America in Mastering GEOGRAPHY

Select: *Physical Features* from the *Physical Environment* menu, then select *Environmental Issues.*

What areas of North America other than Louisiana may be vulnerable to coastal flooding and groundwater depletion?

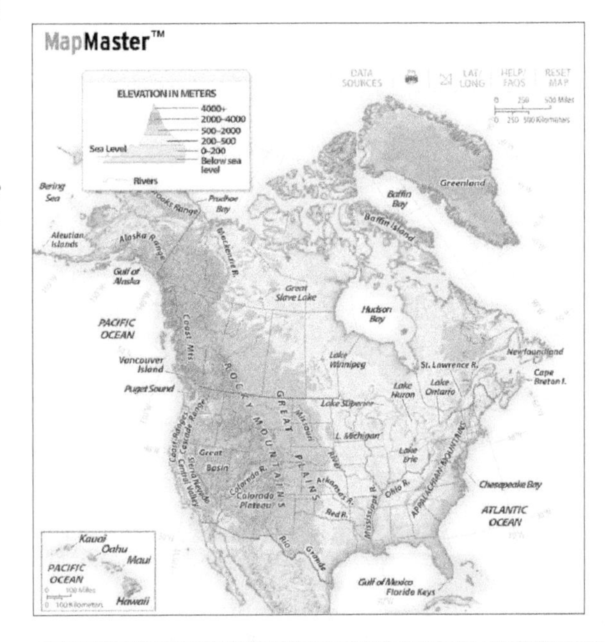

On the Internet

Useful Internet sites for learning about geography in North America are maintained by the major professional organizations, including the Association of American Geographers (**www.aag.org**, or scan the QR on the first page of this chapter), the American Geographical Society **(www.amergeog.org)**, the National Council for Geographic Education **(www.ncge.org)**, and the Canadian Association of Geographers (Association Canadienne des Géographes) **(www.cag-acg.ca)**.

The National Geographic Society offers access to material from its magazine and television programs, as well as online mapping at **www.nationalgeographic.com.**

Explore

NEW ORLEANS

(top) Use Google Earth to explore the impact of Hurricane Katrina on New Orleans. (bottom) Desire residential area in New Orleans. Google Earth provides historical images, in this case August 30, 2005, the day after Hurricane Katrina hit. Move the timeline forward until the floodwaters have disappeared

Fly to: *Desire Neighborhood Development, New Orleans, LA, USA.*

Click icon for: *Show historical imagery.*

Move time line to: *August 30, 2005.*

What was the date, after Katrina, when the buildings were visible again?

The Physical and Political World

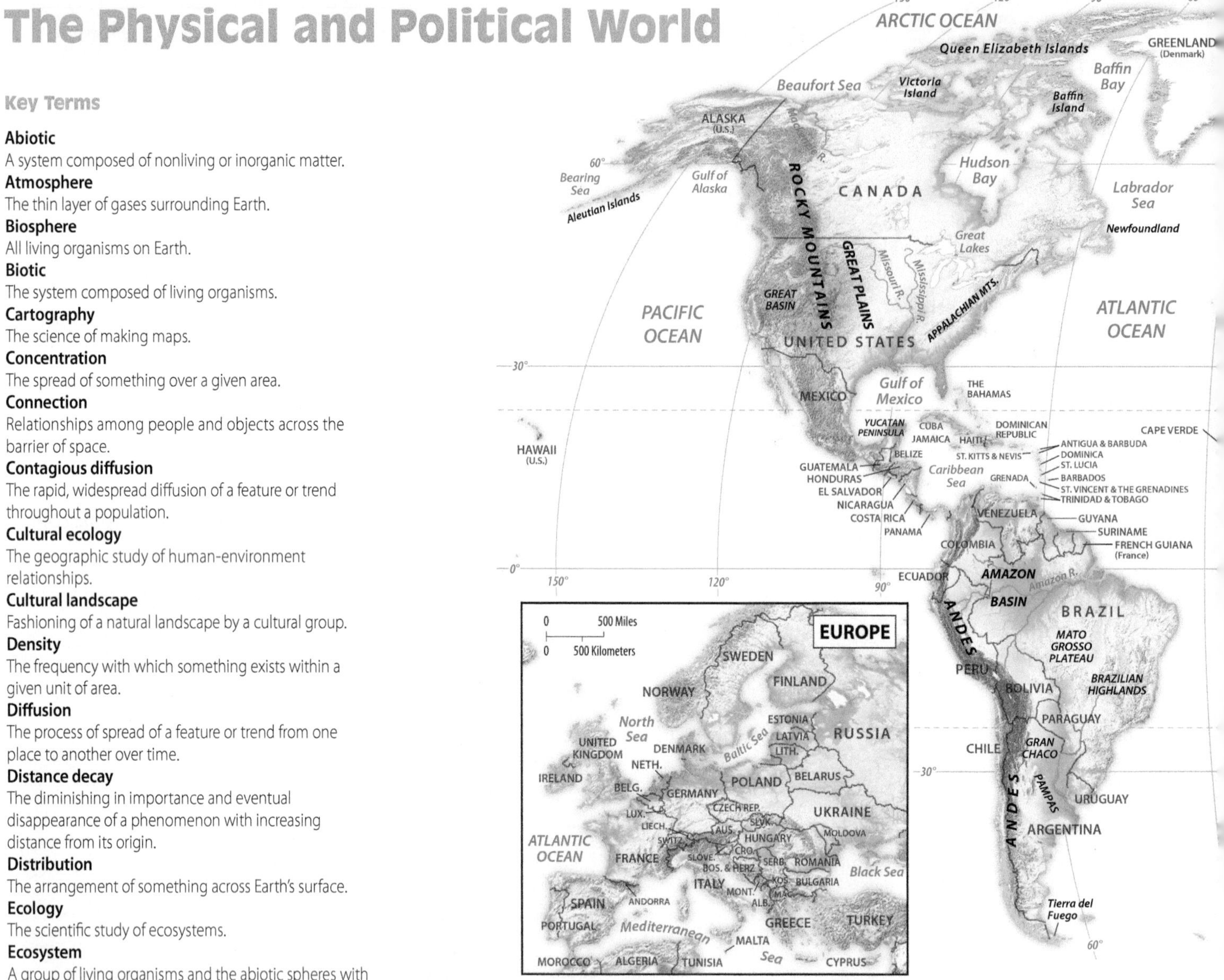

Key Terms

Abiotic
A system composed of nonliving or inorganic matter.

Atmosphere
The thin layer of gases surrounding Earth.

Biosphere
All living organisms on Earth.

Biotic
The system composed of living organisms.

Cartography
The science of making maps.

Concentration
The spread of something over a given area.

Connection
Relationships among people and objects across the barrier of space.

Contagious diffusion
The rapid, widespread diffusion of a feature or trend throughout a population.

Cultural ecology
The geographic study of human-environment relationships.

Cultural landscape
Fashioning of a natural landscape by a cultural group.

Density
The frequency with which something exists within a given unit of area.

Diffusion
The process of spread of a feature or trend from one place to another over time.

Distance decay
The diminishing in importance and eventual disappearance of a phenomenon with increasing distance from its origin.

Distribution
The arrangement of something across Earth's surface.

Ecology
The scientific study of ecosystems.

Ecosystem
A group of living organisms and the abiotic spheres with which they interact.

Environmental determinism
A nineteenth- and early twentieth-century approach to the study of geography that argued that the general laws sought by human geographers could be found in the physical sciences. Geography was therefore the study of how the physical environment caused human activities.

Expansion diffusion
The spread of a feature or trend among people from one area to another in an additive process.

Formal region (or uniform or homogeneous region)
An area in which everyone shares in one or more distinctive characteristics.

Functional region (or nodal region)
An area organized around a node or focal point.

Geographic grid
A system of imaginary arcs drawn in a grid pattern on Earth's surface.

Geographic Information Science (GIScience)
The development and analysis of data about Earth acquired through satellite and other electronic information technologies.

Geographic information system (GIS)
A computer system that stores, organizes, analyzes, and displays geographic data.

Global Positioning System (GPS)
A system that determines the precise position of something on Earth through a series of satellites, tracking stations, and receivers.

Globalization
Actions or processes that involve the entire world and result in making something worldwide in scope.

Greenwich Mean Time (GMT)
The time in that time zone encompassing the prime meridian, or 0° longitude.

Hearth
The region from which innovative ideas originate.

Hierarchical diffusion
The spread of a feature or trend from one key person or node of authority or power to other persons or places.

Hydrosphere
All of the water on and near Earth's surface.

International Date Line
A meridian that for the most part follows 180° longitude. When you cross the International Date Line heading east (toward America), the clock moves back 24 hours (one day), and when you go west (toward Asia), the calendar moves ahead one day.

Latitude
The numbering system used to indicate the location of parallels drawn on a globe and measuring distance north and south of the equator (0°).

Lithosphere
Earth's crust and a portion of the upper mantle directly below the crust.

Location
The position of anything on Earth's surface.

Longitude
The numbering system used to indicate the location of meridians drawn on a globe and measuring distance east and west of the prime meridian (0°).

Map
A two-dimensional, or flat, representation of Earth's surface or a portion of it.

Map scale
The relationship between the size of an object on a map and the size of the actual feature on Earth's surface.

Meridian
An arc drawn on a map between the North and South poles.

Parallel
A circle drawn around the globe parallel to the equator and at right angles to the meridians.

Pattern
The regular arrangement of something in a study area.

Place
A specific point on Earth distinguished by a particular characteristic.

Polder
Land created by the Dutch by draining water from an area.

Possibilism
The theory that the physical environment may set limits on human actions, but people have the ability to adjust to the physical environment and choose a course of action from many alternatives.

Prime meridian
The meridian, designated as 0° longitude, that passes through the Royal Observatory at Greenwich, England.

Projection
The system used to transfer locations from Earth's surface to a flat map.

Region
An area of Earth distinguished by a distinctive combination of cultural and physical features.

Relocation diffusion
The spread of a feature or trend through bodily movement of people from one place to another.

Remote sensing
The acquisition of data about Earth's surface from a satellite orbiting the planet or other long-distance methods.

Scale
The relationship between the portion of Earth being studied and Earth as a whole.

Site
The physical character of a place.

Situation
The location of a place relative to other places.

Space
The physical gap or interval between two objects.

Space-time compression
The reduction in the time it takes to diffuse something to a distant place, as a result of improved communications and transportation systems.

Spatial interaction
The movement of physical processes, human activities, and ideas within and among regions.

Stimulus diffusion
The spread of an underlying principle, even though a specific characteristic is rejected.

Toponym
The name given to a portion of Earth's surface.

Transnational corporation
A company that conducts research, operates factories, and sells products in many countries, not just where its headquarters or shareholders are located.

Vernacular region (or perceptual region)
An area that people believe exists as part of their cultural identity.

▸ LOOKING AHEAD

Earth is a planet in orbit around the Sun. Energy from the Sun drives all four of Earth's physical spheres. The starting point for geographic inquiry is understanding how solar energy reaches Earth's atmosphere.

2 Weather, Climate, and Climate Change

Weather conditions—such as storms, snowfall, clear skies, and warmth or cold—are driven by radiant energy received from the Sun. This solar energy is distributed through the atmosphere, both horizontally and vertically, and eventually back into space. This distribution is enabled by winds that carry heat from warm areas to cool ones, and cool air from colder regions toward warmer climates. Similarly, air carries moisture from the oceans over the land and returns drier air to ocean areas. In the process of air and moisture moving, or circulating, from one place to another, weather is created.

Weather varies from day to day because atmospheric circulation is constantly changing. Despite these changes, circulation tends to follow certain patterns. Weather patterns are one of the fundamental features of Earth's surface that make places different from each other. **Climate** is the totality of weather conditions over a period of several decades or more. Even though Earth's circulation and thus climate is highly complex and regulated by factors largely beyond our control, it can be modified by humans. In this chapter we will explore the processes that govern weather, the distribution of climates that results from those processes, and some of the ways in which human activities are modifying weather and climate.

How does energy move in the Earth-atmosphere system?

THUNDERSTORM, NORMAN, OKLAHOMA

What makes the weather?

What climates are found on Earth, and how is climate changing?

SCAN TO ACCESS WEATHER AND CLIMATE INFORMATION FOR THE UNITED STATES

2.1 Earth-Sun Geometry

- The intensity of solar radiation depends mainly on the angle at which the Sun's rays hit the surface at a particular place.
- Day length is affected by latitude and season of the year.

Energy travels through space as **radiation**. The amount of radiation or **solar energy** intercepted by a particular area of Earth, or **insolation** (incoming solar radiation), depends on two factors:

- the intensity of solar radiation, or the amount arriving per unit of time.
- the number of hours during the day that the solar radiation is striking.

◄ 2.1.1 **ANGLE OF INCIDENCE**

INTENSITY

Daily and seasonal differences in intensity are caused by variations in the **angle of incidence**—the angle at which solar radiation strikes a particular place at any moment in time. This angle varies from place to place, with time of day, and with the seasons (Figure 2.1.1). Throughout the year, the area of Earth's surface where the Sun is overhead at midday shifts due to Earth's tilt and continual revolution around the Sun. The intensity of solar radiation at a given place and time depends on its latitude and season of the year (Figure 2.1.2).

▼ 2.1.2 **VARIATION WITH SEASON**
The tilt of Earth's axis varies with the season, affecting the angle at which solar radiation strikes Earth's surface.

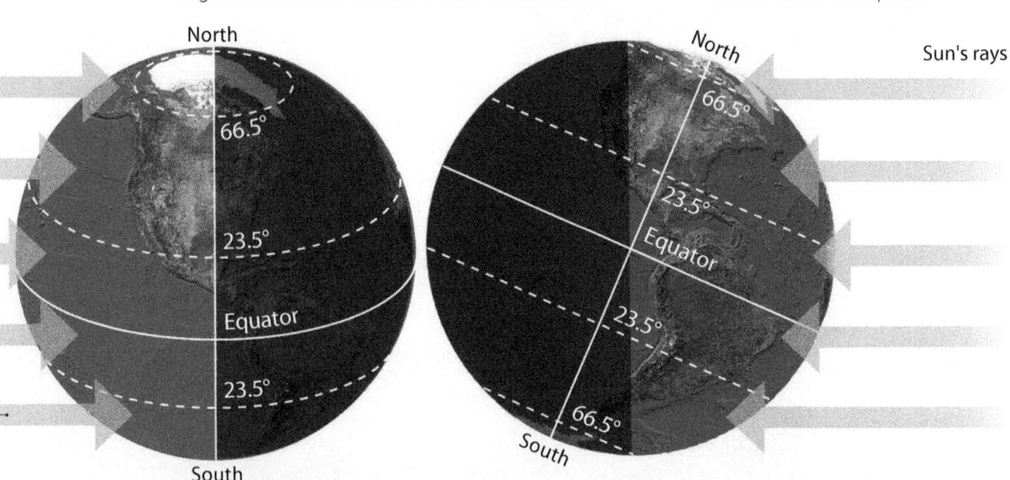

DAY LENGTH

MasteringGEOGRAPHY™
Animations
Earth-Sun Relations

Variations in the length of day from place to place result from the 23.5° tilt of Earth's axis away from a perpendicular relation to the Sun (Figure 2.1.3). Places on the equator always receive 12 hours of sunlight and 12 hours of night. But in higher latitudes, the amount of daylight varies considerably with the seasons. For example, in a 24-hour day at the **summer solstice**, a Northern Hemisphere city like Winnipeg, Manitoba, Canada receives nearly six times as much solar radiation, measured at the top of the atmosphere, as it does at the **winter solstice** (Figure 2.1.4).

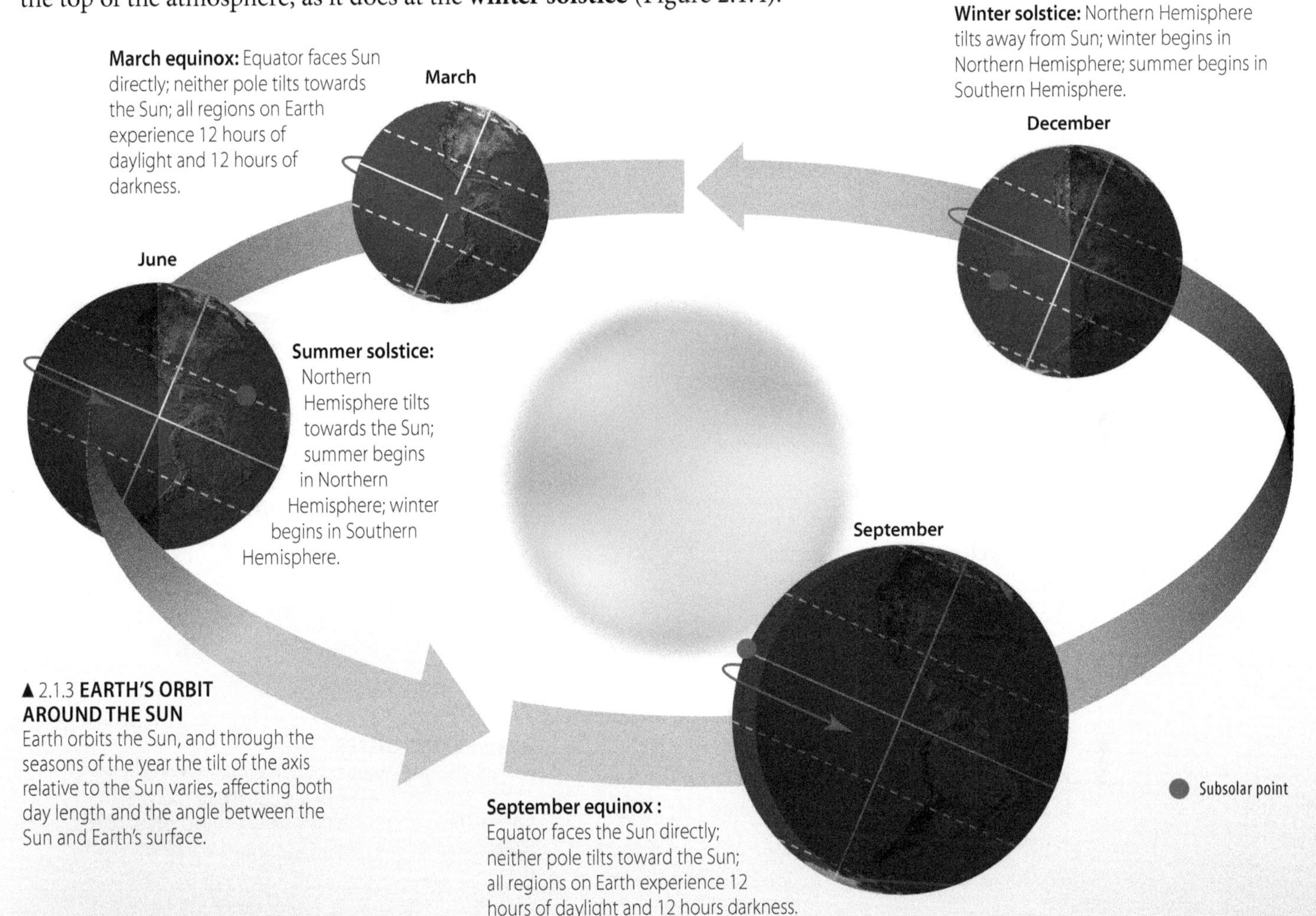

▲ 2.1.3 **EARTH'S ORBIT AROUND THE SUN**
Earth orbits the Sun, and through the seasons of the year the tilt of the axis relative to the Sun varies, affecting both day length and the angle between the Sun and Earth's surface.

▼ 2.1.4 **ON THE DAY OF THE SUMMER SOLSTICE**
Areas north of the Arctic Circle are in full sunlight for the entire 24 hours. This multiple-exposure photo shows the passage of the Sun through the sky at Sigurdarstadavik Bay near Kopasker, Iceland, around midnight in June .

2.2 Energy Exchange Mechanisms

- Energy exchanges occur by radiation, conduction, convection, and latent heat transfer.
- Radiant energy is sent to Earth as shortwave radiation and returned to space as longwave radiation.

Once solar energy enters Earth's atmosphere, a wide variety of energy exchange processes take place, redistributing this energy vertically and around the globe. By constructing a budget of energy exchanges, we can see the relative importance of different parts of this complex system.

ENERGY EXCHANGE MECHANISM

The most important process of heat transfer, a type of energy exchange, in the environment is radiation. Energy transmitted by electromagnetic waves, including radio, television, light, and heat, is radiation, or radiant energy. You feel heat radiating from a burner on the stove without touching it. Heat travels from the burner to your skin, which senses it. Radiation can travel through space and through materials, although materials may restrict radiant energy flow (Figure 2.2.1).

Radiant energy waves have different lengths. The **wavelength** is the distance between successive waves, like waves on a pond. Wavelength affects the behavior of the energy when it strikes matter; some waves are reflected, and some are absorbed. Two ranges of wavelengths—called **shortwave energy** and **longwave energy**—are most important for understanding how solar energy affects the atmosphere. Most energy arriving from the Sun is shortwave, while all energy radiated by Earth is longwave.

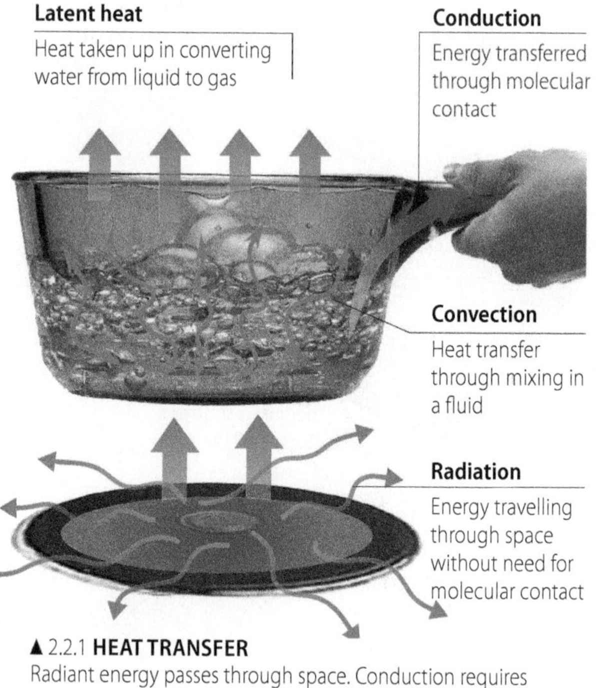

▲ 2.2.1 **HEAT TRANSFER**
Radiant energy passes through space. Conduction requires contact between objects, as through the pot handle. **Convection** involves mixing in a fluid, while heat transfer requires a change of state, as from liquid to vapor.

RADIATION IN THE ATMOSPHERE

As energy from the Sun passes through the atmosphere, some wavelengths are absorbed, warming the atmosphere, while others pass through or are reflected, either to be absorbed elsewhere or to travel back into space. Clouds play a major role in reflecting energy back to space and in this way atmospheric moisture and weather processes can significantly affect the energy budget, an accounting of the major energy exchanges in the Earth-atmosphere-Sun system.

When heat is absorbed by an object, its temperature rises, and heat is stored in the object. When this stored heat is released, an object cools. The ability of an object to store heat depends on what it is made of. Some materials can absorb or release a large amount of heat with only small changes in temperatures, while others heat and cool quickly with only small inputs and releases of energy.

Of all the gases in the atmosphere, only a few allow much of the incoming shortwave solar energy to pass through but still absorb most outgoing longwave radiation. Gases with these properties are called **greenhouse gases**, and they are critical to heat exchange in the atmosphere (Figure 2.2.2). Among the most important ones are water vapor, **carbon dioxide (CO_2)**, **ozone (O_3)**, and **methane (CH_4)**. Although these gases

water vapor
carbon dioxide
ozone
other trace gases
60%
26%
8%
6%

▲ 2.2.2 **GREENHOUSE GAS EFFECTS ON ATMOSPHERIC HEATING**
Gases in the atmosphere vary in their ability to absorb shortwave and longwave radiation. This diagram shows relative contributiuons of water vapor, carbon dioxide, ozone, and other trace gases to heating of a clear (cloud-free) sky.

together constitute a small fraction of 1 percent of the atmosphere, they are the most important in atmospheric heating. Water vapor contributes the most to atmospheric heating. Human activities are increasing the amount of some greenhouse gases in the atmosphere, and this is believed to be the chief cause of global warming. We will return to this topic later in the chapter.

Energy absorbed at ground level is transferred back to the atmosphere and then to space via longwave radiation, convection, and latent heat exchange (evaporation/condensation of water), which will be discussed in more detail in the next section (Figure 2.2.3).

▼ 2.2.3 **EARTH'S ENERGY BUDGET**
(A) About half of the solar radiation that reaches Earth is absorbed by the surface; the rest is either **(B)** absorbed in the atmosphere or **(C)** reflected back to space. The energy absorbed at the surface is transferred upward, mainly to the atmosphere, by **(D)** latent heat exchange, **(E)** convection, and **(F)** longwave radiation. Energy absorbed by the atmosphere is radiated back to space. The energy exchanges shown here are net exchanges; large amounts of energy are sent back and forth between the Earth's surface and the atmosphere by longwave radiation, but the net exchange is upward.

MasteringGEOGRAPHY
Animations
Atmospheric Energy Balance

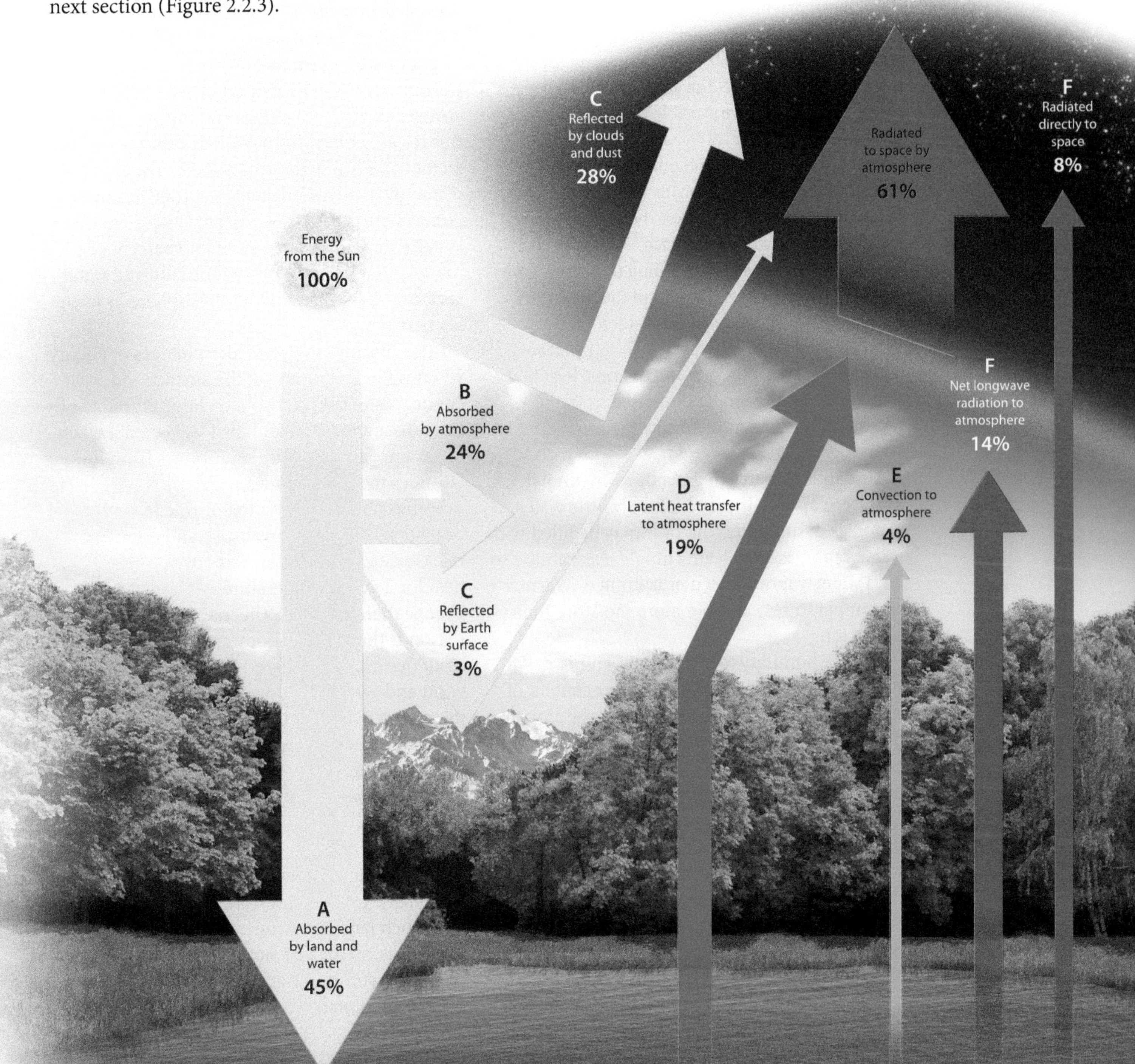

2.3 Latent Heat

- Conversion among vapor, liquid, and solid states involves large energy transfers.
- Latent energy plays a critical role in the movement of energy in the Earth-atmosphere system.

Water plays a central role in Earth's energy budget, through a process called latent heat exchange. This process transfers tremendous amounts of energy from Earth's surface to the atmosphere, from low latitudes to high ones, and it is also the mechanism most influential in causing precipitation.

TWO TYPES OF HEAT

We can distinguish between two types of heat—sensible and latent. **Sensible heat** is detectable by your sense of touch. It is heat you can feel, from warm water or a hot pan, and you can measure it with a thermometer. The atmosphere, oceans, rocks, and soil all have sensible heat. **Latent heat**, on the other hand, is "in storage" in water and water vapor. Latent—which means "hidden"—describes the heat that controls the state of water. When ice melts, it must absorb heat energy from its surroundings. This is why ice melting in your hand feels so cold—it is absorbing heat from your hand. The heat becomes stored in the meltwater as latent heat. Latent heat also is stored in water vapor. If you ever had a finger scalded by steam, you know the startling amount of latent heat that was stored in the vapor and released from the water when it condensed from a vapor to liquid on your finger.

Air contains water in gaseous form, known as **water vapor**. Air may hold very little water vapor, as in dry desert air, or it may be filled with water vapor, as in a steamy jungle. Air's ability to hold water-vapor molecules is limited. Warmer air temperatures can hold more moisture than cooler air.

Relative humidity tells us how wet air is. **Relative humidity** is the actual water content of the air compared to how much water the air could potentially hold, expressed as a percentage. For example, if a 30°C (86°F) sample of air contains half the water vapor that it could hold at that temperature, its relative humidity is 50 percent. But if cooled to 22°C (71°F), that same air sample with the same amount of water vapor would be three-fourths saturated, at 75 percent relative humidity since cooler air has a lower capacity to hold moisture. When air is cooled to the point of 100 percent relative humidity, then condensation occurs and latent heat is released.

LATENT ENERGY TRANSFERS

When water evaporates, as from the ocean or a vegetated land surface, heat is taken up in the evaporating water and the surface is cooled (Figure 2.3.1). The atmosphere, however, is warmed. The water vapor condenses in the atmosphere to form clouds and precipitation, and this release of heat from **condensation** is a major factor in warming the atmosphere. The amount of energy involved in latent heat transfers in the atmosphere is vast, especially for major weather systems and hurricanes. About 40 percent of the solar energy absorbed by the land and water surface is transferred to the atmosphere by latent heat exchange.

In addition to the vertical exchanges of energy when water evaporates at the surface and condenses in the atmosphere, large amounts of heat are carried horizontally ("advected") when, for example, warm humid air blows from low latitudes to high latitudes.

Bodies of water play a major role in heat storage. Water can absorb and release much larger quantities of heat for a given temperature change than can land. The main reason is that a water body such as an ocean can be stirred by the wind and carry heat down below its surface to depths of 10 meters (33 feet) or more seasonally. In addition, water can absorb much more heat than soil per unit mass of these materials. In contrast, land surfaces warm and cool to only about 2 meters (6 feet) depth seasonally. Oceans can store a season's heat in a much larger volume of matter than can land areas, so they do not heat up as quickly in summer nor do they cool down as quickly in winter.

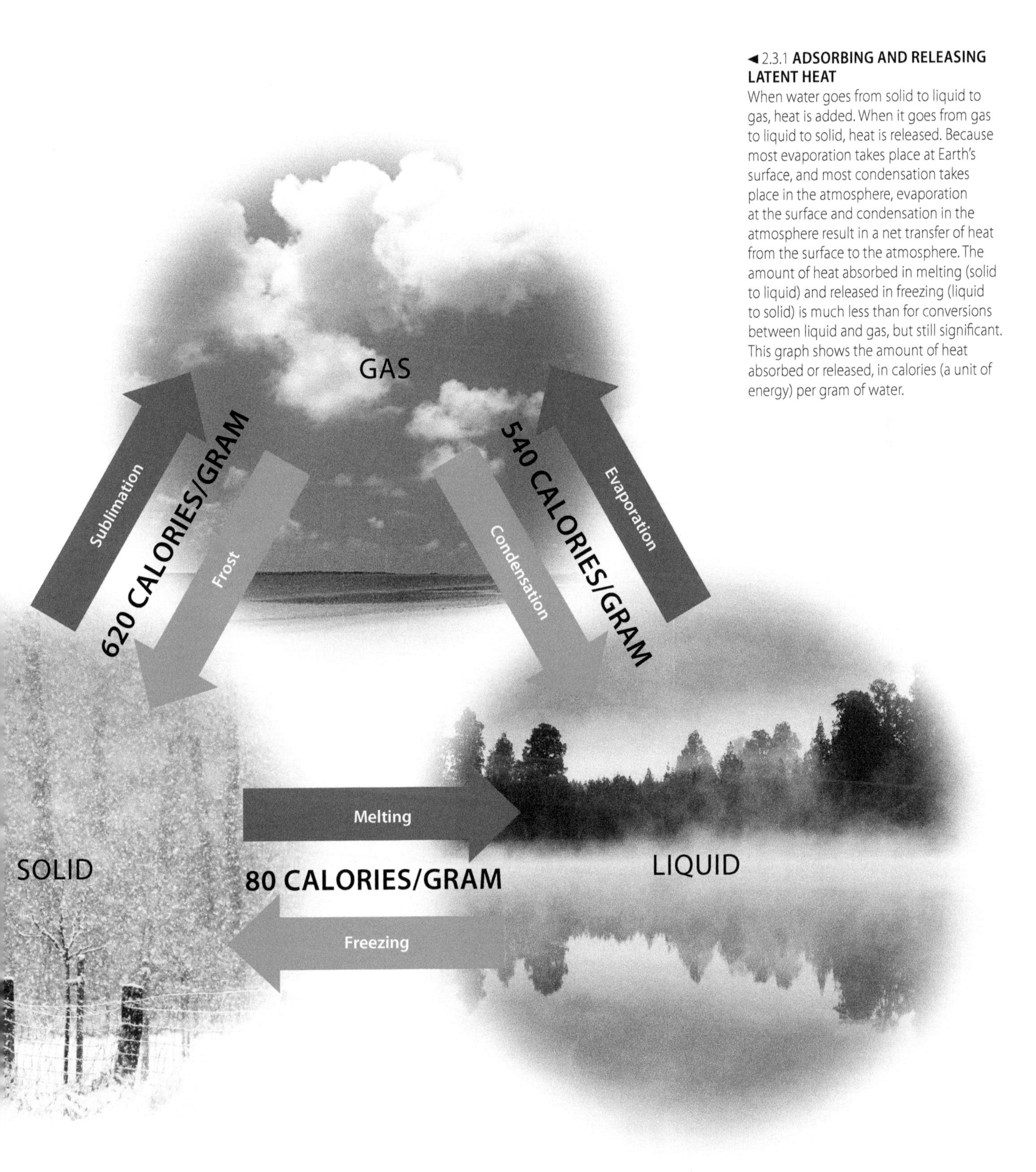

◄ 2.3.1 **ADSORBING AND RELEASING LATENT HEAT**

When water goes from solid to liquid to gas, heat is added. When it goes from gas to liquid to solid, heat is released. Because most evaporation takes place at Earth's surface, and most condensation takes place in the atmosphere, evaporation at the surface and condensation in the atmosphere result in a net transfer of heat from the surface to the atmosphere. The amount of heat absorbed in melting (solid to liquid) and released in freezing (liquid to solid) is much less than for conversions between liquid and gas, but still significant. This graph shows the amount of heat absorbed or released, in calories (a unit of energy) per gram of water.

2.4 Variations in Temperature

- Energy inputs, controlled by latitude and season, are the most important factor influencing temperature of the Earth's surface.
- Heat storage in water and proximity to the oceans also affects temperature.

The temperature of any place is a result of its energy budget, especially the amount of energy that arrives from the Sun. We can see these effects in the world temperature maps.

SEASONAL VARIATIONS IN TEMPERATURE

On average, the highest temperatures are found in low latitudes, because the Sun is highest in the sky in these areas, and therefore the intensity of solar radiation is highest (Figure 2.4.1). High solar elevation angles occur throughout the year, and this makes the tropics (the area between the **Tropic of Cancer** and the **Tropic of Capricorn**) consistently warm (Figure 2.4.2).

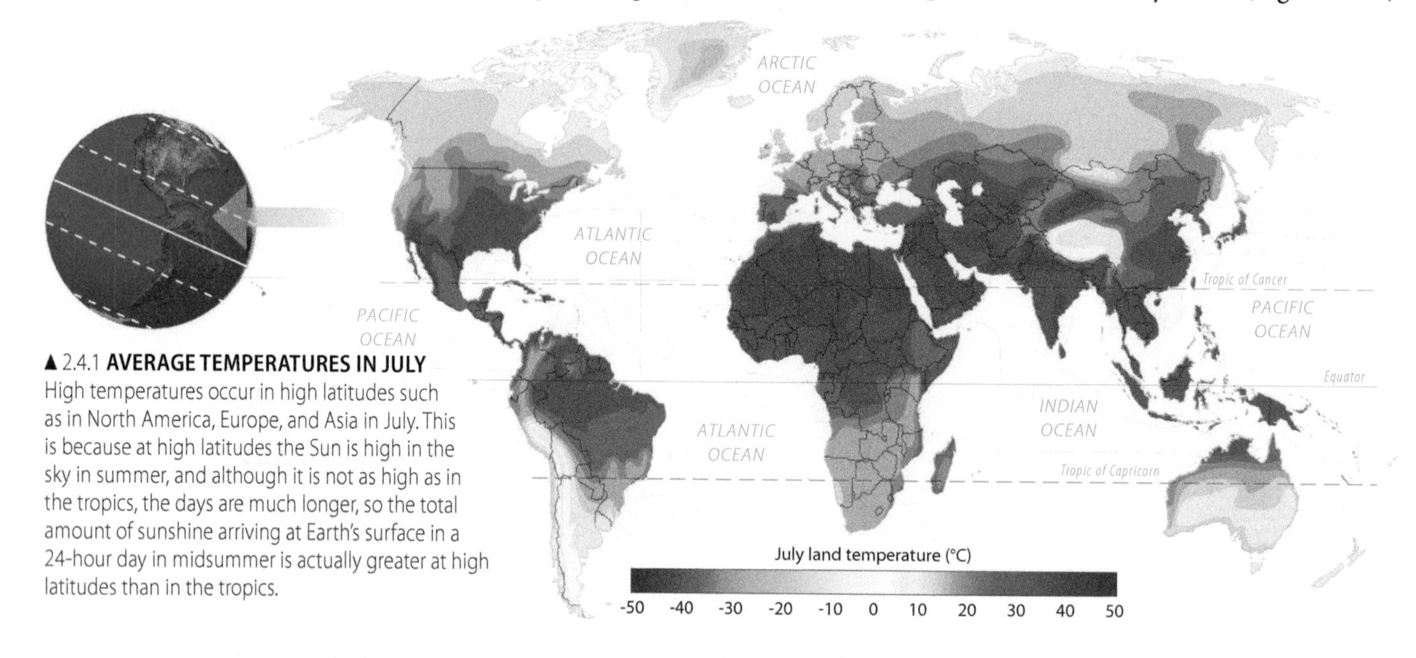

▲ 2.4.1 **AVERAGE TEMPERATURES IN JULY**
High temperatures occur in high latitudes such as in North America, Europe, and Asia in July. This is because at high latitudes the Sun is high in the sky in summer, and although it is not as high as in the tropics, the days are much longer, so the total amount of sunshine arriving at Earth's surface in a 24-hour day in midsummer is actually greater at high latitudes than in the tropics.

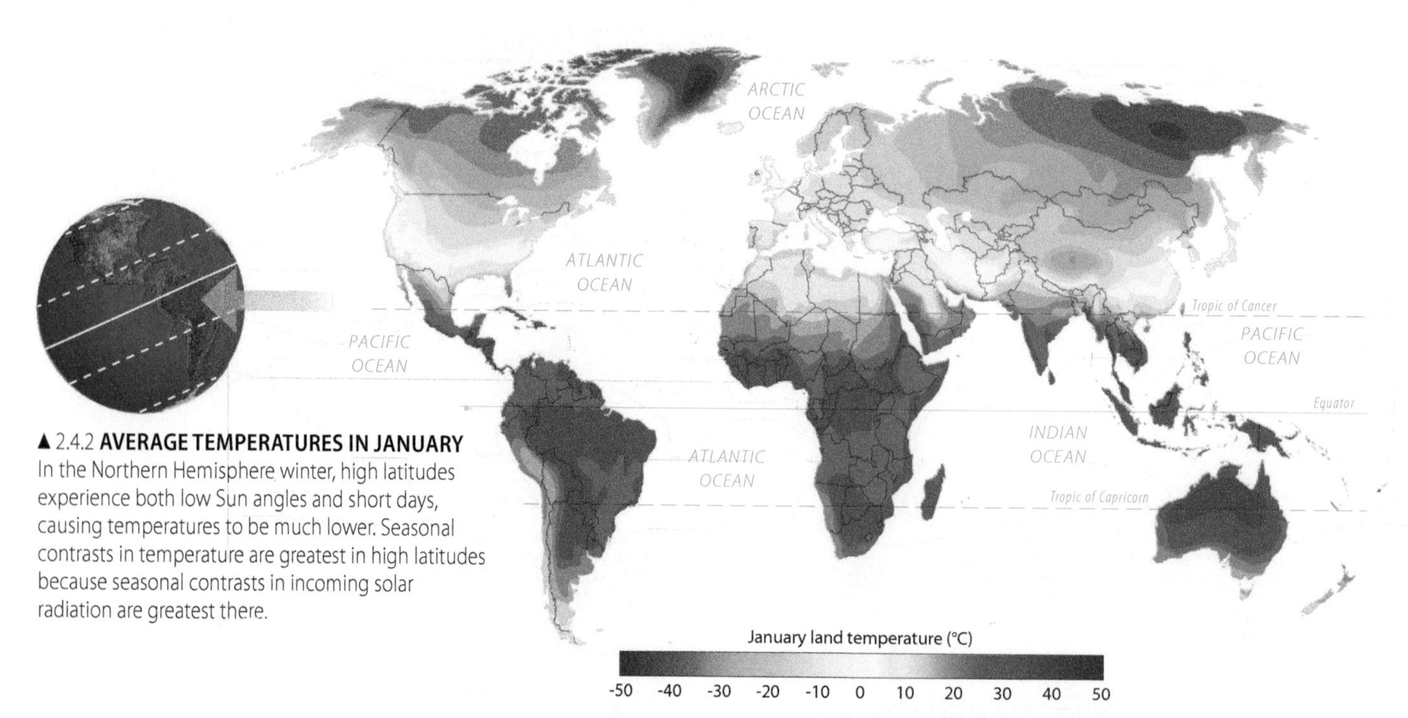

▲ 2.4.2 **AVERAGE TEMPERATURES IN JANUARY**
In the Northern Hemisphere winter, high latitudes experience both low Sun angles and short days, causing temperatures to be much lower. Seasonal contrasts in temperature are greatest in high latitudes because seasonal contrasts in incoming solar radiation are greatest there.

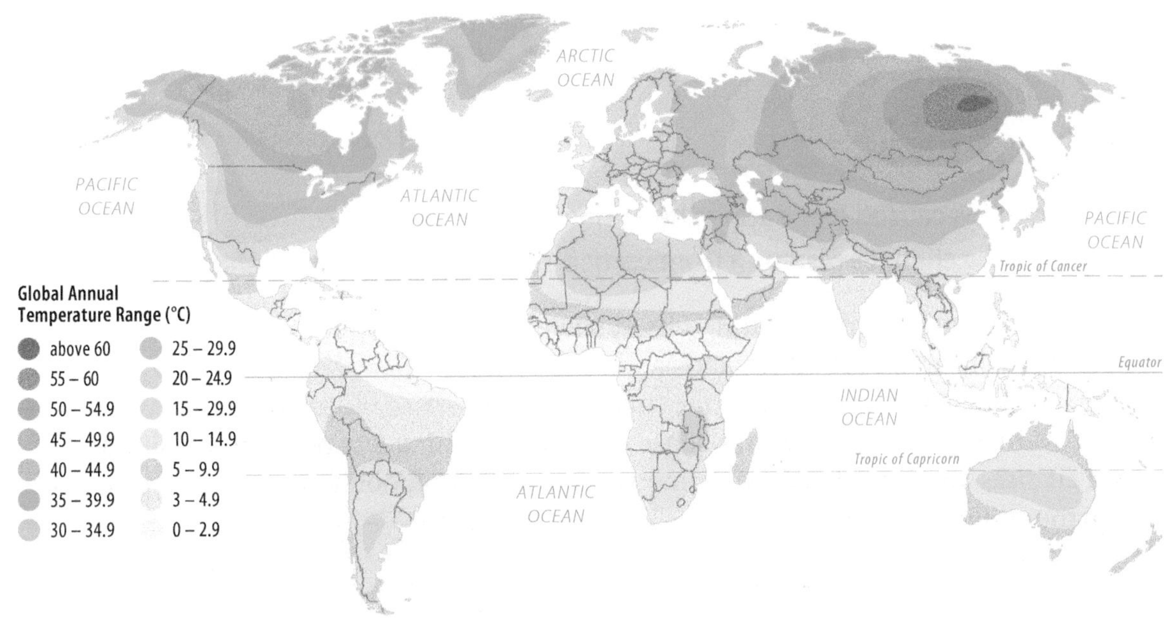

◄ 2.4.3 **TEMPERATURE RANGES**
The difference between summer and winter temperatures is greatest (blue tones) on land areas, and especially toward the eastern parts of continents in high latitudes where seasonal contrasts are large. In this latitude, winds generally blow from west to east, carrying the influence of the sea to the western part of the continents and the influence of the interior to the eastern parts of the continents.

EFFECTS OF LAND AND WATER

Large landmasses in high latitudes, such as North America and Asia, have very large temperature differences between July and January (Figure 2.4.3). Average July temperatures in Siberia reach above 10°C (50°F), whereas average January temperatures reach −40°C (−40°F). The difference between July and January temperatures exceeds 50°C (90°F) in eastern Siberia. In contrast, the January and July temperatures in the Aleutian Islands of the north Pacific, at about the same latitude as Siberia, are about 15°C (59°F) in July and 5°C (41°F) in January, a range of only about 10°C (18°F). The moderate climates of areas near the ocean compared with the climatic extremes of midcontinent regions are a result of this heat storage in water, and the **advection** of heat from ocean areas to adjacent land areas (Figure 2.4.4).

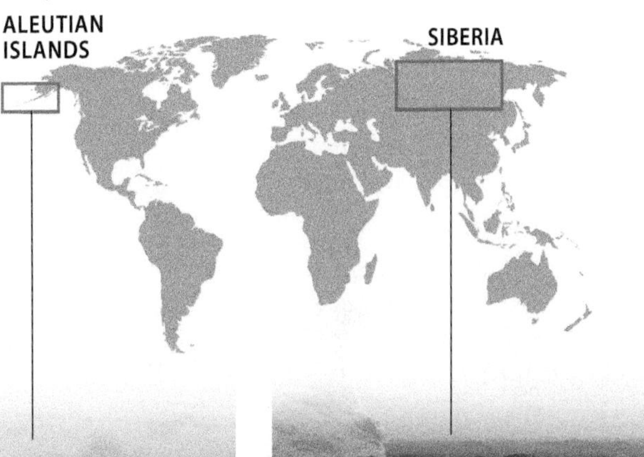

◄▼ 2.4.4 **EFFECTS OF WATER**
Temperature range is greater in Siberia (below) than in the Aleutian Islands (below left), because heat is stored in oceans, moderating temperatures in coastal areas.

2.5 Convection and Adiabatic Processes

- **Convection is caused by heating the atmosphere from below.**
- **Temperature changes resulting from vertical motions create clouds and precipitation.**

Convection is movement in a fluid, caused when part of the fluid (whether gas or liquid) is heated. The heated portion expands and becomes less dense, rising up through the cooler portion. Convection causes the turbulence you see in water and in puffy white clouds overhead. In becoming less dense—that is, weighing less per unit of volume—warm air rises above cooler, denser air, just as a hot-air balloon rises through the cooler air surrounding it.

CONVECTION ON A SUNNY DAY

Think of an island on a sunny day (Figure 2.5.1). Solar energy is absorbed by the sandy surface. As the surface warms, it reradiates longwave energy, some of which is absorbed by the air just above the ground. The water surface warms less than the land does, and the air grows much warmer over the island than over the water (Figure 2.5.2).

▼ 2.5.1 **CONVECTION OVER AN ISLAND**
As the ground surface is heated the warm air above it rises, as above this island.

WARM AIR RISING

COOL AIR SINKING

HORIZONTAL WIND FROM OCEAN TO LAND

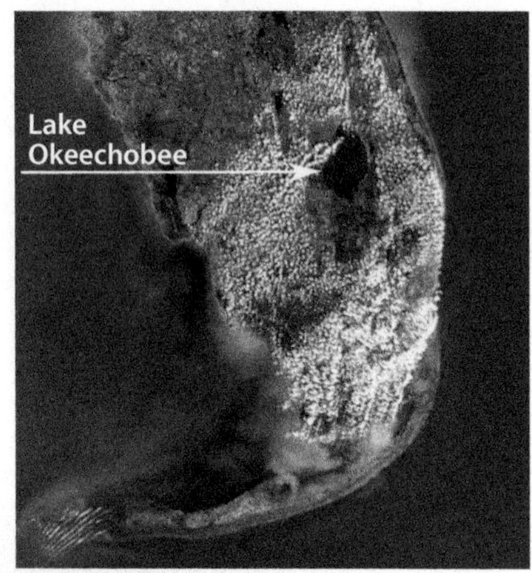

▲ 2.5.2 **CONVECTION AND CLOUDS OVER FLORIDA**
This satellite image of Florida shows clouds forming over the warm land as a result of daytime heating, while the cooler ocean and Lake Okeechobee are cloud free. The cloudy areas are areas of rising air.

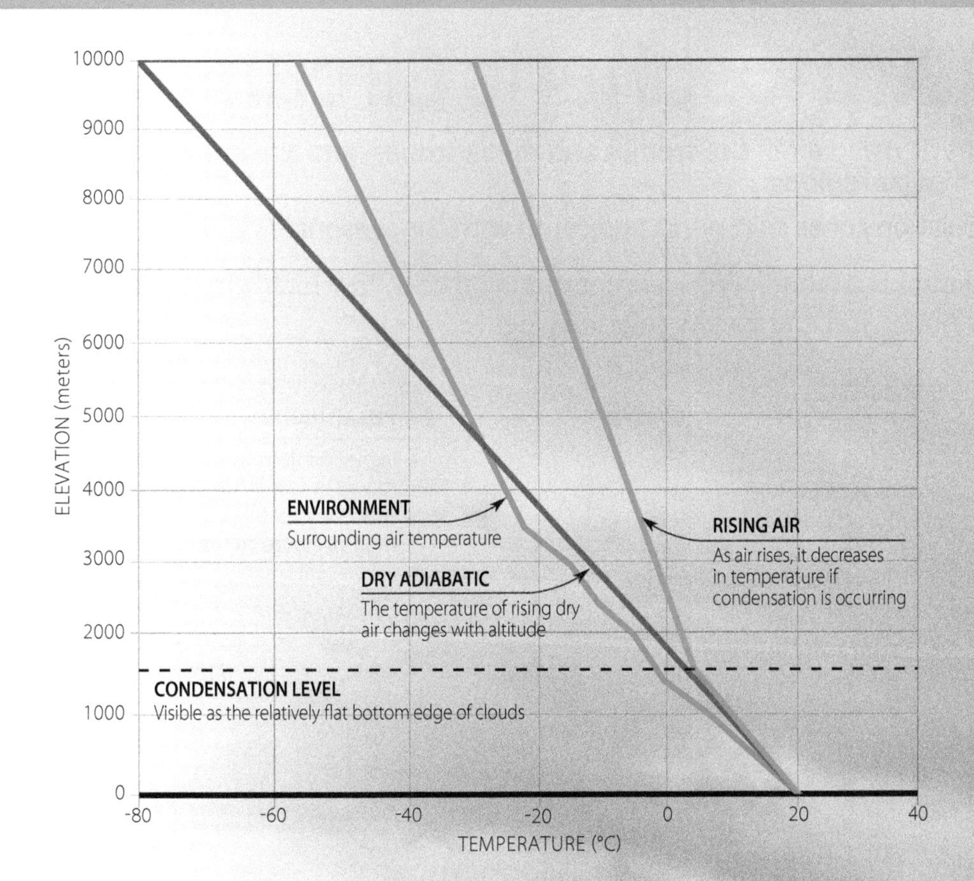

MasteringGEOGRAPHY
Animations
Atmospheric Stability and Coriolis Effect

◄ 2.5.3 **VERTICAL PATTERNS OF AIR TEMPERATURE**
The temperature environment aloft controls vertical air movements.

CONVECTION AND PRECIPITATION

Air is a gas, compressed by the weight of overlying air. When it rises, air has less weight above it, and the lower pressure allows the air to expand. Compressing a gas causes an increase in temperature, whereas expanding it causes a decrease (for example, an aerosol spray can gets cold in your hand when you use it). The decrease in temperature that results from expansion of rising air is called **adiabatic cooling**; the word adiabatic means "without heat being involved."

When convection causes air to rise, the pressure in that air decreases with elevation and this causes the air to cool (Figure 2.5.3). Cooling, in turn, lowers the amount of water vapor it can hold. Condensation occurs when the amount of water the air can hold reaches the actual vapor content—the relative humidity reaches 100%.

As air rises above the condensation level it continues to cool, further reducing its ability to hold water vapor and causing more condensation. This condensation releases latent heat, warming the air. The added heat isn't enough to prevent condensation, but it does make the air warmer than the air around it at the same elevation, and so promotes further rising air.

Because air pressure under a column of rising air is lower than places nearby, on a map this appears as a center of low pressure. Wind blows toward this center of low pressure, in a spiraling path. If Earth did not rotate, winds would simply blow in a straight line from areas of high pressure to areas of low pressure. However, on our real spinning planet, winds follow an indirect, curving path. This deflection of wind (and any other object moving above Earth's rotating surface) is called the **Coriolis effect** (Figure 2.5.4). The spiral is caused by the Coriolis effect, which results from Earth's rotation on its axis.

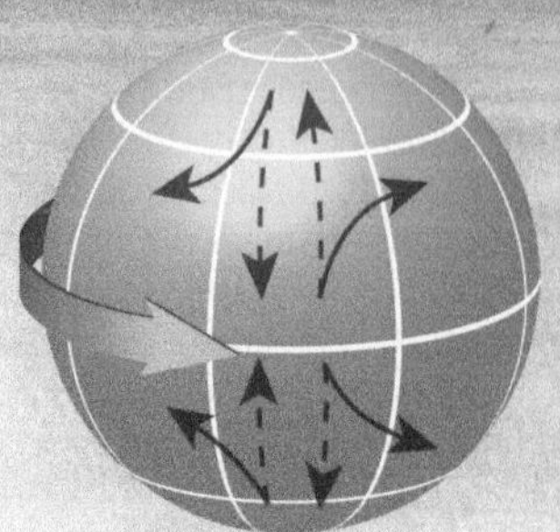

▲ 2.5.4 **CORIOLIS EFFECT**
On the surface of a spinning object the apparent motions of other objects, including moving air (wind), are altered. The result is winds that are deflected to the right in the Northern Hemisphere, and to the left in the Southern Hemisphere. This deflection is called the Coriolis effect.

2.6 Global Atmospheric Circulation

- **There are areas of rising air in the tropics and midlatitudes, and sinking air in the subtropics and polar regions.**
- **Latitudinal circulation zones shift north and south with the seasons.**

The global circulation is a set of large-scale convectional cells, in which wind patterns are influenced by both wind and Earth's rotation (Figure 2.6.1).

▼ 2.6.1 **GENERAL CIRCULATION**
The generalized atmospheric circulation pattern includes east-west trending cells, with rising air in the tropics and midlatitudes, and sinking air in the subtropics and polar regions. Surface winds blow from areas of sinking air to areas of rising air, but are deflected to blow from the east in the tropics and from the west in midlatitudes.

WESTERLIES

On the poleward sides of the subtropical high-pressure cells, circulation is toward the poles. But these winds are deflected by the Coriolis effect, so winds prevail from the southwest in the Northern Hemisphere and from the northwest in the Southern Hemisphere.

POLAR HIGHS

In the polar regions, the intense cold caused by low insolation creates dense air and high pressure. In these **polar high-pressure zones**, the air is so cold, it contains very little moisture, and convection and precipitation are limited.

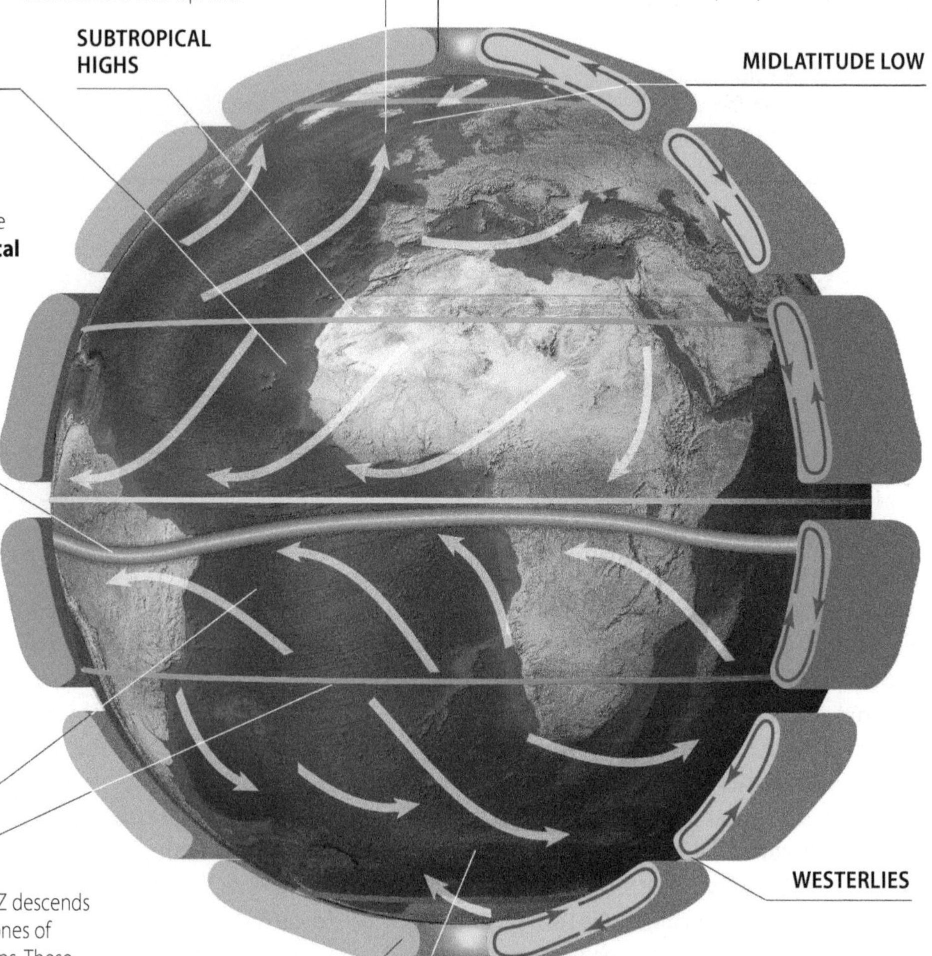

TRADE WINDS

Air converges toward the equator at the surface, replacing the rising air. The Coriolis effect deflects this moving air to the right in the Northern Hemisphere to form the Northeast Trade Winds and to the left in the Southern Hemisphere to form the Southeast Trade Winds. Aloft, air circulates away from the **Intertropical Convergence Zone** (ITCZ) both northward and southward, toward subtropical latitudes. This air has lost most of its moisture in the daily rainfalls, and it is now warm and dry.

INTERTROPICAL CONVERGENCE

In the tropics, dependable year-round inputs of solar energy heat the air, expanding it and creating low pressure. As a result, convectional rising of air occurs daily above Earth's equator. This forms the Intertropical Convergence Zone (ITCZ), so called because it is a zone between the Tropics of Cancer and Capricorn where surface winds converge. Convectional precipitation is common, usually as afternoon thunderstorms.

SUBTROPICAL HIGHS

The warm, dry air that spreads poleward from the ITCZ descends at about 25° north and south latitudes. This creates zones of high pressure that are especially strong over the oceans. These **subtropical high-pressure (STH) zones** are areas of dry air, bright sunshine, and little precipitation.

This descending dry air associated with subtropical high-pressure cells creates an arid climate, so most of the world's major desert regions are on land in this zone, at about 25° north and south latitudes. The STH zones are strongest over the eastern side of oceans, and so major desert areas occur on the western edges of continents at this latitude. The eastern sides of these continents tend to be more humid because the circulation brings in warm, humid tropical air.

MIDLATITUDE LOW

Poleward of the subtropical highpressure zones are the **midlatitude low-pressure zones**. These lower-pressure areas experience convergence of warm air blowing from subtropical latitudes and cold air blowing from polar regions. The warm and cold air masses collide in swirling low-pressure cells that move along the boundary between the two air masses, which is known as the polar front.

In January, the ITCZ is generally south of the equator because solar radiation is greatest there (Figure 2.6.2). In the Northern Hemisphere high pressure develops over land, with low pressure over the seas.

Asia has the world's highest average pressures during January and the world's lowest average pressure during July (Figure 2.6.3). As a result, wind directions in the vicinity are reversed seasonally. This produces a **monsoon circulation**, in which winter winds from the Asian interior produce extremely dry winters in most of south and east Asia, while summer winds blowing inland from the Indian and Pacific oceans result in wet summers. These wet summers are the well-known monsoon season of heavy rains in southern Asia (Figure 2.6.4).

▲ 2.6.4 **MONSOON RAINFALL IN INDIA**

▼ 2.6.2 **JANUARY CIRCULATION PATTERNS**

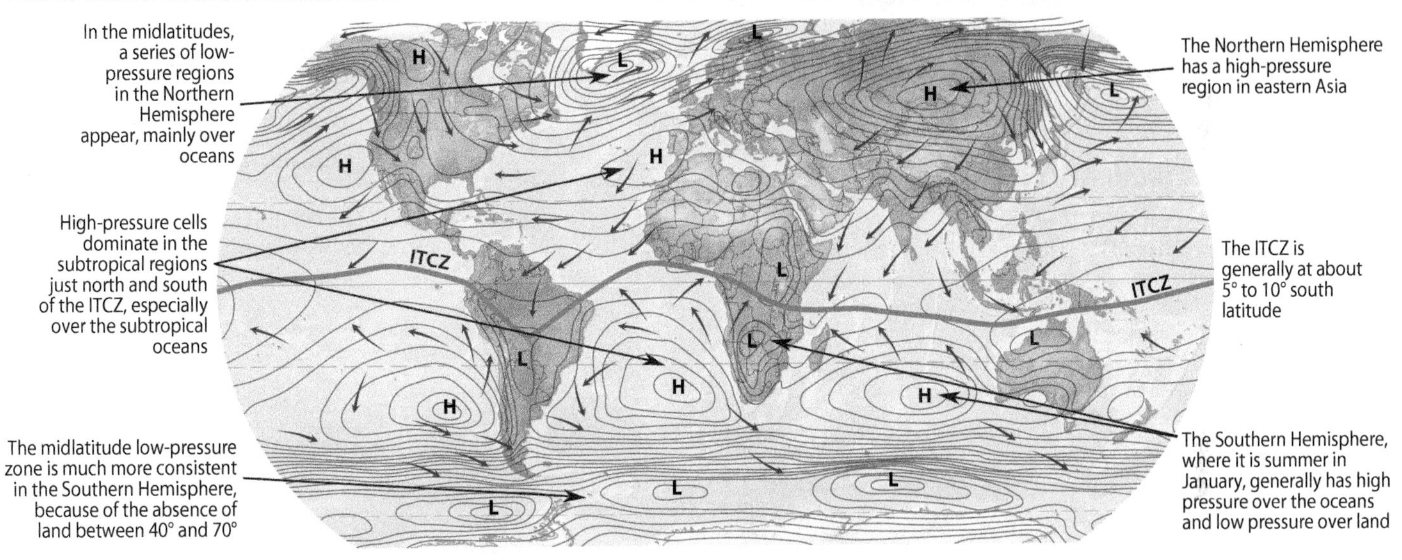

▼ 2.6.3 **JULY CIRCULATION PATTERNS**

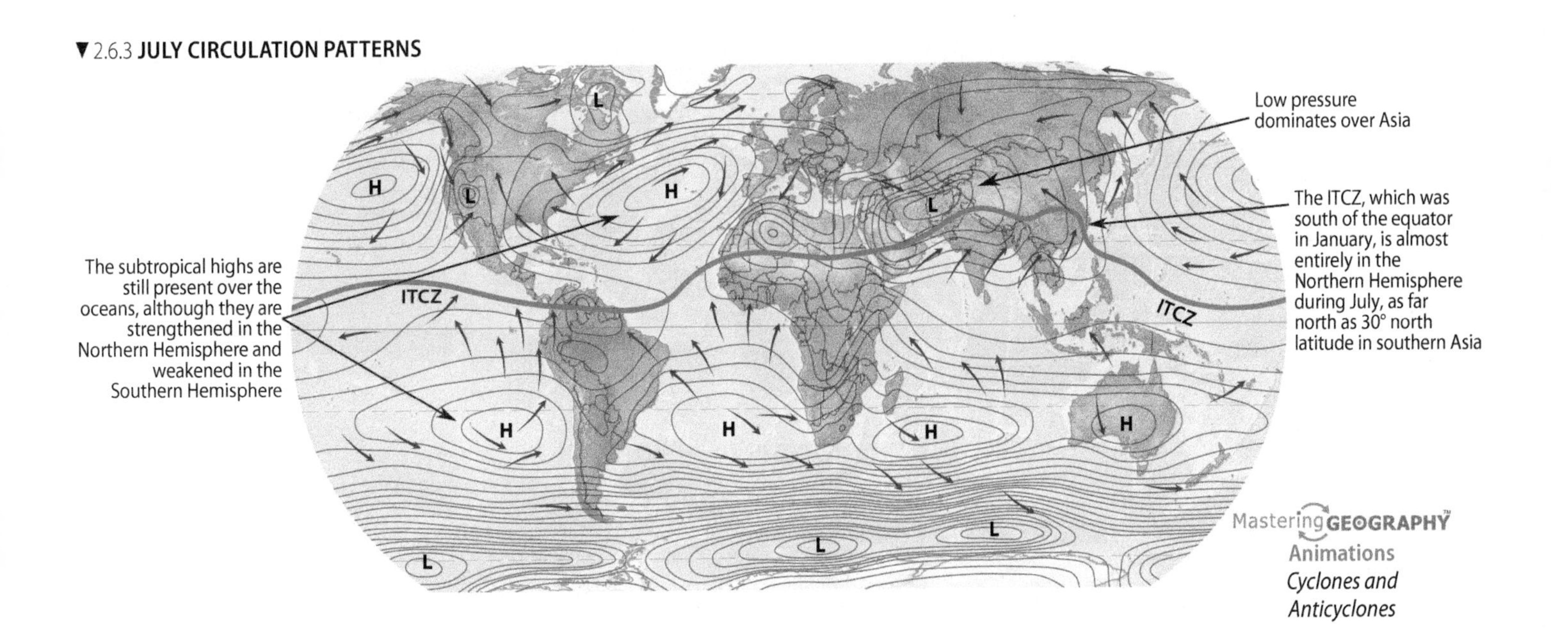

MasteringGEOGRAPHY™
Animations
Cyclones and Anticyclones

2.7 Ocean Circulation

- **Wind direction affects ocean surface currents, while water salinity and temperature affect subsurface circulation.**
- **Ocean circulation patterns vary over periods of years, producing variations in weather.**

When wind blows over the ocean, it drags the sea surface, creating waves and currents (Figure 2.7.1). The continuing drag of prevailing winds also causes broad currents in the ocean's surface layers. In addition to wind, differences in seawater temperature and salinity give water different densities, promoting movement of currents from areas of greater density to those of less density. This is similar to the way a high-pressure area in the atmosphere causes wind currents to blow toward a low-pressure area. Both of these factors—wind and temperature/salinity—are important in creating ocean currents that redistribute heat around Earth (Figure 2.7.2).

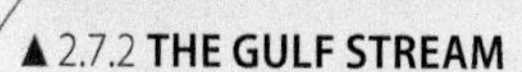

▲ 2.7.2 **THE GULF STREAM**
This image shows warm water in red and cool water in blue. The Gulf Stream is a strong warm current that flows northeastward across the Atlantic. Giant swirling eddies are prominent.

▼ 2.7.1 **GENERAL OCEANIC CIRCULATION**
In this diagram, surface currents are shown in red, while currents at depth are shown in blue.

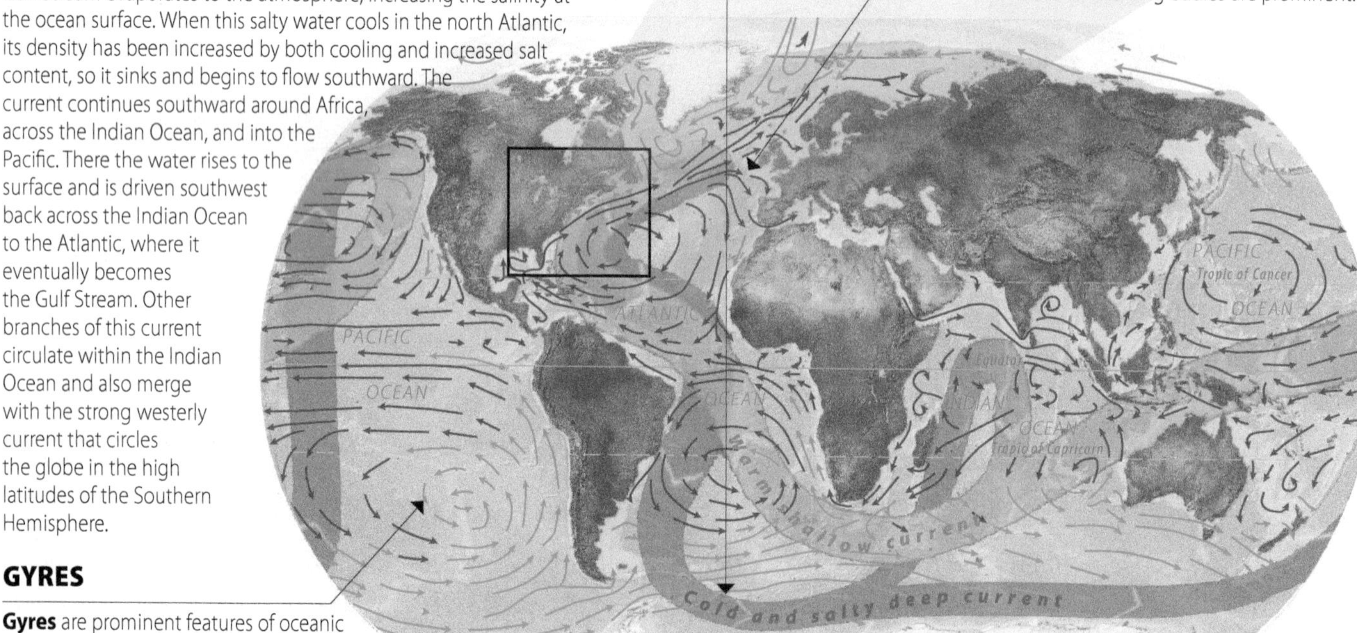

THERMOHALINE CIRCULATION

In addition to wind-driven surface circulation patterns, water also circulates vertically in the oceans. One of the most important of these circulations is called the thermohaline circulation, because it is driven by variations in both temperature and salinity. Warm water in the Gulf Stream evaporates to the atmosphere, increasing the salinity at the ocean surface. When this salty water cools in the north Atlantic, its density has been increased by both cooling and increased salt content, so it sinks and begins to flow southward. The current continues southward around Africa, across the Indian Ocean, and into the Pacific. There the water rises to the surface and is driven southwest back across the Indian Ocean to the Atlantic, where it eventually becomes the Gulf Stream. Other branches of this current circulate within the Indian Ocean and also merge with the strong westerly current that circles the globe in the high latitudes of the Southern Hemisphere.

GYRES

Gyres are prominent features of oceanic circulation. These wind-driven circular oceanic flows mirror the movement of prevailing winds. Gyres form beneath tropical high-pressure cells. The Gulf Stream forms the western limb of the gyre in the North Atlantic. Where ocean currents circulate warm water from low equatorial latitudes to higher latitudes, they are carrying heat poleward by advection. Such flows are balanced by cool currents traveling equatorward, most notably along the west coasts of midlatitude and subtropical land areas. These cold currents cool the lower portions of the atmosphere above them, causing air to sink. Without rising air, adjacent landmasses may be very dry. Some of the driest areas on Earth, most notably the Atacama Desert of Peru and Chile, owe their aridity partly to this effect. The same occurs along the coast of southern California and northwest Mexico, as well as along the Namib desert of southern Africa and much of Australia.

EL NIÑO-SOUTHERN OSCILLATION (ENSO)

The close linkage between oceanic and atmospheric circulation is demonstrated by a phenomenon called **El Niño**. The term is Spanish for "the (male) child," a reference to the Christ child, because the phenomenon occurs around Christmastime. El Niño is a circulation change in the eastern tropical Pacific Ocean that occurs every several years. In this change, the usual cool flow from South America westward is slowed and sometimes reversed, replaced by a warm-water flow from the central Pacific eastward. The counterpart of El Niño, in which especially cool waters are found in this region, is known as La Niña ("the female child").

The typical (La Niña) circulation causes deep ocean water to rise to the surface off the coast of Peru, delivering nutrients that support fish populations (Figure 2.7.3a). The reversed flow of water (Figure 2.7.3b) contributed to the collapse of the Peruvian anchovy industry in the 1970s. El Niño events are far reaching because a modification of circulation in one part of the globe may cause circulation patterns to change elsewhere in North and South America and the Pacific region. For example, El Niño events are linked to flooding in the U.S. Southwest, reduced rainfall in India, and droughts in Australia (Figure 2.7.4). La Niña events often bring wet weather in south and Southeast Asia, and dry conditions in the southern United States.

These large-scale ocean currents have profound effects on weather and climate patterns. In many cases, the strength of the currents changes over time, ranging from a few years to millennia. Variations in the strength of the thermohaline circulation are believed to be correlated with abrupt climate shifts tens of thousands of years ago.

▼ 2.7.3a **NORMAL CLIMATIC CONDITIONS**

Southeast trade winds
Descending warm air
Rising warm moist air
Accumulation of warm water
Upwelling of cold water
South equatorial current

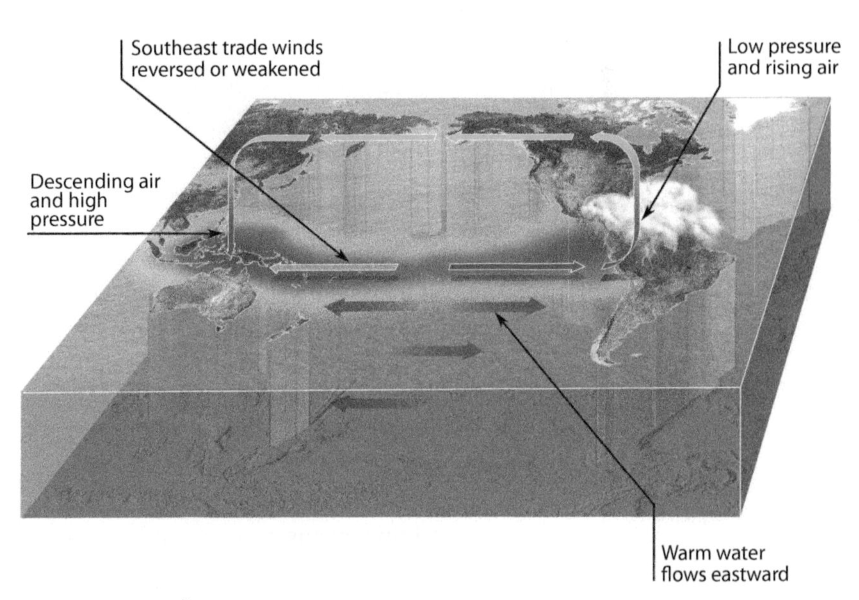

▲ 2.7.3b **THE EL NIÑO CYCLE**
Normal (and also La Niña) conditions in the upper diagram and reversed (El Niño) conditions in the lower diagram.

▼ 2.7.4 **EFFECTS OF EL NIÑO**
Cracked dry reservoir bottom after drought, Lake Burrendong, in southeast Australia.

2.8 Causes of Precipitation

- Precipitation occurs where humid air rises.
- Four general mechanisms cause air to rise.

Precipitation occurs when air rises sufficiently to cause condensation. There are four types of conditions that cause air to rise: convection, orographic uplift, convergence, and fronts.

CONVECTIONAL PRECIPITATION

On a warm, humid summer day, the sky is clear in the morning and the Sun is bright. The Sun warms the ground quickly and the air temperature rises. Most of the warming of the air takes place close to the ground, because the humid air is a good absorber of longwave radiation, which is being reradiated from the ground (Figure 2.8.1).

▲ 2.8.1 **CONVECTION**
As the air near the ground warms, it expands, becomes less dense, and rises through the surrounding cooler air above.

Convectional storms are responsible for a large portion of the world's precipitation. In tropical climates, where strong insolation makes temperatures high, all that is needed for intense daily convectional storms is a source of humidity. In midlatitude climates, such storms occur mostly in the summer because higher temperatures allow the air to hold more moisture, and this means more latent heat can be released, causing strong convection.

Convection works along with other mechanisms that cause air to rise and form precipitation. Often these other mechanisms are the triggers that lead to more intense convection.

OROGRAPHIC UPLIFT

Precipitation sometimes occurs when the horizontal winds move air against mountain ranges, forcing air to rise as it passes over the mountains. This is called **orographic precipitation** (Figure 2.8.2). As the air rises, it cools adiabatically (by expansion); the cooling causes condensation, and precipitation results. After air has moved up the windward side of a mountain and over the top, it then descends on the leeward side. As it does so, its relative humidity drops significantly. The leeward side of a mountain range is often much drier than the rainy windward side.

◄ 2.8.2 **OROGRAPHIC UPLIFT**
Wind is forced to rise over mountains, generating clouds and precipitation.

CONVERGENT PRECIPITATION

Within larger storm systems, large areas of low pressure form, drawing in air that converges from surrounding areas. This rising air causes precipitation (Figure 2.8.3). Such regions of low pressure and precipitation usually move, guided by large-scale circulation patterns described in 2.6. We see them on satellite images of Earth as large areas of clouds, within which precipitation occurs.

► 2.8.3 **CONVERGENT PRECIPITATION**
This satellite image shows an area of low pressure and rising air over the western United States. The colored areas are regions of especially high cloud tops, and thus likely precipitation.

MasteringGEOGRAPHY
Animations
Cold Fronts and Warm Fronts

FRONT

Frontal lifting forms along a **front**, which is a boundary between two air masses. An air mass is a large region of air—hundreds or thousands of square kilometers—with relatively uniform characteristics of temperature and humidity. An air mass acquires these characteristics from the land or water over which it forms.

In North America, air masses that form over central Canada tend to be cool (because of Canada's relatively high latitude) and dry (because of the region's isolation from oceanic moisture sources) (Figure 2.8.4). Air of this type is called continental polar air. In contrast, air masses that form over tropical water, such as over the Gulf of Mexico, tend to be warm and moist and are called maritime tropical air masses.

If these two types of air masses were to meet, a boundary, or front, may form between them. Because cool air is relatively dense, it tends to move under less dense warm air, while warm air tends to rise over cool air. Where air rises along these fronts, precipitation occurs.

▼ 2.8.4 **COLD FRONT**
The cold air advances and drives under the warm air, forcing it upward.

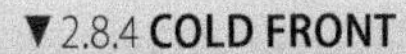

2.9 Storms

- **Tropical cyclones are intense convective systems that form over tropical and subtropical oceans.**
- **Midlatitude cyclones are the dominant precipitation-forming storms of the midlatitudes.**

MasteringGEOGRAPHY
Animations
Midlatitude Cyclones

Storms are areas of concentrated convection that bring precipitation and sometimes strong winds. On a regional scale, large storms can affect areas hundreds to thousands of kilometers across. Storms, also called **cyclones**, are large low-pressure areas in which winds converge in a counterclockwise swirl in the Northern Hemisphere and clockwise in the Southern Hemisphere. There are two types of cyclones, tropical and midlatitude.

TROPICAL CYCLONE

Tropical cyclones are intense, rotating convectional systems that develop over warm ocean areas in the tropics and subtropics, primarily during the warm season. When such storms have wind velocities exceeding 119 kilometers per hour (74 miles per hour), we call them **hurricanes** in North America, **typhoons** in the western Pacific, and cyclones in the Indian Ocean and northern Australia (Figure 2.9.1).

▼ 2.9.1 **ANATOMY OF A HURRICANE**
Hurricanes have multiple spiral bands of very intense convection.

These storms typically develop in the eastern portion of an ocean within the **trade wind** belt. They begin as areas of low pressure (rising air) and converging winds, drawing in warm, moisture-laden air. Tropical storms move with the general circulation over the subtropical Atlantic, Pacific, and Indian oceans, from east to west. They thrive on warm, moist air, so they are most intense over ocean areas during the warm season (Figure 2.9.2). They lose intensity over land because they lose their source of energy. Furthermore, the smooth ocean surface favors development of high winds. In contrast, the hills and trees of land areas slow the wind by friction.

Humid tropical air contains a great deal of energy, both as sensible heat and latent heat in the water vapor, but especially latent heat. As this energy is drawn into the developing storm, condensation and the resulting release of latent heat intensifies the convection. The center of low pressure grows more intense, and as wind speed grows, the Coriolis effect causes a spiraling circulation to develop.

The greatest threat to humans from hurricanes is in tropical and subtropical coastal areas, on the eastern margins of the continents and in southern Asia, where monsoon circulation draws air northward from the Indian Ocean. When a hurricane strikes land, the combination of intense wind and extremely low pressure causes a **storm surge**, an area of elevated sea level in the center of the storm that may be several meters high. The surge carries large waves crashing inland onto low-lying coastal areas, with devastating results. In fact, the majority of hurricane deaths and damage result from storm surges. Adding to this hazard, tornadoes commonly are spawned as a hurricane comes ashore.

▲ 2.9.2 **HURRICANE IGOR**
At its peak, this 2010 Atlantic hurricane had winds of over 155 mph (250 km/h). It struck land in Bermuda and Newfoundland. Off the coast of Newfoundland waves 83 feet (25 m) high were recorded.

MIDLATITUDE CYCLONES

Midlatitude cyclones are centers of low pressure that develop along the **polar front**. They move from west to east along that front, following the general circulation in the midlatitudes. Midlatitude cyclones are usually much less intense than hurricanes, but they are much more common (Figure 2.9.3).

In a midlatitude cyclone, air is drawn toward the center of low pressure from both the warm and the cold sides of the polar front. Where warm air is drawn toward cold, typically on the eastern side of the storm, a **warm front** develops. On the western side, the spiraling motion causes cold air to drive under the warm air, forming a **cold front**. As the center of low pressure moves eastward, these fronts move with it, bringing precipitation to areas over which they pass.

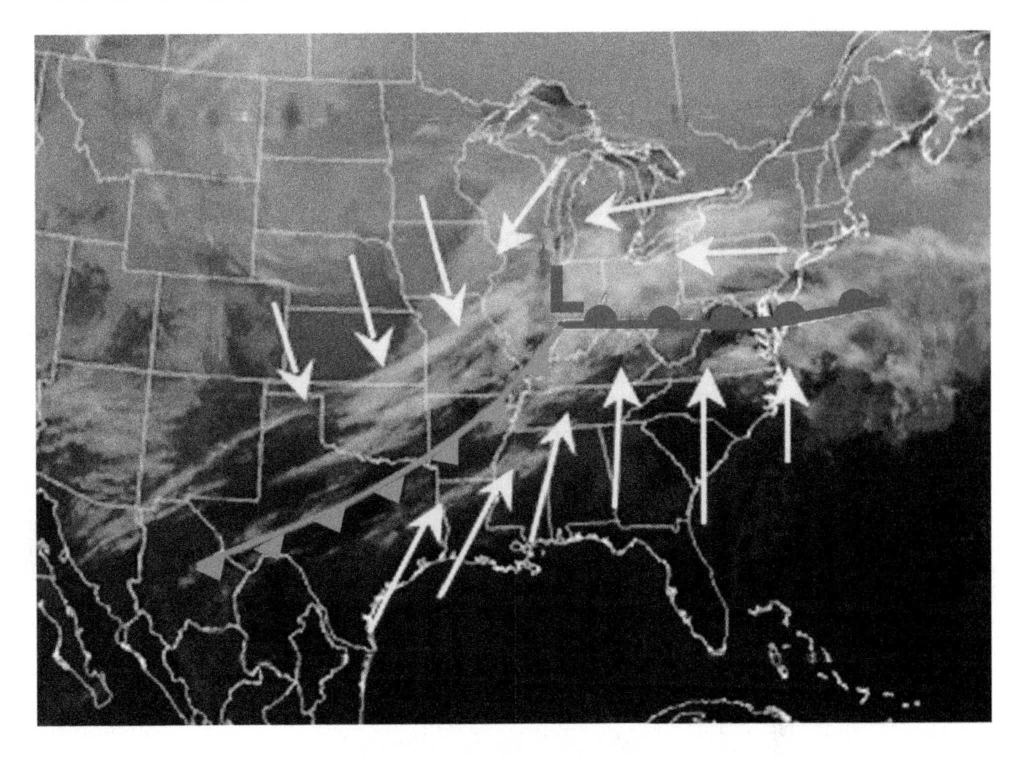

▲ 2.9.4 **MIDLATITUDE CYCLONE**
This storm, shown centered over Indiana on February 21, 2011, produced rain ahead of the warm front (in red), and snow along the cold front (in blue).

The passage of a front at the surface usually is marked by a significant change of temperature, precipitation, and shifting winds. The repeated passage of such storms creates highly variable weather conditions, with alternating cold and warm air as the polar front moves back and forth across the land. Occasionally, especially in North America, strong cold fronts associated with midlatitude cyclones produce **tornadoes** (Figure 2.9.4). Characteristics and patterns of occurrence of tornadoes, midlatitude cyclones, and tropical cyclones are summarized in Figure 2.9.5.

▼ 2.9.3 **LIGHTNING AND THUNDERSTORM SUPERCELL, KANSAS**

▼ 2.9.5 **CHARACTERISTICS OF THREE MAJOR STORM TYPES**

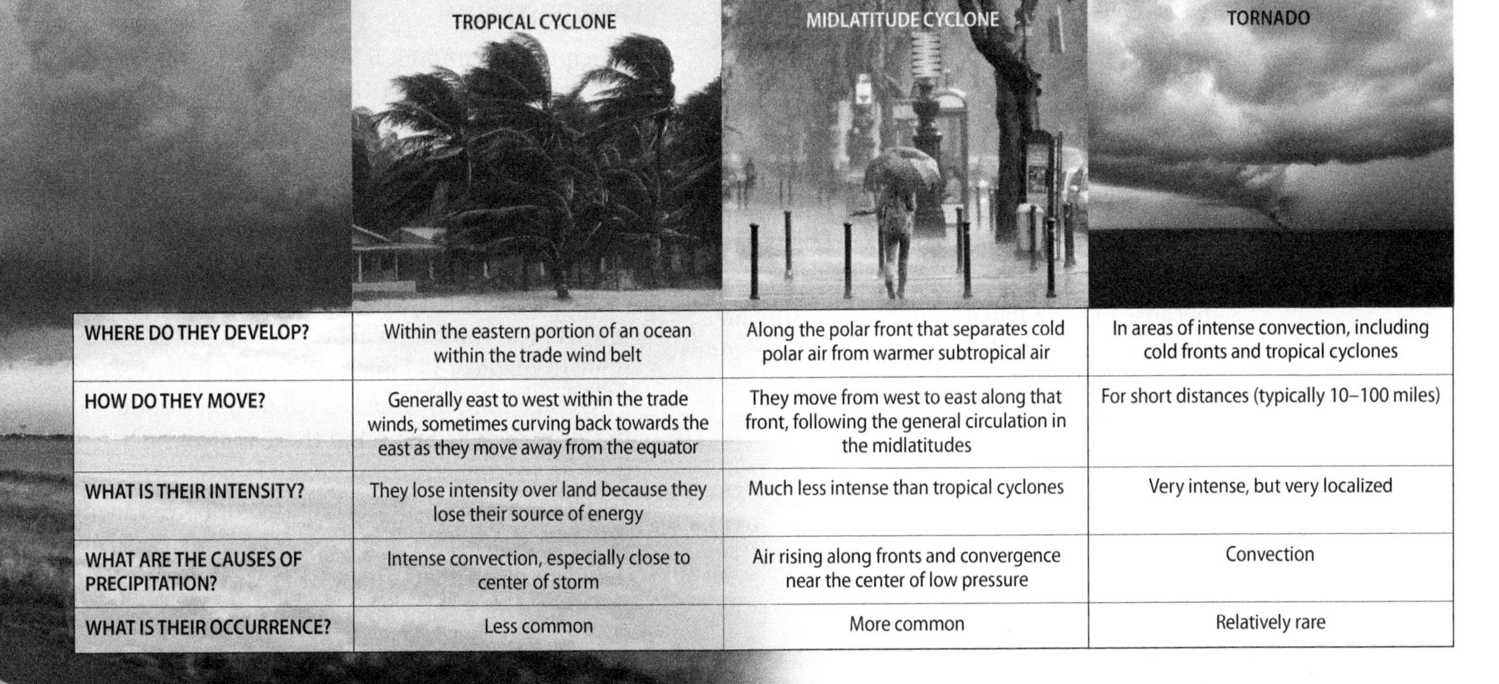

	TROPICAL CYCLONE	MIDLATITUDE CYCLONE	TORNADO
WHERE DO THEY DEVELOP?	Within the eastern portion of an ocean within the trade wind belt	Along the polar front that separates cold polar air from warmer subtropical air	In areas of intense convection, including cold fronts and tropical cyclones
HOW DO THEY MOVE?	Generally east to west within the trade winds, sometimes curving back towards the east as they move away from the equator	They move from west to east along that front, following the general circulation in the midlatitudes	For short distances (typically 10–100 miles)
WHAT IS THEIR INTENSITY?	They lose intensity over land because they lose their source of energy	Much less intense than tropical cyclones	Very intense, but very localized
WHAT ARE THE CAUSES OF PRECIPITATION?	Intense convection, especially close to center of storm	Air rising along fronts and convergence near the center of low pressure	Convection
WHAT IS THEIR OCCURRENCE?	Less common	More common	Relatively rare

2.10 Global Climates

► **Climates are classified on the basis of annual averages and seasonal variations in temperature and precipitation.**

Climate is the summary of weather conditions over several decades or more—a place's weather pattern over time. The vegetation, natural resources, and human activities that characterize a particular region of Earth are heavily influenced by its climate. Geographers have struggled for decades to find an effective classification system for climates. Part of the problem is in deciding where to draw boundaries. It is a little like deciding how to define the difference between two colors, when there is an infinite variation in shades between them. The most commonly used classification system is shown here (Figure 2.10.1).

To understand a region's distinctive climate, the two most important measures are air temperature and precipitation.

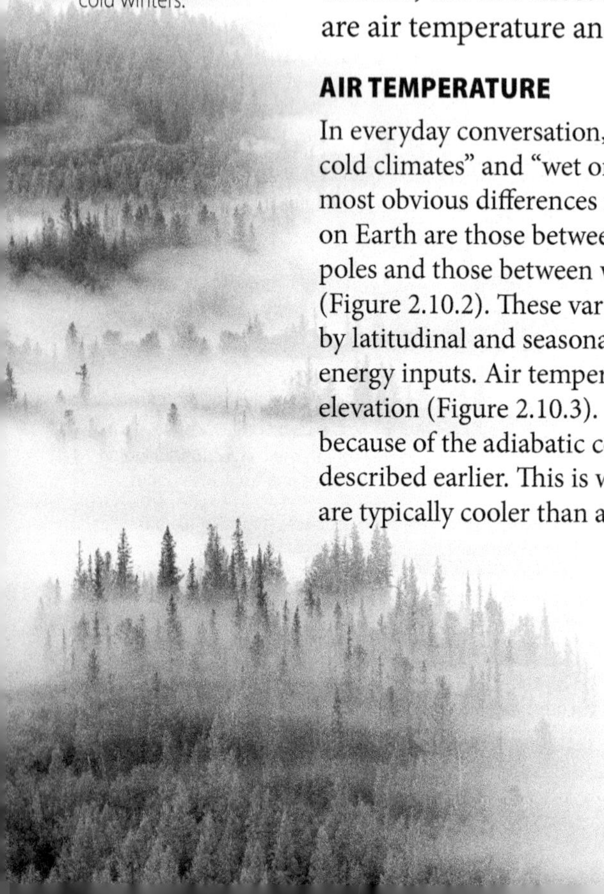

▼ 2.10.2 **FOREST IN SIBERIA**
Coniferous forests are found in areas with short, warm summers and long cold winters.

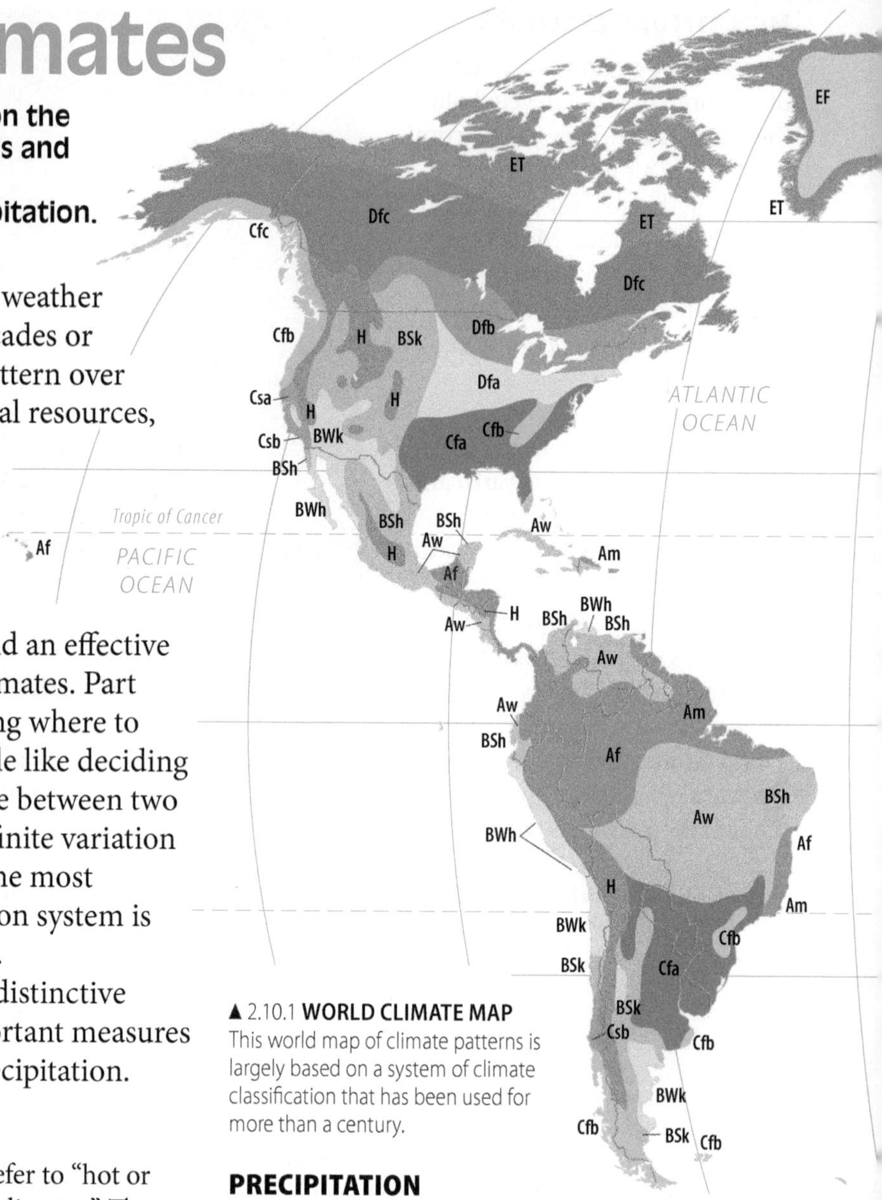

▲ 2.10.1 **WORLD CLIMATE MAP**
This world map of climate patterns is largely based on a system of climate classification that has been used for more than a century.

AIR TEMPERATURE

In everyday conversation, we refer to "hot or cold climates" and "wet or dry climates." The most obvious differences in air temperature on Earth are those between the tropics and the poles and those between winter and summer (Figure 2.10.2). These variations are caused by latitudinal and seasonal variations in solar energy inputs. Air temperature also varies with elevation (Figure 2.10.3). This variation occurs because of the adiabatic cooling of rising air described earlier. This is why mountain regions are typically cooler than adjacent lowlands.

PRECIPITATION

Precipitation amount is extremely variable between places and over time. A thunderstorm can unleash heavy rain in one place and a light sprinkling just a short distance away. Worldwide, annual precipitation generally ranges from virtually none in some desert areas to more than 300 centimeters (120 inches) in wet tropical areas. A few tropical mountain areas have recorded more than 10 meters (396 inches) of rainfall per year.

Average rainfall amounts tell us much about the climate of a region, but there is also much they do not tell. The timing and reliability of rainfall are equally important. In cool areas such as midlatitudes, precipitation frequently takes the form of gentle rainfall, whereas in tropical regions, it may come in torrential downpours. We also experience significant variations in

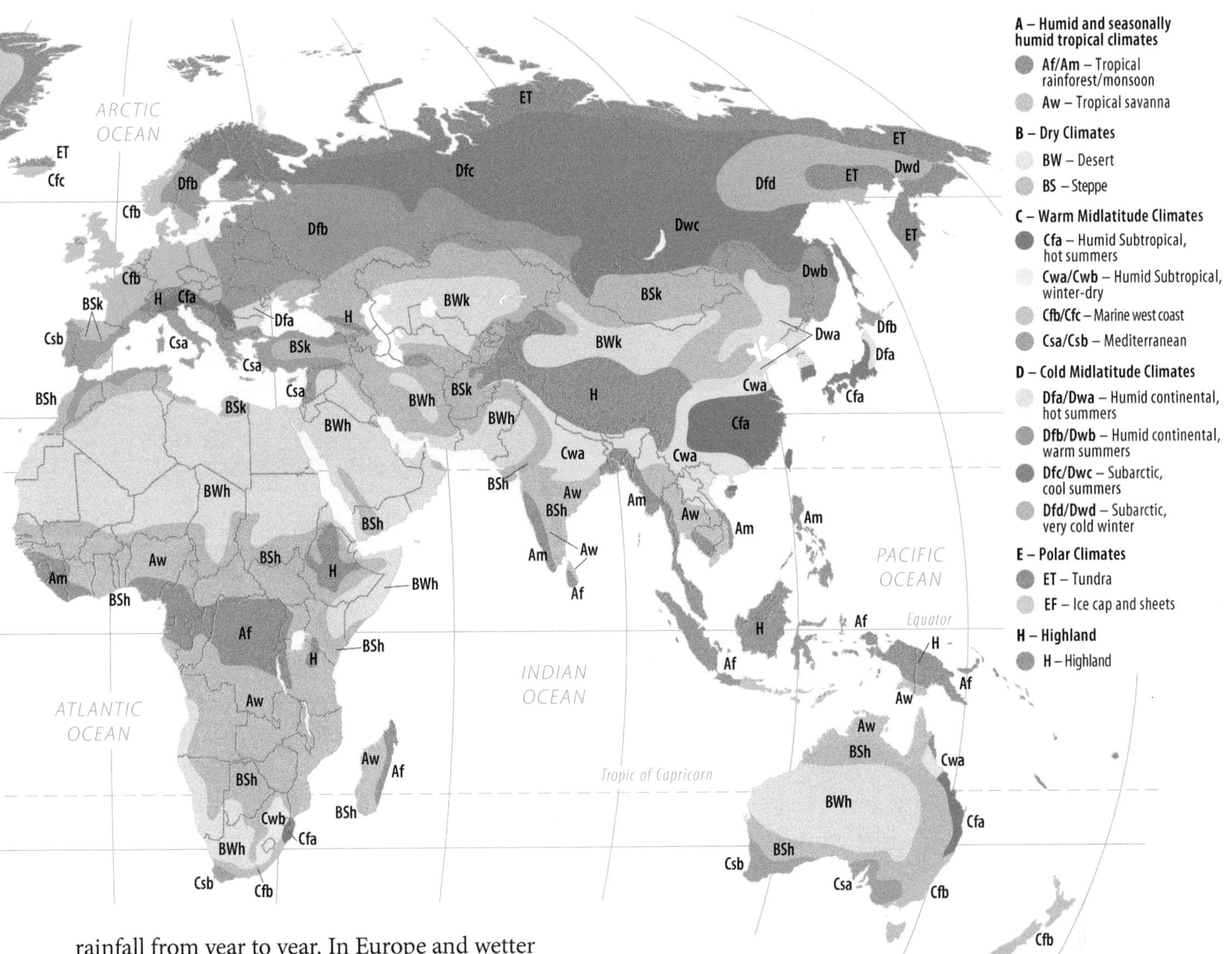

rainfall from year to year. In Europe and wetter portions of North America, the amount of annual rainfall varies by less than 15 percent per year on average. In many of the tropical and subtropical regions, on the other hand, it fluctuates from 15 to 20 percent, and in the semiarid and arid lands, it can fluctuate by 50 percent or more.

Evaluating water availability involves much more than just measuring rainfall. Plants demand water and play a key role in what happens to the precipitation. Most climate classification systems consider precipitation in relation to what a region's vegetation needs. Plants consume very large amounts of water, totaling about two-thirds of all precipitation that falls on Earth's land areas. Because heat energy must be available to evaporate water, warmer climates make possible a greater amount of evaporative water use. Thus temperature is considered in distinguishing between arid and humid climates.

► 2.10.3 **VARIATION IN CLIMATE RELATED TO ELEVATION**

Use Google Earth to explore the Himalayan Mountains and the effects of elevation on climate. Mt Everest is in the distance.

Fly to: *Libang, Nepal*

Zoom to an eye altitude of around 20,000 feet (6,000 m).

Tilt to bring the horizon into view.

Look around and explore the landscape.

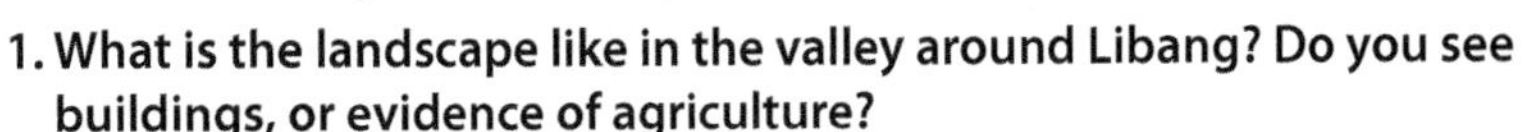

1. **What is the landscape like in the valley around Libang? Do you see buildings, or evidence of agriculture?**
2. **At elevations of 8,000 to 13,000 ft (2500 to 4000 m), there is darker green vegetation. What do you think that is?**
3. **When you look at elevations above 13,000 ft (4,000 m), how does the vegetation cover change? What factors might account for those changes?**

2.11 Diversity of Climates

The most commonly used classification system groups climates into five major categories (Figure 2.11.1):

- Cold midlatitude (Figure 2.11.2).
- Warm midlatitude (Figure 2.11.3).
- Polar (Figure 2.11.4).
- Dry (Figure 2.11.5).
- Humid and seasonally humid tropical (Figure 2.11.6)

COLD MIDLATITUDE CLIMATES

Cold midlatitude climates are typically "continental," so named because they are remote from the ocean and therefore deprived of the sea's input of moisture and moderating influence on temperature. These climates occur between about 35° and 60° latitude in the interior and eastern portions of Northern Hemisphere continents. They occur exclusively in the Northern Hemisphere because very little landmass exists in the Southern Hemisphere at these latitudes.

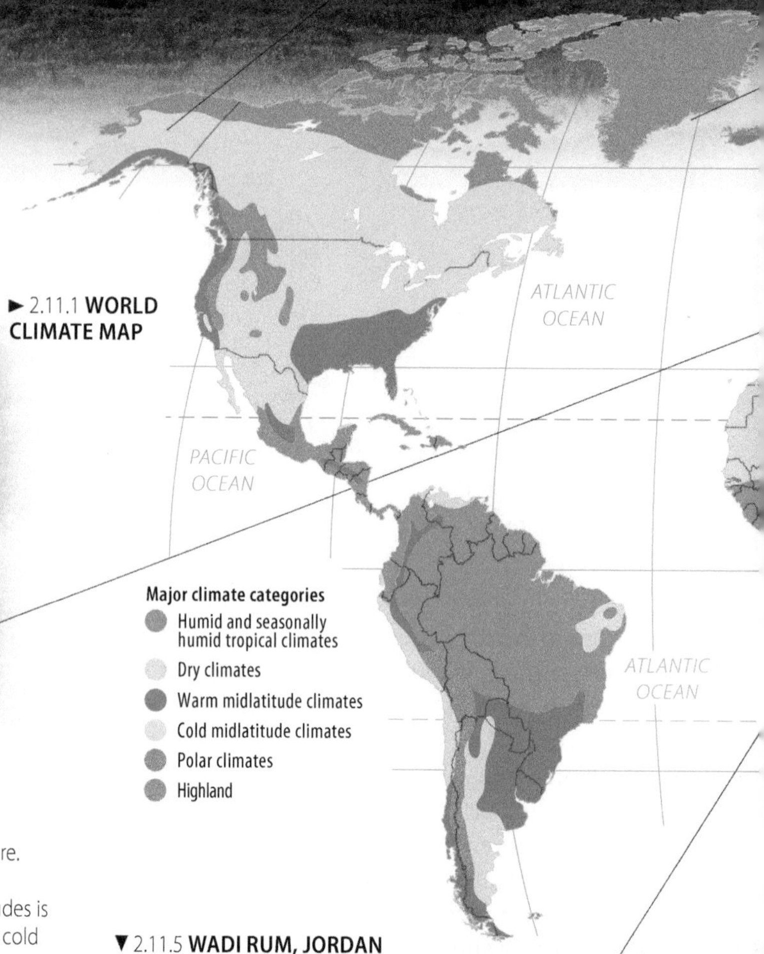

▼ 2.11.2 DENALI NATIONAL PARK, ALASKA

► 2.11.1 WORLD CLIMATE MAP

▲ 2.11.3 TUSCANY, ITALY

▼ 2.11.5 WADI RUM, JORDAN

WARM MIDLATITUDE CLIMATES

In the midlatitudes, seasonal variations of insolation profoundly influence temperature. The midlatitudes experience a distinct cool season (winter). In subtropical locations, winter may be a month or two in which frost can occur. Precipitation in the midlatitudes is heavily influenced by the polar front—the boundary between warm tropical air and cold polar air along which midlatitude cyclones form and travel. Humid subtropical climates occur in latitudes between about 25° and 40° on the eastern sides of continents and between about 35° and 50° on the western sides.

On continental west coasts between about 35° and 65° are mild climates with small temperature variations and plentiful moisture year-round. These areas are cooler than humid subtropical climates, especially in summer. These marine west coast climates are moderated by ocean temperatures. Maritime influence on temperature is so great that winter temperatures normally associated with subtropical latitudes are found as far poleward as 55°.

On the western margins of continents, another distinctively seasonal climate occurs. The dry-summer Mediterranean climate envelops the Mediterranean region of Europe and parts of northern Africa, lending this climate type its familiar name. Precipitation is seasonal and caused by northward and southward movement of the subtropical high-pressure zones. In summer, these zones move poleward and bring aridity; in winter, they move toward the equator and are replaced by more frequent storms of the midlatitude low-pressure zone.

POLAR CLIMATES

High-latitude climates are characterized by two important features: low average temperatures and extreme seasonal variability. These climates occur at and around the poles and are differentiated by temperature alone, because temperature determines whether water is present in liquid form. These climates result from scant insolation, varying from zero during winter at the poles to among the highest daily totals on Earth for a brief period at midsummer. Tundra climates have a brief cool summer in which vegetation grows, while ice cap climates are too cold to support significant vegetation growth at any time of the year.

▲ 2.11.4 **GREENLAND, SOUTHEAST COAST**

ARCTIC OCEAN

INDIAN OCEAN

PACIFIC OCEAN

Tropic of Cancer

Equator

Tropic of Capricorn

HUMID AND SEASONALLY HUMID TROPICAL CLIMATES

These climates lie mostly within 10° north and south of the equator but can extend to 20° north and south. These areas are strongly influenced by the ITCZ and include the world's tropical rain forests. Because tropical areas are warm throughout the year, the variation in temperature in a single day is greater than the difference between the warmest and coolest months of the year. The high temperatures mean that a great deal of energy is available to evaporate water.

In many areas of the humid tropics, rainfall is concentrated in part of the year, allowing for a distinct dry season. This is caused by seasonal shifts in the location of the ITCZ or by monsoonal circulation patterns (for example, in southern and southeastern Asia and to a lesser extent West Africa). The ITCZ moves north in the Northern Hemisphere summer (May) and south between November and April, and rainfall shifts with it.

In southern and southeastern Asia, precipitation is seasonal largely because of the monsoonal circulation pattern. In the summer, the Asian landmass heats and develops an extensive low-pressure region that draws air in from all directions. The low-pressure cell draws over the land a steady flow of moist air from the Indian and Pacific oceans, and the rainfall is intense.

▼ 2.11.6 **RWENZORI MOUNTAINS, UGANDA, EAST AFRICA**

DRY CLIMATES

Dry climates are generally located in bands immediately to the north and south of the low-latitude humid climates. Many dry regions are associated with the subtropical high pressure zone; the Sahara desert in Africa is of this type. Other deserts are caused mainly by mountain ranges that isolate land from ocean sources of moisture. In Asia, for example, a mountain belt stretches east to west from China to the Caucasus, blocking Indian Ocean moisture from the interior. The North American Rockies and South American Andes similarly block Pacific moisture.

Geographers distinguish between arid (desert) and **semiarid climates**. Arid climates have severe moisture shortages year-around, while most semiarid climates have a rainy season in which sufficient moisture is available for abundant, if brief, vegetation growth. Most of the world's grasslands are in this climate type.

2.12 Global Warming

- Climate has changed dramatically over the last few million years and continues to change today.
- Much of our understanding of climate change comes from models, which demonstrate both natural and human influences on climate.

Past climate changes can give us valuable clues in determining whether recent changes are similar to those of the past or whether they are unique because of human actions. Models help us predict future climates.

CLIMATE HISTORY

Viewed over the entire 4.6 billion-year history of Earth, the climate of the past 2 million years has been quite exceptional. This period, which includes our present time, is known to geologists as the **Quaternary Period**. Within the Quaternary Period there have been intervals in which global average temperature was as much as 10°C (18°F) cooler than the present and warm intervals in which the climate was warmer than today (Figure 2.12.1). In the past 1 million years, there have been about 10 cold intervals, occurring fairly regularly about once per 100,000 years. During the cooler periods, great continental ice sheets, like those covering much of Antarctica and Greenland today, extended over much of North America, northern Europe, and northern Asia. Periods of glaciation were also periods of lowered sea level because water was taken out of the sea and stored on land in the form of glacial ice. Even areas not covered by ice were much different in the past than they are today.

Climate has varied within the past 1,000 years also. Between A.D. 800 and 1000, climates were warm enough that seafarers from present-day Scandinavia were able to establish settlements in Greenland. However, from about 1500 to 1750, temperatures were especially cool. In this period, known as the **Little Ice Age**, glaciers advanced in Europe, North America, and Asia. Since the early 1800s, climates have warmed significantly. The 1930s and 1940s were relatively warm, and cooling occurred from about 1945 to 1970. Global mean temperatures have risen dramatically since about 1975 (Figure 2.12.2).

▲ 2.12.1 **415,000 YEARS OF CLIMATE HISTORY**
Major variations of climate, consisting of alternating cold and warm spells, suggest a climate system that is unstable but partly regulated by a "pacemaker" that is related to variations in the geometry of Earth's orbit around the Sun.

▼ 2.12.2 **SHRINKING ICE CAP**
The extent of Arctic sea ice has declined significantly in recent decades. The following images show September ice extent in (left) 1979 and (right) 2011 as well as the median September ice extent for the period 1979-2011.

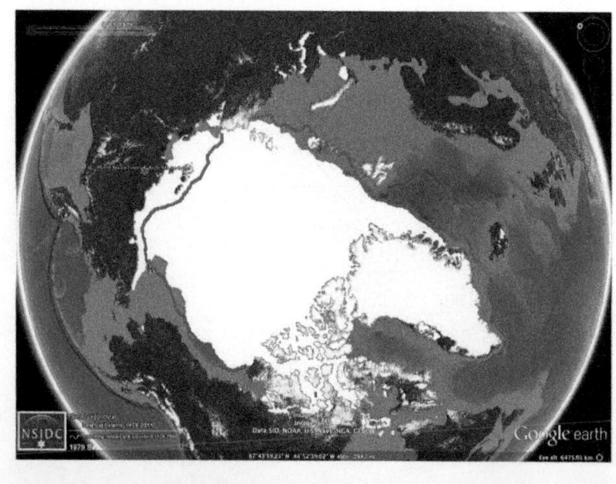

SOME POSSIBLE CAUSES OF CLIMATIC VARIATION

Climate change has many potential causes. Understanding the causes of climatic change is critical because if we learn that climate is affected by human activity then we have the opportunity to limit such effects. If climatic change were entirely governed by natural processes, then there would be no need for concern about potential human impacts on the atmosphere.

Solar Output

Recent studies based on satellite observations of the Sun are beginning to suggest that variations in solar output could be responsible for some of the climatic variations observed in the past few hundred years, but the effects are probably small in comparison to other factors (Figure 2.12.3). In the long run—over tens of thousands of years—we know that the geometry of Earth's revolution around the Sun fluctuates, and these fluctuations appear to function as pacemakers, determining the timing of major climate shifts at scales of about 100,000 years.

Volcanic Eruptions

Volcanic eruptions can influence climate for a few years by injecting large amounts of dust and gases—especially sulfur dioxide—into the upper atmosphere. These gases reduce the amount of solar radiation filtering through the atmosphere to Earth, briefly lowering temperatures.

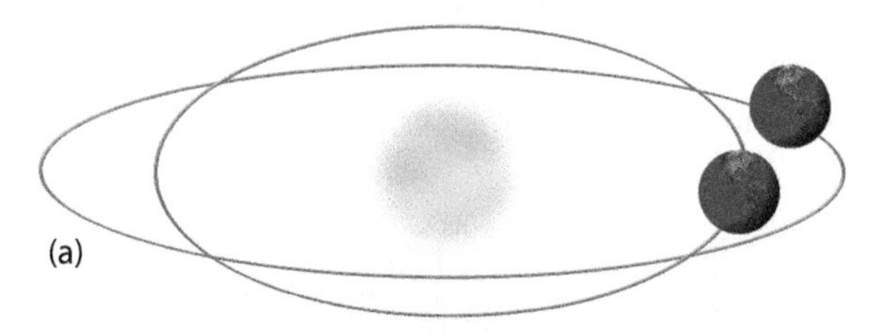

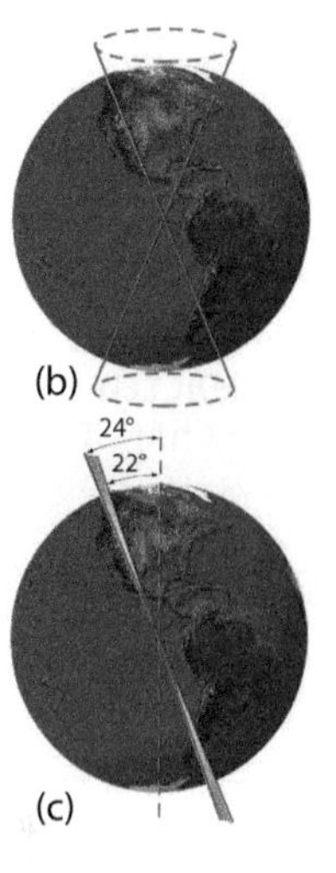

▲ 2.12.3 **CHANGES IN EARTH ORBIT**
Climate may vary with long-term variations in the geometry of Earth's orbit around the Sun. These variations include (a) the annual variation of distance between Earth and the Sun, (b) the orientation of the Earth's axis relative to the shape of Earth's orbit around the Sun, and (c) the degree of tilt of that axis relative to our orbit around the Sun.

HUMAN IMPACTS ON CLIMATE

Processes such as changes in orbital geometry or volcanic eruptions have clearly affected climate in the past, but do not explain the dramatic warming of the last 200 years. We now know that this warming is largely due to humans and their effects on atmospheric carbon dioxide. As described in section 2.2, CO_2 is one of several important greenhouse gases. Human-caused increases in atmospheric CO_2 are the primary factor in recent **global warming**. Computer climate models help to demonstrate this.

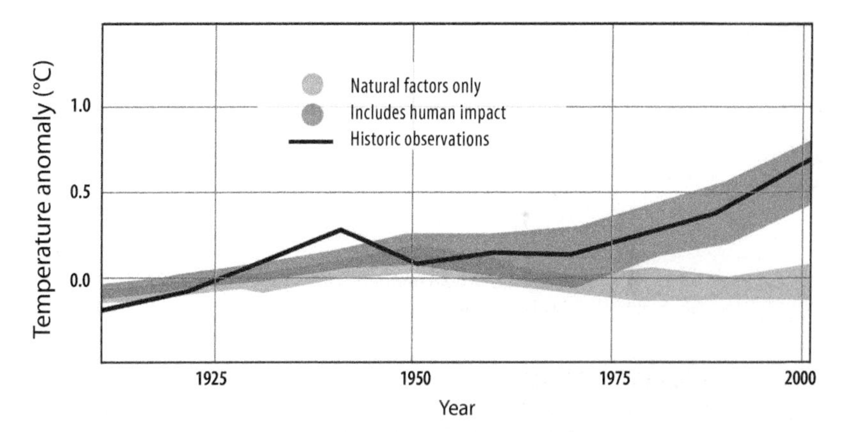

▲ 2.12.4 **GLOBAL TEMPERATURE CHANGES PREDICTED BY MODELS AND OBSERVED**
Just as we predict the weather with computer models, we can predict climate change. Computer models are accurate only if they include human emissions of CO_2 and other pollutants.

Models as evidence of the past and predictions of the future.

The models that we use to predict future global warming are similar to the models that we use to predict the weather two or three days in the future (Figure 2.12.4). We put in a computer program as much as we know about the present state of the atmosphere and land/ocean surface, along with equations that represent how radiant energy modifies temperatures, how temperature and pressure affect precipitation, wind, and so forth. The computer then simulates future atmospheric circulation.

Because these models, and the equations within them, have been built on thousands—even millions—of careful observations (that is, real-world data), they provide some of the most convincing evidence that global warming is caused by human activity. In addition to predicting climate or climates of the future, we can use the models to "predict" historic climates. If we start with the climate of, say, 1910 and add to it what we know about historic changes in natural factors such as solar activity and volcanic eruptions since then, we can predict the temperatures of the past 100 years. These predictions of the past don't match very well with the observations, particularly since about 1950, when observed temperatures are much higher than the predicted ones. However, when we also add to the model human impacts on the atmosphere, such as emissions of carbon dioxide, methane, and other pollutants, the models do a very good job of predicting the temperatures of the past 100 years. The results demonstrate both that the models can predict temperatures with a reasonable degree of accuracy and that humans are causing global warming.

2.13 Global Warming Consequences

- Significant human-caused climate changes are underway, and further changes are expected in the coming decades.
- Several factors contribute to uncertainties regarding future climate changes.

The most significant summary of our understanding of global warming consequences comes from the work of the Intergovernmental Panel on Climate Change (IPCC), a group of scientists appointed by the governments of more than 130 countries around the world. While the process of summarizing and reporting information at the IPCC is not totally free of political influence, and some scientists have criticized the IPCC's objectivity, its reports are generally regarded as credible syntheses of a very complex body of science.

In its 2007 assessment of global climate change, the IPCC identified a number of ways in which climate is changing, and it included statements reflecting the confidence of the participating scientists that these changes (1) began in the twentieth century, (2) are caused at least in part by humans, and (3) will occur in the twenty-first century (Figure 2.13.1).

▼ 2.13.1 **DETECTED AND PREDICTED CLIMATE CHANGE**
A summary of the climate changes detected and predicted by the Intergovernmental Panel on Climate Change.

PHENOMENON AND DIRECTION OF TREND	LIKELIHOOD THAT TREND OCCURRED IN LATE TWENTIETH CENTURY	LIKELIHOOD OF A HUMAN CONTRIBUTION TO OBSERVED TREND	LIKELIHOOD OF FUTURE TRENDS BASED ON PROJECTIONS FOR TWENTY-FIRST CENTURY
Warmer and fewer cold days and nights over most land areas	✔✔✔	✔✔	✔✔✔✔
Warmer and more frequent hot days and nights over most land areas	✔✔✔	✔✔ (nights)	✔✔✔✔
Increased frequency of warm spells/heat waves over most land areas	✔✔	✔	✔✔✔
Increased frequency of heavy precipitation events over most areas	✔✔	✔	✔✔✔
Increases in the areas affected by droughts	✔✔ (in many regions since 1970s)	✔	✔✔
Increased intense tropical cyclone activity	✔✔	✔	✔✔
Increased incidence of extremely high sea level (excludes tsunamis)	✔✔	✔	✔✔

✔✔✔✔ Virtually certain (>99% probability); ✔✔✔ Very likely (>90% probability); ✔✔ Likely (>66% probability); ✔ More likely than not (>50% probability).

Many aspects of expected global warming are highly uncertain. There are three types of major uncertainties:

1. WE HAVE A LIMITED UNDERSTANDING OF THE CLIMATE SYSTEM AND HOW TO MODEL IT ACCURATELY

While the accuracy of the computer models used to predict weather and climate have improved dramatically in recent years, they still have limits. This is in part because of two factors. First, we don't have highly detailed information on the present conditions of temperature, pressure, and so on for every place and altitude. Second, the rules we use to estimate atmospheric flows may not be completely representative of the complexities of the environment. We face similar problems with long-term climate projections. For example, while the models of global climate we have at present are reasonably consistent in predicting future temperature averages, they aren't as consistent for precipitation, in part because of the importance of storms and storm tracks.

Our models also suffer from incomplete understanding of some important atmospheric processes. For example, human-caused global warming can alter the water content of the atmosphere. We believe that atmospheric water content is more likely to increase than decrease. But if it does increase, would the added water trap more outgoing radiation and thus increase warming, or would it instead lead to increased cloud cover that reflects incoming energy? The answer to this question depends on a more detailed understanding of atmospheric circulation and processes of cloud formation than we have at present.

▼ 2.13.2 **COAL BURNING POWER PLANT IN NEW MEXICO**
We have very large reserves of fossil fuels, but using them is likely to increase global warming and related climate changes.

2. IT IS POSSIBLE THAT UNFORESEEN, POTENTIALLY RAPID CHANGES IN CONDITIONS (ALSO KNOWN AS "TIPPING POINTS") COULD AFFECT THE CLIMATE

In recent years, concern has grown that with ongoing climate change, certain parts of the Earth–atmosphere system may be reaching tipping points, or conditions in which the pace of change may increase rapidly and irreversibly. One of these is the decrease in Arctic sea ice. As sea ice decreases, there is a tendency for the ocean to absorb more solar radiation because ice reflects more energy than does open water. This could cause large parts of the Arctic Ocean to become ice free for a much longer period each year. On the other hand, open water loses heat more rapidly through evaporation than does ice-covered water, so this may not be as great a problem as some have argued. Another possible tipping point relates to accelerated melting of ice caps, such as the Greenland ice cap and portions of Antarctica. While we are relatively certain that such mechanisms of instability exist, most of the ones being considered are not well understood, and it is difficult to know whether the rapid changes envisioned are likely or not, and if they are likely, whether they will happen soon or far in the future.

3. FUTURE CONDITIONS OF THINGS LIKE THE RATE OF CARBON DIOXIDE EMISSIONS ARE UNKNOWN

The amount of carbon dioxide added to the atmosphere is controlled mainly by our use of fossil fuels and by land use and land management, such as forest harvesting and planting. Most of our predictions are based on assumptions that people will continue to burn fossil fuels and manage the land in about the same ways as they have in recent years, but of course this could change in the future (Figure 2.13.2).

The amount of solar energy received by Earth and its atmosphere varies by latitude, season, and time of day. This energy determines broad temperature patterns and also drives atmospheric and ocean circulation. The distribution of climate types reflects energy inputs and circulation systems.

Key Questions

How does energy move in the Earth-atmosphere system?

- The geometry of Earth's orbit around the Sun determines the latitudinal, seasonal, and daily variation in solar energy.
- Energy passes among the Earth's surface, atmosphere, and space by radiation, convection, latent heat exchange, and conduction.
- Atmospheric gases that are relatively transparent to incoming shortwave solar energy but absorb outgoing longwave radiation are key to atmospheric heating.
- Latent heat exchange that occurs when water evaporates/condenses or melts/freezes plays a central role in energy exchange and precipitation.

What makes the weather?

- Temperature is controlled mainly by energy inputs, with strong influences from surface characteristics such as land or water.
- Convection causes air to rise, causing clouds and precipitation to form.
- Ocean circulation includes wind-driven surface currents, and deeper circulation related to water temperature and salinity.
- Precipitation results from convection, orographic uplift, convergence, and fronts.
- Tropical cyclones are large convectional storms. Fronts are a key feature of midlatitude cyclones.

What climates are found on Earth, and how is climate changing?

- Climate, or average weather, varies around the globe in relation to prevailing patterns of energy inputs and circulation.
- Major climate types include humid and seasonally humid tropical climates, dry climates, warm and cold midlatitude climates, and polar climates.
- Climate has changed dramatically over the last few million years and is changing today.
- Models help us to understand the causes of climate change and also to predict future climates.
- Current and expected future climate change includes warmer temperatures and increased frequency of intense precipitation.

Thinking Geographically

For a two-week period, keep a daily journal of the weather, including such things as air temperature, wind direction, cloudiness, and precipitation. For the same period, clip the weather map from a daily newspaper.

1. **How did the patterns of high and low pressure systems affect the weather?**
2. **How might an increase in average annual temperature of 5°C (9°F) affect your day-to-day life? Consider both the immediate effects, and the indirect effects that might derive from broader economic or social trends in the world (2.CR.1).**

▼ 2.CR.1 **RISING SEA LEVEL**
Storm-related flooding, as shown here in the Passur River, Ganges delta, Bangladesh, is worsened by rising sea level.

Interactive Mapping

PREDICTED CLIMATE CHANGE

Open: MapMaster World Layered Thematic Analysis Activities in MasteringGEOGRAPHY™

Select: *Global Surface Warming, Worst-Case Projections.*

1. **What parts of the globe are expected to see the greatest warming?**
2. **In general, are ocean or land areas expected to warm more? Why do you think this would be the case?**

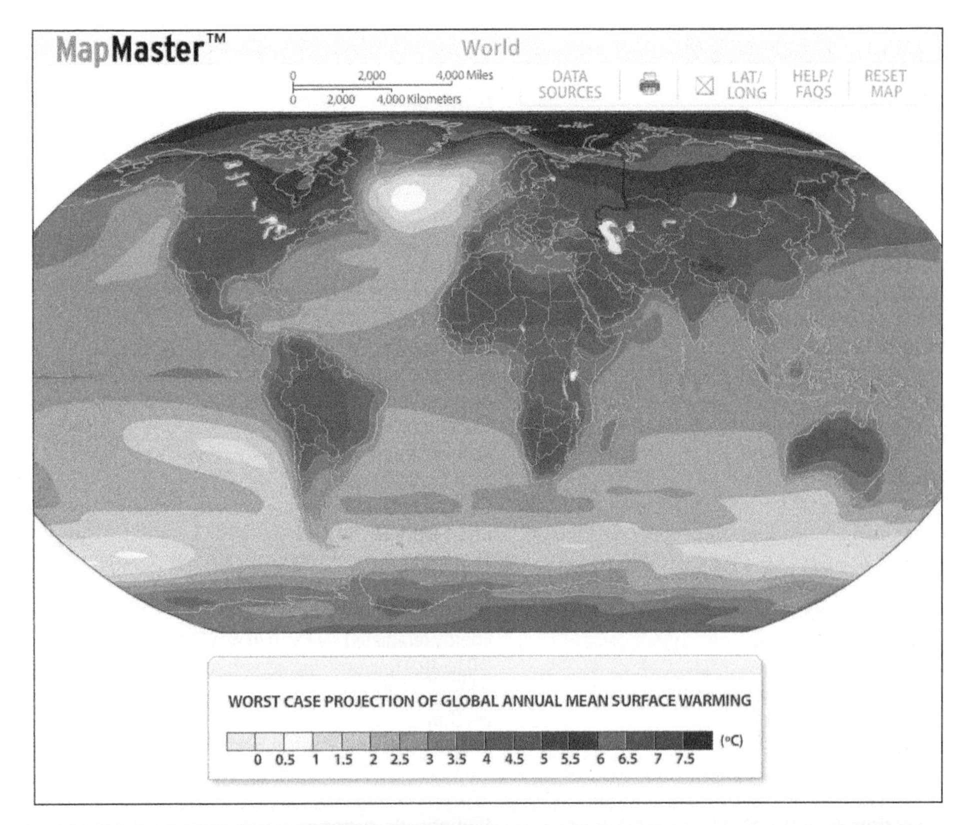

Explore

SEATTLE, WASHINGTON

Use Google Earth to explore the effects of mountains on the weather.

Fly to: *Ames Lake, Washington*

Click to turn on: *Conditions and Forecasts* under the Weather menu.

1. **What is the weather in Ames Lake today? Look around the area. What does the vegetation cover look like?**

Fly to: *Wenatchee, Washington*

Now check the weather about 100 miles east in Wenatchee, Washington.

2. **How is the weather in Wenatchee different from the weather in Ames Lake?**
3. **Is there a link between the differences in weather and the differences in vegetation?**

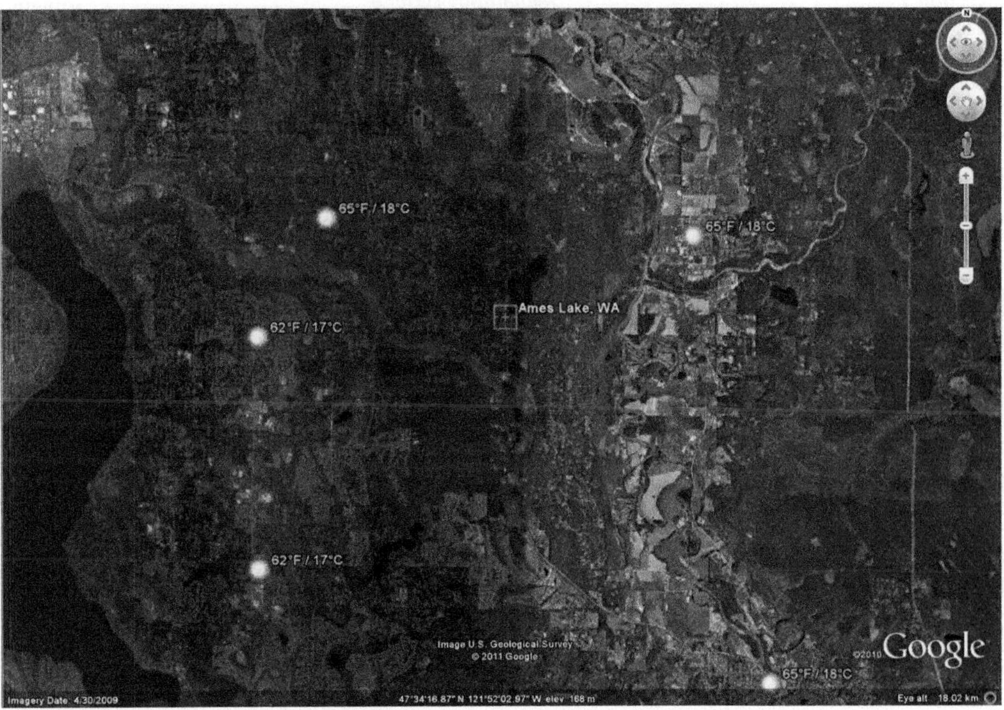

On the Internet

The Intergovernmental Panel on Climate Change (IPCC) brings together the best available scientific information about climate change and policy responses to that change. The IPCC web site is **www.ipcc.ch.** The Carbon Dioxide Information Analysis Center (CDIAC) at Oak Ridge National Laboratory has a great wealth of information on greenhouse gases, climate change, and related issues at **cdiac.ornl.gov.**

Key Terms

Adiabatic cooling
The cooling of air as a result of expansion of rising air; adiabatic means "without heat being involved".

Advection
The horizontal movements of air or substances by wind or ocean currents.

Angle of incidence
The angle at which solar radiation strikes a particular place at a point in time.

Autumnal equinox
In the Northern Hemisphere September 22 or 23, one of two dates when, at noon, the perpendicular rays of the sun strike the equator (meaning that the Sun is directly overhead along the equator).

Carbon dioxide
A trace gas found in the atmosphere with chemical formula CO_2; a major contributor to the greenhouse effect.

Climate
The totality of weather conditions over a period of several decades or more.

Cold front
The boundary formed when a cold air mass advances against a warmer one.

Condensation
Water changing from a gas state (vapor) to a liquid or solid state.

Convection
Circulation in a fluid caused by temperature-induced density differences, such as the rising of warm air in the atmosphere.

Coriolis effect
The tendency of an object moving across Earth's surface to be deflected from its apparent path as a result of Earth's rotation.

Cyclone
Large low-pressure areas in which winds converge in a counterclockwise swirl in the Northern Hemisphere (or clockwise in the Southern Hemisphere).

Desert climate
A climate with low precipitation and temperatures warm enough to cause potential evapotranspiration to be substantially higher than precipitation.

El Niño
A circulation change in the eastern tropical Pacific Ocean, from westward flow to eastward flow, that occurs every few years.

Front
A boundary between warm air and cold air.

Global warming
A general increase in temperatures over a period of at least several decades caused primarily by increased levels of carbon dioxide in Earth's atmosphere.

Greenhouse gases
Trace substances in the atmosphere that contribute to the greenhouse effect; water vapor, carbon dioxide, ozone, methane, and chlorofluorocarbons are important examples.

Gyre
A circular ocean current beneath a subtropical high-pressure cell.

Hurricane
An intense tropical cyclone that develops over warm ocean areas in the tropics and subtropics, primarily during the warm season. Hurricanes in the Pacific Ocean are called typhoons; in the Indian Ocean they are called cyclones.

Insolation
The amount of solar energy intercepted by a particular area of Earth.

Intertropical Convergence Zone (ITCZ)
A low-pressure zone between the Tropic of Cancer and the Tropic of Capricorn where surface winds converge.

Latent heat
Heat stored in water and water vapor, not detectable by people; latent means "hidden".

Little Ice Age
The period between about 1500 to 1750, when climates on Earth were especially cool.

Longwave energy
Energy reradiated by Earth in wavelengths of about 5.0 to 30.0 microns . Includes infrared radiation, which we sense as heat.

Methane
A trace gas found in the atmosphere with chemical formula CH_4; a major contributor to the greenhouse effect.

Midlatitude cyclone
A storm characterized by a center of low pressure in the midlatitudes usually associated with a warm front and a cold front.

Midlatitude low-pressure zones
Regions of low pressure and air converging from the subtropical and polar high-pressure zones.

Monsoon circulation
Seasonal reversal of pressure and wind in Asia, in which winter winds from the Asian interior produce dry winters, and summer winds blowing inland from the Indian and Pacific oceans produce wet summers.

Orographic precipitation
Precipitation caused by air being forced to rise over mountains.

Ozone
A gas composed of molecules with three oxygen atoms; it is a highly corrosive gas at ground level, but in the upper atmosphere essential to protecting life on Earth by absorbing ultraviolet radiation.

Polar front
A boundary between cold polar air and warm subtropical air that circles the globe in the midlatitudes.

Polar high-pressure zones
Regions of high pressure and descending air near the North and South poles.

Quaternary Period
The period of geologic time encompassing approximately the last 3 million years.

Radiation
Energy in the form of electromagnetic waves that radiate in all directions.

Relative humidity
The actual water content of the air compared to how much water the air could potentially hold, expressed as a percentage.

Semiarid climate
A climate with precipitation slightly less than potential evapotranspiration for most of the year.

Sensible heat
Heat detectable by sense of touch, or with a thermometer.

Shortwave energy
Radiant energy emitted by the Sun in wavelengths about 0.2 to 5.0 microns.

Solar energy
The radiant energy from the Sun.

Storm surge
An area of elevated sea level in the center of a hurricane that may be several meters high, and which does most of the damage when a hurricane comes ashore.

Subtropical high-pressure (STH) zones
Regions of high pressure and descending air at about 25° north and south latitudes.

Summer solstice
For places in the Northern Hemisphere, June 20 or 21 is the date when at noon the Sun is directly overhead along the parallel of 23.5° north latitude; for places in the Southern Hemisphere, December 21 or 22 is the date when at noon the Sun is directly overhead at places along the parallel of 23.5° south latitude.

Tornado
A rapidly rotating column of air usually associated with a thunderstorm, often having winds in excess of 300 kilometers/hour (185 miles/hour).

Trade wind
The prevailing wind in subtropical and tropical latitudes that blows toward the Intertropical Convergence Zone, typically from the northeast in the Northern Hemisphere and from the southeast in the Southern Hemisphere.

Tropic of Cancer
The parallel of 23.5° north latitude.

Tropic of Capricorn
The parallel of 23.5° south latitude.

Typhoon
The name applied to a hurricane in the Pacific Ocean.

Vernal (spring) equinox
In the Northern Hemisphere March 20 or 21, one of two dates when at noon the perpendicular rays of the Sun strike the equator (the Sun is directly overhead along the equator).

Warm front
A boundary formed when a warm air mass advances against a cooler one.

Water vapor
Water in the air in gaseous form.

Wavelength
The distance between successive waves of radiant energy, or of successive waves on a water body.

Winter solstice
For places in the Southern Hemisphere, June 20 or 21 is the date when at noon the Sun is directly overhead at places along the parallel of 23.5° north latitude; for places in the Northern Hemisphere, December 21 or 22 is the date when at noon the Sun is directly overhead at places along the parallel of 23.5° south latitude.

Major Climates and Their Köppen Equivalents

CLIMATE TYPE — **CLIMATE CHARACTERISTICS**

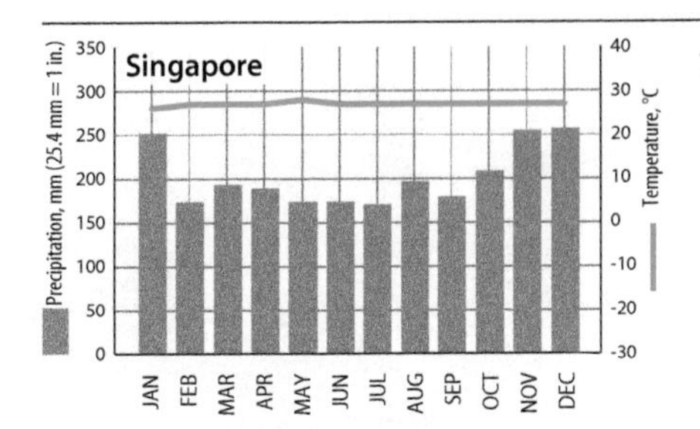

TROPICAL: CLIMATES THAT ARE WARM ALL YEAR

Climate type	Climate characteristics
Humid tropical	
Af	Tropical, constantly warm and humid with no dry season (shown at left)
Am	Tropical, constantly warm and humid, but with a short dry season
Seasonal humid tropical	
Aw	Tropical, constantly warm and humid, but with a pronounced dry low-Sun season and wet high-Sun season

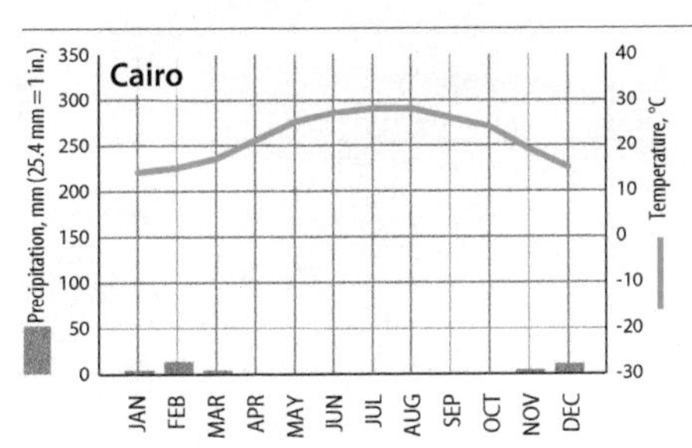

DRY CLIMATES

Climate type	Climate characteristics
Desert	
BWh	Hot **desert climate** (shown at left)
BWk	Cool desert climate
Semiarid	
BSh	Hot semiarid (steppe) climate
BSk	Cool semiarid (steppe) climate

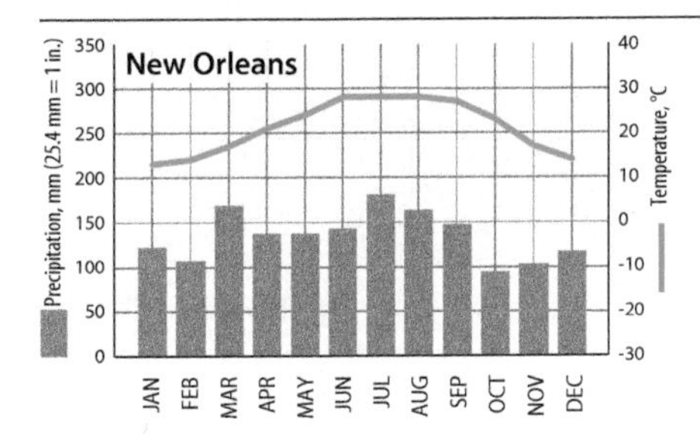

MIDLATITUDE CLIMATES WITH WARM SUMMERS AND COOL WINTERS

Climate type	Climate characteristics
Humid subtropical	
Cfa	Humid, warm subtropical climate, with hot summers and no dry season (shown at left)
Cw	Humid, warm subtropical climate, with hot summers and dry winters
Marine west coast	
Cfb	Marine west coast climate, with warm summers and no dry season
Cfc	Marine west coast climate, with cool summers and no dry season
Mediterranean	
Cs	Mediterranean climate, with dry, warm summers and cool, wet winters

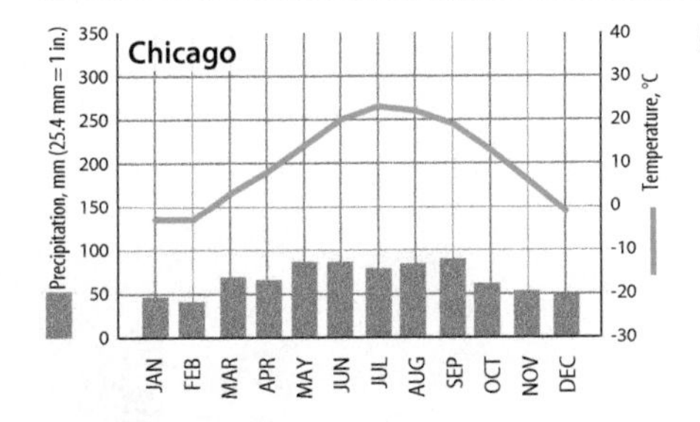

MIDLATITUDE CLIMATES WITH WARM SUMMERS AND COLD WINTERS

Climate type	Climate characteristics
Humid continental	
Dfa	Humid continental climate, with hot summers, cold winters, and no dry season (shown at left)
Dwa	Humid continental climate, with hot summers, and dry, cold winters
Dfb	Humid continental climate, with warm summers, cold winters, and no dry season
Dwb	Humid continental climate, with warm summers, and dry, cold winters
Subarctic	
Dfc	Moist subarctic climate, with cool summers, very cold winters, and no dry season
Dwc	Moist subarctic climate, with cool summers and very cold, dry winters
Dfd	Moist subarctic climate, with cool summers, frigid winters, and no dry season
Dwd	Moist subarctic climate, with cool summers and frigid dry winters

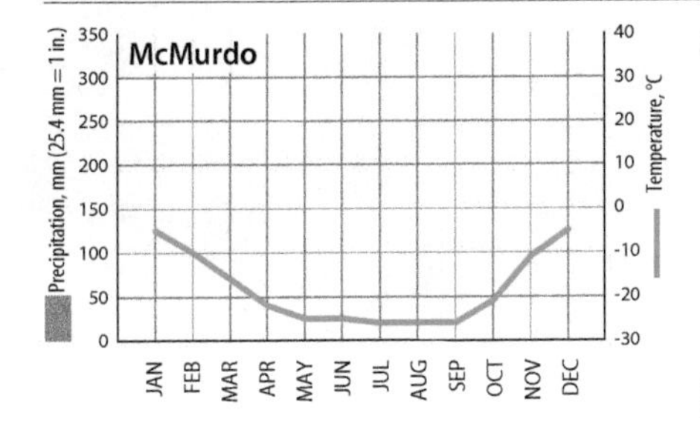

POLAR CLIMATES

Climate type	Climate characteristics
Tundra	
ET	Tundra climate, with very cool, short summers and frigid winters
Ice cap and ice sheets	
EF	Ice cap climate, with temperatures consistently below freezing (shown at left)

▶ LOOKING AHEAD

Earth's crust is composed of tectonic plates that move relative to each other. Relative motion between plates creates large-scale landforms at plate boundaries.

3 Landforms

Earth's surface is dynamic—always in motion. Earthquakes, volcanic eruptions, and landslides are sudden and dramatic, but the rest of the landscape is also slowly changing, year after year.

The landforms we see today—plains, hills, and valleys—reflect the processes that created them. Layers of sedimentary rocks tell us about the rivers and seas in which the sediment accumulated. Mountains are gradually rising in response to movements of Earth's crust. Valleys and floodplains have shapes that derive from the way water flows, forms channels, and deposits sediment. Coasts and beaches have characteristic forms created by the action of waves at the ocean's edge.

In addition to these natural processes, human activity has profoundly changed the solid Earth surface, just as it has changed the atmosphere and the biosphere. Agricultural practices are depleting soil fertility. In semiarid areas, overgrazing by animals has produced desert-like conditions. In many agricultural areas, soil is eroding at rates 10 to 100 times greater than natural rates of erosion. Some estimates of human-caused earth moving suggest that more earth is moved by humans than by all natural processes combined.

LA SAL MOUNTAINS AND ARCHES NATIONAL PARK, UTAH.

How are processes within Earth reflected in the landscape?

How does rock movement shape the land?

SCAN TO ACCESS THE LATEST EARTHQUAKE, VOLCANO, AND LANDSLIDE INFORMATION FROM THE USGS

How do glaciers and waves shape the land?

3.1 Catastrophic Earthquakes

- **An earthquake that occurs under the ocean generates tsunami waves that cause devastation beyond that caused by ground shaking.**
- **Despite enormous damage and loss of life, Japan's preparedness for earthquakes averted a much larger catastrophe.**

On March 11, 2011, the fourth-largest earthquake since reliable estimates of earthquake magnitude began occurred just off the coast of northern Japan.

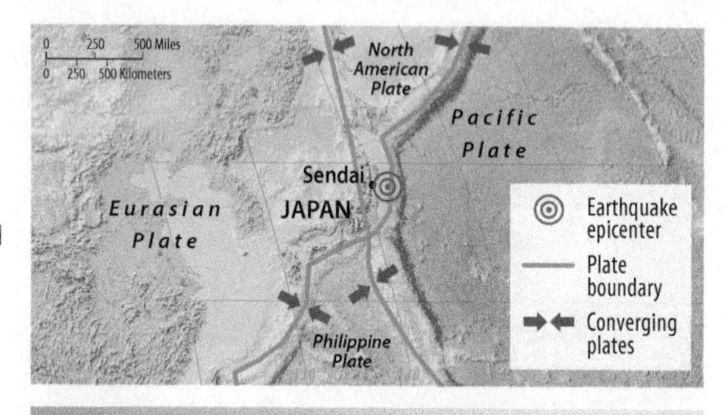

► 3.1.1 **TSUNAMI DESTROYING SENDAI, MARCH 11, 2011**
Much of Japan's population is concentrated near the coast. A seawall protected this area from lesser tsunamis in the past, but not from one of the size that occurred in 2011.

MasteringGEOGRAPHY
Animations
Tsunami

The quake happened at a place where two **tectonic plates**, or large pieces of Earth's rigid crust, are moving toward each other. Such motion occurs in many places around the world, including most of the lands bordering the Pacific Ocean, both on the east side (North and South America) and on the west, including eastern Russia, Japan, the Philippines, Indonesia, and New Zealand. The three larger quakes of which the magnitude is known accurately took place in Chile in 1960, Alaska in 1964, and accurately Indonesia in 2004.

The center of motion in the crust, or **focus**, of the 2011 quake was about 130 km offshore, at a depth of 32 km (20 miles). The motion was strong enough to cause skyscrapers to sway in Tokyo, 373 km (232 miles) away. Damage in coastal cities such as Sendai was significant. However, the most severe damage was caused by a **tsunami**, or earthquake-generated sea wave, which struck the coastal region minutes later. Just as ocean waves rise up and break as they approach a beach, the tsunami rose to heights of 10 meters (30 feet) or more as it reached the shore of Japan. The wave crashed over seawalls built to protect against such events and rushed several kilometers inland, in some places, smashing and washing away nearly everything in its path. Fortunate people had sufficient time to reach high ground, but over 20,000 lost their lives (Figure 3.1.1).

▼ 3.1.2 **FUKUSHIMA DAIICHI NUCLEAR POWER PLANT**
The plant was damaged by the tsunami, leading to meltdown of three reactors.

▲ 3.1.3 **DAMAGE IN PORT-AU-PRINCE, HAITI**
The earthquake of January 12, 2010, devastated the capital city.

Many thousands were left homeless. In addition to this devastation, a group of nuclear reactors at a power plant was severely damaged (Figure 3.1.2), resulting in releases of radiation to the environment and loss of electric generating capacity.

Japan has considerable experience with earthquakes and tsunamis. In 1923, an event that was substantially smaller than the 2011 quake, in terms of the energy of motion, occurred just offshore of the Tokyo-Yokohama region. A resulting fire burned hundreds of thousands of homes and caused a tsunami that was as high as 12 meters (39 feet). Over 140,000 people were killed. But a long history of past earthquakes, combined with Japan's position among the world's wealthy nations, makes it better able to manage earthquakes and tsunamis. The 2011 event is by far the largest to have affected Japan since 1923. The next largest occurred in the city of Kobe in 1995, killing 5,500.

For comparison, the United States Geological Survey estimates that, worldwide, over 700,000 people have been killed by earthquakes and resulting tsunamis since 2000. Two events killed over 200,000 people each: the 2004 earthquake/tsunami in Indonesia and the Indian Ocean, and the 2010 earthquake in Haiti (Figure 3.1.3). These disasters occurred in areas that have not had the capacity to build sophisticated buildings that can withstand strong motion, nor did they have warning systems or response systems that could provide medical care, shelter and supplies to those injured or left homeless. The magnitude of the Haitian quake was much less than that in Japan, but it killed 10 times as many people. The greater vulnerability of people in poorer countries to natural disasters relative to those in richer countries also applies to other hazards, such as storms, floods, and droughts.

MasteringGEOGRAPHY
Animations
Seismographs

3.2 Plate Tectonic Framework

- ► **Earth's crust is composed of tectonic plates that move relative to each other.**
- ► **Relative motion between plates creates large-scale landforms at plate boundaries.**

Earth resembles an egg with a cracked shell. Earth's crust is thin and rigid, averaging 45 kilometers (28 miles) in thickness. The rock just beneath the crust, known as the **mantle**, is fluid enough to move slowly along in convection currents, driven by heat within Earth's core. These currents are analogous to winds in the atmosphere, which carry heat away from Earth's surface. This motion of the mantle causes the tectonic plates that make up Earth's rigid crust to move (Figure 3.2.1).

► 3.2.1 **CHANGES CAUSED BY PLATE MOVEMENTS**

Two hundred million years ago Earth's continents were all joined in one supercontinent known as Pangaea.

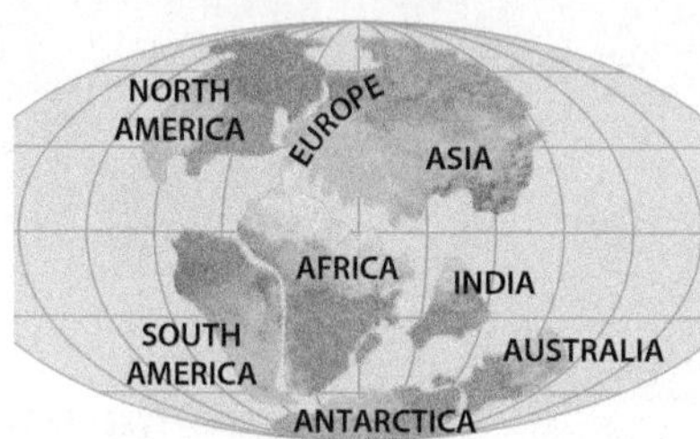

By 135 million years ago the continent had broken up.

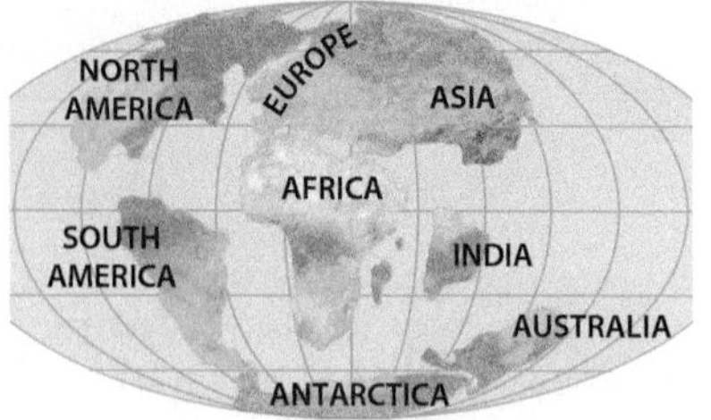

By 65 million years ago, the arrangement was beginning to look like the present, although North America and Europe were still joined.

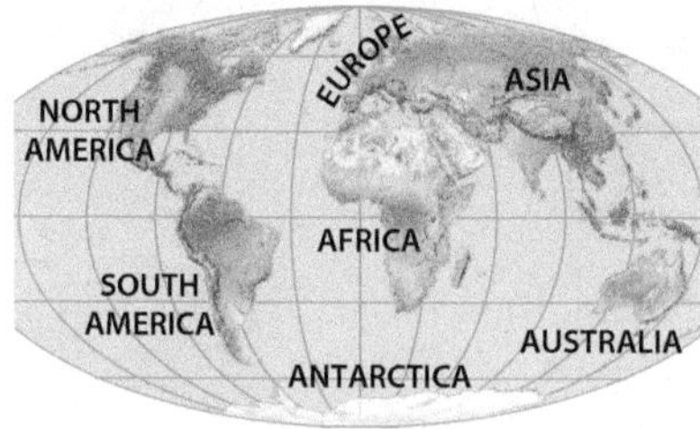

The present arrangement of continents.

MasteringGEOGRAPHY
Animations
Divergent Boundary Formation

Three types of boundaries form between moving plates of Earth's crust depending on whether the plates are spreading apart, pushing into each other, or grinding past each other (Figure 3.2.2). Movement of the plates causes earthquakes to rumble, volcanoes to erupt, and mountains to form (Figures 3.2.3 and 3.2.4). Over periods of hundreds of millions of years, the geography of Earth is thoroughly changed by these movements (Figure 3.2.5).

▼ 3.2.2 **CROSS SECTION OF THE UPPER MANTLE AND CRUST**

DIVERGENT PLATE BOUNDARIES

A boundary where plates are spreading apart is a **divergent plate boundary**. The rates of movement are very slow; typically only a few centimeters per year (2.54 centimeters = 1 inch), and most divergent boundaries are under water. The Mid-Atlantic Ridge is a well-known example, as are the rift valleys of East Africa, which divide sub-plates of the African Plate. These valleys are hundreds of meters deep and extend thousands of kilometers from Mozambique in the south to the Red Sea in the north. Divergent plate boundaries are areas of volcanic activity in which the erupting lava creates new crust.

CONVERGENT PLATE BOUNDARIES

A boundary where plates push together is a **convergent plate boundary**. Material from one plate is slowly forced downward by the collision, back into the mantle. Because seafloor crust is denser than continental crust, when a plate of continental crust collides with a plate of oceanic crust, the denser oceanic plate sinks beneath the lighter continental crust. The oceanic plate is carried into Earth's mantle, where some of it is remelted. This magma then migrates toward the surface, causing volcanic eruptions at sites above the plunging plate. This occurs, for example, to the south and southwest of Indonesia where the Eurasian and Indo-Australian plates converge.

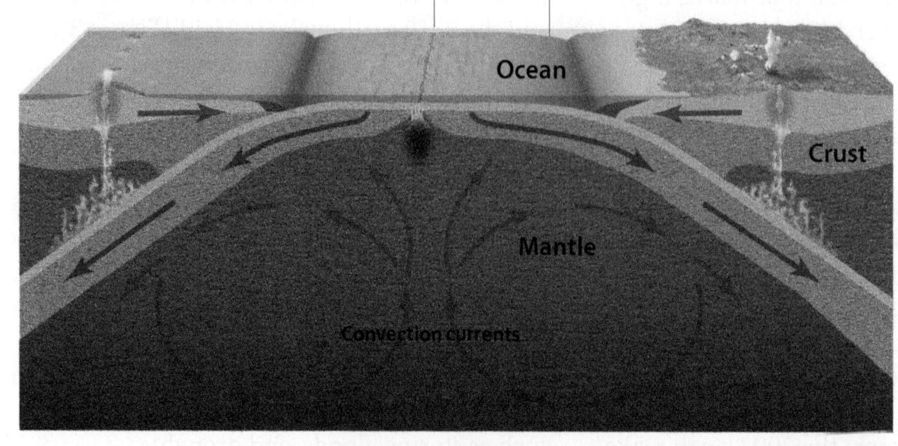

▲ SAN ANDREAS FAULT

▲ HIMALAYAS

▲ 3.2.3 **MOUNTAIN RANGES CREATED BY PLATE MOTION**
The Himalayas (above) formed, and are still growing, because of the Indo-Australian and Eurasian plates coming together in a convergent boundary.

◄ 3.2.4 **TRANSFORM PLATE BOUNDARY**
The Coast Ranges formed between the Pacific and North American Plates in a transform boundary.

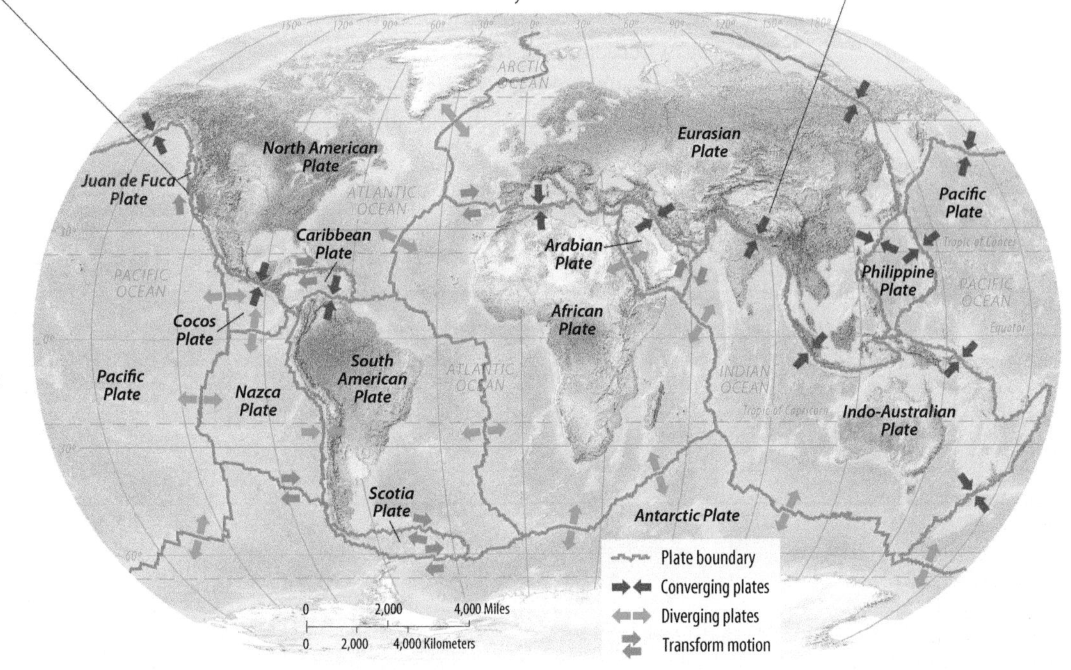

► 3.2.5 **TECTONIC PLATE BOUNDARIES**
This map shows the convergent, divergent, and transform plate boundaries. A boundary where the plates neither converge nor diverge, but grind past each other, is a **transform plate boundary**. California's San Andreas Fault is an example, where the Pacific Plate is moving northwest relative to the North American Plate. The boundary between these plates is not a smooth one, and ridges and mountains are built as the two plates grind against one another. The plates bind for long periods and then abruptly slip, causing the earthquakes that frequently strike California.

VERTICAL MOVEMENTS OF EARTH'S CRUST

Parts of the crust move vertically as well as horizontally. As two plates collide, material may be forced downward into Earth's interior or upward to form mountains. Over millions of years, vertical movements along plate boundaries produce mountain ranges thousands of meters high. Vertical movement of crust also occurs because the crust "floats" on the underlying mantle, much like a boat floats in water. If material is added to the crust, it sinks, and if material is removed, it rises. Deposition of sediment or accumulation of ice in glaciers can cause the crust to sink. These vertical movements caused by loading or unloading the crust are called *isostatic adjustments*.

MasteringGEOGRAPHY™
Animations
Convergent margins: India-Asia collision

3.3 Geological Hazards: Volcanoes and Earthquakes

- Volcanic eruptions and earthquakes occur mainly along plate boundaries.
- Volcanic eruptions and earthquakes are relatively infrequent but can be catastrophic.

DEADLY HAZARDS

Earthquakes and volcanoes present major geologic hazards to populations in geologically active areas, particularly near plate boundaries. These hazards result from the great power of the phenomena, but also because people have chosen to live in places where they occur. If earthquakes and volcanic eruptions were frequent then people would have adapted to them, most likely by living elsewhere. However, major events are infrequent and societies are thus less mindful of the hazards.

Although the magnitude of an earthquake relates to the energy released, it tells us little about the damage it causes. Generally, damage is greater at places closer to the **epicenter** and at places built on ground that is subject to landsliding or collapse. Earthquake damage is also greater where surface rocks or sediments are particularly susceptible to shaking, or where buildings are not designed to absorb the energy.

VOLCANOES

Like earthquakes, volcanoes are clustered along boundaries between tectonic plates. Heat within Earth generates **magma** (molten rock). If this magma reaches the surface and erupts, a **volcano** is formed. The magma may flow over the surface (as **lava**), forming a plain of volcanic rock, or it may build up to form a mountain. The chemistry of the magma/lava determines its texture and therefore the type of landform it builds.

SHIELD VOLCANOES

Shield volcanoes erupt runny lava that cools to form a rock called basalt. They are called **shield volcanoes** because of their shape. Each

▲ 3.3.1 **DEADLY CONSEQUENCES**
Rescuers run while carrying a body bag with a victim as Mount Merapi erupts.

◄ 3.3.2 **ERUPTION OF MT. MERAPI, INDONESIA**
Indonesia's Mount Merapi volcano erupts, sending a plume of ash and smoke about 3,500 meters into the air. Poisonous, superheated clouds called nuees ardentes also rush down mountain slopes during many such eruptions.

of the Hawaiian Islands is a large shield volcano, although the only currently active one is Mauna Loa, on the island of Hawaii (the "Big Island"). These generally sedate volcanoes make news on the rare occasions when they grow more active, and flows of lava threaten settlements. The mid-ocean ridges are formed of similar basaltic lava.

COMPOSITE CONE VOLCANOES

Explosive volcanoes that cause death and destruction are more likely to be **composite cone volcanoes**. Composite cones are made up of a mixture of lava and ash. Their magma is thick and gassy, and it may erupt explosively through a vent. The eruption sends ash and clouds of sulfurous gas high into the atmosphere (Figure 3.3.1 and 3.3.2). It may also pour lethal gas clouds and dangerous mudflows down the volcano's slopes. Repeated eruptions build a cone-shaped mountain, made up of a mixture of lava and ash layers.

Eruptions of composite cone volcanoes have killed tens of thousands of people at a time, but such disasters are much less frequent than severe earthquakes. Thousands of volcanoes stand dormant (inactive, but with the potential to erupt) around the world. About 600 are actively spewing lava, ash, and gas—some daily—but they rarely cause damage because people do not live near them. Others, however, have not erupted in hundreds of years, so people have settled nearby. In some areas earthquake watch centers provide warnings of volcanic eruptions. But when warnings are not available, the danger can be great. In general, predicting volcanic eruptions is more accurate than predicting earthquakes, because volcanoes give many warnings before erupting.

EARTHQUAKES

Thousands of **earthquakes**—sudden movements of Earth's crust—occur every day. They are clustered along plate boundaries (Figure 3.3.3). The place where Earth's crust actually moves is the focus of an earthquake. The focus is generally near the surface but can be as deep as 600 kilometers (372 miles) below Earth's surface. The point on the surface directly above the focus is the epicenter. The energy released at the focus travels worldwide in all directions and at various speeds through different layers of rock. Earthquake intensity is measured on a 0-to-9 logarithmic scale (in which an increment of 1 corresponds to a 10-fold increase in energy) called the moment magnitude scale. Earthquakes with a magnitude of 3 to 4 are minor; magnitude 5 to 6 quakes can break windows and topple weak buildings; and magnitude 7 to 8 quakes are devastating in populated areas. The 2011 earthquake in Japan had a magnitude of 9.0.

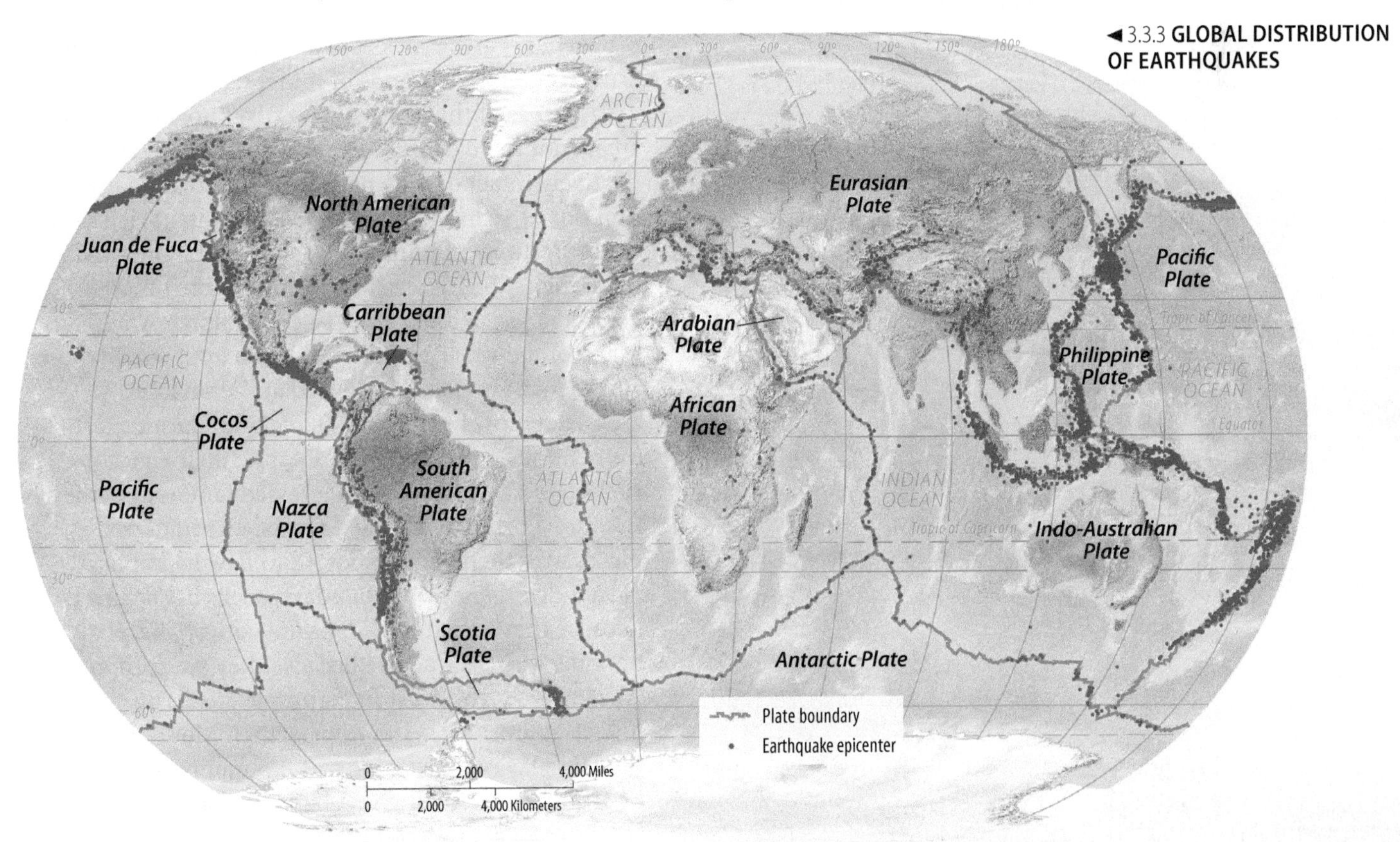

◄ 3.3.3 **GLOBAL DISTRIBUTION OF EARTHQUAKES**

3.4 Bedrock Geologic Settings

- Movements of Earth's crust produce a great diversity of rock types and of geologic structures.
- Variations in rock resistance to erosion are reflected in topography.

Over time, movements of Earth's crust create a great variety of rocks. These can be grouped into three basic categories that reflect how they form (Figure 3.4.1):

MasteringGEOGRAPHY Animations *Folds*

MasteringGEOGRAPHY Animations *Erosion of Deformed Sedimentary Rock*

► 3.4.1 **MAJOR ROCK TYPES**

Rock types	How do they form?	What are some examples?
Igneous Rocks	Molten crustal material (magma) cools and solidifies.	Basalt (common in volcanic areas, including much of the ocean floor) and granite (common in continental areas)
Sedimentary Rocks	Rocks formed from particles that have been transported, deposited, and cemented together.	Sandstone, shale, and limestone form when sand, clay, and calcium-rich sediments, respectively, accumulate and are cemented to bind their grains together
Metamorphic Rocks	Existing rocks are exposed to great pressure and heat, altering them into more compact, crystalline rocks.	Marble (metamorphosed limestone) and slate (metapmorphosed shale)

◄▼ 3.4.2 **MASSANUTTEN MOUNTAIN, IN THE SHENANDOAH VALLEY, VIRGINIA**
The mountain is the top of a U-shaped fold in the rocks, with a resistant sandstone layer forming ridges where it meets the surface. The valley in the top of the mountain sits within the U of the rock folds.

Minerals are the substances that comprise rocks. Each type of mineral has specific chemical and crystalline properties. Earth's rocks are diverse in part because the crust contains thousands of minerals.

Vast areas of the continental crust known as **shield** areas (not to be confused with shield volcanoes) are relatively intact despite partial erosion over many millions of years. Shield areas often contain rich concentrations of minerals, such as metal ores and fossil fuels. Shields are located in the core of large continents such as Africa, Asia, and North America. Many of the world's mining districts exist where these continental shields are exposed at the surface.

Crustal movements along plate boundaries exert tremendous stress on rocks. Despite their rigidity, rocks bend and fold. When stressed far enough, they fracture along cracks called **faults**. The fractured pieces may then be transported to new locations. Near a divergent plate boundary, rocks break apart because they are stretched; near a convergent plate boundary, rocks fracture because they are compressed. Alternatively, the crust may rumple like a rug, creating folds (Figure 3.4.2). The Appalachian Mountains and the Himalayas are examples of mountain ranges created by faulting and folding that happened along convergent plate boundaries. Faulting also occurs along transform boundaries.

Differences in geologic structures and rock types from one place to another are a critical part of the geographic variability of Earth's surface (Figures 3.4.3 and 3.4.4). These geologic features influence the surface in three different ways. Movement of the crust such as that along faults creates landforms such as the mountain ranges seen in Figure 3.2.3. Rocks vary in their resistance to processes that break them down and erode them: weak rocks are removed more rapidly, while more resistant rocks remain in place (Figure 3.2.5). This creates landforms in which the shape of the land reflects the underlying rock structures, with resistant rocks forming high elevations and steep slopes and weaker rocks forming valleys and gentle slopes.

◄ ▲ 3.4.3 **FLAT TOP WILDERNESS, COLORADO**
Mountains created by faulting that lifts one block of crust up relative to another that is dropped down.

MasteringGEOGRAPHY
Animations
Faults

◄ ▼ 3.4.4 **MARBLE CANYON, COLORADO PLATEAU, ARIZONA**
This spectacular landscape is dominated by cliffs formed of relatively resistant near-horizontal sedimentary rocks, separated by gentler slopes formed of more erodible sedimentary rocks.

◄ 3.4.5 **GLACIAL EROSION IN CENTRAL QUÉBEC, CANADA**
Use Google Earth to observe resistant rocks as hills and weak rocks as depressions.

Fly to: *49° 20′ N, 69° W*, and zoom out to an eye altitude of about 30 km.

What do you think may have caused all the straight lines?

3.5 Slopes and Weathering

- Rocks decay upon exposure to air and water, breaking them into moveable pieces.
- Rock particles move downslope and through river systems in a series of steps.

Weathering is the process of breaking rocks into pieces ranging in size from boulders to pebbles, sand grains, and silt, down to microscopic clay particles and dissolved solids. It is the first step in the formation of soil. Without weathering, the force of gravity and the agents of water, wind, and ice would not be able to move rocks to shape the land. Rocks begin to break down the moment they are exposed to the weather at Earth's surface (Figure 3.5.1). They are attacked by water, oxygen, carbon dioxide, and temperature fluctuations. Weathering takes place in two ways, as chemical weathering and as mechanical weathering.

CHEMICAL WEATHERING

Rocks may be broken down as a result of **chemical weathering**, which is a change in the minerals that compose rocks when they are exposed to air and water. Acids released by decaying vegetation also chemically weather rocks. Some of the dissolved products of chemical weathering are carried away by water seeping through soil and rocks (Figure 3.5.2). The water eventually may carry these chemical materials to rivers and then to the sea. This is the source of the salinity (dissolved salt) of the oceans.

One example of chemical weathering is oxidation. Iron is a common element in rocks, and it combines with oxygen in the air to form iron oxide, or rust. Iron oxide has very different properties from the original iron—it is physically weaker and more easily eroded. You can see the effects of oxidation on iron or steel surfaces exposed to the weather; the rusty oxide easily flakes away.

▲ 3.5.1 **WEATHERED ROCK UNDER TROPICAL RAIN FOREST, SOUTH AMERICA**
This material was once solid rock, but has been broken down by chemical and mechanical processes. The weathering continues, reducing larger particles to smaller particles, and in some cases creating new clay minerals from the residue of other rocks.

◄ 3.5.2 **MAMMOTH CAVE NATIONAL PARK, KENTUCKY**
Formed by the chemical weathering of calcium carbonate. This mineral, a major component of limestone and other sedimentary rocks, weathers to soluble ions that are carried by streams to the sea. In some areas of limestone bedrock, underground water may remove large quantities of rock, forming passageways, and even carve out large caverns in the limestone. If the caverns collapse, they create depressions called sinkholes at the surface. Such features are found in many parts of the world, including the Caribbean (especially Puerto Rico, Cuba, and Jamaica), several parts of the southeastern United States (especially Florida, Kentucky, and Missouri), and southeastern China.

MECHANICAL WEATHERING

Rocks are also broken down by physical force. This is called **mechanical weathering**. Rocks expand and contract with frequent changes in temperature, and this causes them to break apart. Also, water can seep into cracks and freezes into ice crystals when the temperature turns colder. The water, which expands about 9 percent when frozen, widens cracks in rocks. Plant roots growing in cracks between rocks also contribute to mechanical weathering; you probably have observed sidewalks that have been heaved by tree roots. Mechanical and chemical weathering work together to break down rocks. Often, mechanical forces open cracks, and water seeps in to weather the rock chemically.

MOVING WEATHERED MATERIAL

Once rocks are weathered, they may be carried from one place to another (Figure 3.5.3). Material most commonly moves downhill by gravity. This happens in two ways: by mass movement or by surface erosion. In **mass movement**, rocks roll, slide, or freefall downhill under the steady pull of gravity. In surface erosion, water, which flows downhill because of gravity, carries solid rock particles with it (Figure 3.5.4). Surface erosion may also result when wind or ice carries material from one place to another.

Material moves faster down steeper hills than down gentler ones. The steepness of a hill is measured through its slope, which is the difference in the elevation between two points (known as the rise) divided by the horizontal distance between the two points (known as the run). The greater the rise and shorter the run between the two points, the faster materials move down the hill. Wherever slopes occur, gravity is available to move material. Even the gentlest slope provides the potential energy necessary to move at least some material downward, either through mass movement or surface erosion. Erosion is usually much more rapid, however, on steep slopes of land than on gentle slopes.

MasteringGEOGRAPHY
Animations
Physical Weathering

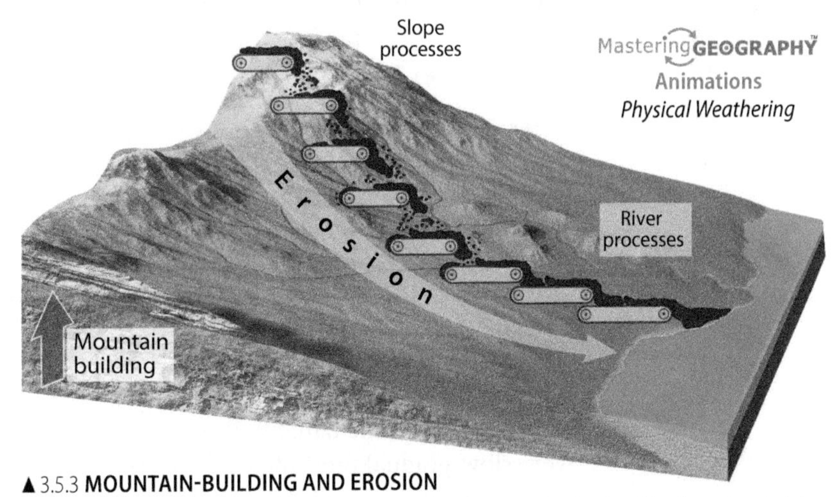

▲ 3.5.3 **MOUNTAIN-BUILDING AND EROSION**
Mountain-building processes lift rocks up, exposing them to weathering. Weathered material is carried in multiple steps, represented by the conveyor belts, downslope to streams, and then along valleys to the sea. Along the way, this weathered material is temporarily stored as soils and river deposits. Storage can be for short time periods, but more likely for hundreds or thousands of years, so the journey of a single grain of sand from source to sea can take a very long time.

► 3.5.4 **A STREAM IN GLEN ETIVE, HIGHLANDS, SCOTLAND**
Slope processes move sediment downhill, and the stream carries it down the valley.

3.6 Mass Movements

- **Mass movements send soil downslope, through the direct effects of gravity on slope materials.**
- **Landslides occur on slopes where rock and soils are weakened to the point that they cannot resist gravitational forces.**

Some **mass movements** are very common and gradual, while others are rare and sudden.

COMMON AND GRADUAL

The most common form of mass movement is **soil creep**. As the name suggests, creep is a very slow, gradual movement of material down the slope of a hill. A tiny movement can cause creep—a rodent digging, a worm burrowing, an insect pushing aside soil. Creep occurs near the surface, in the top 1 to 3 meters (3 to 10 feet) of soil. Even though creep is very slow and usually unnoticed, it is relatively steady and can move a substantial amount of material downslope over time (Figure 3.6.1).

RARE AND SUDDEN

Rockfalls are an example of a mass movement process that is only found in a specific environment—and only occurs occasionally—particularly in steep mountain environments, where resistant rocks form cliffs (Figure 3.6.2). When weathering weakens material to the point gravity can move it, eventually the rock falls.

▲ 3.6.1 **SOIL CREEP, VALENCIA, SPAIN**
Tilted trees and other objects are often indicators of creep.

► 3.6.2 **THREATENING ROCK, PUEBLO BONITO, NEW MEXICO**
The cliff above Pueblo Bonito, an ancient ruin in New Mexico, fell on the pueblo (village) in 1941. The pueblo was built at the base of the cliff around 850 AD—nearly 1100 years before the rock fell, destroying a large part of the pueblo. The pueblo had been abandoned hundreds of years earlier, and scientists had monitored the movement of the rock for years before it fell, so no one was injured.

More dangerous and dramatic mass movements, such as landslides and mudflows, can occur on steep slopes, especially during wet conditions (Figure 3.6.3). In landslides, a relatively intact mass of rock or soil slides over the surface below. Landslides typically follow intense rains, because material with a high water content is weaker and less able to resist the force of gravity. Alternatively, if the slide occurs in weathered rock and soil, it tends to break down into a thick fluid as it moves downhill, becoming a mudflow. In many areas, population growth has pushed development to steeply sloping areas that are vulnerable to such catastrophes (Figure 3.6.4).

► 3.6.3 **LANDSLIDE AT LAKE TAHOE, CALIFORNIA**
Slides and rockfalls at this site have repeatedly caused road closures.

▼ 3.6.4 **LANDSLIDE IN NOVA FRIBURGO, BRAZIL**
Consequences of the heavy rains in January 2011.

3.7 Surface Erosion

- **Soil surface erosion occurs when heavy rains cause excess water to flow across the surface, or when high winds blow on bare ground.**
- **Water and wind erosion are accelerated when the soil surface is exposed.**

The most common form of soil erosion is caused by rainfall. Intense rain sometimes falls faster than soil can absorb it. Water that cannot soak into the ground must run off the surface as overland flow. As it runs off the surface, water picks up soil particles and carries them down the slope. With enough of this **overland flow**, water can carve channels.

The smallest channels eroded by the flow of water—only a few centimeters deep—are called rills (Figure 3.7.1). Rills are small enough that soil creep or a farmer's plow can obliterate them. If channels gather enough water, however, they become larger and permanent carriers of water. As these stream channels deepen, they gather water and eroded soil from adjacent slopes. When the streams gather enough water, they form gullies (Figure 3.7.2) or, ultimately, permanent valleys.

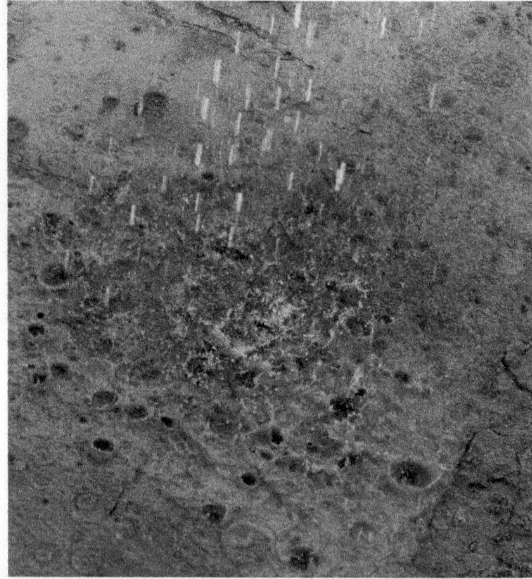

◄▲ 3.7.1 **SOIL EROSION**

Soil surface erosion results from raindrop impact on bare soil combined with water flowing across the soil. Both are favored by intense rain falling on bare soil. The photographs above show overland flow being generated in heavy rain; at left are rills formed by that flow.

ACCELERATED EROSION

Surface erosion by water is relatively slow in most natural environments because the ground is covered by grass and trees. But on large parts of Earth's surface, humans have removed vegetation by clearing forests and plowing fields. Once the vegetative ground cover is removed, slow surface erosion can increase, to become accelerated erosion. Ground where vegetation has been removed can suffer more surface erosion in a few months than it experienced during the previous several thousand years. The eroded soil contributes to water pollution downstream, and the remaining soil may be less productive for agriculture.

People clear natural vegetation because they want to use trees for fuel, lumber, and paper, or they want to use the land for another purpose, especially agriculture and urban development. Erosion has increased as a result of the elimination of the vegetation cover for all these reasons, but it is particularly severe where agriculture has replaced forest. To meet the needs of a growing population, food production expands in two principal ways—by opening up new land for agriculture and by using existing farmland more intensively. Both strategies can result in erosion of the rich soil necessary for productive agriculture.

Opening up new land for agriculture was a major contributor to increased erosion in the United States during the eighteenth and nineteenth centuries. More than 2 million square kilometers (800,000 square miles) of forest were cleared and replaced with plowed fields and pastures in the eastern and midwestern United States. This deforestation probably increased the rate of soil erosion by 10 to 100 times. Erosion of agricultural land is further increasing as farmers use existing fields more intensively. In many agricultural areas, soil is being lost faster than it can be replaced by natural soil formation, and so soil fertility is being lost over time.

Since the 1930s, abandonment of less-productive lands and adoption of soil conservation technologies have significantly reduced erosion on U.S. farms (Figure 3.7.3). In addition, many farmers are finding ways to use fertilizer more efficiently so that less is washed off of fields into streams. Stream pollution from agricultural sources remains a major problem, but it has been much reduced in the last few decades.

▲ 3.7.2 **GULLY EROSION IN KANSAS**
Increases in runoff caused by agricultural land use can result in creation of gullies.

▼ 3.7.3 **SOIL EROSION IN TWO INDIANA SOYBEAN FIELDS**
(left) A field that had been planted using conventional techniques including plowing the soil. The field experienced heavy erosion when intense early-season rains occurred. (right) A field adjacent had been planted using soil conservation techniques that leave much of the soil covered in residue from the previous season's corn crop. Soil erosion was dramatically reduced in this field.

3.8 Streams

- Sediment is carried by running water.
- Streams create floodplains with meandering channels.

Streams collect water from two sources, groundwater and overland flow. When rain falls on the land surface, most of it infiltrates, or soaks into the soil, where it may drain to streams or to groundwater.

SEDIMENT TRANSPORT IN RUNNING WATER

Groundwater migrates slowly through the soil and underlying rocks. Most of the water flowing into streams is supplied not by rainfall directly, but by groundwater. If rain falls intensely, the soil may not be able to absorb it as fast as it falls. **Runoff**, or the total flow in streams, comes from soil water, groundwater, and overland flow.

A stream drains water from an area called its **drainage basin**. Drainage basins may be as small as a farm field, or as large as a major portion of a continent. Smaller basins are nested within larger basins. In general, the greater the area of its drainage basin, the more water a stream must carry. Small rills deliver water and material to larger streams, which join others to form still larger rivers, which flow to the sea. The volume of water that a stream carries per unit of time is its **discharge**. Discharge of any stream usually increases after storms and decreases during dry spells.

When water flows across the land, either on a hillslope or in a channel, it carries small particles of rock with it. Smaller particles are mixed with the water making it look muddy; larger particles may roll or bounce along the bottom of the flow. This movement of material in a stream is called **sediment transport** (Figure 3.8.1). The amount of sediment a stream carries increases as the amount of flowing water increases, so larger streams typically carry more sediment than do smaller ones. The amount of sediment carried when a river is in flood may be hundreds or thousands of times more than the amount carried at ordinary flow levels (Figure 3.8.2). Sediment transport also tends to be greater on steep slopes than on gentle ones.

MasteringGEOGRAPHY
Animations
Sediment Transport by Streams

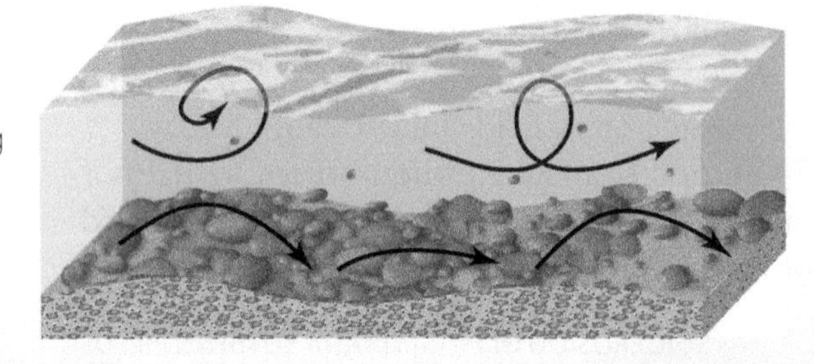

► 3.8.1 **SEDIMENT TRANSPORT**
Sediment is carried both suspended in the flowing water and by rolling or bouncing along the stream bed.

▼ 3.8.2 **COLORADO RIVER NEAR LEE'S FERRY, ARIZONA**
The green water in the distance has very little sediment because the sediment was trapped in a reservoir upstream. The muddy water in the foreground is caused by heavy rain in the drainage area of a local tributary.

FLOODPLAINS

Sediment is carried downstream in a series of steps, with particles being eroded in one place and then deposited farther downstream (Figure 3.8.3). Channel beds and banks are made up of materials transported by streams, temporarily deposited, and then eroded again as streams take **meandering** (winding) courses (Figure 3.8.4). By continually eroding and depositing material in channels and adjacent low-lying surfaces called **floodplains** (Figure 3.8.5), streams tend toward a stable condition, known as **grade**. A graded stream transports exactly as much sediment as it has collected. Streams rarely operate at a condition of grade for long, because daily changes in weather and disturbances from erosion and human activities continually upset the balance. As the stream's condition changes and the transport of sediment increases or decreases, the shape of the stream channel may change. An especially heavy flow resulting from a storm may cause the channel to shift. When increased erosion upstream generates more sediment than a stream can carry, the excess is deposited in the channel or on the floodplain.

Sediment deposition slowly raises the elevation of a stream, which can in turn reduce the difference in elevation between places upstream and downstream and also reduce the stream's slope. Lowering the slope reduces the amount of sediment arriving from upstream.

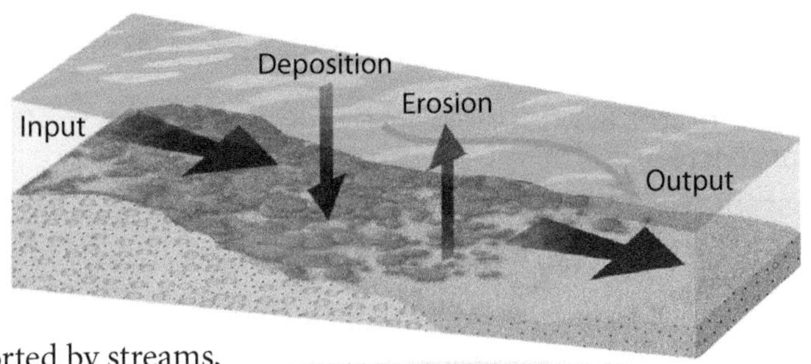

◄ 3.8.3. **SEDIMENT IS CONTINUALLY ERODED, TRANSPORTED, AND DEPOSITED**
In any given part of a river, if more sediment is deposited than eroded (input is greater than output), the channel bed rises. If more is eroded than deposited (output is greater than input), the bed is lowered. The shape of a channel is continually adjusting to the amount of water and sediment moving through it.

MasteringGEOGRAPHY
Animations
Meandering Streams

▲ 3.8.4 **RIVER COLE, UNITED KINGDOM**
Steep channel banks show areas of erosion, while sloping gravel bars in the channel bed are areas of deposition.

▼ 3.8.5 **VYVENKA RIVER, KAMCHATKA PENINSULA, RUSSIA**
Rivers meander from side to side, creating floodplains. The floodplain is an extension of the river channel, and is connected to runoff and erosion processes in the watershed upstream.

3.9 Fluvial Landscapes

- **Stream erosion creates networks of valleys in which water and sediment are carried downstream.**
- **Stream systems respond to changes of land use, climate, and other factors affecting sediment and runoff.**

Streams gather runoff from the land. This runoff becomes concentrated in low areas and forms channels.

Over time, streams erode downward, creating a network of tributary channels and their associated valleys. The streams in these valleys collect the sediment from adjacent hillsides, as well as from upstream sources. Streams generally develop concave longitudinal profiles, with steeper slopes in headwater areas and gentler slopes downstream (Figure 3.9.1). Small streams have small, steeply sloping valleys, and larger streams have larger valleys with gentler slopes.

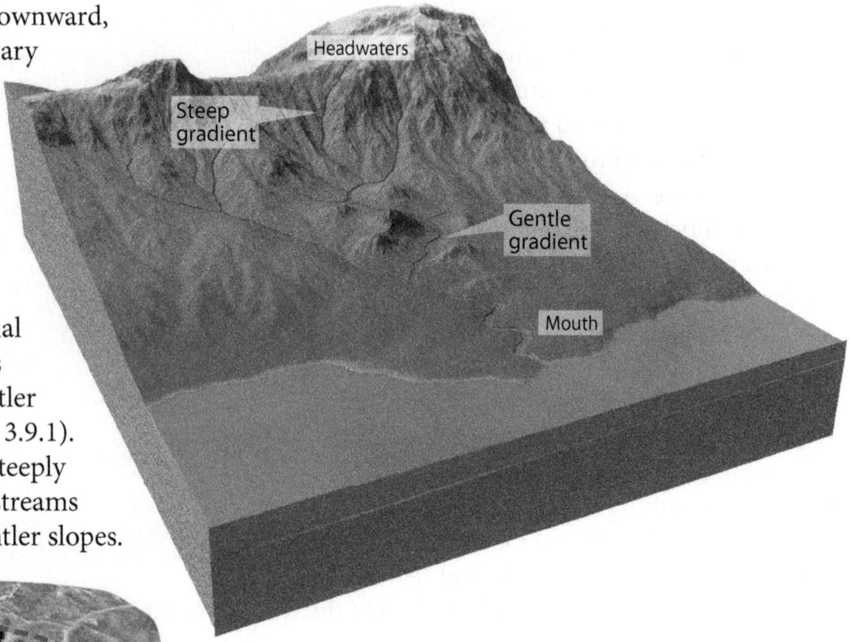

▲ 3.9.1. **AN IDEALIZED PROFILE OF A STREAM AND ITS TRIBUTARIES**
Note the concave-upward shape of the profile, with steeper gradient toward the heads of streams and gentler gradient downstream toward the mouth. As the stream flows downhill, it gathers more and more water, and its erosive power is greatly increased. The gentler gradient downstream slows the flow, balancing the erosive power of the stream with the amount of sediment it must carry.

◄ 3.9.2 **A FLUVIALLY–ERODED LANDSCAPE IN WEST VIRGINIA**
The topography of this landscape is a result of millions of years of streams cutting down and carrying material delivered from the adjacent slopes. The size of valleys increases downstream along with the amount of water and sediment carried.

Fluvial landscapes develop over periods of thousands to millions of years, during which environments may change in a variety of ways (Figure 3.9.2). Tectonic forces can tilt the crust, causing changes in slopes; climates vary and so change amounts of runoff; and vegetation and land use can change, affecting both runoff and sediment inputs. Nearly all streams have thus experienced significant changes in water and sediment inputs, and these changes are often visible in abandoned floodplains, called terraces (Figure 3.9.3), and other features created by streams in the past but no longer part of the active channel system.

In agricultural areas of the United States and many other countries, accelerated erosion led to significant deposition of sediment on floodplains, raising their elevations. In the Midwest, for example, many valleys accumulated layers of sediment 1 to 2 meters deep in the nineteenth and early twentieth centuries. Today, in response to reduction of soil erosion and other factors, many streams are beginning to cut down through this accumulated sediment, leaving terraces isolated above present channels. Streams are almost always changing, and are only rarely stable over long periods of time (Figure 3.9.4).

▲ 3.9.3 **STREAM TERRACES ALONG THE SNAKE RIVER**

These terraces are former floodplains, now found well above the modern channel because the stream has cut down to a new, lower, level.

◄ 3.9.4 **NIGER DELTA**

Use Google Earth to explore a major delta. As the Niger River approaches the Atlantic Ocean, the low stream slope causes water to slow and sediment to be deposited in a vast depositional surface called a **delta**. Like many major deltas, this region contains substantial oil deposits.

Fly to 4° 50′ N, 6° 50′E.

Zoom out to an eye altitude of about 10 km, and move around the landscape of the region.

1. **What are the physical features of the delta?**
2. **What evidence is there of human alteration of the landscape?**

3.10 Coastal Processes and Landforms

- Waves erode coasts, move sediment along shorelines, and create beaches and related features.
- Sea levels rise and fall creating changes in coastal landforms.

A coast is an especially active area because an enormous amount of energy is concentrated on the shorelines from pounding waves. Coastal land may be lost to erosion, or gained through deposition, at rates up to several meters per year. Rising sea level accelerates many coastal processes and increases vulnerability of coastal communities.

MasteringGEOGRAPHY
Animations
Beach Drift and Longshore Currents

Winds blow across the sea surface, transferring their energy to the water by generating waves. Waves on the sea surface are a form of energy, traveling horizontally along the boundary between water and air. As the wind blows harder and longer, it transfers more energy and generates bigger waves. Waves also grow larger with an increase in the expanse of water across which the wind blows.

The speed at which a wave travels is affected by its wavelength—the distance from one wave crest to the next. A small ripple travels very slowly, whereas an ocean wave may travel 10 to 50 kilometers per hour (6 to 30 miles per hour). A tsunami, which is an extremely long wave created by an underwater earthquake, may travel hundreds of kilometers per hour. In the open ocean it may be only a few tens of centimeters high, and passes under ships without their detecting it. When waves reach shallow water, their motion is affected by the sea bottom and the wave shape changes (Figure 3.10.1). We see this as ordinary wind waves break on a beach or as a tsunami can rise to heights of several meters or more.

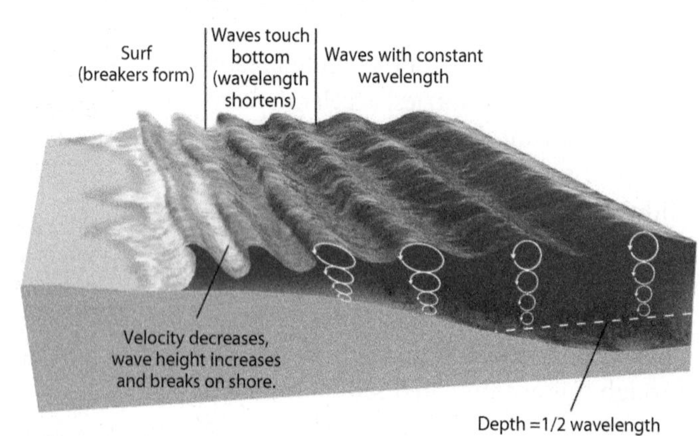

▲ 3.10.1 **WAVES APPROACHING A SHORELINE**
A wave can travel thousands of kilometers across deep ocean water relatively unchanged, but when it nears the shore, the shallow bottom restricts water motion and distorts the wave's shape. As it slows, the top portion rushes forward and breaks.

◄ 3.10.2. **A CLIFFED SHORELINE, UNITED KINGDOM**
Erosive power is concentrated where waves break along the shore. The energy of a wave is released as a tremendous erosive force of rushing water on the shore. If the water immediately offshore is shallow enough, a beach will form. Beach pebbles and granules of sand are rolled back and forth, constantly being ground ever finer. The finest particles are carried into deep water and settle on the seafloor. The larger sand- and gravel-size particles are left behind on the shore to form a beach, which is a surface on which waves constantly break and move material.

When you watch waves break on a **beach**, the most obvious motion of the water is perpendicular to the shoreline, the waves move up the beach and then recede, causing erosion in the process (Fig. 3.10.2). But when the waves approach at an angle to the shore, their energy gives a push to the water in a direction parallel to the shore, and the repeated breaking of many waves generates a **longshore current** traveling parallel to the shore. The longshore current is like a river, carrying sediment through **longshore transport** from areas where it is eroded by waves and depositing it where breaking waves lose the energy to carry it, usually in deep water. Longshore currents can carry enormous amounts of sediment great distances.

Like rivers, **landforms** along shorelines are shaped by the balance between sediment arriving in a portion of the shore and then being removed from it. Distinctive landforms develop, such as the beaches, spits (peninsulas that grow across the mouth of a bay), and barrier islands (long, narrow islands parallel to the mainland). If more sediment is removed than arrives, the coast is eroded. This is the condition of most of the world's shorelines. But in some areas, more sediment arrives than is removed, and the land area grows.

HUMAN IMPACT ON COASTAL PROCESSES

Because of the high rates of change that can occur in coastal environments, and because many commercial and recreational activities are focused there, humans actively manage coastal processes by building breakwaters that parallel the shoreline, jetties and groins that are perpendicular to it, and similar structures. A jetty constructed in one location along the shore to interrupt the movement of sand will cause less sand to be deposited on a beach farther along the shore, worsening the erosion problem there. People also build sea walls parallel to the shore, but the constant pounding of waves removes sand from around the sea walls and ultimately undermines them.

► 3.10.3 **COASTAL SEDIMENT TRANSPORT**

Use Google Earth to explore the coast of southern California. The direction of longshore transport here is north to south. The beach to the north of the harbor is wide because of deposition. The harbor has interrupted the flow of sand, resulting in erosion and a very narrow beach to the south.

Fly to *Redondo Beach Harbor, California.*

Zoom out to an eye altitude of about 8 km. Note the submarine canyon offshore. Sand is moving south toward the harbor but does not reach the beaches to the south.

Where do you think it goes?

MasteringGEOGRAPHY
Animations
Coastal Stabilization Structures

SEA-LEVEL CHANGE

The average elevation of the sea at any location is called **sea level**. Over the past few hundred years, sea level as a whole has risen on Earth, at about 2 millimeters per year along coasts that are otherwise stable. The rate of sea level rise appears to be increasing; over the last 20 years it was about 3 mm per year. Sea level has risen because the volume of seawater has increased by the melting of glaciers, as well as from expansion caused by warming of the sea itself. The recent change in sea level is small compared to the sea-level rise of about 85 meters (280 feet) that occurred at the end of the most recent Ice Age but over periods of decades it threatens to wipe out many low-lying Pacific Islands and threaten coastal settlements around the world.

Sea-level changes are significant at two different time scales. In the short term (a few decades) the direction of sea-level change affects how a shoreline erodes. If sea level rises, the water offshore becomes deeper and waves break closer to the land, causing more erosion. If sea level falls, the shallower water causes waves to break further offshore, dissipating their energy and reducing shoreline erosion. Because many shorelines have gentle slopes, a minor increase in sea level can translate into a much larger landward migration of the shoreline (Figure 3.10.4).

In the long term—over thousands of years—large sea-level changes can reshape shorelines. During Pleistocene glaciation, the sea was substantially lower and rivers in coastal areas cut deep valleys as they approached the sea. When the glaciers melted, sea level rose worldwide, between about 10,000 to 20,000 years ago, drowning the river valleys and creating large bays (Figure 3.10.5).

In areas where tectonic activity raises the land, sea level has fallen rather than risen relative to the adjacent land. This has left inland soils, animals, and vegetation that are typical of beaches. Much of the U.S. West Coast has been tectonically raised in the last few million years. As a result, the region lacks the many deep river mouths of the U.S. East Coast, but it does have former shorelines that are now above sea level, forming **marine terraces**.

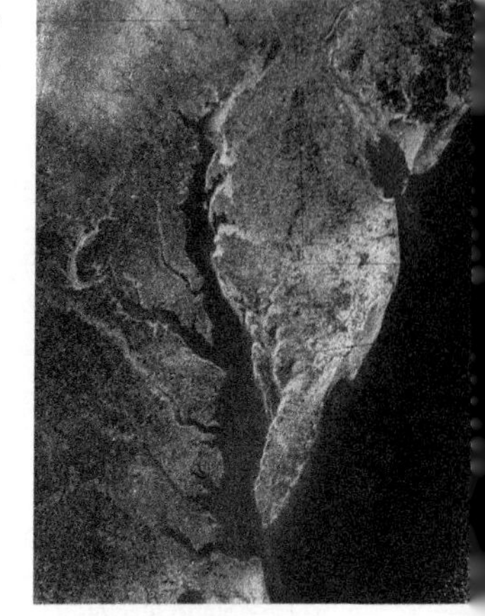

▲ 3.10.5 **DELAWARE AND CHESAPEAKE BAYS**

These deep embayments are the drowned lower portions of the Delaware and Susquehanna river valleys, inundated by sea level rise at the end of the last glacial period.

► 3.10.4 **MIAMI BEACH**

The city is built on a barrier island that was built by wave action and longshore transport. Under natural conditions, without any human impact, sediment would be eroded from the seaward side of the island and deposited on the mainland side, and the island would move landward over time. The city can't move, however, and sand is imported to replace what is lost. Today the shorefront is highly vulnerable to major storms. Sea level rise increases both erosion and the threat of inundation in storms.

3.11 Glacial Processes

- **Glacial formation is controlled by climate and depends on how much snow falls per year in relation to the amount of melt.**
- **Glacial flow patterns are also linked to climate, moving from areas of ice accumulation to areas of melt.**

MasteringGEOGRAPHY
Animations
Glacial Processes

Glaciers are rivers of ice that flow from places where snow accumulates yearly to warmer places where the ice melts (Figure 3.11.1). Water enters the head of the glacier as snow. The glacier flows downhill until the ice eventually leaves the glacier through ablation—melting, evaporation, or drifting away as icebergs (Figure 3.11.2).

▼ 3.11.1 **GLACIAL BUDGETS**
In the upper portion of the glacier more snow falls than melts, causing accumulation. The ice flows from the accumulation zone to locations where there is net ablation (loss of ice to melting, icebergs, etc.). Over time, the glacier responds to variations in climate that affect both accumulation and loss of ice.

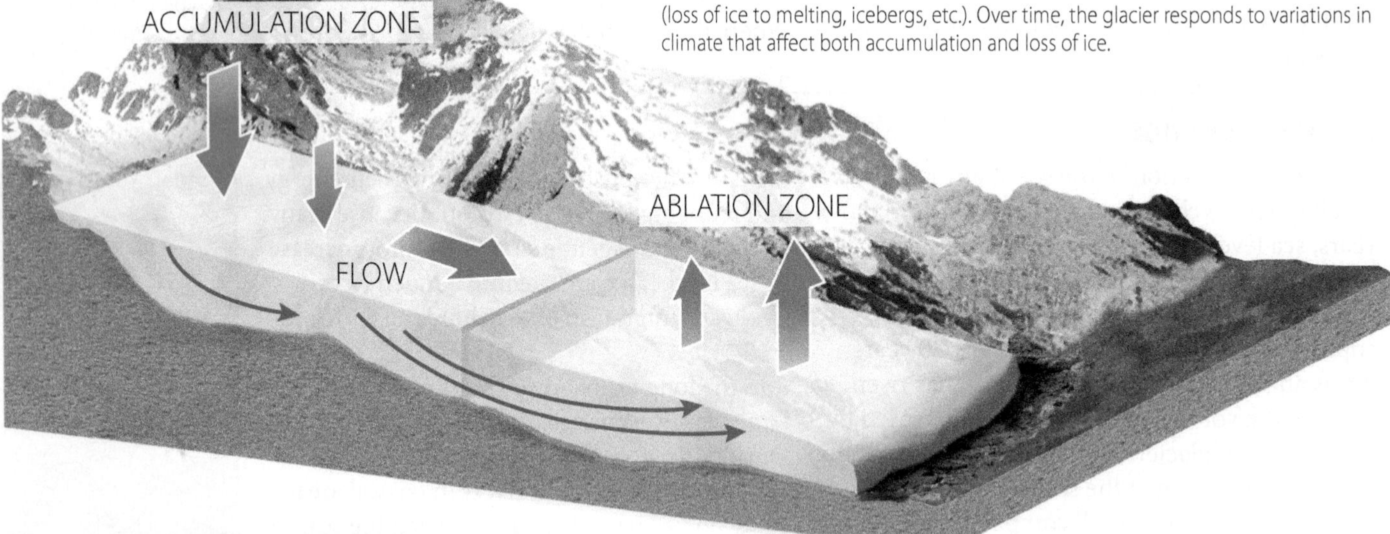

▼ 3.11.2 **MELTWATER FROM RUSSELL GLACIER**
Meltwater from the Russell Glacier that drains the Greenland Ice Sheet. Like most Greenland glaciers it is both receding and speeding up as a result of global warming, and the rivers that drain it are becoming more swollen by greater quantities of meltwater.

GLACIAL FLOW

Glaciers flow very slowly, usually at rates of a few meters to a few hundred meters per year. Glaciers may change size from one year to the next, depending on variations in weather and climate. They grow when they receive more snowfall at their source areas, and they shrink if warm temperatures increase the melting at the terminus. Some of the fastest-flowing glaciers in the world are in climates that are not the coldest, but rather have high rates of accumulation and ablation. For example, some glaciers in New Zealand flow so fast that they reach into elevations with subtropical climate before completely melting. On the other hand, glaciers in Antarctica tend to flow very slowly. Glaciers can also act erratically. For many years, a glacier may move only a few meters per year, then for a few years it may suddenly surge forward at several hundred meters (hundreds to thousands of feet) per year, before slowing again.

SHRINKING GLACIERS

For the past 200 years, most of the glaciers of the world have been shrinking as a result of climatic warming (Figure 3.11.3). Scientists have monitored the largest ice masses in the world—Greenland and Antarctica—particularly intensely because of their potential effects on global sea level. The data show net losses of ice in both areas, although the full picture is complex and uncertainties remain. In Greenland, there has been thickening of ice in high-altitude regions but substantial thinning near the margins. In Antarctica, the West Antarctic ice sheet has lost significant mass, while data for East Antarctica are mixed, with indicators of net gain in some areas and net loss in others.

▲▼ 3.11.3 **RETREAT OF THE MUIR GLACIER , ALASKA**
The Muir glacier, like many glaciers around the world, has retreated dramatically in the last two hundred to three hundred years. Above left is as it was in 1941, above right in 1950, and below in 2005.

3.12 Glacial Landforms

- Glacial erosion creates bowl-shaped depressions in mountains and U-shaped valleys.
- Glacial deposition has left large amounts of material across much of the Northern Hemisphere.

A glacier is like a conveyor belt: it picks up sediment from areas of erosion and drops it in depositional areas. As ice accumulates and begins to flow, the glacier picks up more material, and then where the glacier melts sediment is deposited (Figure 3.12.1).

The accumulation area is the erosion site, because ice is eroding the land by moving material, and the ablation area is the deposition site. Areas where glaciers form typically leave behind deep, bowl-shaped depressions and U-shaped valleys. Areas that have been heavily eroded by glacial ice often have very thin soils or bare rock. In depositional areas, we find materials that have been shaped by flowing ice, materials that were carried in meltwater streams and so bear the marks of flowing water, and sediments left behind in lakes and seas that received glacial meltwater. Glacial deposits play an especially important role in shaping landforms in places with rapid melting because large quantities of material are dropped in these places, forming moraines.

Much of the northcentral United States is actually overlain by ice-deposited material called till. At the place of maximum ice extent (the end of the conveyor belt) we find accumulations of ice-borne material, mainly till, called terminal moraines. Some of these are quite large; in fact, Long Island, New York, and Cape Cod, Massachusetts, are the tops of terminal **moraines** from past glaciations. Flowing meltwater leaves the glacier and deposits debris close to the glacier in a broad, gently sloping plain, known as an **outwash plain**. The outwash plain contains a thick layer of rocks deposited close to the glacier in a layer of sand and gravel that can exceed a thickness of 100 meters (330 feet). The finer silt and clay materials are usually carried much farther and may be deposited in lakes, seas, or distant valleys.

▼ 3.12.1 **LANDFORMS OF GLACIAL EROSION AND DEPOSITION**
Turquoise Lake, Colorado, shown here, lies in a glacial valley with a large terminal moraine that impounds the lake.

Throughout much of the last 3 million years, glaciers covered much of North America, Europe, and northern Asia as well as large parts of the Andes Mountains in South America (Figure 3.12.2). This time in Earth history, known as the **Pleistocene Epoch**, included many periods of glaciation separated by relatively ice-free intervals. In each of these glacial episodes, ice shaped landforms as it flowed across the land as **continental glaciers**, often obliterating the evidence of previous environmental conditions. The most recent glacial advance reached its maximum around 20,000 years ago, and by about 9,000 years ago, most areas that had been overlain by ice sheets were exposed. Because relatively little time has elapsed since the glaciers melted away, the areas that had been glaciated bear very fresh marks of glacial processes (Figure 3.12.3).

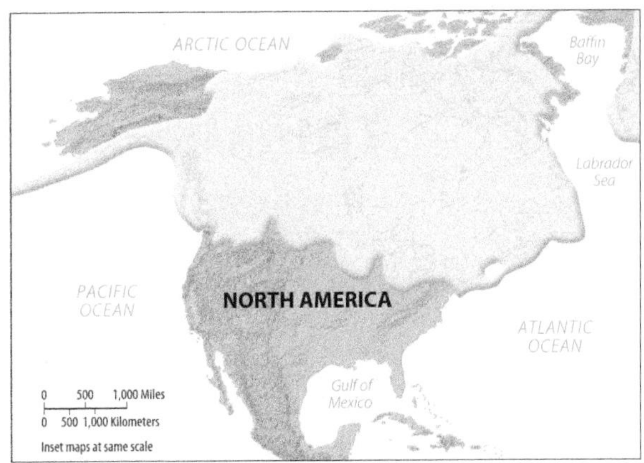

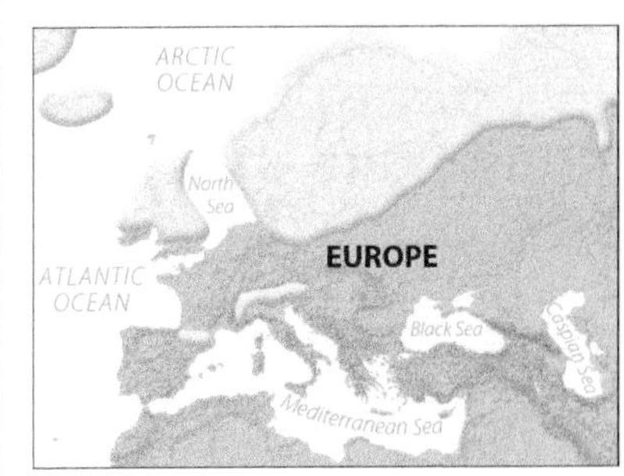

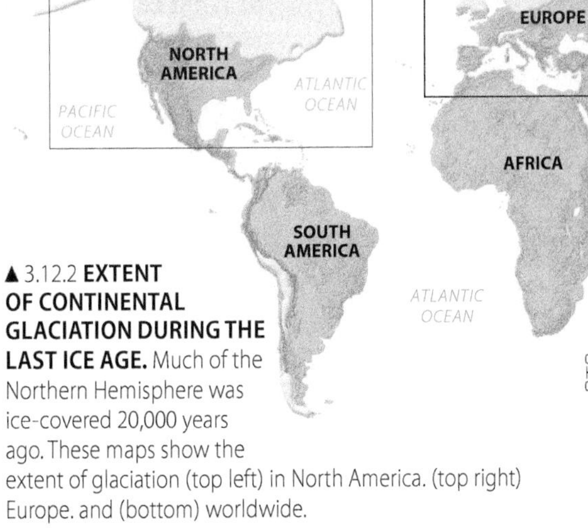

▲ 3.12.2 **EXTENT OF CONTINENTAL GLACIATION DURING THE LAST ICE AGE.** Much of the Northern Hemisphere was ice-covered 20,000 years ago. These maps show the extent of glaciation (top left) in North America. (top right) Europe. and (bottom) worldwide.

◄▼ 3.12.3 **GLACIER'S EFFECTS ON THE LANDSCAPE** (above) Aerial view of the braided Waiho River and part of Waiho Loop, an 11,000 year old glacial terminal moraine, on the west coast of South Island, New Zealand. (below) Glacial U-shaped valley, Steen's Mountain, Oregon.

Earth's surface is continually changing, both in sudden and dramatic movements and gradually over millions of years. Landforms reflect both processes operating within the solid Earth, and on the surface.

Key Questions

How does plate tectonic movement shape Earth's surface?

- Movement along plate boundaries creates mountain ranges.
- Most earthquakes and volcanic eruptions are associated with plate tectonic movement.

How does rock weathering and sediment transport shape the landscape?

- Rocks are broken down and become mobile.
- Gravity and erosional processes carry sediment downslope to rivers.
- Rivers carry sediment downstream, temporarily depositing it in floodplains along the way.
- Most landscapes are composed of slopes and valleys, with the sizes and shapes of these features reflecting the processes that form them.

What landforms are created by glaciers and waves?

- Coastal regions experience rapid erosion and deposition due to the concentration of wave energy on the shore.
- Glaciation during the Pleistocene Epoch dramatically altered landscapes in much of the world.
- Climate change is causing glaciers to shrink and sea level to rise.

▼ 3.CR.1 **MOUNT ST. HELENS**

Thinking Geographically

Mount St. Helens in Washington state erupted violently in 1980. The Loma Prieta earthquake in California occurred in 1989 (Figure 3.CR.1).

1. **How far apart are these two points on a map?**
2. **Was there a possible connection between these two events?**
3. **List reasons why there might be a connection and why there might not.**
4. **What is the most significant natural disaster that has occurred near where you live, or that might occur in your region?**
5. **What are the factors that contribute to this hazard?**
6. **What can people in your region do to reduce their vulnerability to that hazard?**

Interactive Mapping

EARTHQUAKES ARE NOT UNIFORMLY DISTRIBUTED

Launch MapMaster World in MasteringGEOGRAPHY™

Turn on the *Physical Features* layer from the *Physical Environment* menu.

Locate the major mountain ranges.

Now turn on the earthquakes layer and compare the distribution of earthquakes with the positions of the major mountain ranges. You may have to turn the earthquake layer on and off a few times to see both the mountains and the earthquakes.

Is there a relation between earthquakes and major mountain ranges? If so, why would that be the case?

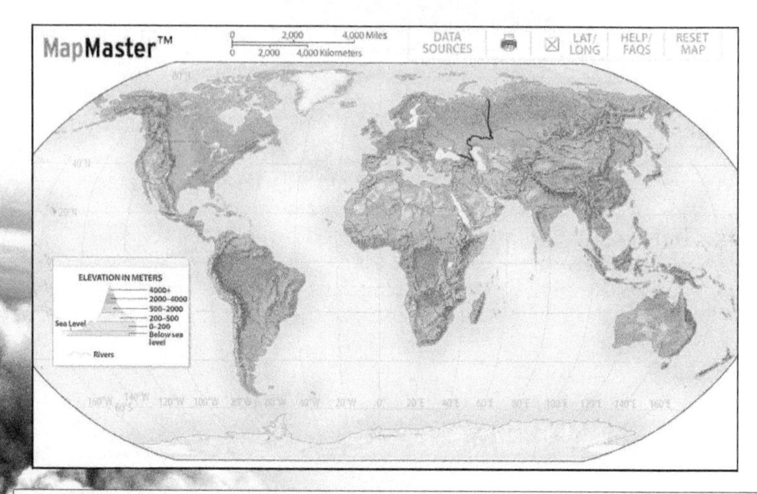

On the Internet

The United States Geological Survey has excellent and up-to-the-minute information on streamflow and water resource conditions at **http://waterwatch.usgs.gov/.**

NASA's Earth Observatory site, **http://earthobservatory.nasa.gov/**, has excellent imagery on its "Imagery" page showing recent events and conditions around the world.

Explore

EXPLORE MEANDERING RIVERS

Open Google Earth and find a meandering river channel. Here are some coordinates of good examples:

Fly to: *31° 20′ N 91° 4′ W or* Fly to: *36° 9′ N 89° 36′ W or* Fly to: *69° 53′ N 159° 28′ W*

Zoom to an eye altitude of 10 to 50 km to see the river meanders.

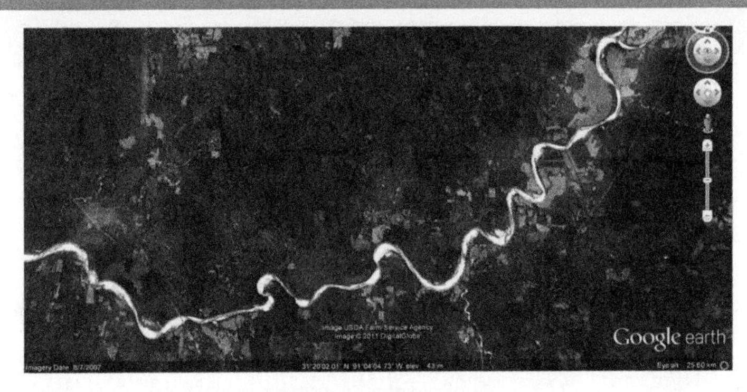

Can you identify the meanders of the channel?

Use the ruler tool to measure the width of the river, including the whole channel and not just the portion with water in it at the time of the image. Now measure a distance along the channel that is long enough to include several bends of the river (click on path in the dialog box to measure a curved line). Count the number of right-hand bends in the measured portion of the river and divide that number into the distance. The result is the meander wavelength, which is normally 10 to 15 times the river width.

Do these channels conform to that relationship?

Key Terms

Beach
A deposit of wave-carried sediment along a shoreline, on which waves break.

Chemical weathering
The breakdown of rocks or minerals through chemical reactions at Earth's surface.

Composite cone volcano
A volcano formed by a mixture of lava eruptions and more explosive ash eruptions.

Continental glacier
A thick glacier hundreds to thousands of kilometers across, large enough to be only partly guided by underlying topography.

Convergent plate boundary
A boundary between tectonic plates in which the two plates move toward one another, destroying or thickening the crust.

Delta
A deposit of sediment formed where a river enters a lake or an ocean.

Discharge
The quantity of water flowing past a point on a stream per unit time.

Divergent plate boundary
A boundary between tectonic plates in which the two plates move away from each other, and new crust is created between them.

Drainage basin
The geographic area that contributes runoff to a particular stream, defined with respect to a specific location along that stream --the runoff from the drainage basin passes that point on the stream.

Earthquake
A sudden release of energy within Earth, producing a shaking of the crust.

Epicenter
The location on Earth's surface immediately above the focus of an earthquake.

Fault
A fracture in Earth's crust along which displacement of rocks has occurred.

Floodplain
A low-lying surface adjacent to a stream channel and formed by materials deposited by the stream.

Focus (of an earthquake)
The location in Earth where motion originates in an earthquake.

Glacier
A large mass of flowing, perennial ice.

Grade
A condition in which a stream's ability to transport sediment is balanced by the amount of sediment delivered to it.

Igneous rock
Rock formed by crystallization of magma.

Landform
A characteristic shape of the land surface, such as a hill, valley, or floodplain.

Lava
Magma that reaches Earth's surface.

Longshore current
A current in the surf zone along a shoreline, parallel to the shore.

Longshore transport
Sediment transport by a longshore current.

Magma
Molten rock beneath Earth's surface.

Mantle
The portion of Earth above the core and below the crust.

Marine terrace
A nearly level surface along a shoreline, elevated above present sea level, formed by coastal erosion at a time when sea level at the location was higher than at present.

Mass movement
Downslope movement of rock and soil at Earth's surface, driven mainly by the force of gravity acting on those materials.

Meandering
The tendency of flowing water to follow a sinuous course with alternating right- and left-hand bends.

Mechanical weathering
The breakdown of rocks into smaller particles caused by application of physical or mechanical forces.

Metamorphic rock
Rock formed by modification of other rock types, usually by heat and/or pressure.

Moraine
An accumulation of rock and sediment deposited by a glacier, usually in or near the melting area.

Outwash plain
An accumulation of sand and gravel carried by meltwater streams from a glacier, usually deposited immediately beyond the terminal moraine from the glacier.

Overland flow
Water flowing across the soil surface on a hillslope, usually resulting from precipitation falling faster than the ground can absorb it.

Pleistocene Epoch
A period of geologic time consisting of the first part of the Quaternary Period beginning about 3 million years ago and ending about 12,000 years ago.

Runoff
Flow of water from the land, either on the soil surface or in streams.

Sea level
The general elevation of the sea surface, averaging out variations caused by waves, storms, and tides.

Sediment transport
The movement of rock particles by surface erosional processes.

Sedimentary rock
Rock formed through accumulation and fusing of many small rock fragments at Earth's surface.

Shield
The ancient core of a continent.

Shield volcano
A volcano with relatively gentle slopes formed by eruption of relatively fluid lavas.

Soil creep
The slow downslope movement of soil caused by many individual, near-random particle movements such as those caused by burrowing animals or freeze and thaw.

Tectonic plates
Large pieces of Earth's crust that move relative to one another.

Transform plate boundary
A boundary between tectonic plates in which the two plates pass one another in a direction parallel to the plate boundary.

Tsunami
An extremely long wave created by an underwater earthquake; the wave may travel hundreds of kilometers per hour.

Volcano
A vent in Earth's surface where magma erupts as lava.

▸ LOOKING AHEAD

Vegetation and soils are linked to climate and landforms via the movements of water, carbon, and nutrients between the atmosphere, hydrosphere, lithosphere, and biosphere. These links—biogeochemical cycles—and the resulting patterns of vegetation and soils are explored in the next chapter.

4 Biosphere

All of Earth's physical systems—the hydrosphere, biosphere, lithosphere, and atmosphere—are closely connected. Water, carbon, oxygen, and a variety of nutrients move steadily among these systems. These exchanges play critical roles in regulating environmental processes, determining spatial patterns of vegetation and soil on Earth's surface, and linking all of Earth's subsystems. Climate, through its influence on water availability and movement, plays a central role in these exchanges. In the last few centuries, the human role in these physical processes has grown enormously, to the extent that not only are we modifying ecosystems directly but also indirectly through modifying climate itself.

How do water and nutrients cycle through ecosystems?

BROADLEAF DECIDUOUS FOREST (SOME CLEARED FOR AGRICULTURE), SLOVAKIA

How are matter and energy cycled through ecosystems?

What is the distribution of life forms and ecological communities?

SCAN TO VIEW SATELLITE IMAGERY OF THE BIOSPHERE FROM NASA

4.1 Biogeochemical Cycles and Ecosystems

- **Water, nutrients, and carbon move among the atmosphere, hydrosphere, lithosphere, and biosphere.**
- **Humans play a major role in many biogeochemical cycles through their everyday activities.**

Biosphere processes, the growth and decay of plants and animals and, indeed, all life processes, depend on exchanges of energy and matter. Earth receives a constant supply of energy in the form of light from the sun. Matter, including the essential substances water and carbon, is available in the atmosphere, hydrosphere, and lithosphere, as well as in the biosphere itself.

BIOGEOCHEMICAL CYCLES

Biogeochemical cycles are recycling processes that supply essential substances such as carbon, nitrogen, and other nutrients to the biosphere (Figure 4.1.1). Biogeochemical cycles also connect the lithosphere, hydrosphere, biosphere and atmosphere. The **hydrologic cycle** in which water flows among these subsystems is an example of a biogeochemical cycle. Other important examples include the **carbon cycle**, in which that element moves between the atmosphere (as CO_2 and other substances), the biosphere (as living and formerly matter), the hydrosphere (as dissolved compounds), and the lithosphere (as rock and fossil fuels).

The law of conservation of energy states that energy is not created or destroyed under ordinary conditions, but it may be changed from one form to another. Similarly, the law of conservation of matter states that matter may be changed from one form to another under ordinary conditions but cannot be created or destroyed, except in nuclear reactions. Therefore, a change in the quantities of a substance stored or moving in one part of the system can have consequences throughout the system. The Mississippi River example at right illustrates these effects.

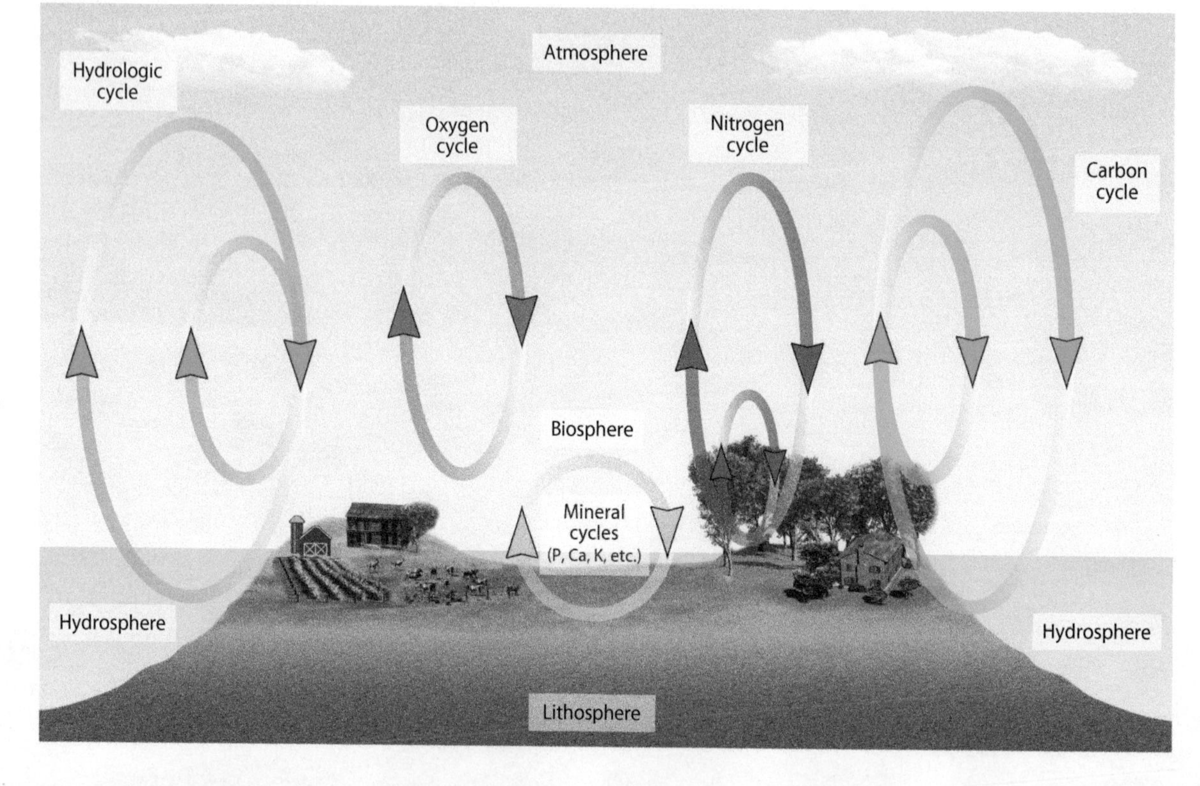

► 4.1.1
BIOGEOCHEMICAL CYCLES
These cycles transfer matter among the atmosphere, biosphere, hydrosphere, and lithosphere. The cycles represented here are shown in greatly simplified form. Important minerals cycling in the environment include phosphorus (P), calcium (Ca), and potassium (K), among others.

▲ 4.1.2 **THE MISSISSIPPI DELTA**
This is a unique environment. With its extensive wetlands, rich fisheries, and mineral resources, it is different from any other environment in North America. The delta is under severe environmental threat, but not just from local problems like oil spills. It is also affected by distant activities like fertilizer use in Illinois and reservoir construction in North Dakota.

HUMAN INFLUENCES

The Mississippi River collects water, sediment, and dissolved substances from 4.76 million km^2 of land area in the central United States and southern Canada (Figure 4.1.2). Its major tributaries include the Ohio River, which provides most of the water, and the Missouri, which provides much sediment. The Mississippi collects this water and its substances and carries it southward. Along the way, sediment is temporarily stored in floodplains, nutrients are taken up and released in aquatic ecosystems, and the water is withdrawn to serve cities and farms along the river and then discharged back to the Mississippi as wastewater.

The Mississippi Delta is a vast area of over 50,000 km^2 in Louisiana and Mississippi, largely consisting of wetlands near sea level. These wetlands have been created and modified through millions of years of sediment accumulation and support a highly diverse ecological community. Offshore, in the waters of the Gulf of Mexico, nutrients from inland delivered to the gulf help support a similarly rich and diverse marine community. Huge accumulations of oil and gas in both the delta and offshore areas have been formed from nutrients and organic matter delivered by the river.

In the last century, two factors related to land use and river management have transformed these environments.

- First, millions of large and small dams have been built throughout the watershed, both on the main stem and its many tributaries. These dams, combined with navigation structures on the main stem, have decreased the supply of sediment to the vast wetlands of the delta. These wetlands are naturally sinking, and the lack of addition of sediment from inland is leading to a loss of about 50 km^2 (20 mi^2) of wetlands per year.
- Second, the use of large amounts of fertilizer on cropland has caused very large amounts of nitrogen to be delivered to the waters of the gulf. This nitrogen stimulates growth of floating plants, which sink to bottom waters when they die. Decay of this plant matter depletes oxygen, creating a "dead zone" in the Gulf of Mexico.

4.2 Hydrologic Cycle

- At any given time, most of the world's water is in the oceans.
- Budgets that account for water exchanges help us to understand the hydrologic cycle and water resources.

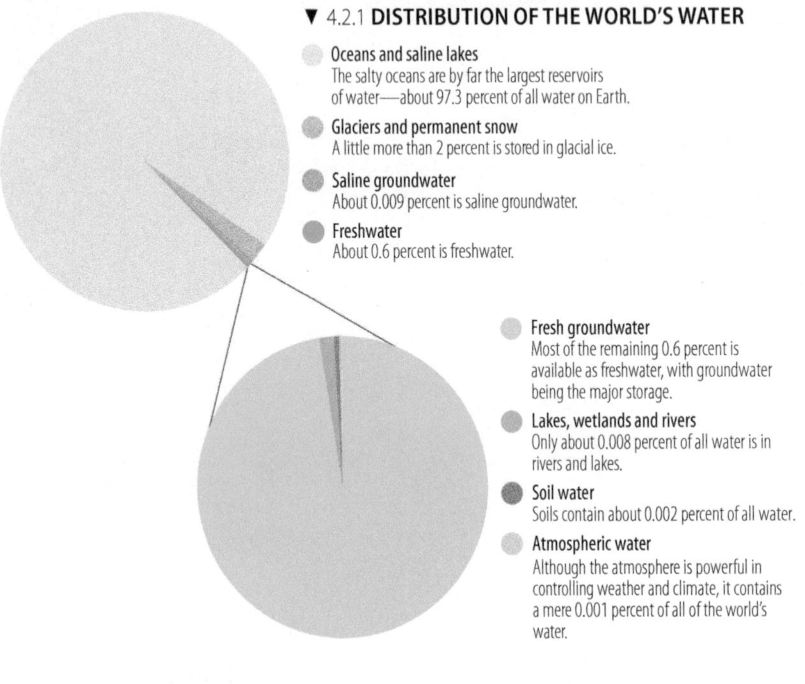

▼ 4.2.1 **DISTRIBUTION OF THE WORLD'S WATER**

Water is stored in the atmosphere, biosphere, hydrosphere, and lithosphere in solid, gaseous, and liquid forms (Figure 4.2.1). Water cycles through the atmosphere, lithosphere, and hydrosphere by means of evaporation, condensation, precipitation, and runoff (Figure 4.2.2). Evaporation converts liquid water in lakes and oceans into vapor, returning it into the atmosphere. Water falls from the atmosphere to the ground and ocean through condensation and precipitation. Runoff carries water from the land to the sea, most of it temporarily stored as **groundwater**. This flow is the hydrologic cycle.

MasteringGEOGRAPHY
Animations
Earth's Water and Hydrologic Cycle.

▼ 4.2.2 **THE HYDROLOGIC CYCLE**
Numbers in parentheses refer to quantities of water transferred on an average annual basis.

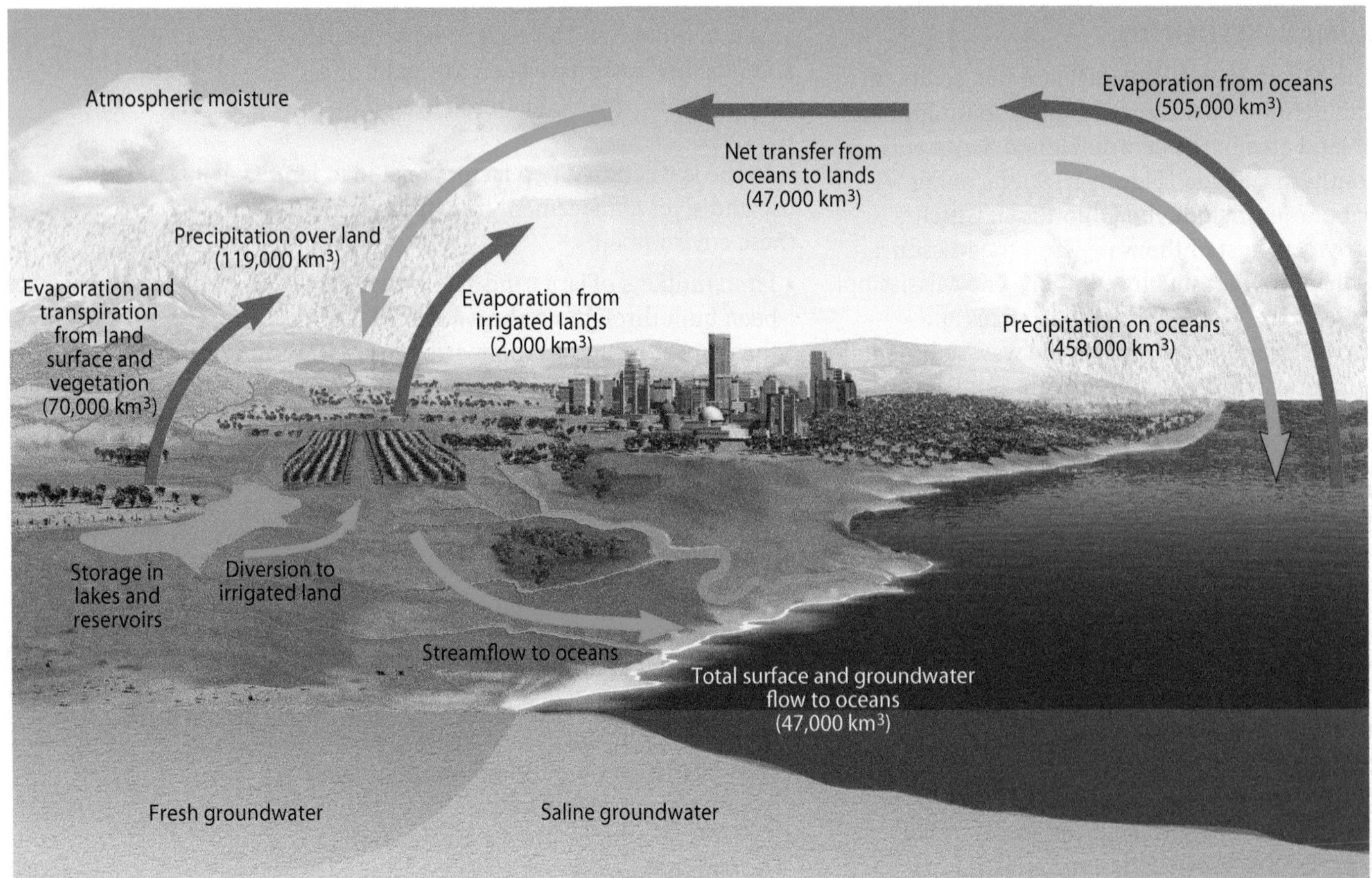

Estimates of the global average rates of evaporation, condensation, precipitation, and runoff are shown in the global water budget. Each process varies geographically with the amount of water available. Excess water evaporated into the atmosphere over the oceans is carried by wind over land areas, where it condenses into clouds and falls as precipitation (Figure 4.2.3). About two-thirds of the water that falls on land areas evaporates there, and the remaining one-third drains into rivers, which return this excess to the sea as runoff. A significant portion of river flow (an average of about 5 percent worldwide) is returned to the atmosphere via plant-water use on irrigated lands rather than flowing to the ocean as liquid water (Figure 4.2.4).

▲▼ 4.2.3 **WATER CYCLING IN THE AMAZON BASIN**
Water from the Atlantic feeds precipitation over the Amazon, which is cut off from Pacific moisture by the Andes mountains. As the trade winds carry moisture across South America from east to west, evapotranspiration supplies water to the atmosphere, which then falls out again as precipitation. In the eastern part of the Amazon, precipitation is mainly derived from evaporation from the Atlantic, but in the west, recycled water that has been used by vegetation forms a significant portion of total rainfall.

Moisture from the Atlantic
Atmospheric transport and precipitation
Andes
Mouth of the Amazon
Atlantic
Soil water storage and release
Amazon
Pacific

1940

2009

◄ 4.2.4 **LAKE MEAD**
This reservoir, along the Nevada-Arizona border, supplies water and electricity to cities and irrigated agriculture in Nevada, Arizona, and California. Water levels in the reservoir have been low since about 2000 as a result of prolonged drought.

4.3 Local Water Budgets

- Climate and weather control local water movements.
- Local water budgets explain seasonal variations in plant growth, streamflow, and water availability.

Water budgets are critical for water management. They tell us how much water is available in any given area, how water resources are related to climate, and how human activities change water resources (Figure 4.3.1).

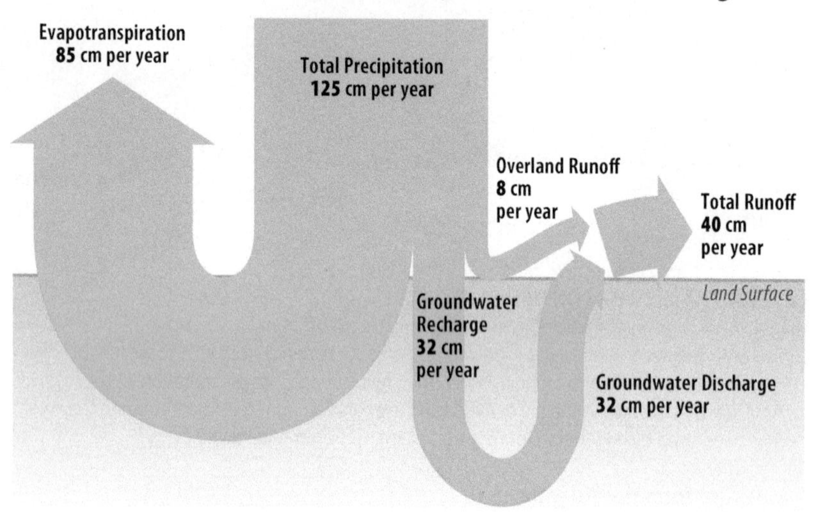

▲ 4.3.1 **LOCAL WATER BUDGET FOR A TYPICAL MIDLATITUDE SITE**
The quantities in this budget are typical of a midlatitude climate. Note that most rainfall goes to evapotranspiration.

Water budget diagrams illustrate the annual course of the water budget (Figure 4.3.2). In well-vegetated areas, much of the conversion of water from liquid to vapor takes place in plants' leaves. This water is replaced by water drawn from the soil through plant roots. We call this plant mechanism **transpiration**, and when combined with evaporation we call the process **evapotranspiration (ET)**.

Although ET rates vary tremendously over time and from place to place, it generally depends on two factors.

- Water must be available for plants to transpire.
- The air cannot be saturated with humidity.

Energy is necessary to evaporate water. When water vaporizes from a liquid to a gas, energy is absorbed (it becomes latent heat in the water vapor), which cools the plant leaves. Thus, the rate of evapotranspiration depends mostly on energy availability. ET occurs fastest under warm conditions and it virtually halts

▼ 4.3.2 **WATER BUDGET DIAGRAM FOR CHICAGO, ILLINOIS, A TYPICAL HUMID MIDLATITUDE SITE**

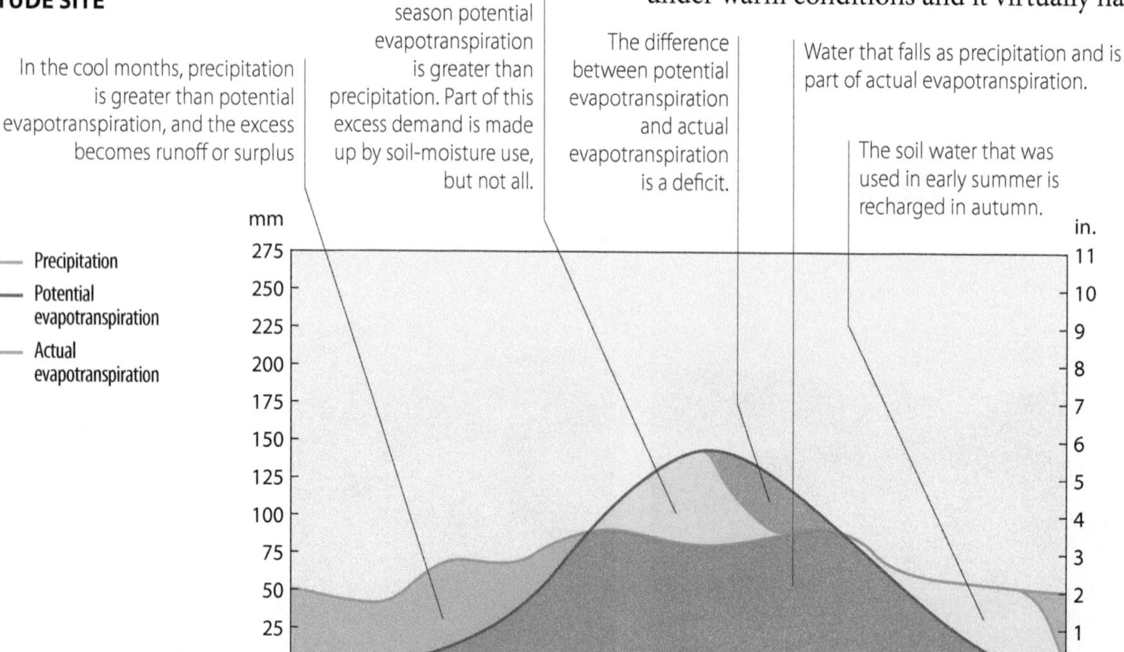

below freezing. Atmospheric humidity and wind speed also are significant: ET is faster on dry, windy days than on humid, calm days. Evapotranspiration is low in winter and high in summer, reflecting active transpiration by plants.

In warm weather, conditions may favor high ET, but the soil may be too dry to supply this demand (Figure 4.3.3). Therefore, we distinguish between potential ET and actual ET. **Potential evapotranspiration (POTET)** is the amount of water that would be evaporated if it were available. **Actual evapotranspiration (ACTET)** is the amount that actually is evaporated under existing conditions. If water is plentiful, then ACTET equals POTET. But if water is in short supply, ACTET is less than POTET.

ACTET never exceeds POTET. Comparing POTET with precipitation is a useful way to describe water availability in various climates. For example, if precipitation always is greater than POTET, plants have plenty of water, and the climate is quite humid. If precipitation is less than POTET most of the time, then plants never get as much water as they need, the natural vegetation is adapted to dryness, and the climate is arid.

Water budgets vary tremendously from one climate to another (Figure 4.3.4). In some humid climates, precipitation almost always meets demand, and stored soil moisture provides the small additional amount of water needed during brief dry spells. In semiarid and arid climates, the demand for water considerably exceeds precipitation during the year, so soil moisture is never fully replenished and plants must withstand severe moisture deficits.

▲ 4.3.3 **IRRIGATING RECREATIONAL AREAS**
In climates like Chicago's, people water lawns and golf courses to meet the high summer evapotranspiration needs of the grass.

► 4.3.4 **WATER BUDGETS FOR FOUR PLACES**

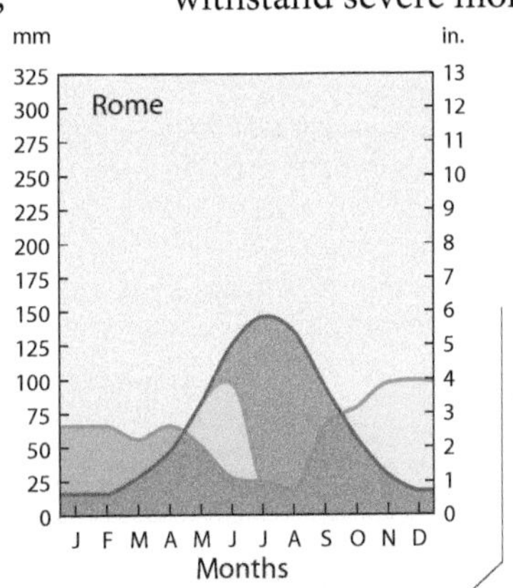

In Rome, precipitation is low in summer while potential evapotranspiration is high, causing a moisture deficit.

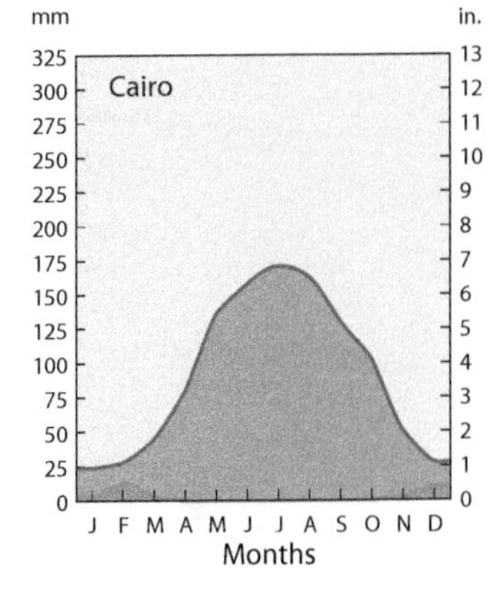

Cairo has a desert climate, in which precipitation is less than potential evapotranspiration all the year.

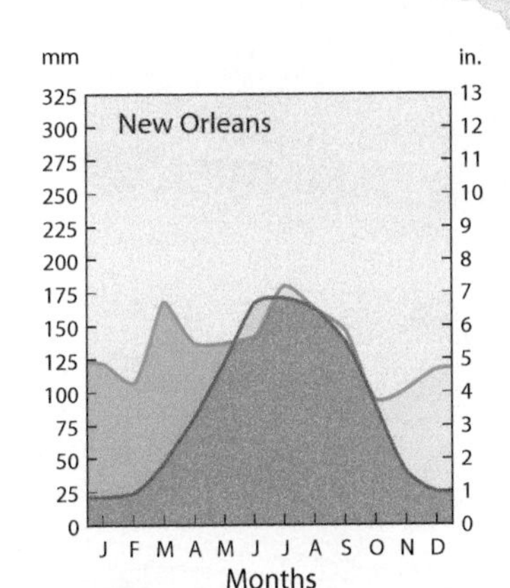

New Orleans is a humid climate in which precipitation is sufficient to meet potential evapotranspiration all year. Potential evapotranspiration is much higher in summer than winter.

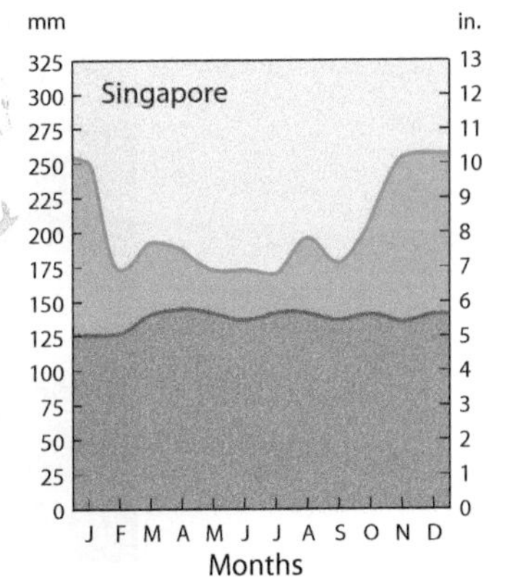

Singapore has high amounts of precipitation that are greater than potential evapotranspiration all year, despite high temperatures.

4.4 Carbon Cycle

- Plant growth and decomposition drive the carbon cycle.
- Increasing atmospheric CO_2 is causing climate change.

In the carbon cycle, carbon in the form of atmospheric carbon dioxide (CO_2) is incorporated into carbohydrates in plant tissues through photosynthesis (Figure 4.4.1). Animals consume plants and use the plant carbohydrates for their own life processes. Carbon is stored in **biomass**—living and formerly living plants and animals, and some of this is deposited in sediments. Through respiration, which occurs in almost all living organisms, carbon is returned to the atmosphere as CO_2. Carbon dioxide is also added to the atmosphere by the combustion of fossil fuels (Figure 4.4.2).

► 4.4.2 **CO_2 CONCENTRATIONS IN THE ATMOSPHERE.**
The concentration of CO_2 measured at Mauna Loa, Hawaii, goes up and down seasonally, reflecting the annual cycle of photosynthesis in the Northern Hemisphere. Concentrations fall from spring through autumn as carbon is taken up in growing plants, and rise through the winter as organic matter decays. In addition, each year the average concentration rises by about 2 parts per million, primarily as a result of fossil fuel combustion.

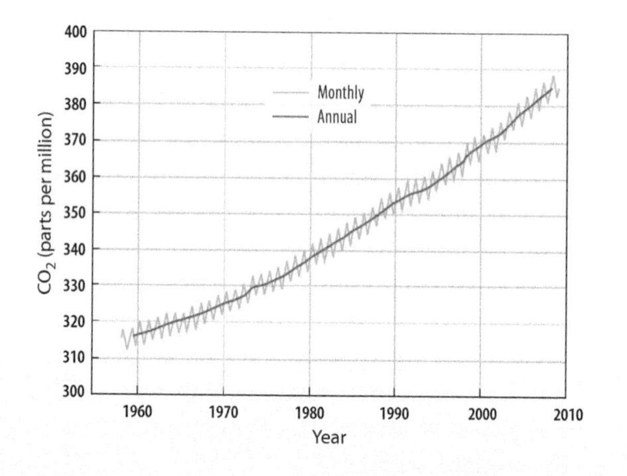

▼ 4.4.1 **THE CARBON CYCLE**
Carbon is taken from the atmosphere and stored in biomass through photosynthesis. It is returned to the atmosphere through respiration. Combustion of fossil fuels and manufacturing of cement release vast quantities of carbon to the atmosphere from long-term storage in rocks. Large quantities are also exchanged between the atmosphere and oceans via gas exchange. Major storages of carbon are present in the soil, plants, the atmosphere, oceans, rocks, and fossil fuels.

PHOTOSYNTHESIS

Photosynthesis, sedimentation, respiration, and combustion cycle carbon and oxygen back and forth between living things and the environment. **Photosynthesis** can be shown by the following equation:

carbon dioxide + water + energy →
carbohydrates + oxygen

For this reaction, land plants obtain carbon dioxide from the air, water from the soil, and energy from solar radiation. They store carbohydrates in tissue for later use and release oxygen to the atmosphere. Plants are the source of atmospheric oxygen, without which animals could not exist.

Respiration involves the opposite reaction to photosynthesis:

carbohydrates + oxygen →
carbon dioxide + water + energy (heat)

In respiration, carbohydrates are broken down when they combine with atmospheric oxygen to CO_2 and water. Energy is released in the process. Some of this energy is lost as heat and some is stored in chemical compounds for later use in other life processes.

When plants are growing, they take carbon out of the atmosphere; when forests are cut down, most of the carbon in the trees is released to the atmosphere, through decomposition or perhaps fire.

HUMAN IMPACTS ON THE CARBON CYCLE RELEASE AND STORAGE

In many parts of the world, especially the humid tropics, deforestation and urban development are reducing carbon storage in the biosphere and sending the carbon to the atmosphere (Figure 4.4.3). But in other areas, such as the eastern United States, forest area is actually increasing, and young forests are growing, storing carbon in the process. Similarly, when undeveloped lands are first converted to agriculture, there is usually a loss of organic matter and thus carbon from the soil, but later that carbon can be replaced if the soils are allowed to recover. Presently, evidence indicates that deforestation is causing large transfers of carbon to the atmosphere, but we also know that in some parts of the world there are large amounts of carbon being absorbed by the biosphere. The uncertainties are large.

Although many processes are involved, clearly combustion of fossil fuels is the most important factor driving the global increase in atmospheric CO_2 and thus global warming. The greatest emissions of CO_2 have historically come from industrialized regions, such as Europe and North America, but recent economic growth has caused a dramatic increase in fossil carbon emissions. From 2000–2009, global fossil carbon emissions grew at a rate of 3.2 percent per year, with China's emissions more than doubling in that period. China is now the world's greatest CO_2 emitter, although much of those emissions are the result of manufacturing products that are consumed in the United States, Europe, and Japan.

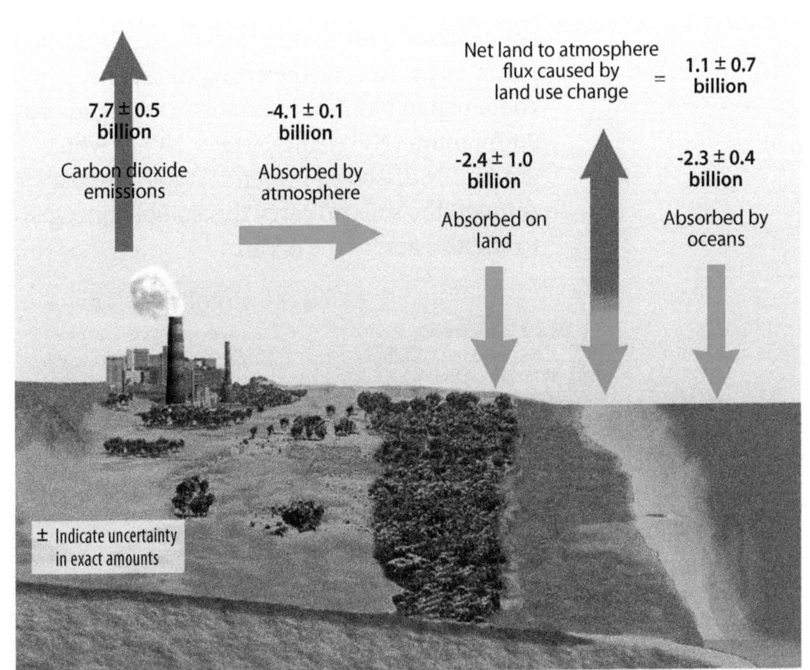

▲ 4.4.3 **GLOBAL CARBON BUDGET FOR 2000–2009**
The total amount of carbon in the atmosphere increases by about 4.1 gigatons (Gt) of carbon per year (a gigaton is a billion metric tons). This increase is caused primarily by fossil fuel combustion and cement manufacturing, which totaled about 7.7 Gt per year in 2000–2009, and to a lesser extent by net land use change, which sends about 1.1 Gt per year to the atmosphere. These additions total much more than the rate of accumulation in the atmosphere, which is only 4.1 Gt per year. The ocean absorbs about 2.3 Gt, and it is believed that the remainder of about 2.4 Gt per year is absorbed in the biosphere and soils (data from Global Carbon Project, 2010).

4.5 Nutrient Cycles

- **Nitrogen and phosphorus cycle among Earth's systems, providing essential nutrients to ecosystems.**
- **Human modifications of nutrient cycles have dramatically altered global ecosystems.**

In addition to the hydrologic and carbon cycles, which pass water and carbon between the atmosphere, lithosphere, hydrosphere, and biosphere, essential nutrients also move through Earth's systems. Among the most important of these are nitrogen and phosphorus. Both the nitrogen and phosphorus cycles have been heavily modified by humans, with profound effects on ecosystems—plants and animals and the physical environment with which they interact.

NITROGEN CYCLE

In the nitrogen cycle, nitrogen is taken from the atmosphere and fixed, or converted to forms that can be used by plants and animals to build proteins and other molecules (Figure 4.5.1). Nitrogen fixation occurs by both natural and human processes.

- Most natural nitrogen fixation is carried on in the soil, by bacteria, and is then taken up in plants and incorporated in plant tissue. This nitrogen then passes from plants to the animals that consume them, before being broken down and released back to the atmosphere. Bacteria play a major role in converting nitrogen contained in biomass into mobile forms such as ammonium (NH_4) and nitrate (NO_3), which are readily dissolved in water. Water moving through the soil can carry this soluble nitrogen to streams and to the ocean.
- Humans fix nitrogen through fertilizer manufacture. This nitrogen is added to agricultural systems from which it is spread through the environment. Motor vehicles and fossil-fuel fired power plants also emit large amounts of nitrogen oxides that are returned to the surface in precipitation in bioavailable forms, particularly nitrate.

Today more nitrogen is fixed by humans than by natural processes.

PHOSPHOROUS CYCLE

While the major store of nitrogen is in the atmosphere, the major storage in the phosphorus cycle is in the lithosphere (Figure 4.5.2). It is released to soils by rock weathering, taken up by plants, and passed through the biosphere. Phosphorus is released back to soils and water

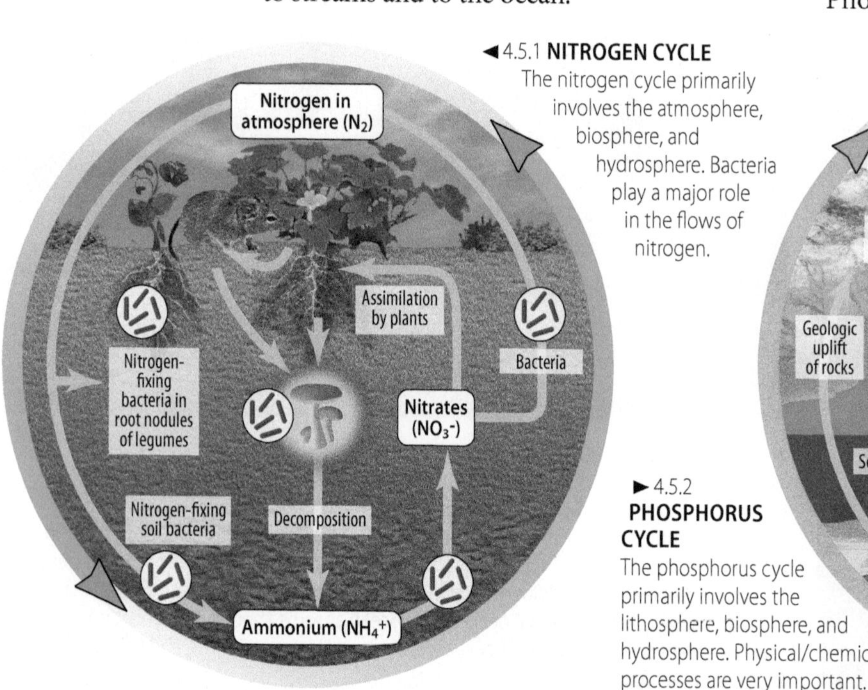

◄ 4.5.1 **NITROGEN CYCLE** The nitrogen cycle primarily involves the atmosphere, biosphere, and hydrosphere. Bacteria play a major role in the flows of nitrogen.

► 4.5.2 **PHOSPHORUS CYCLE** The phosphorus cycle primarily involves the lithosphere, biosphere, and hydrosphere. Physical/chemical processes are very important.

by decomposition of waste and dead biomass. Like nitrogen, phosphorus is an essential nutrient for plant growth. When we harvest crops, the phosphorus they contain is removed from farm fields and this, along with removal in water that runs off the fields, depletes soil phosphorus stocks. We replace this lost phosphorus, and stimulate crop growth, by adding phosphorus fertilizers to the soil. These fertilizers are produced by mining phosphate-rich rock, and by mining accumulations of bird droppings in some coastal and island areas.

EUTROPHICATION

The addition of large amounts of bioavailable nitrogen and phosphorus to the environment, particularly streams, lakes, and coastal areas, has dramatically altered aquatic ecosystems. Just as adding these nutrients to soils stimulates crop production, adding them to water stimulates growth of algae in the water—a process called **eutrophication** (Figure 4.5.3). The added algal growth in surface waters stimulates biological activity and increases populations of zooplankton—small animals that feed on algae. But as the algae and zooplankton die and settle into deeper, darker water, they decompose, consuming oxygen. This causes the deeper water to have low oxygen content, and some organisms that live on the sea bed or near it cannot survive, hence the term "dead zone" (Figure 4.5.4).

To combat these problems, farmers are encouraged to use less fertilizer, or to use techniques that maximize the efficiency of fertilizer use. One example of such a technique is precision agriculture, in which a digital map of a farm field is created showing variations in fertilizer needs. A GPS receiver mounted on a tractor is linked to a GIS system that controls a fertilizer applicator towed behind the tractor. As the tractor moves across a field, the rate of fertilizer application is adjusted to match the needs in each part of the field, so that no more fertilizer is applied than is necessary. This can both reduce costs of fertilizer for the farmer, and improve downstream water quality.

▲ 4.5.3 **ALGAL BLOOM**
Haozhou, China.

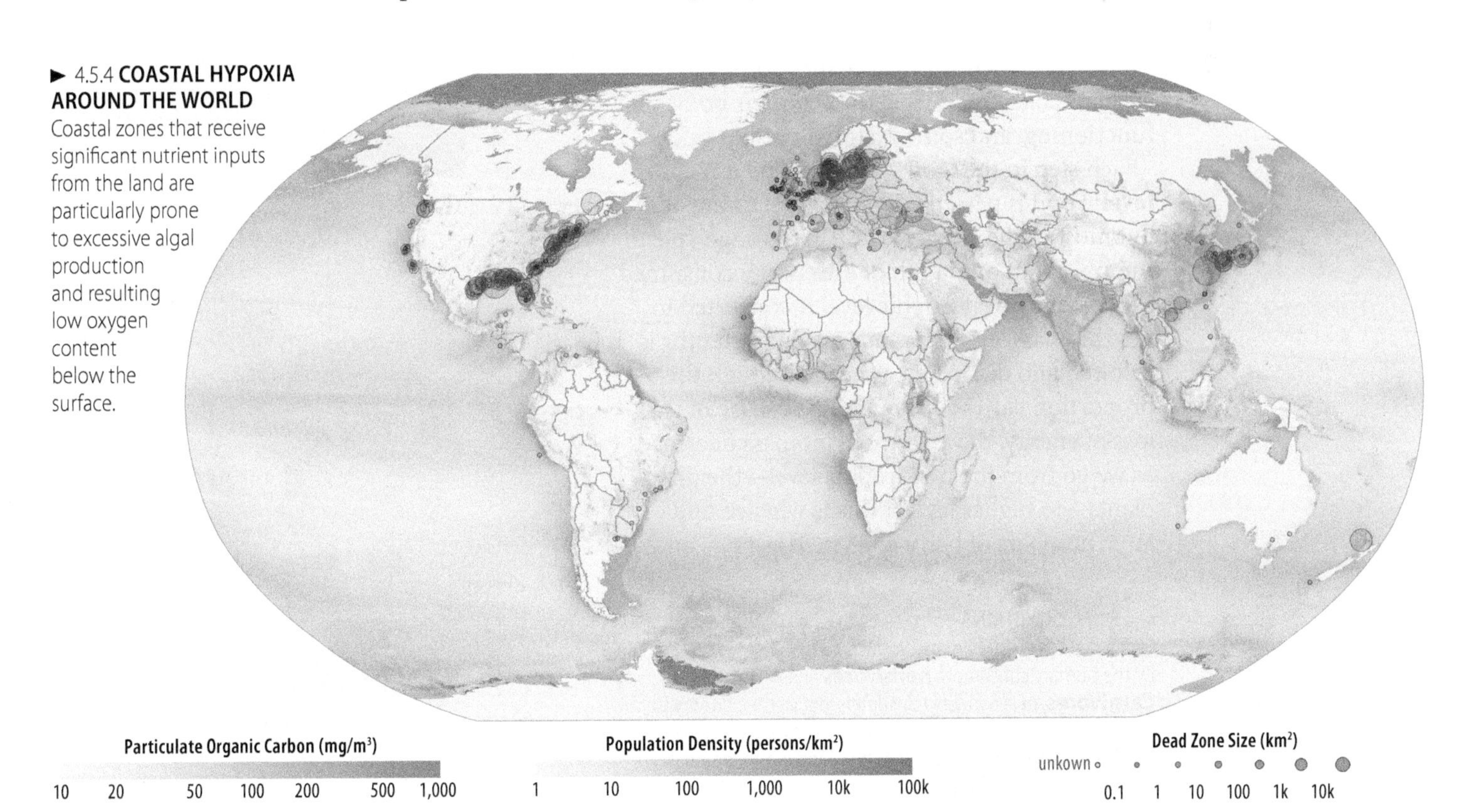

► 4.5.4 **COASTAL HYPOXIA AROUND THE WORLD**
Coastal zones that receive significant nutrient inputs from the land are particularly prone to excessive algal production and resulting low oxygen content below the surface.

4.6 Food Chains and Webs

- Food chains link plant photosynthesis to herbivores and carnivores and ultimately decomposers.
- Complex interactions tie all members of an ecological community to each other.

An ecosystem includes all living organisms in an area and the physical environment with which they interact. An ecosystem can cover an area as small as a field or a pond. On any scale of analysis, large or small, certain fundamental ecosystem elements exist:

- Nonliving matter and energy necessary for production and consumption to occur—water, mineral nutrients, gases such as oxygen and carbon dioxide, and energy (light and heat).
- Producers—green plants and other organisms that produce food for themselves and for consumers that eat them.
- Consumers—organisms that eat producers, other consumers, or both.
- Decomposers—small organisms, such as bacteria, fungi, insects, and worms, that digest and recycle dead organisms.

FOOD CHAIN

Green plants produce food in the form of carbohydrates. The food is distributed through an ecosystem by way of a food chain (Figure 4.6.1). Most of the food that animals consume is used to keep their bodies functioning, and some is stored in their bodies.

Each step in the food chain is called a **trophic level**. Food is passed from one level to the next, but most of the energy is lost. As a rough rule of thumb, about one-tenth of the energy consumed as food at a given trophic level is converted to new biomass, and the remaining nine-tenths is respired and dissipated as heat, although this proportion varies substantially. Because of this loss of energy, the amount of biomass decreases as we go from the first trophic level—the green plants—to higher levels. This is why we find large numbers of herbivores such as mice and

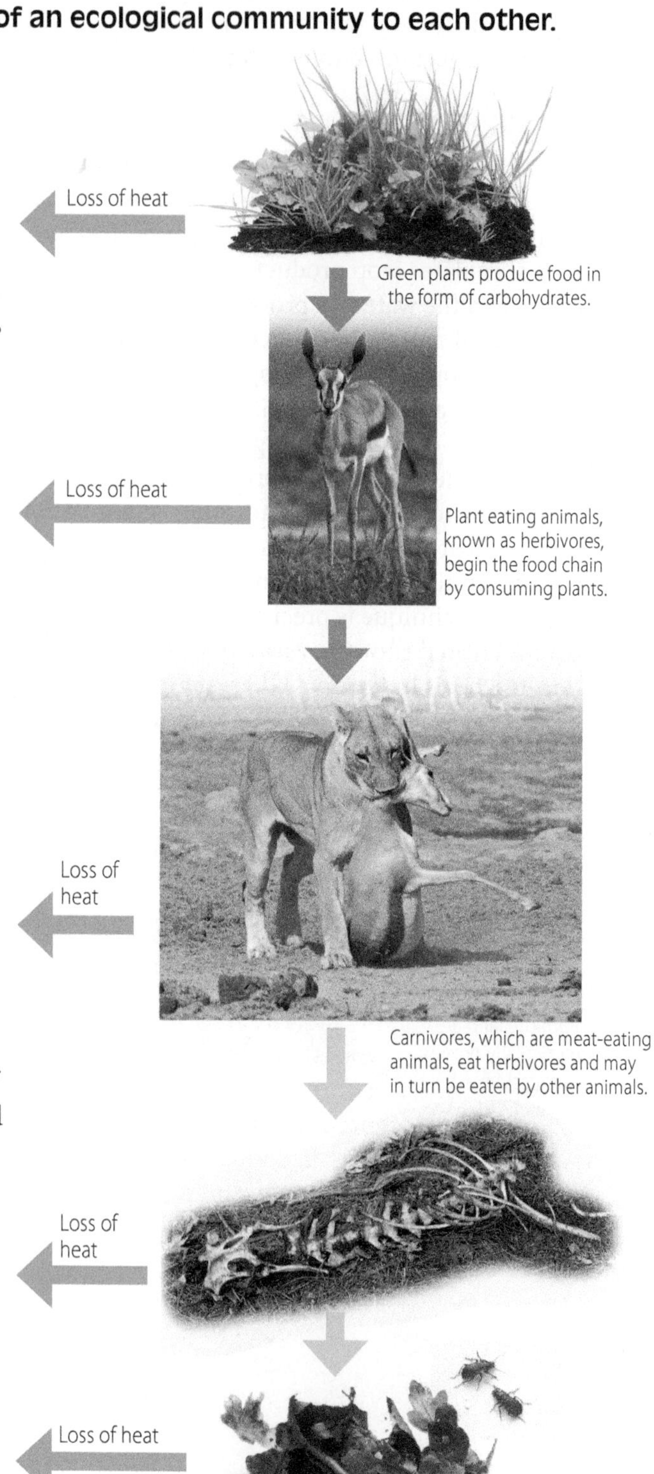

► 4.6.1 **FOOD CHAINS**
Green plants are the primary producers of food, which is eaten by the primary consumers: **herbivores** and **omnivores**. **Carnivores**, or secondary consumers, derive their energy from other animals rather than directly from plants. At each step of this **food chain**, energy is lost as heat, and waste materials are broken down by decomposers.

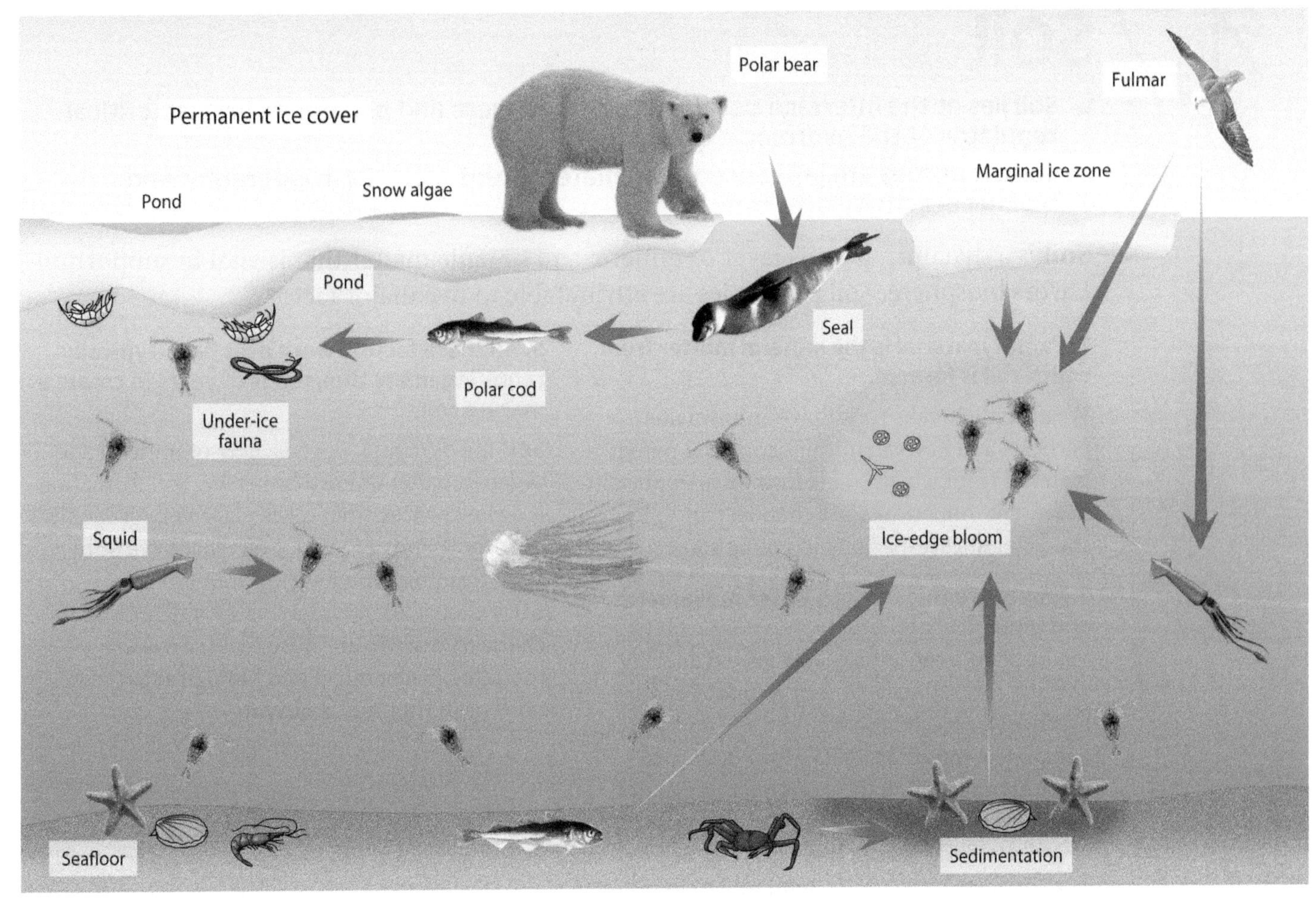

▲ 4.6.2 **ARCTIC SEA ICE FOOD WEB**
Food webs are complex relationships among many species in a community. The food web associated with the Arctic sea ice environment combines marine organisms such as algae, crustaceans, fish, and marine mammals; birds such as ducks and gulls; terrestrial mammals such as arctic fox and polar bears; and a wide range of decomposers that feed on waste and organisms not otherwise consumed from all portions of the web.

rabbits, but relatively few large carnivores like wolves and lions.

One consequence of food chains is that some chemicals in the environment, especially persistent pesticides such as DDT, can accumulate in animal tissues as they pass up the chain. If a substance accumulates in an animal rather than being broken down, then at each trophic level the concentration of that pesticide increases in a process called **biomagnification**. This process was a major factor in the decline of some bird populations, including the bald eagle, in the mid-twentieth century. Because of these effects, DDT and similar pesticides have been banned or severely restricted in much of the world, and the pesticides used today break down relatively rapidly to reduce this problem.

COMPLEX RELATIONSHIPS

Although trophic levels are a one-directional, ladder-like view of energy flow through ecosystems, the reality is more complex. We use the term *food web* to describe this complex system (Figure 4.6.2). Individual organisms may feed at multiple tropic levels, and may rely on different food sources at different times. For example, black bears are omnivores–they eat both plants and animals–but their diet varies through the year and may include shoots of young plants in the spring, roots, insects (which may be either herbivores or carnivores), some mammals such as young deer (herbivores), and fish (again, either herbivores or carnivores).

The multiplicity of producer–consumer and predator–prey relations in a food web means that there is a high degree of linkage through all parts of an ecosystem. A change in the amount of primary production, or in the population of a particular species, may have positive or negative impacts on numerous other species.

4.7 Soil

- Soil lies at the interface between the lithosphere and biosphere, and is a critical regulator of the hydrologic cycle.
- Soil properties are influenced by climate, parent material, topography, and biological activity over time.

Soil is a dynamic, porous layer of mineral and organic matter that is vital in supporting Earth's biosphere. Soil properties are attributable to five major factors:

1. Parent material is the mineral matter from which soil is formed.

Weathering breaks rock down into smaller particles and new chemical forms. The parent material from which soil is formed is important because it influences soil's chemical and physical characteristics, especially in young soils.

2. Climate regulates both water movements and biological activity.

Water plays a central role in rock weathering and soil formation. In a very humid climate, much water passes through the soil and leaches out soluble minerals on its way. Because of this leaching, soils in humid climates generally have lower amounts of soluble minerals such as sodium and calcium compared to soils in dry climates. However, in semiarid areas, water enters the soil and picks up soluble minerals, which are drawn toward the surface as water is evapotranspired; soils of semiarid and arid climates often have a layer rich in relatively soluble minerals near the surface.

3. Biological activity among plants and animals moves minerals and adds organic matter to the soil.

Plants produce organic matter that accumulates on the soil surface, and animals redistribute this organic matter through the soil. Plants and animals also play a role in weathering processes.

4. Topography affects water movement and erosion rates.

Topography also affects the amount of water present in the soil, largely through controlling drainage and erosion. Steeply sloping areas generally have better drainage than flat or low-lying areas, and they are often more eroded.

5. All these factors work over time, typically requiring many thousands of years to create a mature soil.

Soil is a dynamic, porous layer of mineral and organic matter at Earth's surface. Soil formation is a slow process that takes place very gradually over thousands of years. Soils that have only been forming for a few hundred or even a few thousand years have very different characteristics from those that have been modified by chemical and biological processes for tens of thousands of years.

▼ 4.7.2 **WORLD SOIL MAP**

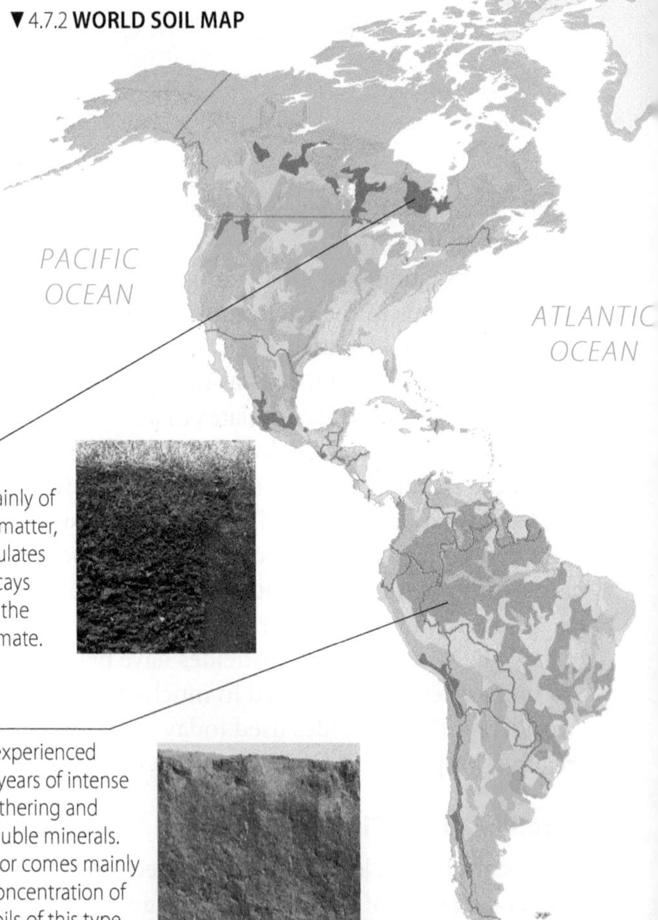

HISTOSOL

This soil is composed mainly of dead organic matter, which accumulates because it decays very slowly in the cold Arctic climate.

OXISOL

This soil has experienced thousands of years of intense chemical weathering and removal of soluble minerals. Its reddish color comes mainly from a high concentration of iron oxides. Soils of this type are often low in nutrients as a result of leaching of soluble minerals by water.

Soil properties vary in layers called soil **horizons** (Figure 4.7.1). Soil horizons are formed through the vertical movement of water, minerals, and organic matter in the soil and also by variations in biological and chemical activity at different depths. The characteristics of soil horizons vary greatly from one soil type to another.

Not all soils contain the typical A, B, and C horizons, but many do. The presence of certain horizons and the characteristics of those horizons are key to identifying distinct soil types. Five examples of soil profiles are shown on this page; these represent just a portion of the great variability in soil characteristics that is seen around the world (Figure 4.7.2).

◄ 4.7.1 **A GENERALIZED SOIL PROFILE**

O HORIZON

Litter—leaves, twigs, dead insects, and other organic matter—accumulates to form a horizon at the surface known as the O (organic) horizon.

A HORIZON

As litter decays, insects, worms, and bacteria consume it and carry it underground, where it helps form the A horizon. Waste from these burrowing animals as well as their dead bodies add more organic matter to the A horizon. In many soils, the A horizon contains much of the nutrients that support plant life. Water may erode materials from the soil surface

B HORIZON

Organisms and water move materials between the A and B horizons. Clay minerals formed from chemical weathering often accumulate in the B horizon, and in dry regions soluble minerals such as calcium accumulate in the B horizon.

C HORIZON

The C horizon contains weathered parent materials that have not been altered as completely by soil-forming processes as materials above it have.

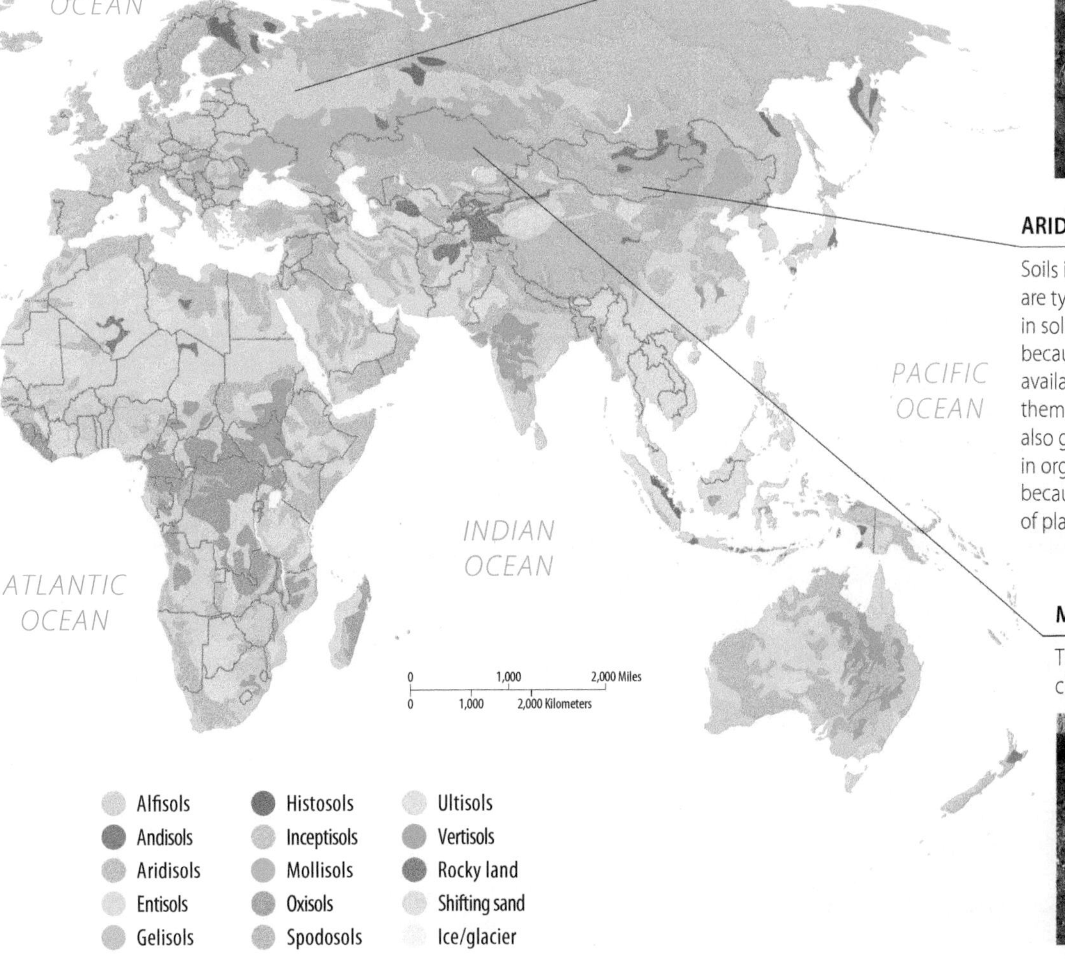

ALFISOL

This soil has a brownish color reflecting moderate organic matter content. It formed under a forest cover and has moderately high fertility.

ARIDISOL

Soils in arid climates are typically very rich in soluble minerals because water is not available to remove them. They are also generally low in organic matter because of low rates of plant growth.

MOLLISOL

This rich, black soil formed in a semarid climate with grassland vegetation. It is high in organic matter and nutrients.

4.8 Diversity of Biomes

- **Photosynthesis rates are greatest in areas that have ample moisture, warm temperatures, and abundant nutrients.**
- **The diversity of life forms is a response to diversity of habitats.**

Photosynthesis requires sunlight, water, nutrients, and a suitable substrate for plant growth. Different climates have different amounts of sunlight and water available, while at the global scale nutrient and substrate conditions are mainly determined by geology and landforms. Land areas with ample sunlight and moisture support the highest rates of photosynthesis (Figure 4.8.1). We also see very high rates of productivity in most wetland and shallow coastal areas. Lower rates of photosynthesis occur in dry climates, and midlatitude climates, which experience limited sunlight for part of the year. Large parts of the oceans have low rates of productivity compared to the land because of limited nutrient availability.

FOOD PRODUCTION AND COMPETITION

Within any particular ecosystem, living things compete for resources such as nutrients and water. The more successful plants and animals in this competition will dominate that environment. In the case of plants, this competition is for light, water, nutrients, and space. Although plants require all these factors to grow, in any ecosystem one factor usually is restricted, which forces competition and adaptation. For example, in an arid environment, plants compete for scant water but do not need to compete for the abundant sunlight. In a humid environment, water is abundant, but plants compete for sunlight. An area with adequate water and light may have poor soils, so plants must compete for nutrients. Through evolution, the plants that have adapted their life forms,

► 4.8.1 **MAP OF GLOBAL PLANT PRODUCTIVITY**
This image is a composite of nearly two years of satellite imagery that uses the color of light reflected from the Earth, to provide information on plant growth. Green tones on land and lighter blues in ocean areas are indicative of plant growth.

physiological characteristics, and reproductive mechanisms that allow them to succeed in particular environments have survived. The plants that are best suited to compete for the resources that are most limiting in a given environment tend to dominate the ground cover there.

BIOMES

Earth's ecosystems are grouped into **biomes** characterized by particular plant and animal types, usually named for a region's climate or dominant vegetation type (Figure 4.8.2). Biomes typically contain many ecosystems. A terrestrial biome has two especially visible features: climate and vegetation. Underlying a biome's label is a diverse community of characteristic plants and animals.

The world vegetation map closely mirrors the world climate map. Ecologically diverse and complex forests occupy humid environments, storing most nutrients in their biomass. In arid and semiarid regions, sparse vegetation is adapted to moisture stress. Forests adapted to winter cold are found in humid midlatitude climates, developing as broadleaf forests in warmer areas and coniferous forests in subarctic latitudes. In high-latitude climates, cold-tolerant short vegetation occupies areas that have a mild summer season. Vegetation is absent in icebound polar climates. The map on this page shows the vegetation types that would occur naturally; the actual distribution of vegetation is somewhat different than what is shown because of human disturbance, especially deforestation and agriculture.

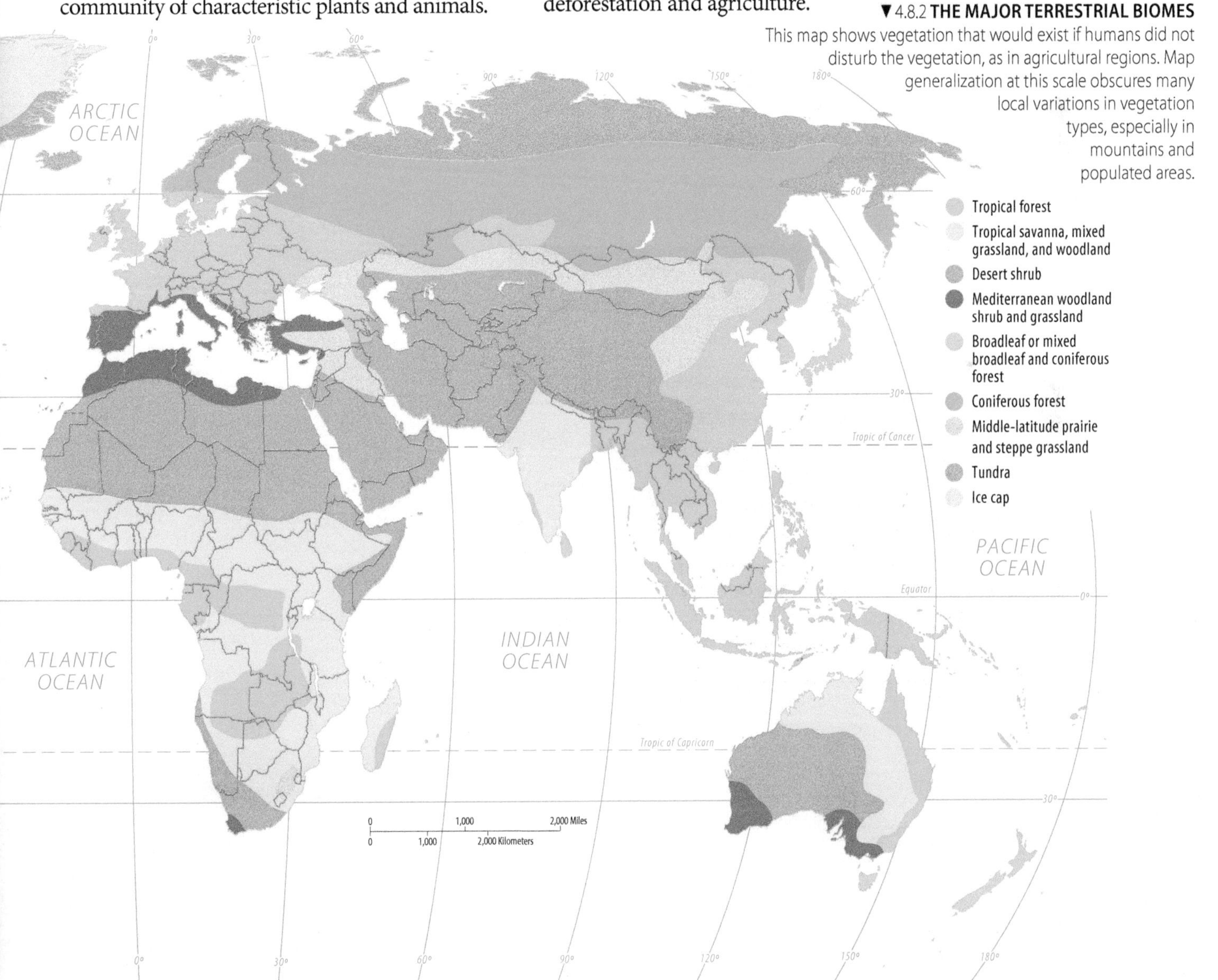

▼4.8.2 **THE MAJOR TERRESTRIAL BIOMES**
This map shows vegetation that would exist if humans did not disturb the vegetation, as in agricultural regions. Map generalization at this scale obscures many local variations in vegetation types, especially in mountains and populated areas.

4.9 Major Biomes

- Forest and woodland biomes occur in areas of moisture surplus.
- Grasslands, deserts, and tundra are found in areas that have significant limitations on plant growth.

Earth contains eight major biomes (Figure 4.9.1).

▼ 4.9.1 **MAJOR BIOMES**

- Tropical forest
- Tropical savanna, mixed grassland, and woodland
- Desert and desert shrub
- Mediterranean woodland shrub and grassland
- Broadleaf or mixed broadleaf and coniferous forest
- Coniferous forest
- Mid-latitude prairie and steppe grassland
- Tundra
- Ice cap

ATLANTIC OCEAN

PACIFIC OCEAN

TUNDRA

Tundra vegetation is dominated by low, tender-stemmed plants and low, woody shrubs (Figure 4.9.2). These survive the cold by lying dormant below the wind, often buried in snow, growing only in the short, cool summer. Tundra vegetation grows very slowly but also decays very slowly, leading to accumulation of organic-rich material.

▲ 4.9.2 **TUNDRA IN HAREFJORD, GREENLAND**

MIDLATITUDE PRAIRIE AND GRASSLANDS

Grasslands dominate semiarid midlatitude areas with hot summers, cold winters, and moderate rainfall (Figure 4.9.3). Grasses are well suited to this climate because they grow rapidly in the short season when temperature and moisture are favorable (generally spring and early summer). During dry or cold periods, above-ground parts of these plants die back, but the roots become dormant and survive. This also allows grasses to survive fire and grow back rapidly, using available moisture at the expense of trees or shrubs that might invade. Many grassland areas have been converted to agriculture, particularly to produce wheat, corn, soybeans, and other small grains.

▲ 4.9.3 **MIDLATITUDE PRAIRIE IN SASKATCHEWAN, CANADA**

▼ 4.9.4 **TROPICAL FOREST IN COSTA RICA**

FORESTS: TROPICAL

In tropical forests tall, broad-leaved trees retain their leaves all year (Figure 4.9.4). A tropical rain forest has a top layer, or canopy, and two more layers beneath. Each layer has different dominant species and associated animal communities. Tropical rain forests are noted for biodiversity, or the variety of living things found in an area. The tropical diversity of trees is paralleled by a great variety of animals. The complex vertical structure adds a diversity of habitats and species. The wide diversity of life in the tropical rain forest places this biome at the center of controversy over deforestation, species extinctions, and biodiversity.

FORESTS: BROADLEAF DECIDUOUS

The **broadleaf deciduous forest**, in which trees lose their leaves for a portion of the year, exists in environments where seasonally cold conditions limit plant growth (Figure 4.9.5). During the summer growing season, long days and a high solar angle promote rapid growth, so that in a year plants may attain 60 to 75 percent of the growth of plants in the tropics, even though the growing season is only 5 to 7 months. These plants have evolved wide, flat leaves that capture as much sunlight as possible. Broadleaf deciduous forests are less diverse than tropical rain forests.

▲ 4.9.5 **BROADLEAF DECIDUOUS FOREST IN THURINGIA, GERMANY**

▲ 4.9.6 **MEDITERRANEAN WOODLAND IN ANDALUCIA, SPAIN**

MEDITERRANEAN WOODLAND SHRUB AND GRASSLAND

In areas of Mediterranean climate we find a mixed woodland and grassland that consists of relatively small trees (Figure 4.9.6). Fire is common in this biome, caused by both humans and lightning. Many plant species are adapted to frequent fire, and some even require fire to maintain their continued presence in the landscape.

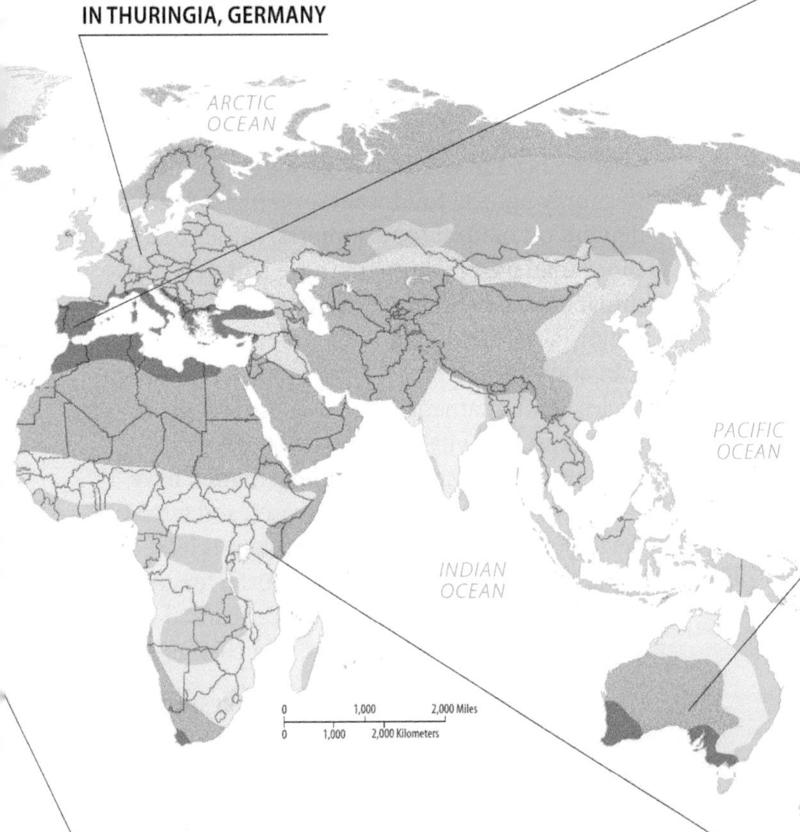

DESERT

In **deserts**, moisture is so scarce that large areas of bare ground exist, and the sparse vegetation is entirely adapted to moisture stress (Figure 4.9.7). Some desert plants are drought-tolerant varieties of types common in more humid areas, such as grasses. Others, such as cacti, are almost exclusive to deserts.

▲ 4.9.7 **DESERT IN SOUTH AUSTRALIA**

TROPICAL SAVANNA AND WOODLAND

Savannas and open woodlands have trees spread widely enough for sunlight to support dense grasses and shrubs beneath them (Figure 4.9.9). This vegetation is common in climates that have a pronounced dry season, where even the trees may lose their leaves, and in some areas dry-season deciduous trees are common.

▼ 4.9.8 **CONIFEROUS FOREST, MAINE**

FORESTS: CONIFEROUS

Coniferous, or **boreal, forest** biome (Figure 4.9.8). During cold winters, low humidity and frozen ground cause moisture stress, but needleleaved trees survive because the leaves have low surface area in relation to their volume, and they are covered with a waxy coating to reduce water loss. Extensive boreal forests are restricted to the Northern Hemisphere, since there is little land in the Southern Hemisphere in latitudes 50° to 70°. Temperate coniferous forests are found in areas of marine west coast climate, where moderate temperatures and ample rainfall allow year-round plant growth. The coniferous forest is much less diverse than the tropical rain forest.

▼ 4.9.9 **TROPICAL SAVANNA IN KENYA, EAST AFRICA**

4.10 Human-dominated Systems

- **Human activity has affected every part of the global biosphere, totally obliterating some biomes.**
- **Even where the major components of biomes remain, species diversity and biogeochemical cycles have been altered.**

Plants, soils, climate, and human activity each reflect the influence of the others. Without humans, climate would have the strongest control on vegetation. Climate controls soil formation and plant growth so strongly that, with the exception of agricultural areas, world vegetation and soil patterns correspond closely to world climates. Humans, however, have had profound influences on ecosystems in most of the world's land areas, including the 37 percent of world land area (excluding Antarctica) that is cropland or permanent pasture.

BIOGEOCHEMICAL CYCLING

In some cases, humans do more processing of biomass and nutrients than nature does.

- We have already seen that humans are playing such a major role in the global carbon cycle that the quantity of CO_2 in the atmosphere is steadily increasing, despite the vast amounts of carbon processed each year by the biosphere.
- Food production through photosynthesis is another critical biogeochemical process in which humans play a major role. About 8 percent of global photosynthesis takes place in agricultural lands.
- Humans play an even larger role in consuming plant matter, either directly as food for themselves and their domestic animals and as fiber for material such as lumber, or indirectly by controlling its characteristics or fate, such as grass grown on golf courses or crop residues left in the field.

Estimates of the amount of biological production that is "appropriated" by humans range from about 3 to 40 percent, depending on what human uses are included in the estimate (Figure 4.10.1). Because this production goes to serve human needs, it is no longer available to supply natural food webs; organisms that depend on natural food supplies, rather than those managed by humans, have suffered. This impact, measured largely as habitat change, is a major factor in loss of biodiversity.

In addition to transforming food webs, nutrient cycles are fundamentally altered. Intensive agricultural activity has meant that in many areas nutrients, especially organic matter, have been depleted in soils. In such cases, soil fertility declines over decades, so the problem is less apparent than if it occurred over just a few years. Variations in yield from year to year caused by weather, insects, plant diseases, and changing

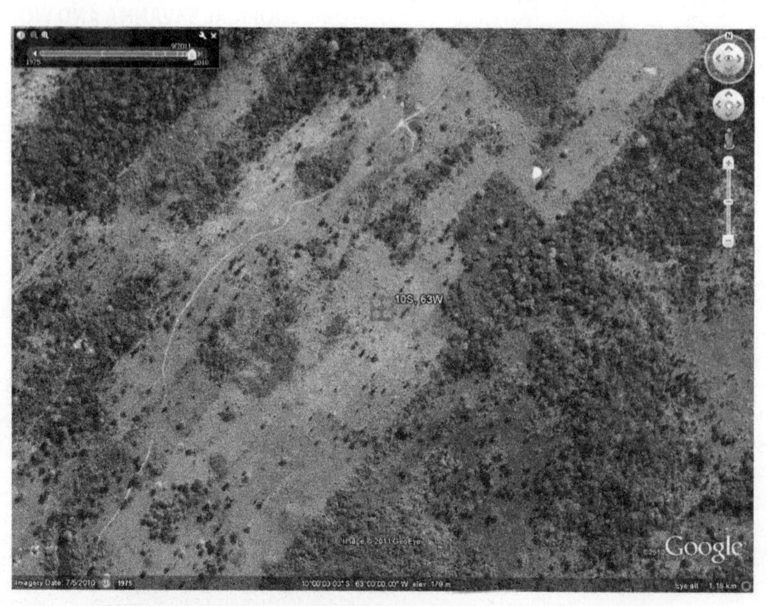

◄ 4.10.1 **AMAZON DEFORESTATION**
Use Google Earth to explore vegetation change in Brazil.

Fly to *10° S, 63° W*, which is in Rondonia, Brazil.

This area has experienced much deforestation in recent decades.

Can you tell what the land is used for?

Compare this with Figure 10.10.2, which shows a similar area.

Zoom out a little, to an eye altitude of 10 km.

Click on *View* and *Historical imagery*. Use the time slider to see the landscape change.

When did most of the deforestation take place?

1935

2005

◀ 4.10.2 **AGRICULTURAL LANDSCAPES REPLACE NATURAL VEGETATION**
In the midlatitudes of North America, much agricultural land was created over the last two centuries. Since the 1930s agricultural land area in the United States has been declining, and former farmland is being replaced by growing forest. At left is an agricultural landscape in Ohio, shown in 1935 and 2005.

technology mask the effects of long-term soil degradation. In some areas reduced cultivation is helping to replace organic matter (Figure 4.10.2), but a long-term decline in soil quality has taken place in many parts of the world.

SOIL AND VEGETATION

Soil has also been transformed by erosion, and accelerated by cultivation. Erosion removes the uppermost part of the soil, which is usually the most fertile part. This removes both nutrients and the ability to store them. In many agricultural regions, topsoil loss has ranged from several centimeters (1 to 2 inches) to tens of centimeters (4 inches to a foot or more). Often, the depth of erosion caused by long-term exposure of the soil surface amounts to a significant part of the entire A horizon.

In semiarid areas, the process can degrade the soil so that only plants adapted to lower soil moisture and nutrient availability, as in deserts, can survive. **Desertification**, a process by which semiarid vegetation and soil become more desert-like as a result of human use, has occurred in semiarid lands around the world (Figure 4.10.3). As a result, many fewer animals can be supported by the available forage. Even though the climate today is not fundamentally altered, the land appears more desert-like than it did before excessive exploitation by humans.

Worldwide, large amounts of organic carbon are stored in soils, but agricultural activities have contributed to the loss of that organic carbon from soils. Some of this carbon has been buried with sediments eroded from farm fields, and some has been broken down and may have contributed to increased atmospheric CO_2. Improved farm management practices that conserve soil also help restore soil organic matter, and are being promoted as a way to offset emissions of carbon to the atmosphere.

▼ 4.10.3 **RANGELAND IN ARIZONA**
The land shown here is well-managed, but in many arid areas overgrazing has damaged soils and reduced the capacity of the land to support grazing animals.

This chapter has examined the biosphere and its role in processing flows of critical substances among Earth's systems.

Key Questions

How do water and nutrients cycle through ecosystems?

- Biogeochemical cycles, including the hydrologic cycle, nutrient cycles, and the carbon cycle connect all Earth systems.
- The hydrologic cycle is principally governed by climate, and regulates many aspects of plant growth as well as water resources.

How is living material created and cycled through ecosystems?

- Plants create food through photosynthesis, and this food is consumed by ecosystems through the opposite reaction of respiration.
- Food energy and nutrients are passed from producers to consumers along food chains. Organisms in ecosystems are linked by food webs.
- Soil is created by physical and biological processes, and plays a central role in biogeochemical cycles.

What is the distribution of life forms and ecological communities?

- Earth's ecological communities are diverse. The world map of biomes corresponds closely to the world climate map.
- Human activity has transformed many ecological communities, and plays a dominant role in biogeochemical cycles.

Thinking Geographically

We often use water without thinking about where it comes from and where it goes after we use it.

1. **Find out where your local drinking water comes from. Is it groundwater or surface water (Figure 4.CR.1)? Where does it go after you use it?**

Emissions of nitrogen oxides caused by fossil fuel combustion in cars and power plants have dramatically increased the availability of nitrogen in ecosystems worldwide.

2. **How do you think this might affect global rates of plant growth, and thus the carbon cycle?**

Global warming is expected to cause the greatest temperature changes in high latitudes, with more modest warming in the tropics.

On the Internet

The International Geosphere-Biosphere Programme is a research organization that studies global change and produces a variety of reports describing recent trends. Visit them at **http://www.igbp.net/**.

The International Union for Conservation of Nature monitors environmental problems worldwide and searches for sustainable solutions to those problems. Their website, **http://www.iucn.org/**, has a wealth of information on biodiversity and global change.

▼ 4.CR.1 **PUEBLO DAM AND RESERVOIR, COLORADO**
Will supply water to Colorado Springs, Pueblo, La Junta, Lamar, and other municipalities in southeastern Colorado.

Interactive Mapping

GLOBAL WARMING

Global warming is expected to have significant effects on the biosphere.

Open: MapMaster Layered World Thematics in

MasteringGEOGRAPHY

Select: *Global Surface Warming Worst Case Projections* from the *Physical Environment* menu.

Where is warming expected to be the greatest?

Then select *Vegetation* from the Physical Environment menu.

What biomes are likely to be most altered by global warming?

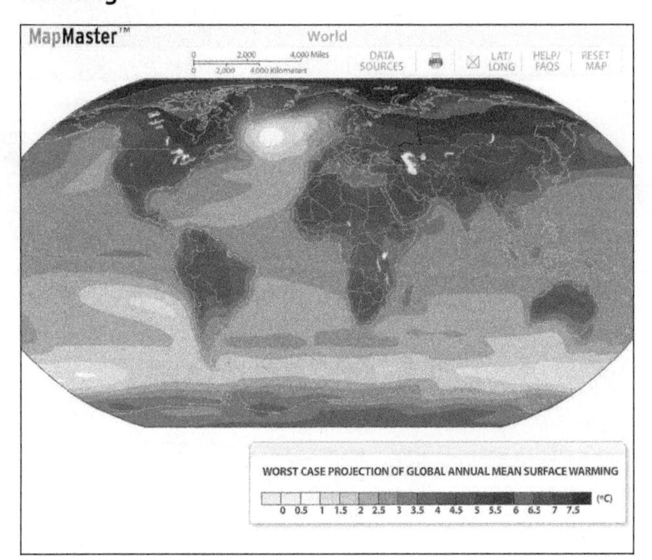

Explore

ARAL SEA

The Aral Sea has changed dramatically in recent decades as a result of upstream use of water for irrigation.

In Google Earth, fly to the Aral Sea.

Click *View*, and select *Historical Imagery*.

Move the slider over the years, from the earliest to the most recent imagery.

What has happened?

Key Terms

Actual evapotranspiration (ACTET)
The amount of water evaporated and/or transpired in a given environment.

Biogeochemical cycle
The environmental recycling process that supplies essential substances such as carbon, nitrogen, and other nutrients to the biosphere.

Biomagnification
The tendency for substances that accumulate in body tissues to increase in concentration as they are passed to higher levels in a food chain.

Biomass
The dry mass of living or formerly living matter in a given environment.

Biome
A large grouping of ecosystems characterized by particular plant and animal types.

Boreal forest
An evergreen needleleaf forest characteristic of cold continental climates.

Broadleaf deciduous forest
A forest with broadleaved trees that lose their leaves in the winter; characteristic of humid midlatitude environments.

Carbon cycle
The movement of carbon among the atmosphere, hydrosphere, biosphere, and lithosphere as a result of processes such as photosynthesis and respiration, sedimentation, weathering, and fossil-fuel combustion.

Carnivore
An animal whose primary food supply is other animals.

Desert
A vegetation type with sparsely distributed plants, specifically adapted for moisture gathering and moisture retention.

Desertification
The process of a region's soil and vegetation cover becoming more desertlike as a result of human land use, usually by overgrazing or cultivation.

Eutrophication
A process in which water bodies receive excess nutrients that stimulate excessive plant growth.

Evapotranspiration
The sum of evaporation and transpiration.

Food chain
The sequential consumption of food in an ecosystem, beginning with green plants, followed by herbivores and carnivores, and ending with decomposers.

Groundwater
The water beneath Earth's surface at a depth where rocks and/or soils are saturated with water.

Herbivore
An animal whose primary food supply is plants.

Horizon
A layer in the soil with distinctive characteristics derived from soil-forming processes.

Hydrologic cycle
The movement of water from the atmosphere to Earth's surface, across that surface, and back to the atmosphere.

Omnivore
An animal that feeds on both plants and other animals.

Photosynthesis
A chemical reaction that occurs in green plants in which carbon dioxide and water are converted to carbohydrates and oxygen.

Potential evapotranspiration (POTET)
The amount of evapotranspiration that would occur if water were available.

Respiration
A chemical reaction that occurs in plants and animals in which carbohydrates and oxygen are combined, releasing water, carbon dioxide, and heat.

Soil
A dynamic, porous layer of mineral and organic matter at Earth's surface.

Transpiration
The use of water by plants, normally drawing it from the soil via their roots, evaporating it in their leaves, and releasing it to the atmosphere.

Trophic level
A position in the food chain relative to other organisms, such as producer, herbivore, or carnivore.

Tundra
A low, slow-growing vegetation type found in high-latitude and high-altitude conditions in which snow covers the ground most of the year.

▶ LOOKING AHEAD

As the world's population grows, human dependence on the biosphere will increase. We anticipate a population of about 9 billion by the middle of the twenty-first century. Population distribution and change are discussed in the next chapter.

5 Population

More humans are alive at this time—about 7 billion—than at any point in Earth's long history. Most of these people live in developing countries, and nearly all of the world's population growth is concentrated in developing countries.

Is the world overpopulated? Will it become so in the years ahead? Geographic approaches are well suited to address these fears. Geographers argue that **overpopulation** is not simply a matter of the total number of people on Earth; rather it depends on the relationship between the number of people and the availability of resources.

Overpopulation is a threat where an area's population exceeds the capacity of the environment to support it at an acceptable standard of living. The capacity of Earth as a whole to support human life may be high, but some regions have a favorable balance between people and available resources, whereas others do not. Further, the regions with the most people are not necessarily the same as the regions with an unfavorable balance between population and resources.

Where is the world's population distributed?

NEW PARENTS WATCH THEIR BABIES THROUGH HOSPITAL WINDOW

Why does population growth vary among countries?

SCAN FOR UPDATED POPULATION DATA

How might population change in the future?

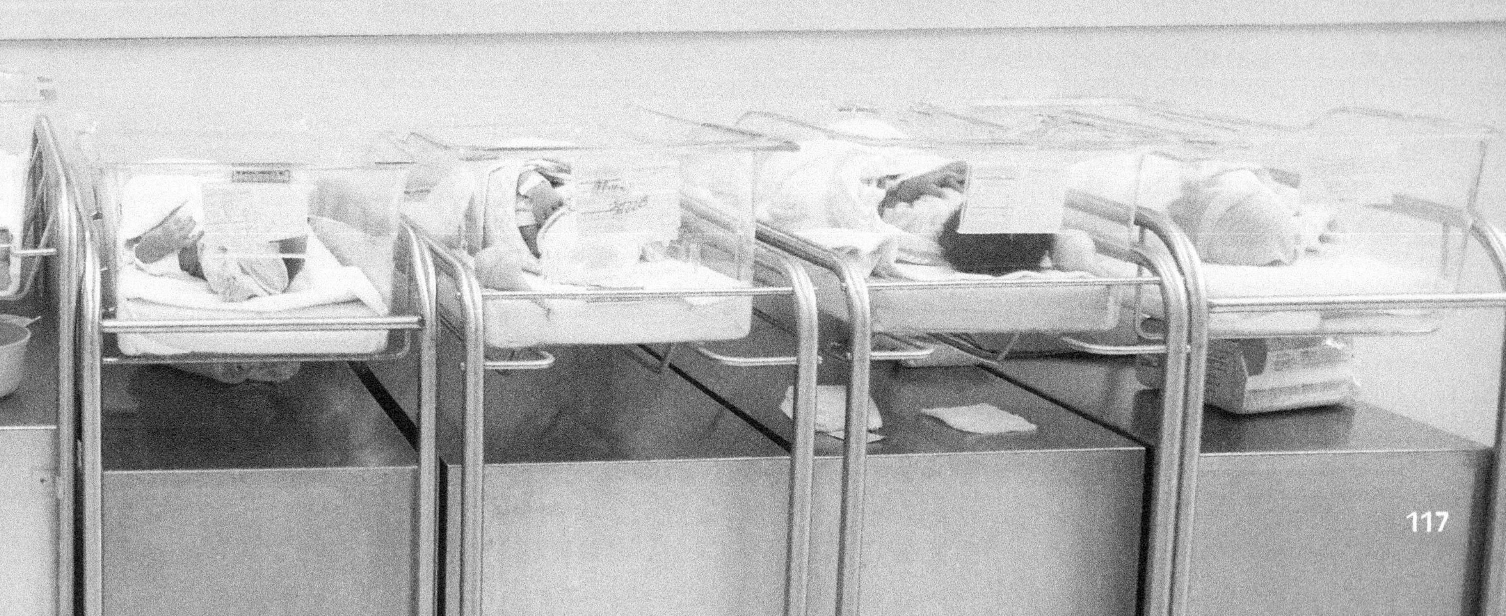

5.1 Population Concentrations

- Two-thirds of the world's inhabitants are clustered in four regions.
- Humans avoid clustering in harsh environments.

Human beings are not distributed uniformly across Earth's surface (Figure 5.1.1). Human beings avoid clustering in certain physical environments, especially those that are too dry, too wet, too cold, or too mountainous for activities such as agriculture (Figure 5.1.2).

The clustering of the world's population can be displayed on a cartogram, which depicts the size of countries according to population rather than land area, as is the case with most maps (Figure 5.1.3). Two-thirds of the world's inhabitants are clustered in four regions—East Asia, South Asia, Southeast Asia, and Europe (Figure 5.1.4).

▲ RORAIMA, BRAZIL

▶ 5.1.2 **SPARSELY POPULATED REGIONS**
Human beings do not live in large numbers in certain physical environments.

5.1.2A COLD LANDS

Much of the land near the North and South poles is perpetually covered with ice or the ground is permanently frozen (permafrost). The polar regions are unsuitable for planting crops, few animals can survive the extreme cold, and few human beings live there.

5.1.2B DRY LANDS

Areas too dry for farming cover approximately 20 percent of Earth's land surface. Deserts generally lack sufficient water to grow crops that could feed a large population, although some people survive there by raising animals, such as camels, that are adapted to the climate. Although dry lands are generally inhospitable to intensive agriculture, they may contain natural resources useful to people—notably, much of the world's oil reserves.

5.1.1C WET LANDS

Lands that receive very high levels of precipitation, such as near Brazil's Amazon River shown in the image, may also be sparsely inhabited. The combination of rain and heat rapidly depletes nutrients from the soil and thus hinders agriculture.

5.1.2D HIGH LANDS

The highest mountains in the world are steep, snow covered, and sparsely settled. However, some high-altitude plateaus and mountain regions are more densely populated, especially at low latitudes (near the equator) where agriculture is possible at high elevations.

▼ 5.1.1 **POPULATION DISTRIBUTION**

Persons per square kilometer
- 1,000 and above
- 250–999
- 25–249
- 5–24
- 1–4
- below 1

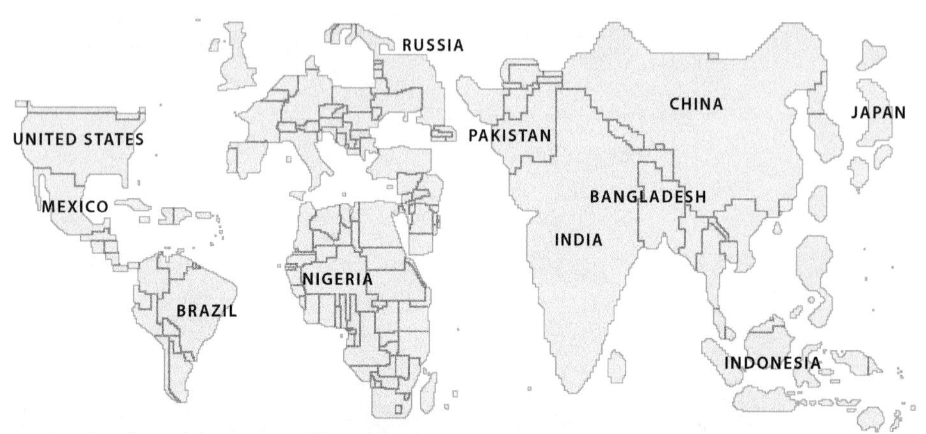

Countries with populations over 100 million are labeled.

▲ 5.1.3 **POPULATION CARTOGRAM**
The population cartogram displays the major population clusters of Europe and East, South, and Southeast Asia as much larger, and Africa and the Western Hemisphere as much smaller, than on a more typical equal-area map, such as the large one in the middle of these two pages.

▼ 5.1.4 FOUR POPULATION CLUSTERS

The four regions display some similarities. Most of the people in these regions live near an ocean or near a river with easy access to an ocean, rather than in the interior of major landmasses. The four population clusters occupy generally low-lying areas, with fertile soil and temperate climate.

5.1.4A **EUROPE**

Europe contains one-ninth of the world's people. The region includes four dozen countries, ranging from Monaco, with 1 square kilometer (0.7 square mile) and a population of 32,000, to Russia, the world's largest country in land area when its Asian land portion is included.

Three-fourths of Europe's inhabitants live in cities. A dense network of road and rail lines links settlements. Europe's highest population concentrations are near the major rivers and coalfields of Germany and Belgium, as well as historic capital cities like London and Paris.

The region's temperate climate permits cultivation of a variety of crops, yet Europeans do not produce enough food for themselves. Instead, they import food and other resources from elsewhere in the world. The search for additional resources was a major incentive for Europeans to explore and colonize other parts of the world during the previous six centuries. Today, Europeans turn many of these resources into manufactured products.

5.1.4B **EAST ASIA**

One-fifth of the world's people live in East Asia. This concentration includes the world's most populous country, the People's Republic of China. The Chinese population is clustered near the Pacific Coast and in several fertile river valleys that extend inland, such as the Huang and the Yangtze. Much of China's interior is sparsely inhabited mountains and deserts. Although China has 25 urban areas with more than 2 million inhabitants and 61 with more than 1 million, more than one-half of the people live in rural areas where they work as farmers.

In Japan and South Korea, population is not distributed uniformly either. Forty percent of the people live in three large metropolitan areas—Tokyo and Osaka in Japan, and Seoul in South Korea—that cover less than 3 percent of the two countries' land area. In sharp contrast to China, more than three-fourths of all Japanese and Koreans live in urban areas and work at industrial or service jobs.

5.1.4C **SOUTHEAST ASIA**

A third important Asian population cluster is in Southeast Asia. A half billion people live in Southeast Asia, mostly on a series of islands that lie between the Indian and Pacific oceans. These islands include Java, Sumatra, Borneo, Papua New Guinea, and the Philippines. The largest concentration is on the island of Java, inhabited by more than 100 million people. Indonesia, which consists of 13,677 islands, including Java, is the world's fourth most populous country.

Several islands that belong to the Philippines contain high population concentrations, and people are also clustered along several river valleys and deltas at the southeastern tip of the Asian mainland, known as Indochina. Like China and South Asia, the Southeast Asia concentration is characterized by a high percentage of people working as farmers in rural areas.

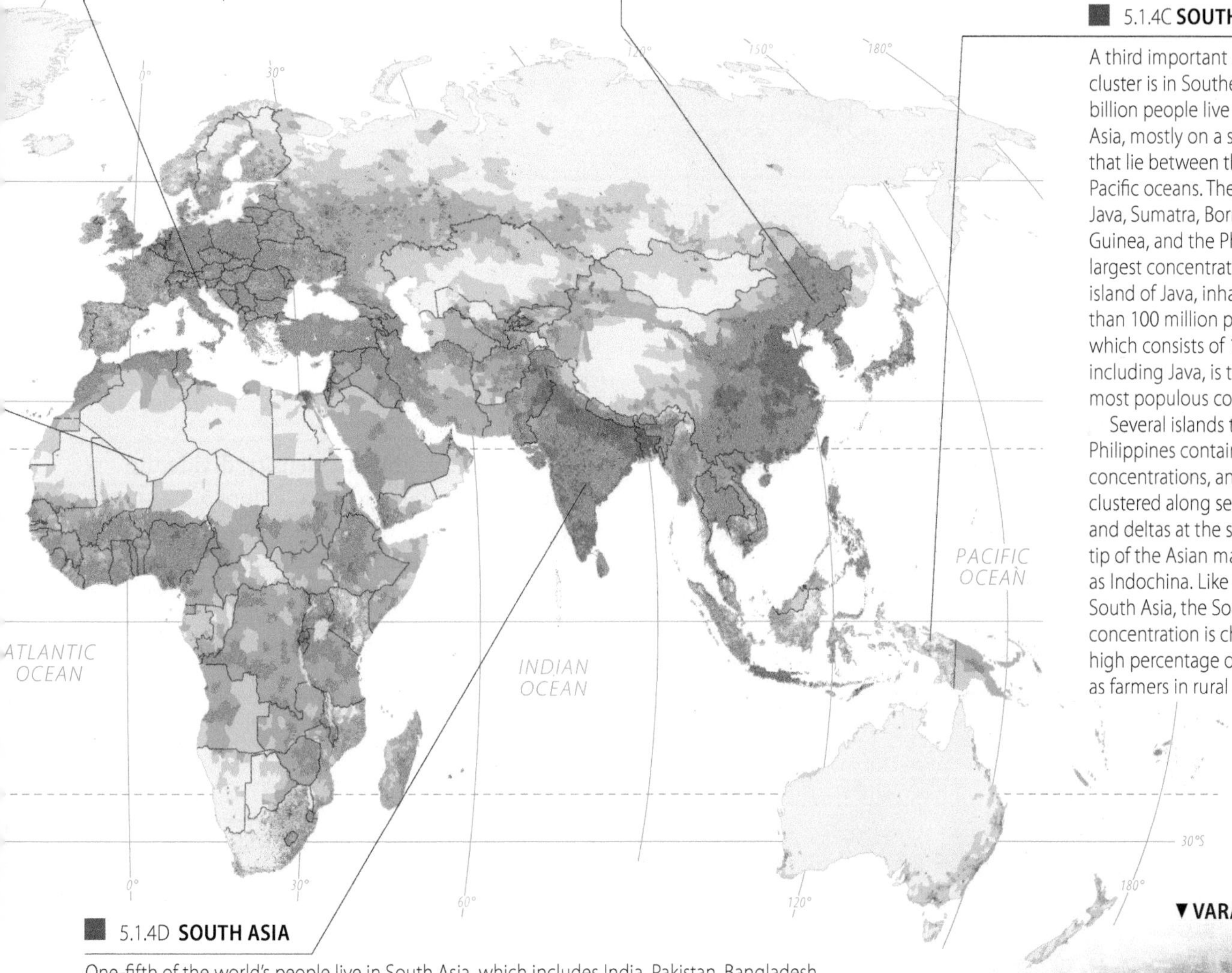

5.1.4D **SOUTH ASIA**

One-fifth of the world's people live in South Asia, which includes India, Pakistan, Bangladesh, and the island of Sri Lanka. The largest concentration of people within South Asia lives along a 1,500-kilometer (900-mile) corridor from Lahore, Pakistan, through India and Bangladesh to the Bay of Bengal. Much of this area's population is concentrated along the plains of the Indus and Ganges rivers. People are also heavily concentrated near India's two long coastlines—the Arabian Sea to the west and the Bay of Bengal to the east.

To an even greater extent than the Chinese, most people in South Asia are farmers living in rural areas. The region contains 18 urban areas with more than 2 million inhabitants and 46 with more than 1 million, but only one-fourth of the total population lives in an urban area.

▼ **VARANASI, INDIA**

5.2 Population Density

- Arithmetic density measures the total number of people living in an area.
- Physiological density and agricultural density show spatial relationships between people and resources.

Density, defined in Chapter 1 as the number of people occupying an area of land, can be computed in several ways, including arithmetic density, physiological density, and agricultural density. These measures of density help geographers to describe the distribution of people in comparison to available resources.

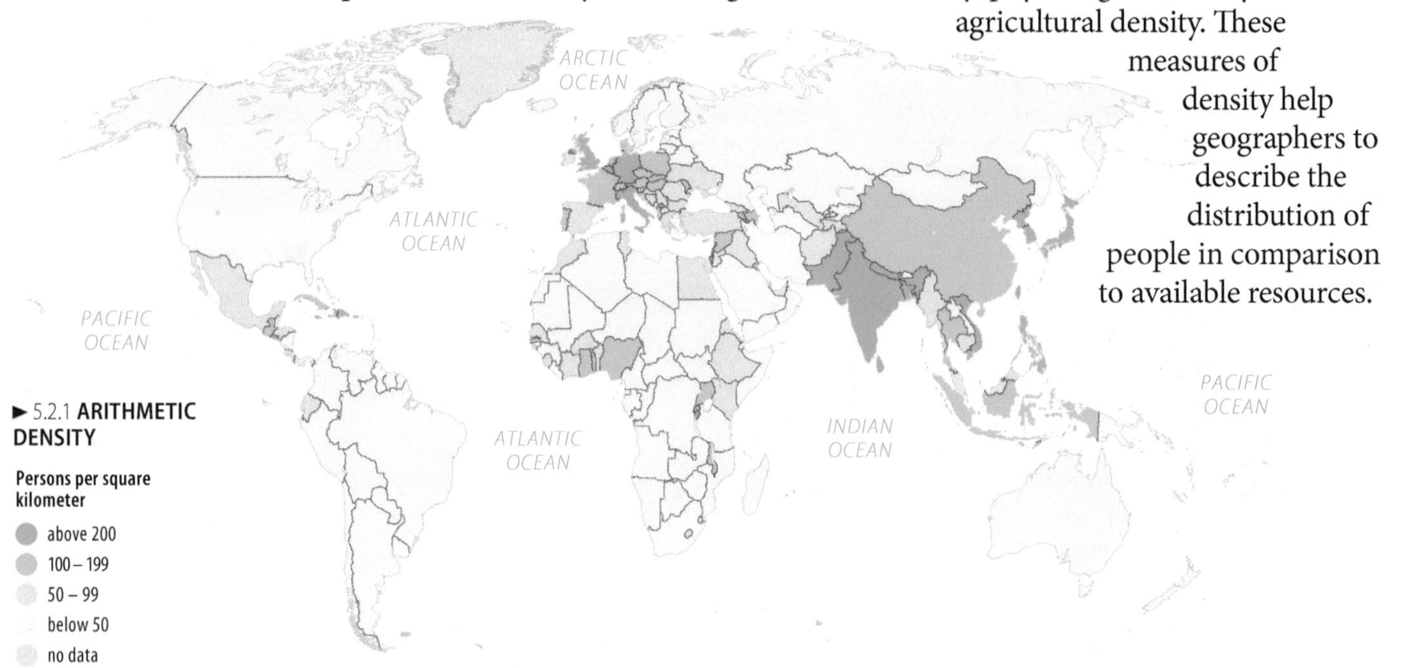

► 5.2.1 **ARITHMETIC DENSITY**

ARITHMETIC DENSITY

Geographers most frequently use **arithmetic density**, which is the total number of people divided by total land area (Figure 5.2.1). Geographers rely on the arithmetic density (also known as *population density*) to compare conditions in different countries because the two pieces of information needed to calculate the measure—total population and total land area—are easy to obtain.

To compute arithmetic density, divide the population by the land area. Figure 5.2.2 shows several examples.

	ARITHMETIC DENSITY (population per square kilometer)	POPULATION 2010 (million people)	LAND AREA (million square kilometers)
Canada	3	34	10.0
United States	32	310	9.6
Netherlands	400	17	0.04
Egypt	80	80	1.0

▲ 5.2.2 **ARITHMETIC DENSITY OF FOUR COUNTRIES**

Compared to the United States, the arithmetic density is much higher in the Netherlands and Egypt and much lower in Canada.

Arithmetic density enables geographers to compare the number of people trying to live on a given piece of land in different regions of the world. Thus, arithmetic density addresses the "where" question. However, to explain why people are not uniformly distributed across Earth's surface, other density measures are more useful. (Figure 5.2.3)

▼ 5.2.3 **HIGH PHYSIOLOGICAL AND AGRICULTURAL DENSITY: EGYPT**
Weekly market at Daraw, Egypt.

PHYSIOLOGICAL DENSITY

A more meaningful population measure is afforded by looking at the number of people per area of **arable land**, which is land suited for agriculture. The number of people supported by a unit area of arable land is called the **physiological density** (Figure 5.2.4). The higher the physiological density, the greater the pressure that people may place on the land to produce enough food.

Physiological density provides insights into the relationship between the size of a population and the availability of resources in a region (Figure 5.2.5). The relatively large physiological densities of Egypt and the Netherlands demonstrate that crops grown on a hectare of land in these two countries must feed far more people than in the United States or Canada, which have much lower physiological densities.

Comparing physiological and arithmetic densities helps geographers to understand the capacity of the land to yield enough food for the needs of the people. In Egypt, for example, the large difference between the physiological density and arithmetic density indicates that most of the country's land is unsuitable for intensive agriculture. In fact, all but 5 percent of Egyptians live in the Nile River valley and delta, because it is the only area in the country that receives enough moisture (by irrigation from the river) to allow intensive cultivation of crops.

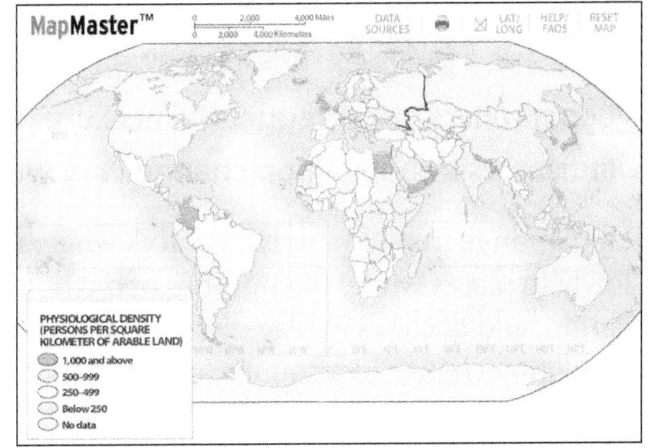

◄ 5.2.4 **PHYSIOLOGICAL DENSITY**

Open MapMaster World in Mastering GEOGRAPHY

Select: *Population* then *Physiological Density.*

What countries other than Egypt and the Netherlands have very high physiological densities?

	PHYSIOLOGICAL DENSITY (population per square kilometer of arable land)	ARABLE LAND (million square kilometers)
Canada	65	0.5
United States	175	1.7
Netherlands	1,748	0.01
Egypt	2,296	0.03

◄ 5.2.5 **PHYSIOLOGICAL DENSITY OF FOUR COUNTRIES**

AGRICULTURAL DENSITY

Two countries can have similar physiological densities, but they may produce significantly different amounts of food because of different economic conditions. **Agricultural density** is the ratio of the number of farmers to the amount of arable land (Figure 5.2.6).

Measuring agricultural density helps account for economic differences. Egypt has a much higher agricultural density than do Canada, the United States, and the Netherlands (Figure 5.2.7). Developed countries have lower agricultural densities because technology and finance allow a few people to farm extensive land areas and feed many people. This frees most of the population in developed countries to work in factories, offices, or shops rather than in the fields.

To understand relationships between population and resources in a country, geographers examine a country's physiological and agricultural densities together. For example, the physiological densities of both Egypt and the Netherlands are high, but the Dutch have a much lower agricultural density than the Egyptians. Geographers conclude that both the Dutch and Egyptians put heavy pressure on the land to produce food, but the more efficient Dutch agricultural system requires fewer farmers than does the Egyptian system.

ARCTIC OCEAN
ATLANTIC OCEAN
PACIFIC OCEAN
ATLANTIC OCEAN
INDIAN OCEAN
PACIFIC OCEAN

▲ 5.2.6 **AGRICULTURAL DENSITY**

Farmers per square kilometer of arable land
- above 100
- 50 – 99
- 25 – 49
- below 25
- no data

	AGRICULTURAL DENSITY (farmers per square kilometer of arable land)	PERCENT FARMERS
Canada	1	2
United States	2	2
Netherlands	23	3
Egypt	251	31

▲ 5.2.7 **AGRICULTURAL DENSITY OF FOUR COUNTRIES**

5.3 Components of Change

- ► Geographers most frequently measure population change through three indicators.
- ► Indicators of population change vary widely among regions.

Population increases rapidly in places where many more people are born than die, increases slowly in places where the number of births exceeds the number of deaths by only a small margin, and declines in places where deaths outnumber births. Geographers measure population change in a country or the world as a whole through three measures—crude birth rate, crude death rate, and natural increase rate.

The population of a place also increases when people move in and decreases when people move out. This element of population change—migration—is discussed in the next chapter.

NATURAL INCREASE RATE

The **natural increase rate (NIR)** is the percentage by which a population grows in a year. The term natural means that a country's growth rate excludes migration. The world NIR during the early twenty-first century has been 1.2, meaning that the population of the world has been growing each year by 1.2 percent.

About 82 million people are being added to the population of the world annually. That number represents a slight decline from the historic high of 87 million in 1989. The world NIR, though, is considerably lower today than its historic peak of 2.2 percent in 1963. The number of people added each year has declined much more slowly than the NIR because the population base is much higher now than in the past. World population reached 1 billion around 1800. The time needed to add each additional billion has declined (Figure 5.3.1).

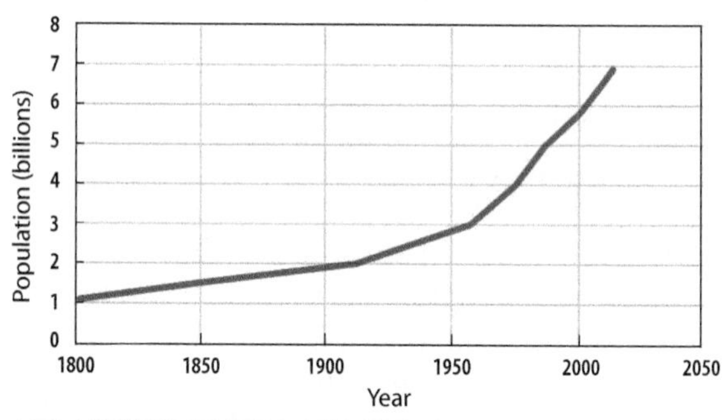

▲ 5.3.1 **WORLD POPULATION GROWTH**

The rate of natural increase affects the **doubling time**, which is the number of years needed to double a population, assuming a constant rate of natural increase. At the early twenty-first-century NIR rate of 1.2 percent per year, world population would double in about 54 years. Should the same NIR continue through the twenty-first century, global population in the year 2100 would reach 24 billion. Should the NIR immediately decline to 1.0, doubling time would stretch out to 70 years, and world population in 2100 would be only 15 billion.

More than 97 percent of the natural increase is clustered in developing countries (Figure 5.3.2). The NIR exceeds 2.0 percent in most countries of sub-Saharan Africa and Southwest Asia & North Africa, whereas it is negative in Europe, meaning that in the absence of immigrants, population actually is declining. About one-third of the world's population growth during the past decade has been in South Asia, one-fourth in sub-Saharan Africa, and the remainder divided about equally among East Asia, Southeast Asia, Latin America, Southwest Asia & North Africa. Regional differences in NIRs show that most of the world's additional people live in the countries that are least able to maintain them.

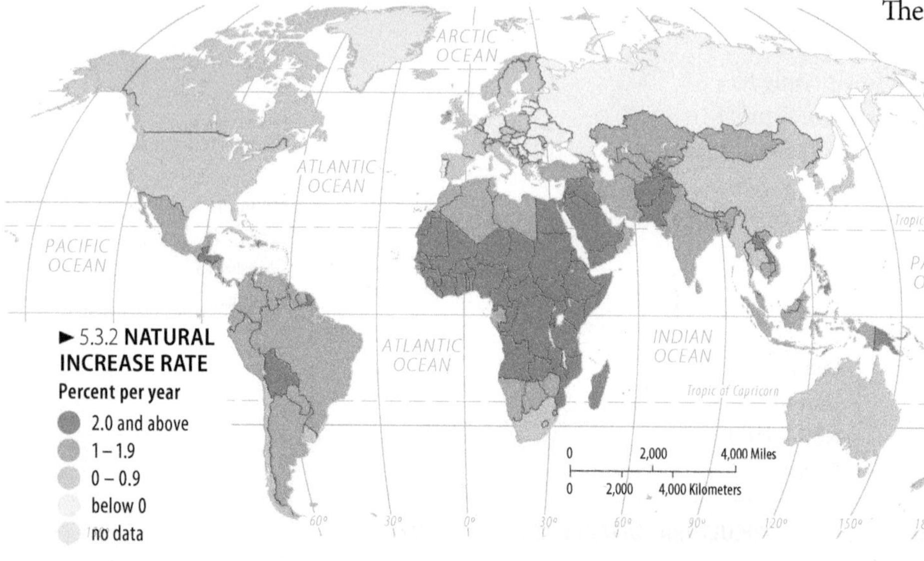

► 5.3.2 **NATURAL INCREASE RATE**

▲ 5.3.3 **HIGH CBR: MALAWI**
Malawi, in sub-Saharan Africa, has one of the world's highest CBRs.

▲ 5.3.4 **CRUDE BIRTH RATE**

Per 1,000 persons
- 40 and above
- 30 – 39
- 20 – 29
- 10 – 19
- below 10
- no data

CRUDE BIRTH RATE

The **crude birth rate (CBR)** is the total number of live births in a year for every 1,000 people alive in the society. A CBR of 20 means that for every 1,000 people in a country, 20 babies are born over a 1-year period.

The world map of CBRs mirrors the distribution of NIRs. As was the case with NIRs, the highest CBRs are in sub-Saharan Africa, and the lowest are in Europe (Figure 5.3.3). Many sub-Saharan African countries have a CBR over 40, whereas many European countries have a CBR below 10 (Figure 5.3.4).

CRUDE DEATH RATE

The **crude death rate** (CDR) is the total number of deaths in a year for every 1,000 people alive in the society. Comparable to the CBR, the CDR is expressed as the annual number of deaths per 1,000 population.

The NIR is computed by subtracting CDR from CBR, after first converting the two measures from numbers per 1,000 to percentages (numbers per 100). Thus if the CBR is 20 and the CDR is 5 (both per 1,000), then the NIR is 15 per 1,000, or 1.5 percent.

The CDR does not display the same regional pattern as the NIR and CBR (Figure 5.3.5). The combined CDR for all developing countries is lower than the combined rate for all developed countries. Furthermore, the variation between the world's highest and lowest CDRs is much less extreme than the variation in CBRs. The highest CDR in the world is 17 per 1,000, and the lowest is 1—a difference of 16—whereas CBRs for individual countries range from 7 per 1,000 to 52, a spread of 45.

Why does Denmark, one of the world's wealthiest countries, have a higher CDR than Cape Verde, one of the poorest? Why does the United States, with its extensive system of hospitals and physicians, have a higher CDR than Mexico and every country in Central America? The answer is that the populations of different countries are at various stages in an important process known as the demographic transition (see section 5.5).

▼ 5.3.5 **CRUDE DEATH RATE**

Per 1,000 persons
- 20 and above
- 15 – 19
- 10 – 14
- 5 – 9
- below 5
- no data

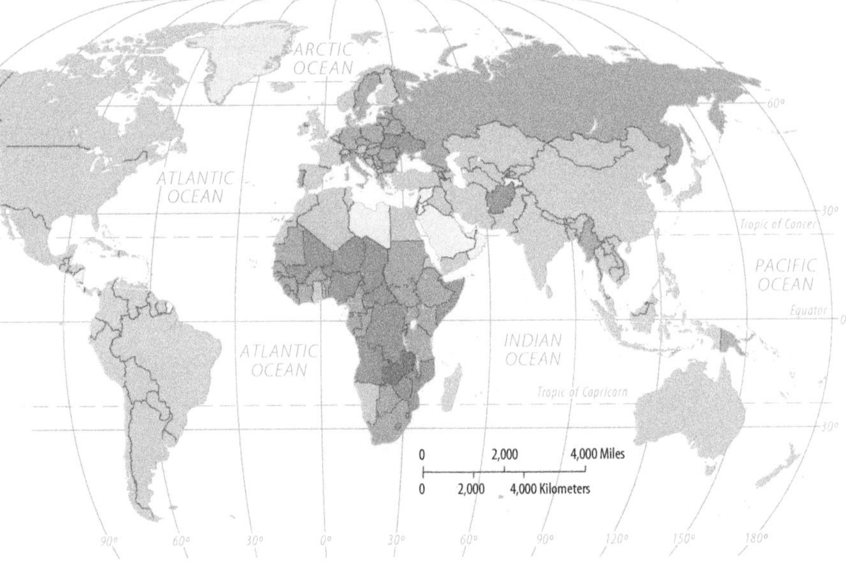

5.4 Population Structure

- Change in a country's population is influenced by rates of fertility and infant mortality.
- Patterns of births and deaths result in distinctive ratios of young and old.

In addition to the CBR discussed in the previous section, total fertility rate also measures the number of births in a country. In addition to the CDR, the infant mortality rate is another measure of a country's deaths. As a result of a combination of births and deaths, a country will display distinctive percentages of young and old people.

TOTAL FERTILITY RATE

The **total fertility rate (TFR)** is the average number of children a woman will have throughout her childbearing years (roughly ages 15 through 49). To compute the TFR, scientists must assume that a woman reaching a particular age in the future will be just as likely to have a child as are women of that age today.

The TFR for the world as a whole is 2.5; it exceeds 5 in many countries of sub-Saharan Africa, compared to 2 or less in nearly all European countries (Figure 5.4.1). The TFR attempts to predict the future behavior of individual women in a world of rapid cultural change, whereas the CBR provides a picture of a society as a whole in a given year.

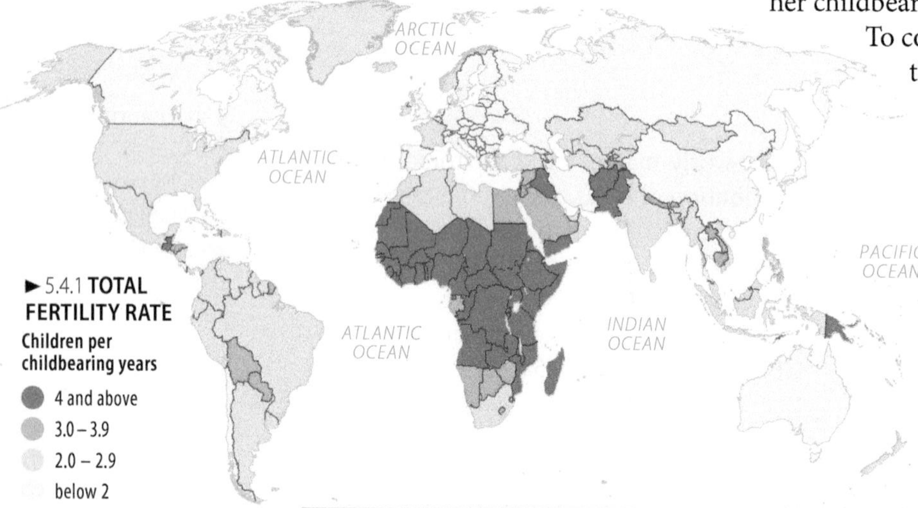

► 5.4.1 **TOTAL FERTILITY RATE**
Children per childbearing years
- 4 and above
- 3.0 – 3.9
- 2.0 – 2.9
- below 2
- no data

INFANT MORTALITY RATE

The **infant mortality rate (IMR)** is the annual number of deaths of infants under 1 year of age for every 1,000 live births. As was the case with the CBR and CDR, the IMR is usually expressed as the number of deaths among infants per 1,000 births rather than as a percentage (per 100).

The highest IMRs are in the poorer countries of sub-Saharan Africa, whereas the lowest rates are in Europe (Figure 5.4.2). IMRs exceed 80 through sub-Saharan Africa, compared to less than 5 in Europe. Otherwise stated, more than 1 in 12 babies die before reaching their first birthday in sub-Saharan Africa, compared to less than 1 in 200 in Europe.

In general, the IMR reflects a country's health-care system. Lower IMRs are found in countries with well-trained doctors and nurses, modern hospitals, and large supplies of medicine.

Although the United States is well endowed with medical facilities, it suffers from a higher IMR than Canada and every country in Europe. African Americans and other minorities in the United States have IMRs that are twice as high as the national average, comparable to levels in Latin America and Asia. Some health experts attribute this to the fact that many poor people in the United States, especially minorities, cannot afford good health care during pregnancy or for their infants.

▼ 5.4.2 **INFANT MORTALITY RATE**
Per 1,000 live births
- 100 and above
- 50 – 99
- 25 – 49
- 10 – 24
- below 10
- no data

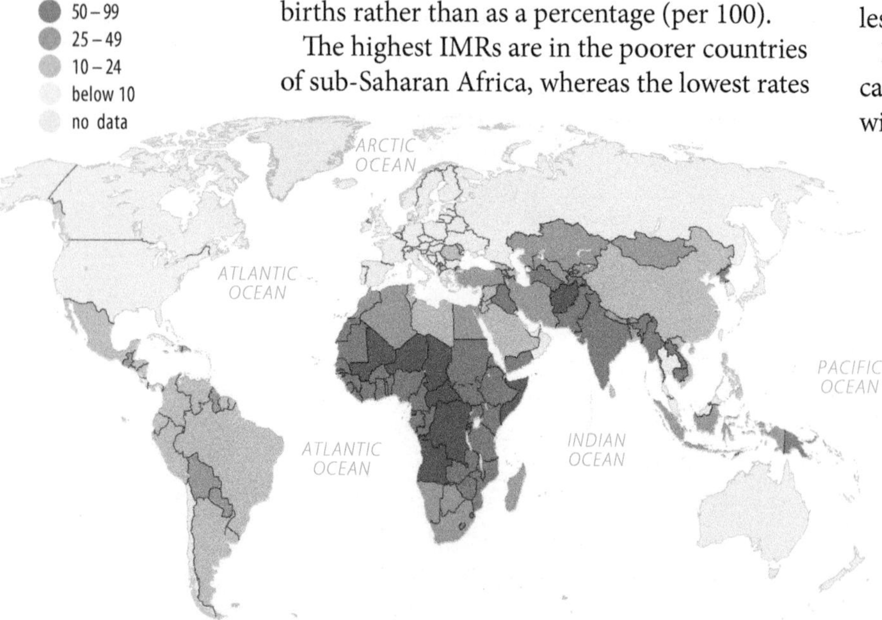

LIFE EXPECTANCY

Life expectancy at birth measures the average number of years a newborn infant can expect to live, assuming current mortality levels. Life expectancy is most favorable in the wealthy countries of Europe and least favorable in the poor countries of sub-Saharan Africa. Babies born today can expect to live into their 80s in much of Europe but only into their 40s in much of sub-Saharan Africa (Figure 5.4.3).

▼ 5.4.3 **LIFE EXPECTANCY AT BIRTH**

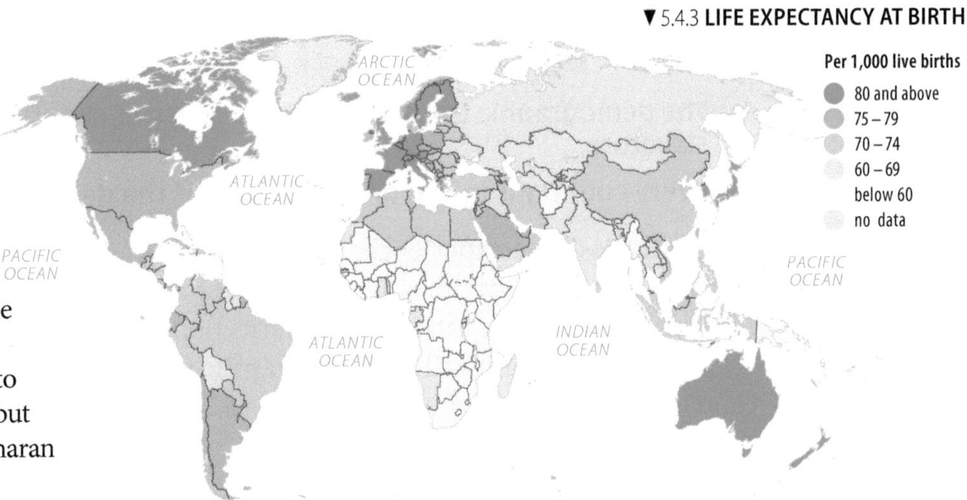

YOUNG AND OLD

One-third of the people in the developing countries are under age 15 compared to only one-sixth in developed countries (Figure 5.4.4). The large percentage of children in developing countries strains their ability to provide needed services such as schools, hospitals, and day-care centers. When children reach the age of leaving school, jobs must be found for them, but the government must continue to allocate scarce resources to meet the needs of the still growing number of young people.

In contrast, developed countries face increasing percentages of older people, who must receive adequate levels of income and medical care after they retire from their jobs. The "graying" of the population places a burden on European and North American governments to meet these needs. More than one-fourth of all government expenditures in the United States, Canada, Japan, and many European countries go to Social Security, health care, and other programs for the older population.

The **dependency ratio** is the number of people who are too young or too old to work, compared to the number of people in their productive years. The larger the percentage of dependents, the greater the financial burden on those who are working to support those who cannot. People who are 0–14 years of age and 65-plus are normally classified as dependents.

A **population pyramid** is a bar graph that displays the percentage of a place's population for each age and gender (Figure 5.4.5). The shape of a country's pyramid is determined primarily by the CBR. A country with a high CBR has a relatively large number of young children, making the base of the pyramid very broad, whereas a country with a relatively large number of older people has a graph with a wider top that looks more like a rectangle than a pyramid. A variety of population pyramids appear on the next page.

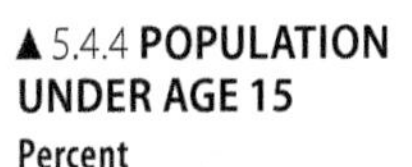
▲ 5.4.4 **POPULATION UNDER AGE 15**

Percent

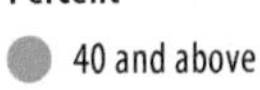

40 and above
30–39
20–29
below 20
no data

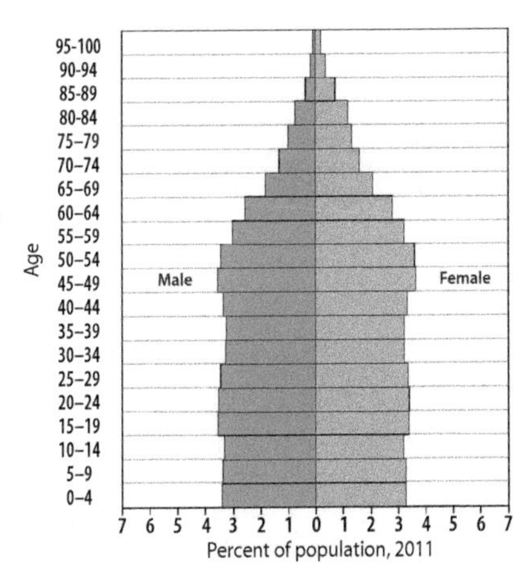

► 5.4.5 **POPULATION PYRAMID OF THE UNITED STATES**

A population pyramid shows the percentage of the total population in 5-year age groups, with the youngest (0 to 4 years old) at the base of the pyramid and the oldest at the top. The length of the bar represents the percentage of the total population contained in that group. By convention, males are usually shown on the left side of the pyramid and females on the right.

5.5 The Demographic Transition

- **The demographic transition is the process of change of a country's population structure.**
- **Every country is in one of four stages of the demographic transition.**

All countries have experienced some changes in natural increase, fertility, and mortality rates, but at different times and at different rates. Although rates vary among countries, a similar process of change in a society's population, known as the demographic transition, is operating. The **demographic transition** is a process with several stages, and every country is in one of them.

FOUR STAGES OF DEMOGRAPHIC TRANSITION

The demographic transition is a process with several stages. Countries move from one stage to the next. At a given moment, we can identify the stage that each country is in.

STAGE 1
- **Very high CBR**
- **Very high CDR**
- **Very low NIR**

The stage for most of human history, because of unpredictable food supply, as well as war and disease.

During most of stage 1, people depended on hunting and gathering for food. A region's population increased when food was easily obtained and declined when it was not. No country remains in stage 1 today.

STAGE 2
- **Still high CBR**
- **Rapidly declining CDR**
- **Very high NIR**

In developed countries 200 years ago, because the Industrial Revolution generated wealth and technology, some of which was used to make communities healthier places to live.

In developing countries 50 years ago, because transfer of penicillin, vaccines, insecticides, and other medicines from developed countries controlled infectious diseases such as malaria and tuberculosis (Figure 5.5.1).

STAGE 3
- **Rapidly declining CBR**
- **Moderately declining CDR**
- **Moderate NIR**

In developed countries 100 years ago. People choosing to have fewer children, in part a delayed reaction to the decline in mortality in stage 2, and in part because a large family is no longer an economic asset when families move from farms to cities.

Some developing countries have moved into stage 3 in recent years, especially where government policies strongly discourage large families.

STAGE 4
- **Very low CBR**
- **Low, slightly increasing CDR**
- **0 or negative NIR**

In some developed countries in recent years. Increased access to birth control methods, as well as increased number of women working in the labor force outside the home, induce families to choose to have fewer children.

As fewer women remain at home as full-time homemakers, they are less likely to be available for full-time care of young children. People who have access to a wider variety of birth-control methods are more likely to use some of them.

▼5.5.1 **STAGE 2: SIERRA LEONE**

A country that has passed through all four stages of the demographic transition has completed a process from little or no natural increase in stage 1, to little or no natural increase in stage 4 (Figure 5.5.2). Two crucial differences:

1. CBR and CDR are high in stage 1 and low in stage 4.
2. Total population is much higher in stage 4 than in stage 1.

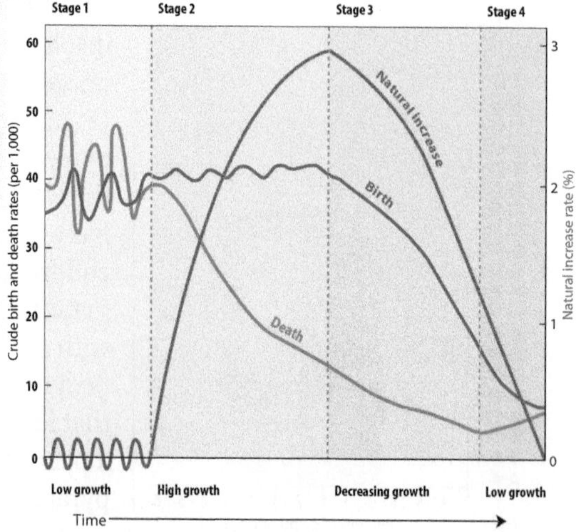

◄5.5.2 **DEMOGRAPHIC TRANSITION**

STAGE 2 (HIGH GROWTH): CAPE VERDE

Cape Verde, a collection of 12 small islands in the Atlantic Ocean off the western coast of Africa, moved from stage 1 to stage 2 about 1950 (Figure 5.5.3). Cape Verde was a colony of Portugal until it became independent in 1975, and the Portuguese administrators left better records of births and deaths than are typical for a colony in stage 1.

Cape Verde's population actually declined during the first half of the twentieth century because of several severe famines, an indication that the country was still in stage 1. Suddenly, in 1950, Cape Verde moved to stage 2. The reason: an anti-malarial campaign launched that year caused the CDR to sharply decline.

Cape Verde's population pyramid shows a large number of females nearing their prime childbearing years. For Cape Verde to enter stage 3, these females must bear considerably fewer children than did their mothers.

STAGE 3 (MODERATE GROWTH): CHILE

Chile's CDR declined sharply in the 1930s, moving the country into stage 2 of the demographic transition. As elsewhere in Latin America, Chile's CDR was lowered by the infusion of medical technology from MDCs such as the United States.

Chile has been in stage 3 of the demographic transition since the 1960s. It moved to stage 3 of the demographic transition primarily because of a vigorous government family-planning policy, initiated in 1966.

Chile's government reversed its policy and renounced support for family planning during the 1970s. Further reduction in the CBR is also hindered by the fact that most Chileans belong to the Roman Catholic Church, which opposes the use of what it calls artificial birth-control techniques. Therefore, the country is unlikely to move into stage 4 of the demographic transition in the near future.

STAGE 4 (LOW GROWTH): DENMARK

Denmark, like most European countries, has reached stage 4 of the demographic transition. The country entered stage 2 of the demographic transition in the nineteenth century, when the CDR began its permanent decline. The CBR then dropped in the late nineteenth century, and the country moved on to stage 3.

Since the 1970s, Denmark has been in stage 4, with roughly equal CBR and CDR. Denmark's CDR has actually increased somewhat in recent years because of the increasing percentage of elderly people. The CDR is unlikely to decline unless another medical revolution, such as a cure for cancer, keeps older elderly people alive much longer.

Denmark's population pyramid shows the impact of the demographic transition. Instead of a classic pyramid shape, Denmark has a column, demonstrating that the percentages of young and elderly people are nearly the same.

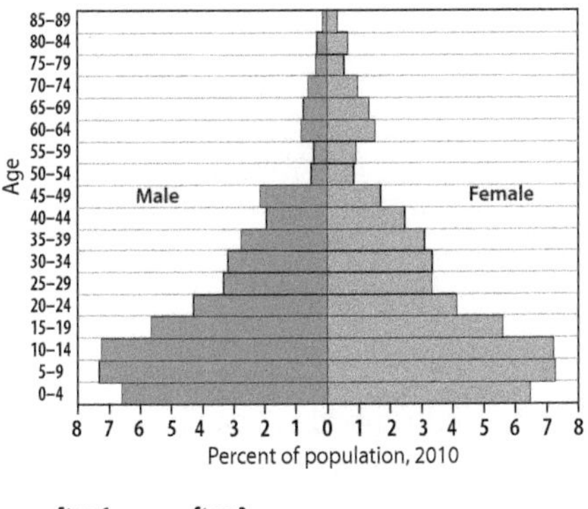

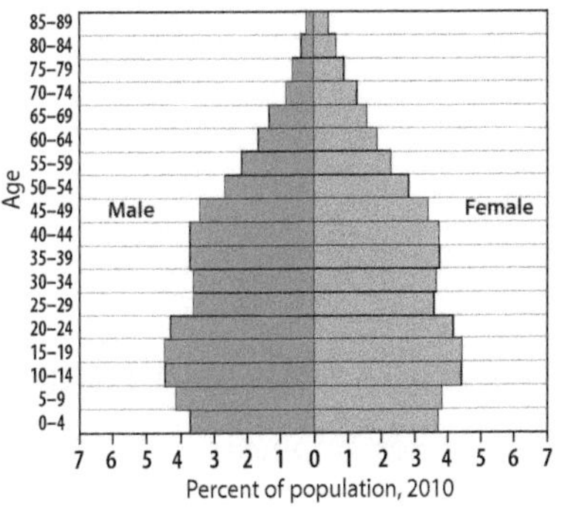

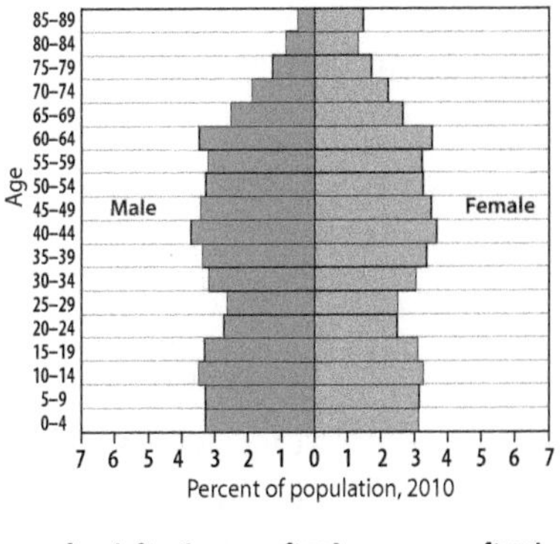

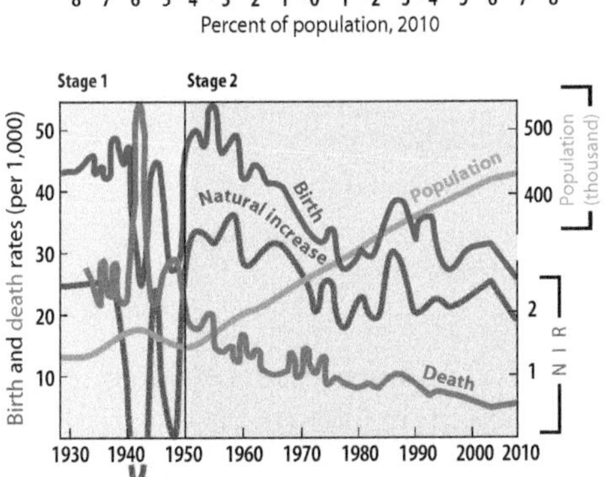

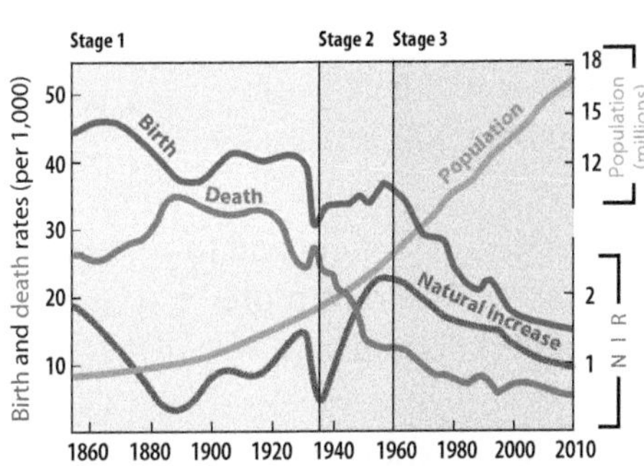

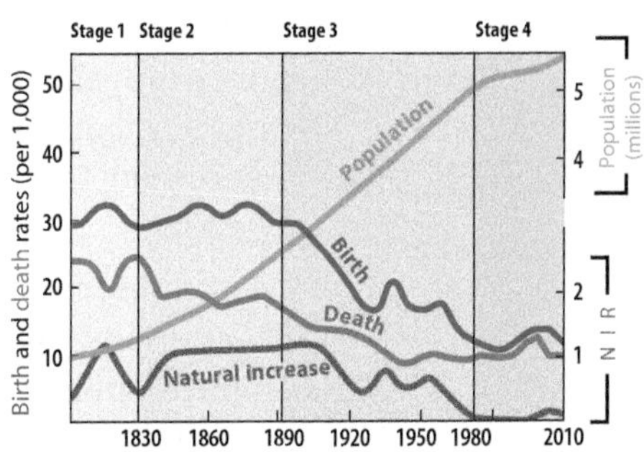

▲ 5.5.3 **POPULATION PYRAMID AND DEMOGRAPHIC TRANSITION FOR CAPE VERDE (left), CHILE (center), DENMARK (right)**

5.6 Declining Birth Rates

- Some developing countries have lowered birth rates through improved education and health care.
- In other developing countries, distribution of contraceptives has reduced birth rates.

Population has been increasing at a much slower rate since the mid-twentieth century. After hitting a peak around 1970, the world NIR has been declining steadily through the late twentieth century and the early twenty-first. The NIR declined beginning in the 1960s in developed countries and beginning in the 1970s in developing ones (Figure 5.6.1).

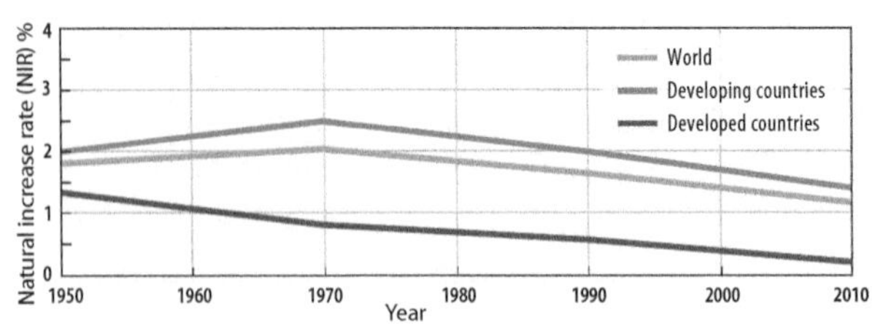

▲ 5.6.1 NIR 1950–2010

In most countries, the decline in the NIR has occurred because of lower birth rates (Figure 5.6.2). Between 1980 and 2010, the CBR declined in every country except for three in Northern Europe—Denmark, Norway, and Sweden—where the CBR increased by only 1.

Two strategies have been successful in reducing birth rates. One alternative emphasizes reliance on education and health care, the other on distribution of contraceptives. Because of varied economic and cultural conditions, the most effective method varies among countries.

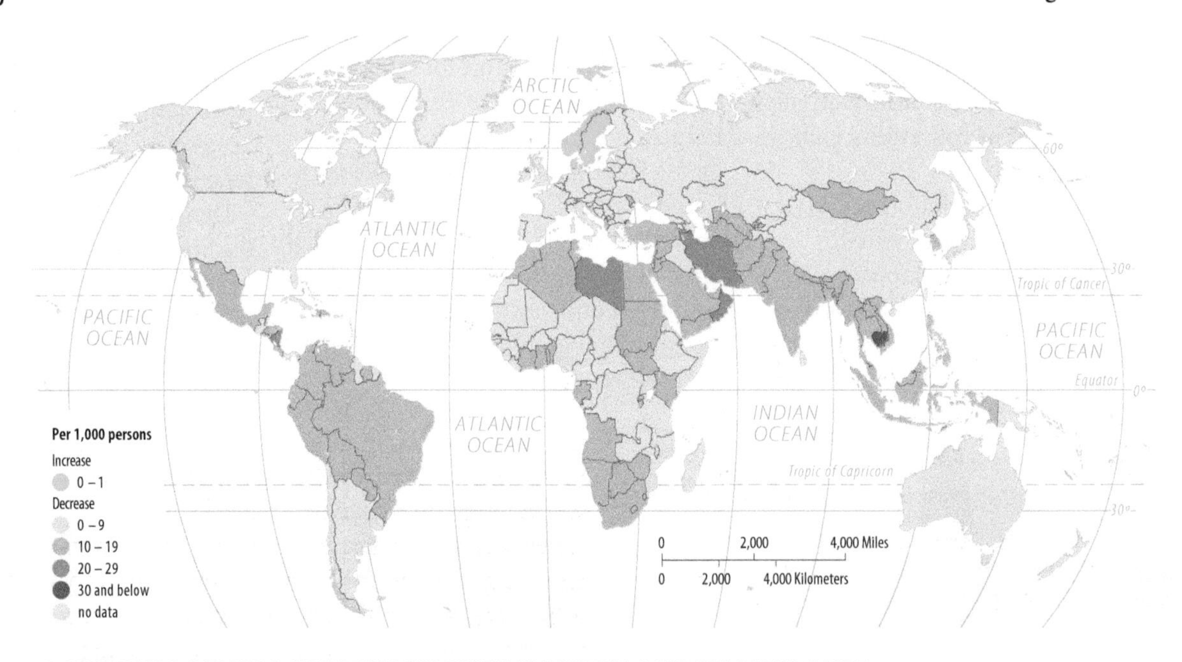

► 5.6.2 CRUDE BIRTH RATE CHANGE 1980–2010

LOWERING BIRTH RATES THROUGH EDUCATION AND HEALTH CARE

One approach to lowering birth rates emphasizes the importance of improving local economic conditions. A wealthier community has more money to spend on education and health-care programs that would promote lower birth rates.

According to this approach:

- With more women able to attend school and to remain in school longer, they would be more likely to learn employment skills and gain more economic control over their lives.
- With better education, women would better understand their reproductive rights, make more informed reproductive choices, and select more effective methods of contraception.
- With improved health-care programs, IMRs would decline through such programs as improved prenatal care, counseling about sexually transmitted diseases, and child immunization.
- With the survival of more infants ensured, women would be more likely to choose to make more effective use of contraceptives to limit the number of children.

LOWERING BIRTH RATES THROUGH CONTRACEPTION

The other approach to lowering birth rates emphasizes the importance of rapid diffusion of modern contraceptive methods. Economic development may promote lower birth rates in the long run, but according to this approach the world cannot wait around for that alternative to take effect. Putting resources into family-planning programs can reduce birth rates much more rapidly.

In developing countries, demand for contraceptive devices is greater than the available supply. Therefore, the most effective way to increase their use is to distribute more of them, cheaply and quickly. According to this approach, contraceptives are the best method for lowering the birth rate.

- Bangladesh is an example of a country that has had little improvement in the wealth and literacy of its people, but 56 percent of the women in the country used contraceptives in 2010 compared to 6 percent two decades earlier. Similar growth in the use of contraceptives has occurred in other developing countries, including Colombia, Morocco, and Thailand.
- The percentage of women using contraceptives is especially low in Africa, so the alternative of distributing contraceptives could have an especially strong impact there (Figure 5.6.3). About one-fourth of African women employ contraceptives, compared to three-fourths in Latin America and two-thirds in Asia (Figures 5.6.4 and 5.6.5). The reason for this is partly economics, religion, and education.
- Very high birth rates in Africa and southwestern Asia also reflect the relatively low status of women there. In societies where women receive less formal education and hold fewer legal rights than do men, having a large family is expected of women, and men regard it as a sign of their own virility.

Regardless of which alternative is more successful, many oppose birth-control programs for religious and political reasons. Adherents of several religions, including Roman Catholics, fundamentalist Protestants, Muslims, and Hindus, have religious convictions that prevent them from using some or all birth-control devices. Opposition is strong within the United States to terminating pregnancy by abortion, and the U.S. government has at times withheld aid to countries and family-planning organizations that advise abortion, even when such advice is only a small part of the overall aid program.

▲ 5.6.3 **WOMEN USING FAMILY PLANNING**

Percent

- 75 and above
- 50–74
- 25–49
- below 25
- no data

▼ 5.6.4 **PROMOTING FEWER CHILDREN**
China's government has errected billboards around the country to encourage families to have fewer children.

▼ 5.6.5 **FAMILY PLANNING METHODS**

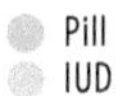

- Pill
- IUD
- Condom
- Female sterilization
- Male sterilization
- Periodic abstinence and withdrawal
- Other
- Not using a method

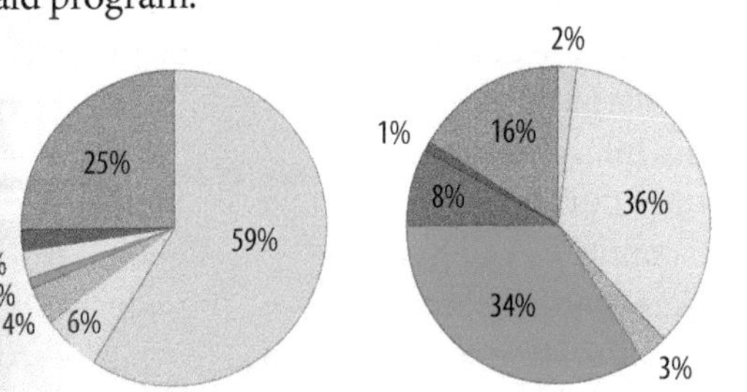

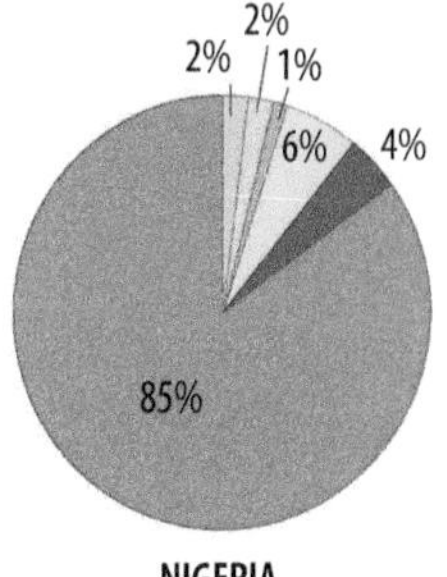

5.7 Population Futures

- World population will still increase but at a slower rate than in the past.
- Some developed countries may move into a possible stage 5 of the demographic transition.

Though NIR is forecast to be much slower in twenty-first century than in the twentieth, world population with continue to grow. Virtually all growth will be in developing countries. The size of the world's population in the twenty-first century depends heavily on what happens in China and India, the two most populous countries.

COMPONENTS OF FUTURE POPULATION GROWTH

Future population depends primarily on fertility. The Population Reference Bureau forecasts world population will be 9.5 billion in 2050 (Figure 5.7.1). The United Nations forecasts that if the current TFR of 2.5 remains unchanged, world population would be even higher in 2050, approximately 12 billion. On the other hand, if TFR declines in the next few years to 1.5, world population would increase to only 8 billion in 2050.

Under all forecasts of total population, the world's future population will definitely have a higher percentage of older persons. The **elderly support ratio** is the number of working-age people (ages 15–64) divided by the number of persons 65 or older (Figure 5.7.2). A small number means that relatively few workers are available to contribute to pensions, health care, and other support that older people need.

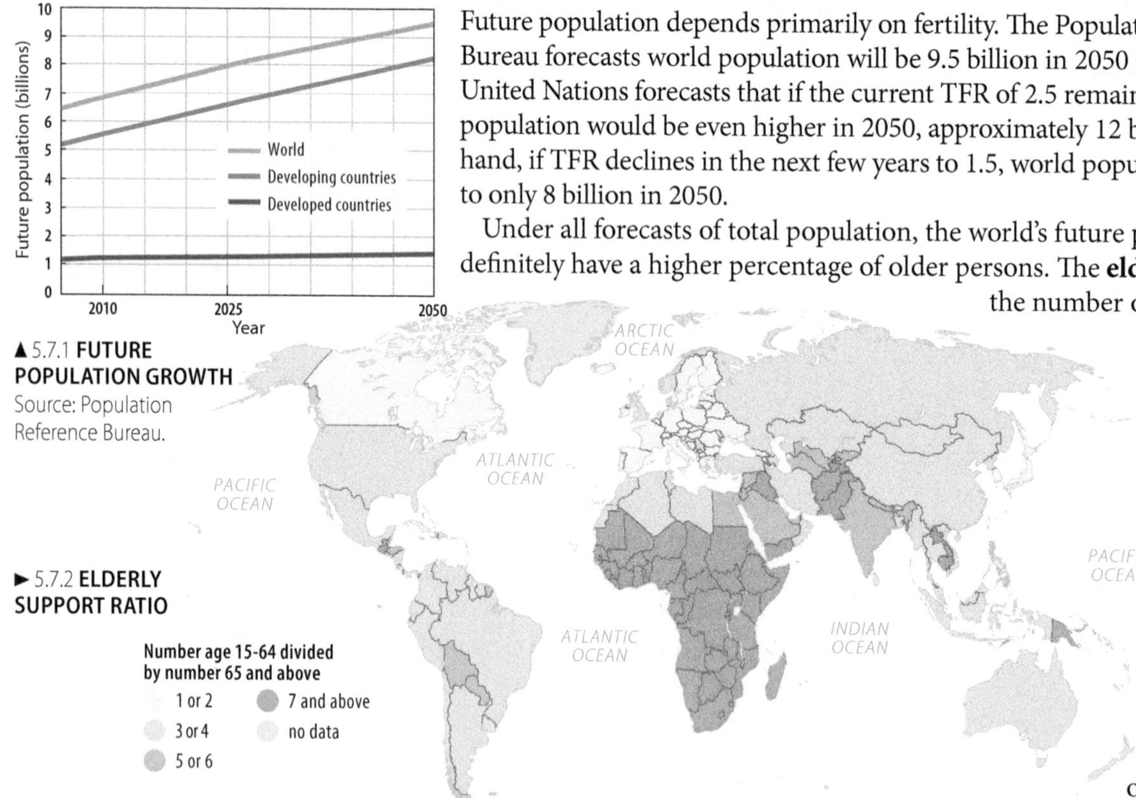

▲ 5.7.1 **FUTURE POPULATION GROWTH**
Source: Population Reference Bureau.

► 5.7.2 **ELDERLY SUPPORT RATIO**

DEMOGRAPHIC TRANSITION POSSIBLE STAGE 5

A possible stage 5 of the demographic transition is predicted by demographers for some developed countries in the twenty-first century. With more elderly people than children, many developed countries will experience population declines during the twenty-first century. Death rates will increase because of high mortality among the relatively large percentage of elderly people.

Meanwhile, after several decades of very low birth rates during the late twentieth century, stage 5 countries will have fewer young women who will eventually bear children. As many of the women within this smaller population choose to have fewer children, birth rates will continue to fall even lower than in stage 4 (Figure 5.7.3).

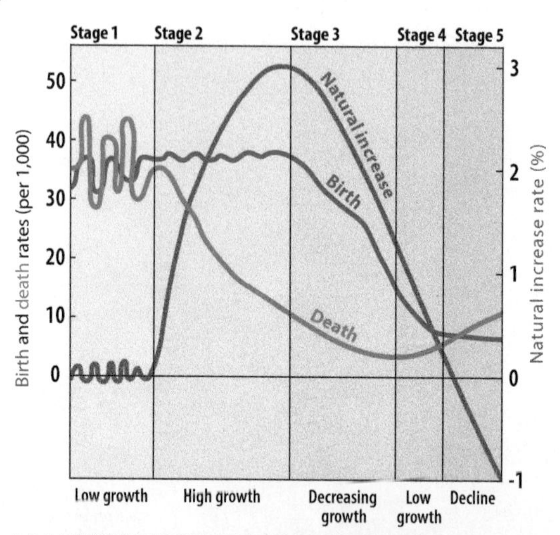

▲ 5.7.3 **POSSIBLE STAGE 5 OF DEMOGRAPHIC TRANSITION**

STAGE 5 (DECLINE): JAPAN

- Very low CBR
- Increasing CDR
- Negative NIR

If the demographic transition is to include a stage 5, Japan is one of the world's first countries to have reached it. Japan's population is expected to decline from an all-time peak of 128 million in 2006 to 119 million in 2025 and 95 million in 2050. With the population decline will come an increasing percentage of elderly people, as reflected in its changing population pyramid (Figure 5.7.4). Japan is forecast to be the first country to have an elderly support ratio of only 1, that is an equal number of workers and retirees.

Japan faces a severe shortage of workers. Rather than increasing immigration, Japan is addressing its labor force shortage primarily by encouraging more Japanese people to work. Programs make it more attractive for older people to continue working, to receive more health-care services at home instead of in hospitals, and to borrow against the value of their homes to pay for health care.

Rather than combine work with child rearing, Japanese women are expected to make a stark choice: either marry and raise children or remain single and work. According to the Japan's most recent census, the majority has chosen to work: More than half of women in the prime childbearing years of 20 to 34 are not married.

INDIA VERSUS CHINA

The world's two most populous countries, China and India, will heavily influence future prospects for global overpopulation. These two countries—together encompassing more than one-third of the world's population—have adopted different family-planning programs.

▲ 5.7.5 **FAMILY PLANNING OFFICE IN KOLKATA (FORMERLY CALCUTTA) INDIA**

INDIA'S POPULATION POLICIES

India was one of the first countries to embark on a national family-planning program, in 1952. Birth-control devices have been distributed for free or at subsidized prices. Abortions, legalized in 1972, have been performed at a rate of several million per year.

India's most controversial family-planning program was the establishment of camps in 1971 to perform sterilizations, surgical procedures by which people were made incapable of reproduction. A sterilized person was paid the equivalent of roughly one month's salary in India. But public opposition grew, because people feared that they would be forcibly sterilized. The government no longer regards birth control as a top policy priority. Government-sponsored family-planning programs have instead emphasized education, including ads on national radio and television networks and information distributed through local health centers (Figure 5.7.5). Given the cultural diversity of the Indian people, the national campaign has had only limited success.

▲ 5.7.6 **POSTER IN SHANGHAI, CHINA, PROMOTES ONE-CHILD POLICY**

CHINA'S POPULATION POLICIES

China has made substantial progress in reducing its rate of growth. The core of the Chinese government's family-planning program has been the One Child Policy, adopted in 1980 (Figure 5.7.6). Couples receive financial subsidies, a long maternity leave, better housing, and (in rural areas) more land if they agree to have just one child. To further discourage births, people receive free contraceptives, abortions, and sterilizations.

As China moves toward a market economy and Chinese families become wealthier, the One Child Policy has been relaxed, especially in urban areas. Clinics provide counseling on a wider range of family-planning options. Instead of fines, Chinese couples wishing a second child pay a "family-planning fee" to cover the cost to the government of supporting the additional person.

Fears that relaxing the One Child Policy would produce a large increase in the birth rate have been unfounded.

▼ 5.7.4 **JAPAN'S POPULATION PYRAMIDS**
1950 (left), 2000 (center), and forecast for 2050 (right).

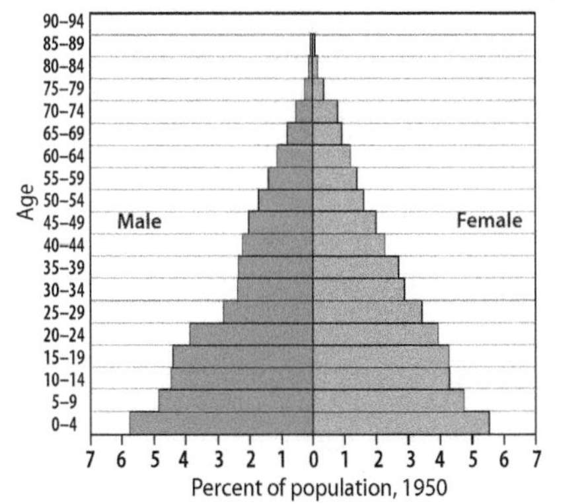

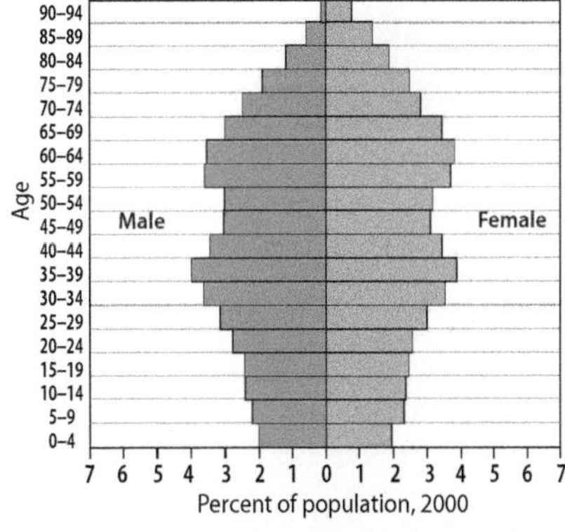

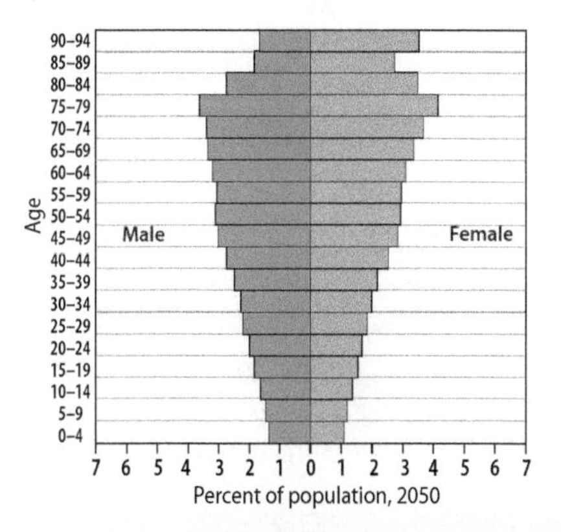

5.8 Malthus's Grim Forecast

- **Malthus predicted that population would increase faster than resources.**
- **Contemporary geographers are divided on the validity of Malthus's thesis.**

English economist Thomas Malthus (1766–1834) was one of the first to argue that the world's rate of population increase was far outrunning the development of food supplies. In *An Essay on the Principle of Population*, published in 1798, Malthus claimed that the population was growing much more rapidly than Earth's food supply because population increased *geometrically*, whereas food supply increased *arithmetically*. Malthus's views remain influential today (Figure 5.8.1).

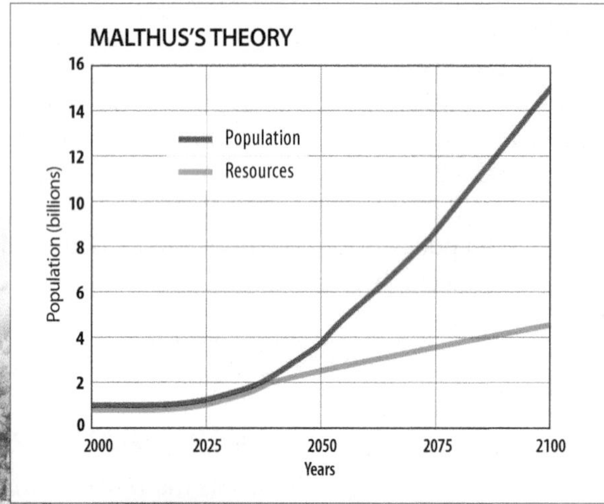

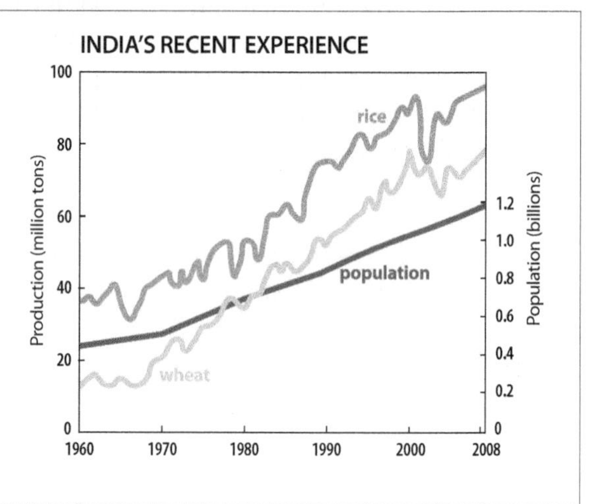

► 5.8.1 **MALTHUS THEORY AND REALITY** (left) Malthus expected population to grow more rapidly than food production. (right) In India, food production has actually increased more rapidly than population.

Supporters of Malthus argue that two characteristics of recent population growth make Malthus's thesis more frightening than when it was first written more than 200 years ago.

1. In Malthus's time only a few relatively wealthy countries had entered stage 2 of the demographic transition. Malthus failed to anticipate that relatively poor countries would have the most rapid population growth because of transfer of medical technology (but not wealth) from developed countries. As a result, the gap between population growth and resources is wider in some countries than even Malthus anticipated.
2. World population growth is outstripping a wide variety of resources, not just food production. According to contemporary supporters of Malthus, wars and civil violence will increase in the coming years because of scarcities of clean water and air, suitable farmland, and fuel, as well as of food.

◄ 5.8.2 **FOOD AVAILABILITY IN INDIA** Market in Kerala.

▲ 5.8.3 **FOOD AVAILABILITY IN INDIA**
Farming potatoes near Bangalore.

MALTHUS'S CRITICS

Many geographers criticize Malthus's theory that population growth depletes resources. To the contrary, a larger population could stimulate economic growth and, therefore, production of more food. Population growth could generate more customers and more ideas for improving technology.

Some theorists maintain that poverty, hunger, and other social welfare problems associated with lack of economic development are a result of unjust social and economic institutions, not population growth. The world possesses sufficient resources to eliminate global hunger and poverty, if only these resources were shared equally.

Some political leaders, especially in Africa, argue that high population growth is good for a country because more people will result in greater power. Population growth is desired in order to increase the supply of young men who could serve in the armed forces. At the same time, developed countries are viewed as pushing for lower population growth as a means of preventing further expansion in the percentage of the world's population living in poorer countries.

MALTHUS'S THEORY AND REALITY

On a global scale, conditions during the past half-century have not supported Malthus's theory. Even though the human population has grown at its most rapid rate ever, world food production has grown at a faster rate than the NIR since 1950, according to geographer Vaclav Smil. Malthus was close to the mark on food production but much too pessimistic on population growth.

Food production increased during the last half of the twentieth century somewhat more rapidly than Malthus would have predicted (Figure 5.8.2). Better growing techniques, higher-yielding seeds, and cultivation of more land all contributed to the expansion in food supply (Figure 5.8.3). Many people in the world cannot afford to buy food or do not have access to sources of food, but these are problems of unequal distribution of wealth rather than an insufficient global production of food, as Malthus theorized.

Following Malthus's model, world population should have quadrupled between 1950 and 2000, from 2.5 billion to 10 billion people, but world population actually grew during this period to only 6 billion. Malthus did not foresee critical cultural, economic, and technological changes that would induce societies sooner or later to move on to stages 3 and 4 of the demographic transition.

5.9 The Epidemiologic Transition

- Each stage of the demographic transition has distinctive causes of death.
- The leading causes of death shift through the demographic transition.

Medical researchers have identified an **epidemiologic transition** that focuses on distinctive causes of death in each stage of the demographic transition. Epidemiologists rely heavily on geographic concepts such as scale and connection, because measures to control and prevent an epidemic derive from understanding its distinctive distribution and method of diffusion. The term epidemiologic transition comes from **epidemiology**, which is the branch of medical science concerned with the incidence, distribution, and control of diseases that affect large numbers of people.

STAGE 1: PESTILENCE AND FAMINE (HIGH CDR)

Stage 1 of the epidemiologic transition was titled the stage of pestilence and famine by epidemiologist Abdel Omran in 1971. Infectious and parasitic diseases were the principal causes of human deaths, along with accidents and attacks by animals and other humans. Malthus called these causes of deaths "natural checks" on the growth of the human population in stage 1 of the demographic transition.

History's most violent stage 1 epidemic was the Black Plague (bubonic plague), which was probably transmitted to humans by fleas attached to migrating infected rats:

- The Black Plague originated among Tatars in present-day Kyrgyzstan.
- It diffused to present-day Ukraine when the Tatar army attacked an Italian trading post on the Black Sea.
- Italians fleeing the Black Sea trading post carried the infected rats on ships west to the major coastal cities of southeastern Europe in 1347.
- The plague diffused from the coast to inland towns and then to rural areas.
- It reached western Europe in 1348 and northern Europe in 1349.

About 25 million Europeans—more than half of the continent's population—died between 1347 and 1350. The Black Plague also diffused east to China, where 13 million died in a single year, 1380.

The plague wiped out entire villages and families, leaving farms with no workers and estates with no heirs. Churches were left without priests and parishioners, schools without teachers and students. Ships drifted aimlessly at sea after entire crews succumbed to the plague.

▼ 5.9.1
EPIDEMIOLOGIC TRANSITION STAGE 2
Cholera, a stage 2 disease, has been a threat in Iraq, such as this location in the Baghdad suburb of Fdailiyah, where drinking water is being drawn from a water pipe that crosses a canal carrying raw sewage.

STAGE 2: RECEDING PANDEMICS (RAPIDLY DECLINING CDR)

Stage 2 of the epidemiologic transition is known as the stage of receding pandemics. A **pandemic** is disease that occurs over a wide geographic area and affects a very high proportion of the population.

In stage 2, improved sanitation, nutrition, and medicine during the Industrial Revolution reduced the spread of infectious diseases (Figure 5.9.1). But death rates did not decline immediately and universally. Poor people crowded into rapidly growing industrial cities had especially high death rates during the Industrial Revolution.

Construction of water and sewer systems were thought to have eradicated cholera by the late nineteenth century. However, cholera reappeared a century later in rapidly growing cities of developing countries as they moved into stage 2 of the demographic transition.

STAGE 3: DEGENERATIVE DISEASES (MODERATELY DECLINING CDR)

Stage 3 is characterized by a decrease in deaths from infectious diseases and an increase in chronic disorders associated with aging. The two especially important chronic disorders in stage 3 are cardiovascular diseases, such as heart attacks, and various forms of cancer.

The decline in infectious diseases such as polio and measles has been rapid in stage 3 countries. Effective vaccines were responsible for these declines (Figure 5.9.2).

STAGE 4: DELAYED DEGENERATIVE DISEASES (LOW BUT INCREASING CDR)

The epidemiologic transition was extended by S. Jay Olshansky and Brian Ault to stage 4, the stage of delayed degenerative diseases. The major degenerative causes of death—cardiovascular diseases and cancers—linger, but the life expectancy of older people is extended through medical advances.

Through medicine, cancers spread more slowly or are removed altogether. Operations such as bypasses repair deficiencies in the cardiovascular system. Also improving health are behavior changes such as better diet, reduced use of tobacco and alcohol, and exercise.

▼ 5.9.2 **EPIDEMIOLOGIC TRANSITION STAGE 3**
Mother holding child receiving vaccination by injection in Zimbabwe.

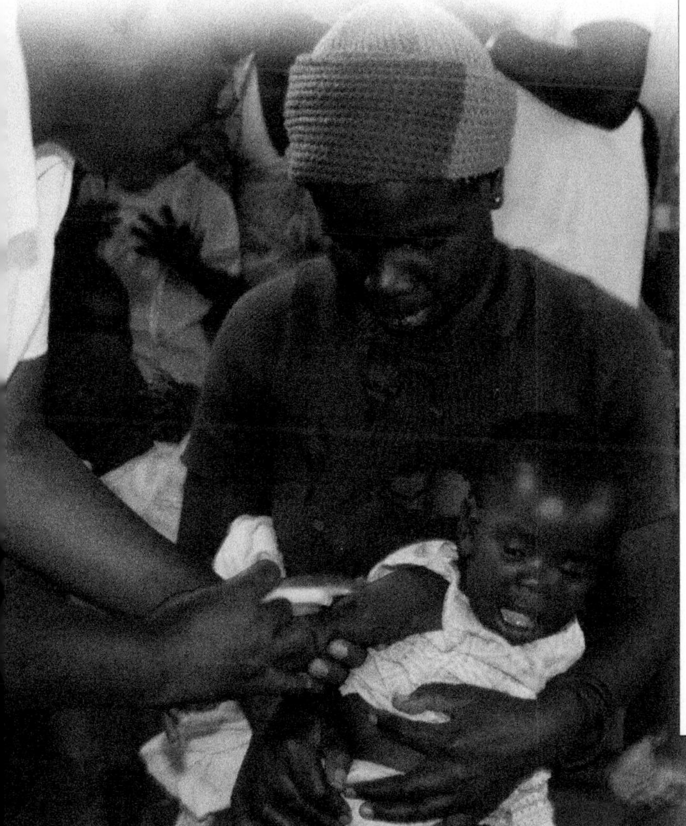

EARLY "GIS" MAPPED CHOLERA DISTRIBUTION

Dr. John Snow (1813–1858) was a British physician not a geographer. To fight one of the worst nineteenth century pandemics, cholera, Snow created a handmade GIS in 1854. On a map of London's Soho neighborhood, Snow overlaid two other maps, one showing the addresses of cholera victims and the other the location of water pumps—for the poor residents of Soho the principal source of water for drinking, cleaning, and cooking (Figure 5.9.3).

The overlay map showed that cholera victims were not distributed uniformly through Soho. Dr. Snow showed that a large percentage of cholera victims were clustered around one pump, on Broad Street (today known as Broadwick Street). Tests at the Broad Street pump subsequently proved that the water there was contaminated. Further investigation revealed that sewage was contaminating the water supply near the pump.

Before Dr. Snow's geographic analysis, many believed that epidemic victims were being punished for sinful behavior and that most victims were poor because poverty was considered a sin. Now we understand that cholera affects the poor because they are more likely to have to use contaminated water.

► 5.9.3 **BIRTH OF GIS**
(top) Dr. John Snow's map of the distribution of cholera in Soho, London, 1854. (bottom) Use Google Earth to see memories of Dr. Snow and the cholera epidemic in modern-day London.

Fly to: *39 Broadwick Street, London, England.*
Drag to *Street view* at 39 Broadwick Street.
Move the compass so that south faces top (north faces bottom).
Move the compass so that east faces top (north faces left).
Click on icons for the *Broad Street pump* and the *Soho Cholera Epidemic.*

1. **What is the current use of the building at 39 Broadwick Street bearing John Snow's name?**
2. **What other evidence of the cholera epidemic can be seen in Broadwick Street?**

5.10 Global Reemergence of Infectious Diseases

- **Some infectious diseases have returned and new ones have emerged.**
- **The most lethal global-scale epidemic has been AIDS.**

Recall that in the possible stage 5 of the demographic transition, CDR rises because more of the population is elderly. Some medical researchers think there is also a stage 5 of the epidemiologic transition, brought about by a reemergence of infectious and parasitic diseases. A consequence of stage 5 would be higher CDRs. Other epidemiologists dismiss recent trends as a temporary setback in a long process of controlling infectious diseases.

In a possible stage 5, infectious diseases thought to have been eradicated or controlled have returned, and new ones have emerged. Three reasons help to explain the possible emergence of a stage 5 of the epidemiologic transition: poverty, evolution, and increased connections.

REASONS FOR POSSIBLE STAGE 5: POVERTY

Infectious diseases are more prevalent in poor areas because:

- Unsanitary conditions may persist.
- Most people can't afford the drugs needed for treatment.

Tuberculosis (TB) is an example of an infectious disease that has been largely controlled in developed countries like the United States but remains a major cause of death in developing countries. An airborne disease often called "consumption," TB spreads principally through coughing and sneezing, damaging lungs.

The death rate from TB declined in the United States from 200 per 100,000 in 1900 to 60 in 1940 and 0.5 today. However, in developing countries, the TB rate is more than ten times higher than in developed countries, and nearly 2 million worldwide die from it annually (Figure 5.10.1).

REASONS FOR POSSIBLE STAGE 5: EVOLUTION

Infectious disease microbes continuously evolve and change in response to environmental pressures by developing resistance to drugs and insecticides. Antibiotics and genetic engineering contribute to the emergence of new strains of viruses and bacteria.

Malaria was nearly eradicated in the mid-twentieth century by spraying DDT in areas infested with the mosquito that carried the parasite. For example, new malaria cases in Sri Lanka fell from 1 million in 1955 to 18 in 1963. The disease returned after 1963, however, and now causes more than 1 million deaths worldwide annually. A major reason was the evolution of DDT-resistant mosquitoes.

▼ 5.10.1
TUBERCULOSIS (TB) DEATHS, 2009

ARCTIC OCEAN
ATLANTIC OCEAN
PACIFIC OCEAN
ATLANTIC OCEAN
INDIAN OCEAN

Tuberculosis death rate per 100,000
- above 50
- 10–49
- 3–9
- below 3
- no data

IMPROVED TRAVEL AND AIDS IN THE UNITED STATES

Airports in New York, California, and Florida are the major ports of entry for visitors arriving in the United States (Figure 5.10.2). Consequently, residents of these states are exposed first to infectious diseases that have reemerged in an age of improved travel.

Not by coincidence, New York, California, and Florida were the nodes of origin for AIDS within the United States during the early 1980s (Figure 5.10.3, left). Though AIDS diffused to every state during the 1980s, these three states, plus Texas (also a major port of entry), accounted for half of the nation's new AIDS cases in the peak year of 1993 (Figure 5.10.3, right). The rapid decline in new cases thereafter resulted from rapid diffusion of preventive methods and medicines such as AZT.

San Francisco 3.5
Las Vegas 1.7
Los Angeles 6.5
Chicago 2.8
Boston 2.1
New York 13.3
Washington 2.2
Atlanta 1.3
Orlando 2.2
Miami 5.3
Honolulu 1.7
Houston 1.5

▲ 5.10.2 **INTERNATIONAL PASSENGER ARRIVALS**
International passenger arrivals to U.S. airports (excluding passengers coming from Canada), 2007 (million passengers)

REASONS FOR POSSIBLE STAGE 5: INCREASED CONNECTIONS

As they travel, people carry diseases with them and are exposed to the diseases of others. Motor vehicles allow rural residents to easily reach urban areas and urban residents to reach rural areas. Airplanes allow residents of one country to easily reach another.

The most lethal epidemic in recent years has been AIDS (acquired immunodeficiency syndrome). Worldwide, 25 million people died of AIDS as of 2007. Another 33 million were living with HIV (human immunodeficiency virus, the cause of AIDS).

The impact of AIDS has been felt most strongly in sub-Saharan Africa. With one-tenth of the world's population, sub-Saharan Africa had two-thirds of the world's total HIV-positive population and nine-tenths of the world's infected children (Figure 5.10.4).

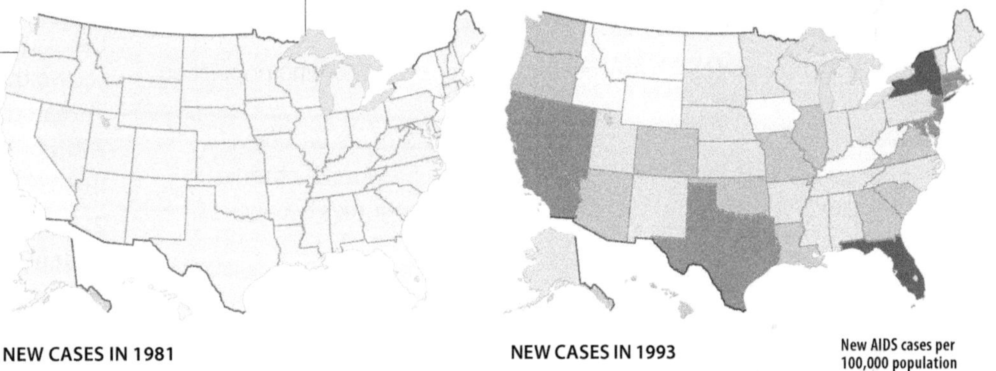

▲ 5.10.3 **DIFFUSION OF AIDS IN THE UNITED STATES**

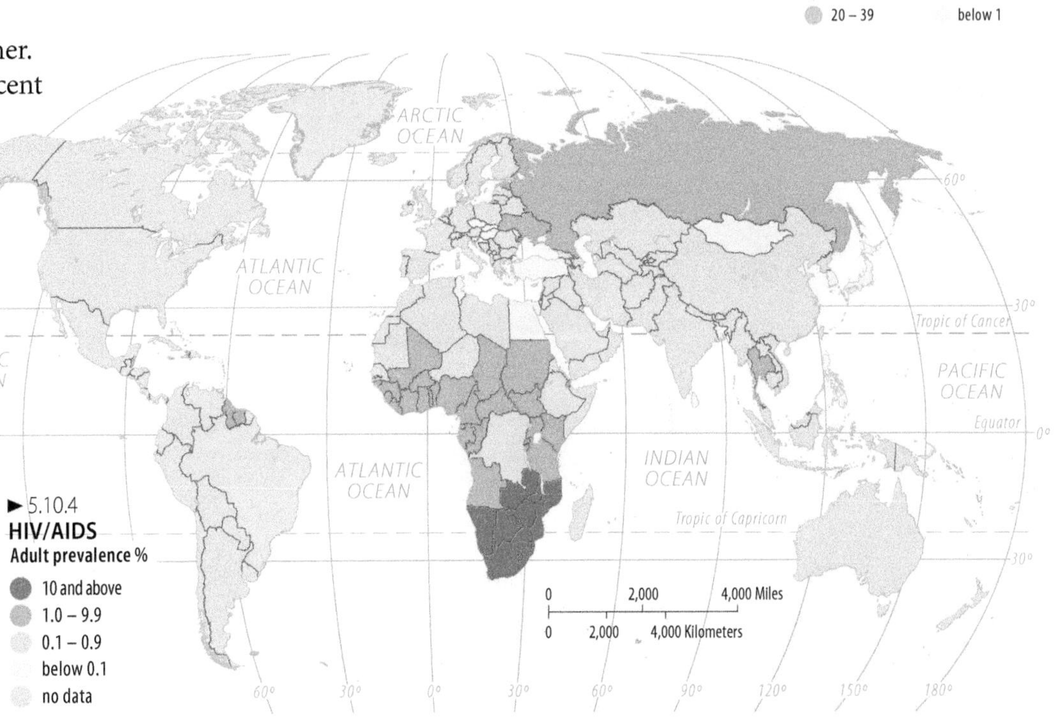

► 5.10.4 **HIV/AIDS**

This chapter has introduced ways in which geographers think about the world, as well as key concepts in understanding geography.

Key Questions

Where is the world's population distributed?

- Global population is highly concentrated; two-thirds of the world's people live in four clusters (Europe, East Asia, Southeast Asia, and South Asia).
- Population density varies around the world partly in response to resources.

Why does population growth vary among countries?

- A population increases because of fertility and decreases because of mortality.
- The demographic transition is a process of change in a country's population from a condition of high birth and death rates, with little population growth, to a condition of low birth and death rates, with low population growth.
- More than 200 years ago, Thomas Malthus argued that population was increasing more rapidly than the food supply; some contemporary analysts believe that Malthus' prediction is accurate in some regions.

How might population change in the future?

- Most countries in Europe and North America face slow or even declining population in the future.
- World population growth is slowing in part because birth rates are declining.
- Meanwhile, death rates are increasing in some countries because of chronic disorders associated with aging and in some developing countries because of infectious diseases.

▼ 5.CR.1 **VERY HIGH ARITHMETIC DENSITY: MARKET, DARAW, EGYPT**
What is the evidence that human behavior is affected by a high population density?

Thinking Geographically

The U.S. Census Bureau is allowed to utilize statistical sampling to determine much of the information about the people of the United States, such as age and gender. However, for determining the total population of each state and congressional district, the Census Bureau is required to count only the people for whom a census form was completed.

1. **What are the advantages of using each of the two approaches to counting the population?**

Some humans live at very high density (Figure 5.CR.1). Scientists disagree about the effects of high density on human behavior. Some laboratory tests have shown that rats display evidence of increased aggressiveness, competition, and violence when very large numbers of them are placed in a box.

2. **Is there any evidence that high density might cause humans to behave especially violently or less aggressively?**

Members of the baby-boom generation – people born between 1946 and 1964 – constitute nearly one-third of the U.S. population.

3. **As they grow older, what impact will baby boomers have on the entire American population in the years ahead?**

On the Internet

The Population Reference Bureau (PRB) provides authoritative demographic information for every country and world region at its website **www.prb.org.**

The Population Division of the United Nations Department of Economic and Social Affairs provides tables on population, births, and deaths for every country, at **http://esa.un.org/unpd/wpp/unpp/panel_population.htm**, or scan the QR code at the beginning of the chapter.

Interactive Mapping:

POPULATION DISTRIBUTION IN SOUTHWEST ASIA AND NORTH AFRICA

Population is highly clustered within Southwest Asia and Northern Africa.

Open: MapMaster Southeast Asia and North Africa in MasteringGEOGRAPHY

Select: *Population Density* from the *Population* menu, adjust opacity to 60%, then select *Physical Features* from the *Physical Environment* menu.

Most people live near what type of physical feature?

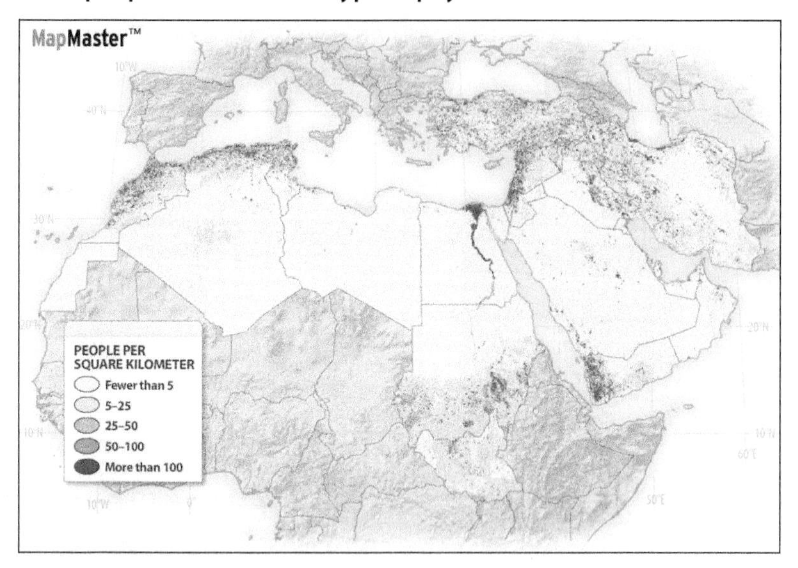

Explore

MAHĀMĪD, EGYPT

Use Google Earth to explore Mahāmīd, a town of 45,000 near the banks of the Nile River.

Fly to: *Mahāmīd, Luxor, Egypt.* Zoom in.

1. What color is most of the land immediately in and around the town? Does this indicate that the land is used for agriculture or is it desert?

Zoom out until you see the entire band of green surrounded by tan.

2. How wide is the green strip? What does the tan color represent? What feature is in the middle of the green strip?

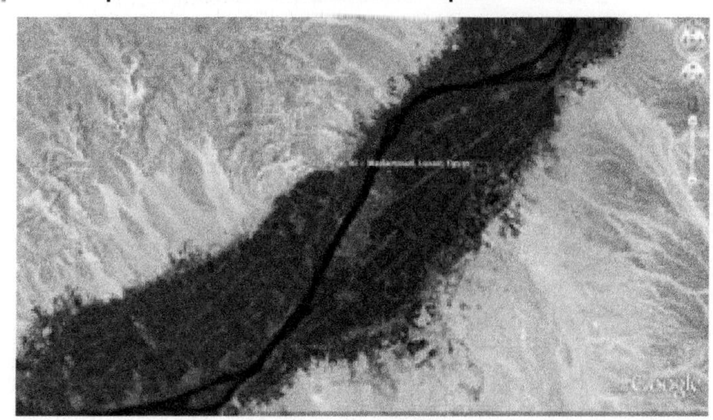

Key Terms

Agricultural density
The ratio of the number of farmers to the total amount of land suitable for agriculture.

Arable land
Land suited for agriculture.

Arithmetic density
The total number of people divided by the total land area.

Crude birth rate (CBR)
The total number of live births in a year for every 1,000 people alive in the society.

Crude death rate (CDR)
The total number of deaths in a year for every 1,000 people alive in the society.

Demographic transition
The process of change in a society's population from a condition of high crude birth and death rates and low rate of natural increase to a condition of low crude birth and death rates, low rate of natural increase, and a higher total population.

Dependency ratio
The number of people who are considered too young or too old to work (under age 15 or over age 64), compared to the number of people in their productive years.

Doubling time
The number of years needed to double a population, assuming a constant rate of natural increase.

Elderly support ratio
The number of working-age people (ages 15–64) divided by the number of persons 65 or older.

Epidemiologic transition
Distinctive causes of death in each stage of the demographic transition.

Epidemiology
Branch of medical science concerned with the incidence, distribution, and control of diseases that affect large numbers of people.

Infant mortality rate (IMR)
The total number of deaths in a year among infants under 1 year old for every 1,000 live births in a society.

Life expectancy
The average number of years an individual can be expected to live, given current social, economic, and medical conditions. Life expectancy at birth is the average number of years a newborn infant can expect to live.

Natural increase rate (NIR)
The percentage growth of a population in a year, computed as the crude birth rate minus the crude death rate.

Overpopulation
The number of people in an area exceeds the capacity of the environment to support life at a decent standard of living.

Pandemic
Disease that occurs over a wide geographic area and affects a very high proportion of the population.

Physiological density
The number of people per unit of area of arable land, which is land suitable for agriculture.

Population pyramid
A bar graph that displays the percentage of a place's population for each age and gender.

Total fertility rate (TFR)
The average number of children a woman will have throughout her childbearing years.

▶ LOOKING AHEAD

Population increases because of births and decreases because of deaths. The population of a place also increases when people move in and decreases when people move out. This element of population change—migration—is discussed in the next chapter.

6 Migration

Humans have always been on the go, whether moving in search of their basic needs or to find vacation spots. The historical distribution of people on Earth owed much to the distribution of food resources. During the last 500 years, the distribution of humans has changed enormously as technology allowed large numbers of people to move longer distances than ever before. The places that migrants come from have changed over time, as have the places they are headed for. Today, the largest international migrations are from poorer countries to wealthier ones.

There are many reasons why a person migrates. Today, the main reason for migration is job-related, as people leave regions with very few good paying jobs in search of places with greater job opportunities and higher pay. Another reason people migrate is that war, violence, and oppression often push people to flee for safer places to live. A temporary form of migration called tourism also changes the distribution of humans on Earth.

Countries are affected by migration in important ways. Internal migration, when people move from one part of a country to another, can result in completely new population patterns over time. Governments respond to the pressures of international migration by controlling who can cross the border and become a citizen of that country.

What are the historical patterns of human migration?

AFRICANS INTERCEPTED BY SPANISH COAST GUARD ATTEMPTING TO IMMIGRATE TO SPAIN IN SMALL BOATS

Why do people migrate?

How does migration change population characteristics?

SCAN FOR ANALYSIS ABOUT INTERNATIONAL MIGRATION

How do states deal with increasing immigration?

6.1 Human Origins

- Humans have always migrated, beginning with our early human ancestors 200,000 years ago.
- The diffusion of humans around the globe can be traced with archaeological, linguistic, and genetic evidence.

Migration has been a constant part of human history. Like any species, humans have moved near and far for many reasons.

REASONS FOR EARLY HUMAN MIGRATIONS

Demic diffusion is the relocation of people themselves from one place to another. The growth of early human populations sent people in search of new food resources. The onset and retreat of glaciers, changing sea levels or escape from dangerous species also spurred their movements. So, too, their migration may have fulfilled a sense of adventure and curiosity. These earliest waves of migration are how humans spread around the globe (Figure 6.1.1).

MIGRATION AND HUMAN ORIGINS

The early migrations that peopled Earth are important to us today because they answer the questions about **human origins**: Where did we come from? What makes us human? By unraveling prehistoric migration patterns, we can identify the location of our origins and track changes that occurred among humans as they fanned out across the planet. Most scientists agree that humans evolved from an ancestral population of early hominids in central East Africa several hundred thousand years ago. The evidence for this view of human origins is the frequency of early hominid skeletons in that region.

EVIDENCE OF EARLY MIGRATIONS

The primary evidence of human diffusion has come from the archaeological record, the examination of skeletal remains and signs of early human presence (Figures 6.1.2, 6.1.3, and 6.1.4).

Recent advances in the recovery and analysis of human genetics have begun to fill in the map of early human migration. The scientific study of languages and how they developed also provides insight on later prehistoric migration. These sources have tended to confirm our basic scientific understanding about human origins and early migration.

► 6.1.1 **THE PEOPLING OF EARTH**
The diffusion of humans out of Africa is the scientifically accepted hypothesis for how humans settled Earth. Accurately dating evidence of human occupation is difficult and controversial in some sites.

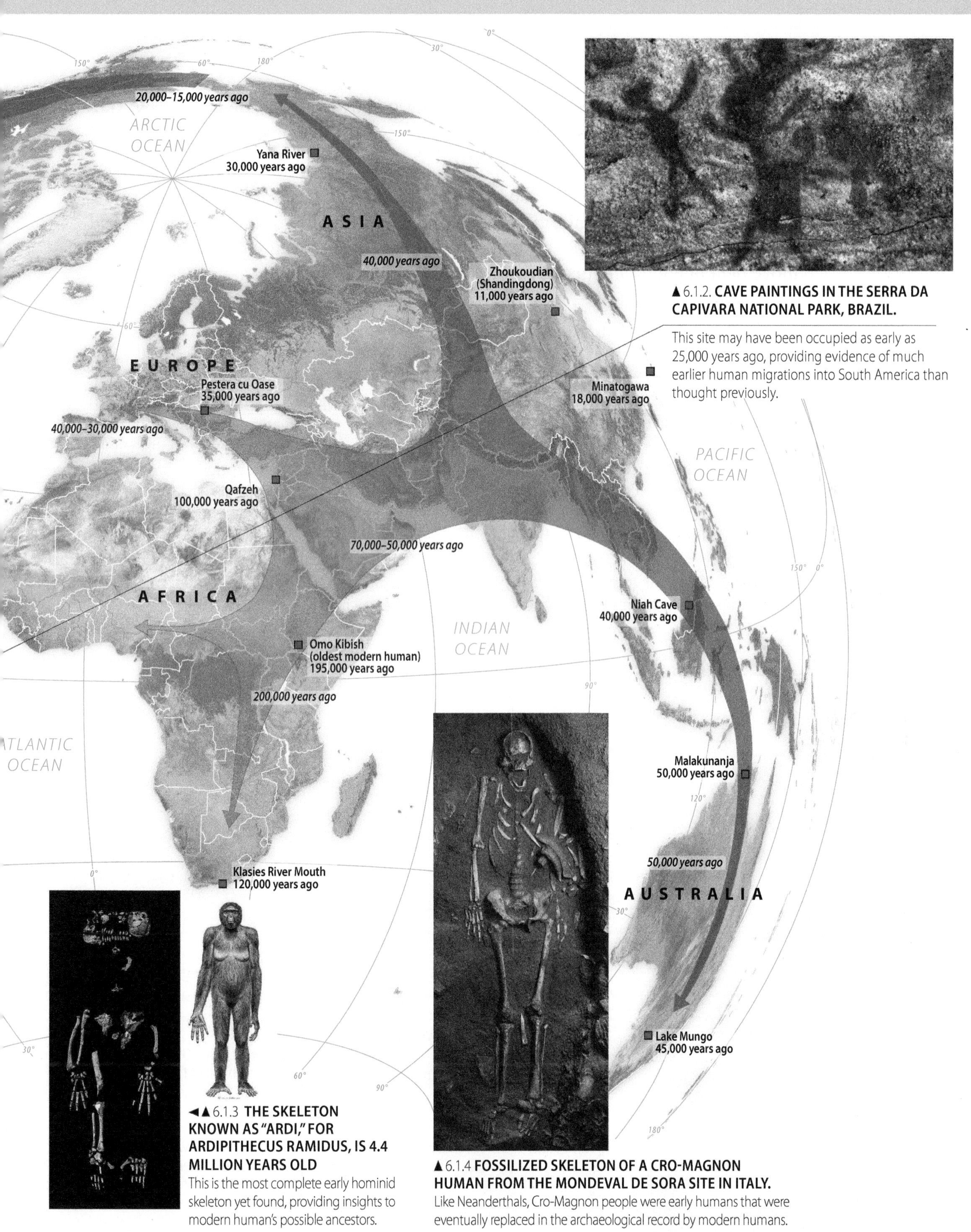

▲ 6.1.2. **CAVE PAINTINGS IN THE SERRA DA CAPIVARA NATIONAL PARK, BRAZIL.**

This site may have been occupied as early as 25,000 years ago, providing evidence of much earlier human migrations into South America than thought previously.

◄▲ 6.1.3 **THE SKELETON KNOWN AS "ARDI," FOR ARDIPITHECUS RAMIDUS, IS 4.4 MILLION YEARS OLD**

This is the most complete early hominid skeleton yet found, providing insights to modern human's possible ancestors.

▲ 6.1.4 **FOSSILIZED SKELETON OF A CRO-MAGNON HUMAN FROM THE MONDEVAL DE SORA SITE IN ITALY.**

Like Neanderthals, Cro-Magnon people were early humans that were eventually replaced in the archaeological record by modern humans.

6.2 Modern Mass Migration

- **Large movements of people around the globe have shaped the modern world.**
- **The number of long-distance migrants has increased over time.**

The modern era, roughly the last 500 years, is characterized by the movement of large numbers of peoples over much longer distances than ever before. The reasons for this relate to the forces of globalization that began with Europe's voyages of conquest and the establishment of colonies around the world. New ship technologies accelerated the movement of goods and people.

FROM EUROPE

During the modern period, Spain, Portugal, Britain, France, and the Netherlands were the primary countries to establish colonies around the world (Figure 6.2.1). **Settler migration** in this period entailed the relocation of European populations to overseas colonies.

The primary activity of settler colonies was extracting natural wealth and trading with their home countries. The European powers established about 150 colonies worldwide between the fifteenth and eighteenth centuries. During the nineteenth century, the largest **migration stream** was that between northwestern Europe and the United States, involving more than 33 million people between 1821 and 1920 (Figure 6.2.2, see also section 6.8 on the Changing Origin of U.S. Immigrants).

In many places, settlement changed the make-up of populations from native to almost wholly European. Settlers to the Americas brought diseases new to the native population, which decimated their number by as much as 90 percent. Violence reduced their numbers further. Even where Europeans remained a minority, as in southern Africa, they retained political and economic power until well into the twentieth century.

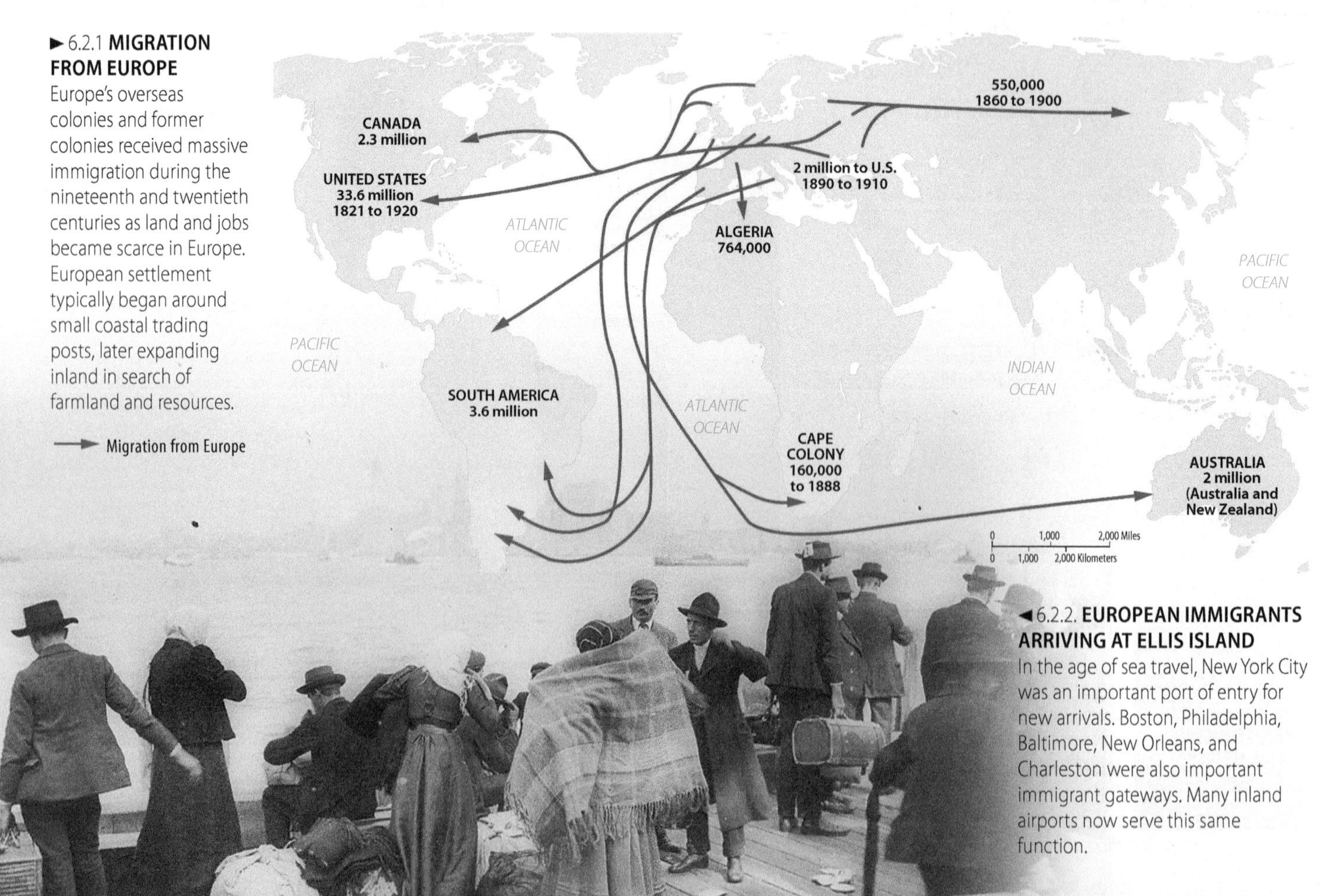

► 6.2.1 **MIGRATION FROM EUROPE**
Europe's overseas colonies and former colonies received massive immigration during the nineteenth and twentieth centuries as land and jobs became scarce in Europe. European settlement typically began around small coastal trading posts, later expanding inland in search of farmland and resources.

◄ 6.2.2. **EUROPEAN IMMIGRANTS ARRIVING AT ELLIS ISLAND**
In the age of sea travel, New York City was an important port of entry for new arrivals. Boston, Philadelphia, Baltimore, New Orleans, and Charleston were also important immigrant gateways. Many inland airports now serve this same function.

FROM AFRICA

The colonial economies built on agriculture and mining typically sought more laborers than were available in the European settler community or the native population. The **forced migration** of enslaved persons, primarily from western and central Africa, to the Western Hemisphere, began with the earliest colonies. Between the seventeenth and eighteenth centuries, about 14.5 million Africans were taken to the new world (Figure 6.2.3). As many as three times this number may have been captured but many did not survive the inhuman conditions on ships crossing the Atlantic.

The slave economies of the New World enabled the production of inexpensive raw materials for European factories and consumers. In the colonies, only the European elite benefited from this arrangement while slave populations were often treated as animals, traded, raped, beaten, or murdered with impunity by the landowner. The legacies of this violence loom large over former slave societies, such as the United States, Brazil, and many Latin American countries. Long after slavery was abolished in the nineteenth century, the descendants of enslaved persons have suffered from discrimination.

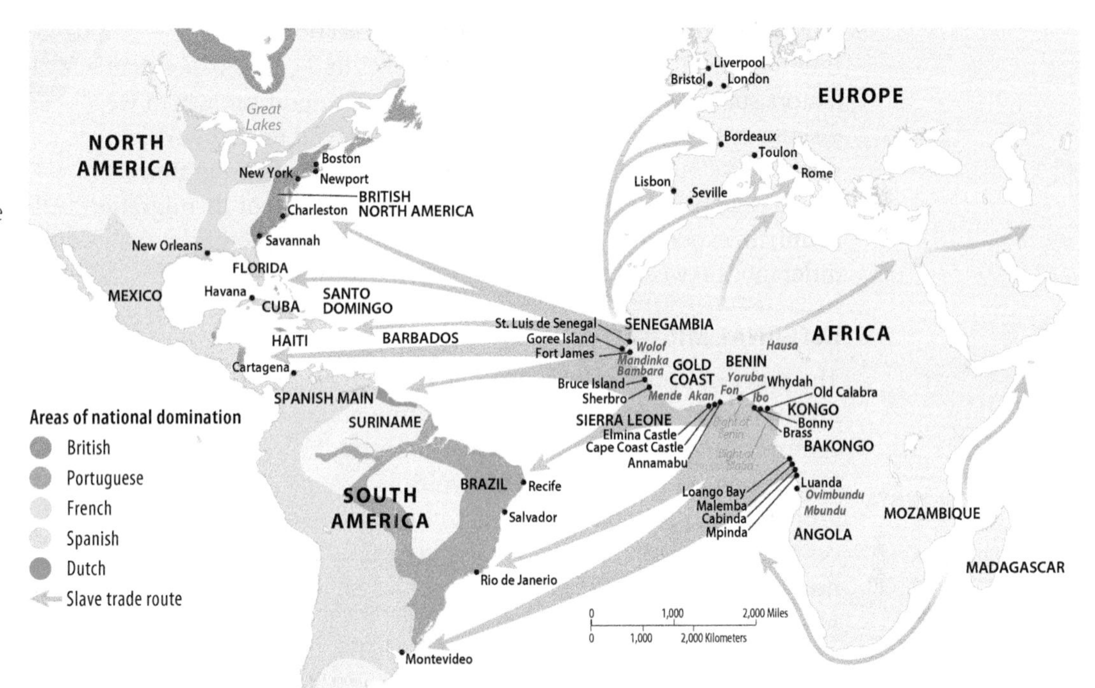

► 6.2.3 **FORCED MIGRATION OF ENSLAVED AFRICANS**
The movement of African slaves to the new world was part of a triangular trade scheme. Captured Africans were traded for goods manufactured in Europe. They were sold in the New World for slave labor and the profits on their sale purchased raw materials produced on plantations. Raw materials such as sugar or cotton were shipped to Europe and sold. The profits were then used to buy manufactured goods, then used to obtain more captured Africans.

FROM ASIA

After slavery was abolished, continued demand for cheap labor began to draw on destitute and hungry Chinese. The Chinese **diaspora**, literally a "scattering" of people to different destinations, began with the large wave of labor migration to southeast Asia and the Americas, as well as most other colonies. Some were coerced to leave China because of poor living conditions. Others were indentured to pay off debt, a form of temporary enslavement. During this time, Chinese traders also expanded their activities throughout East and Southeast Asia and ethnic Chinese trading communities became economically important (Figure 6.2.4). For Chinese, life in these societies has not always been easy. In many places, their economic success is viewed suspiciously and they are regarded as outsiders.

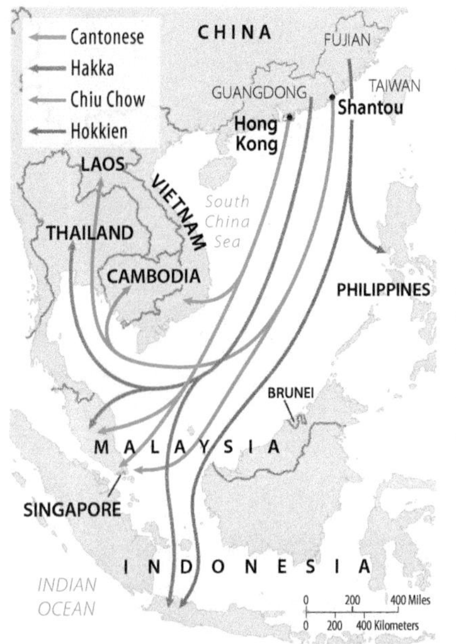

◄ 6.2.4 **EMIGRATION FROM CHINA**
The Chinese diaspora numbers about 40 million people and Chinese migrants can be found in all world regions. Some 30 million live in Southeast Asia (shown here), 5 million in North America, and 2 million in Europe. Chinese traders are increasingly common in poorer countries where they import and sell inexpensive consumer goods in local markets. Their recent success in African countries has come at the expense of local traders.

6.3 Sources and Destinations

- **A combination of push and pull factors influence migration decisions.**
- **Most international migration today is from poor to wealthier countries.**

People migrate because of **push factors** and **pull factors**. A push factor induces people to move out of their present location, whereas a pull factor induces people to move into a new location.

ROLE OF PUSH AND PULL FACTORS IN MIGRATION

The primary set of push and pull factors shaping global migration are economic, as we describe in section 6.4. Environmental factors such as earthquakes and cultural-political factors such as civil war often lead to forced migration, which we discuss in section 6.5. Push and pull factors tend to shape which regions lose population through **emigration**, or out-migration, and which regions gain population from **immigration**, or in-migration. All countries experience at least some immigration and emigration, and the difference between them is called **net migration.**

INTERNAL MIGRATION

The nineteenth century geographer E.G. Ravenstein noted that most migration involves short-distance relocations within countries. Since the eighteenth century, internal migration in most countries has been a process of rural to urban migration (see section 6.7). This typically occurs when rural areas cannot support growing rural populations, some of whom instead seek new livelihoods in urban centers. We will discuss the world's urban population in section 12.9 but we should note here that only half of Earth's population lives in urban areas. Rural to urban migration may well continue for centuries (Figure 6.3.1).

▼ 6.3.1 **NAIROBI, KENYA** The population of this fast growing east African city has more than tripled since the 1980s. Many of the newcomers to the city live in cramped slums.

GLOBAL MIGRATION

International migration has increased considerably in recent decades as economic disparities grow between wealthy and poor regions (Figure 6.3.2). During this latest round of global mass migration, improved communications, information about destinations, and transportation technologies have accelerated the flow of economic migrants between world regions.

At a global scale, Asia, Latin America, and Africa have net out-migration, whereas North America, Europe, and Oceania have net in-migration. The three largest flows of migrants are:

- From Asia to Europe.
- From Asia to North America.
- From Latin America to North America.

Substantial in-migration also occurs from Europe to North America and from Asia to

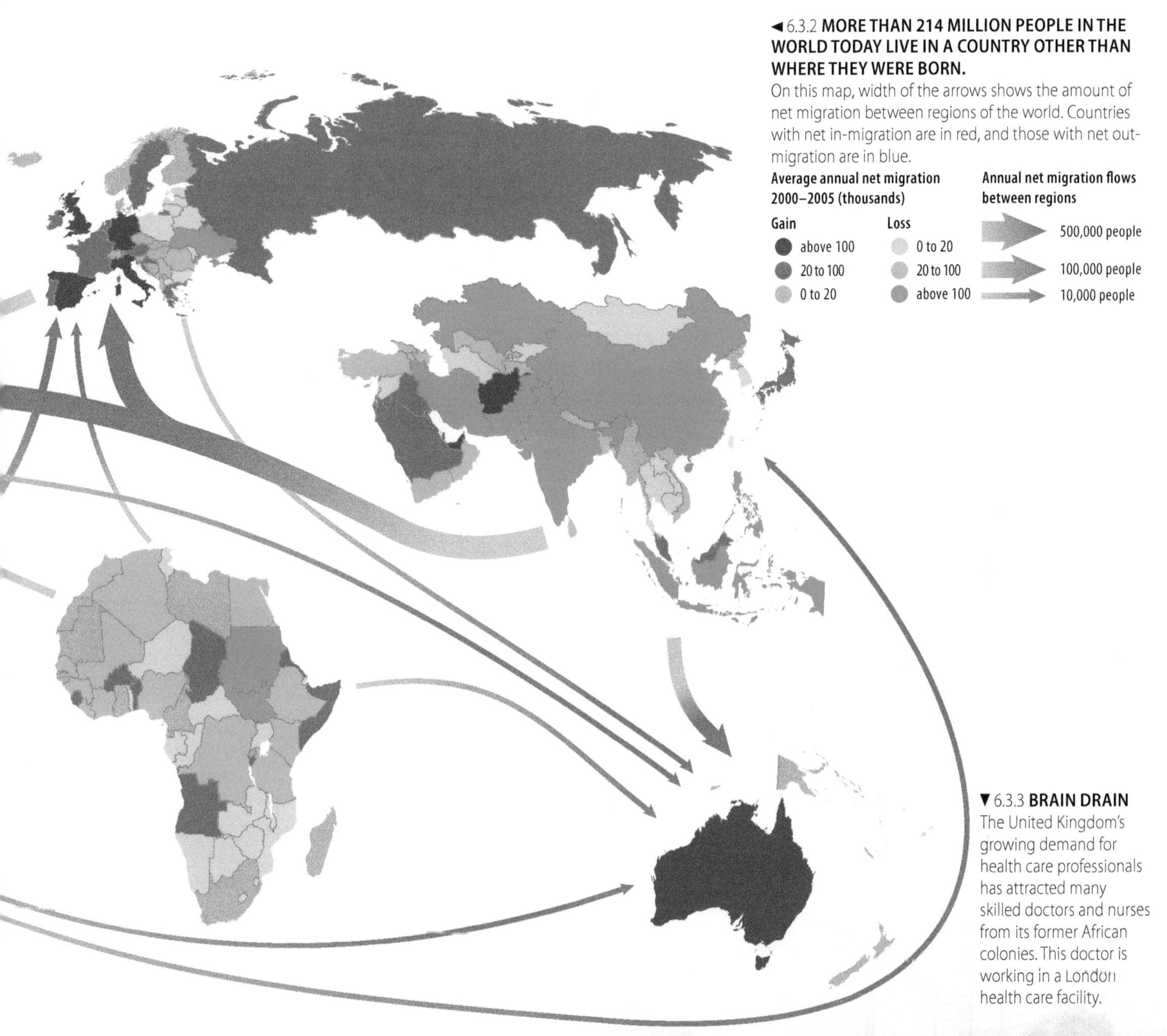

◄ 6.3.2 **MORE THAN 214 MILLION PEOPLE IN THE WORLD TODAY LIVE IN A COUNTRY OTHER THAN WHERE THEY WERE BORN.**
On this map, width of the arrows shows the amount of net migration between regions of the world. Countries with net in-migration are in red, and those with net out-migration are in blue.

Oceania. Lower levels of net migration occur from Latin America to Oceania and from Africa to other regions.

The global pattern reflects the importance of migration from developing countries to developed countries. Migrants from countries with relatively low incomes and high natural increase rates head for relatively wealthy countries, where job prospects are brighter. It is important to remember that regions experiencing net emigration are losing their population to wealthier countries. These emigrants often include the best and brightest in their country, so this process is called **brain drain** (Figure 6.3.3). Because brain drain is common in some critical professions, such as health care, those left behind in the source country acutely feel the loss.

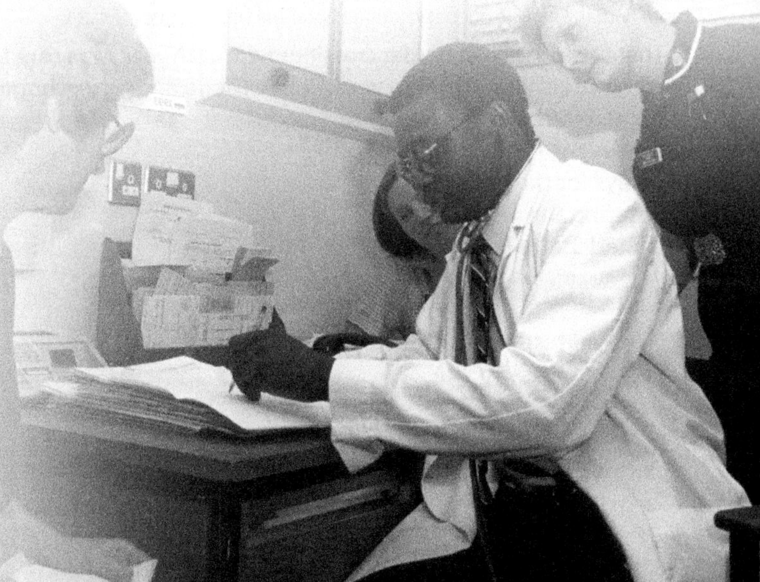

▼ 6.3.3 **BRAIN DRAIN**
The United Kingdom's growing demand for health care professionals has attracted many skilled doctors and nurses from its former African colonies. This doctor is working in a London health care facility.

6.4 International Labor Migration

- ► **Most international migrants are workers looking for jobs.**
- ► **Labor migrants seek countries with growing economies.**

Most people move to find work and to improve their lives. The geographer E.G. Ravenstein considered it a "law of migration" that longer distance migrations would gravitate towards centers of economic strength (Figure 6.4.1). People emigrate from places that have few job opportunities, and they immigrate to places where the jobs seem to be available (Figure 6.4.2).

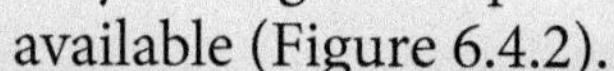

► 6.4.1 **INTERNATIONAL MIGRANTS**
Persons born in a different country other than where they live make up a large percentage of the population in some countries. Nearly all of these people are labor migrants and all of them must find a source of income in their new country.

▲ 6.4.2 **MIGRANT LABOR FROM INDIA**
The wealthy oil-based economies of the Persian Gulf countries have attracted large numbers of laborers from the poor countries of South and Southeast Asia. These Indian workers are part of the construction crews who have turned Doha, Qatar into a global business center.

▼ 6.4.3 **BORDER CROSSING, SAN YSIDRO, CALIFORNIA**
Use Google Earth to explore one of the busiest border crossing points in the world on the Mexico-United States border.

Fly to: *San Ysidro, California*, follow the converging highways south to the San Ysidro Land Port of Entry.

Zoom in to the vehicle queue waiting to enter the United States.

1. **There are 50 million border crossings here each year. Many are Mexican commuters who travel to work in San Diego and return home each night. What causes the long delays for those entering the United States?**

Zoom in and explore the border.

2. **What features are used to funnel border crossings to the Port of Entry?**

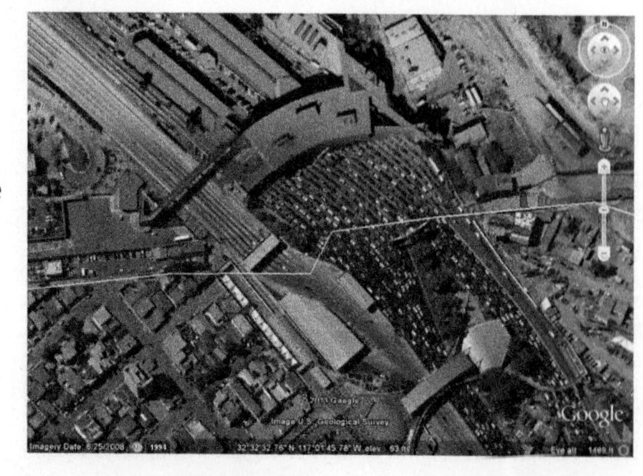

DESTINATION: NORTH AMERICA

The United States and Canada have been especially prominent destinations for economic migrants. Many European immigrants to North America in the nineteenth century truly expected to find streets paved with gold. While not literally so gilded, the United States and Canada did offer Europeans prospects for economic advancement. This same perception of economic plenty now lures people to the United States and Canada from Latin America and Asia (Figure 6.4.3). Some of these are **temporary labor migrants** who work to save enough money to return home and establish new households. **Seasonal migrants** work part of the year tending to certain crops or activities.

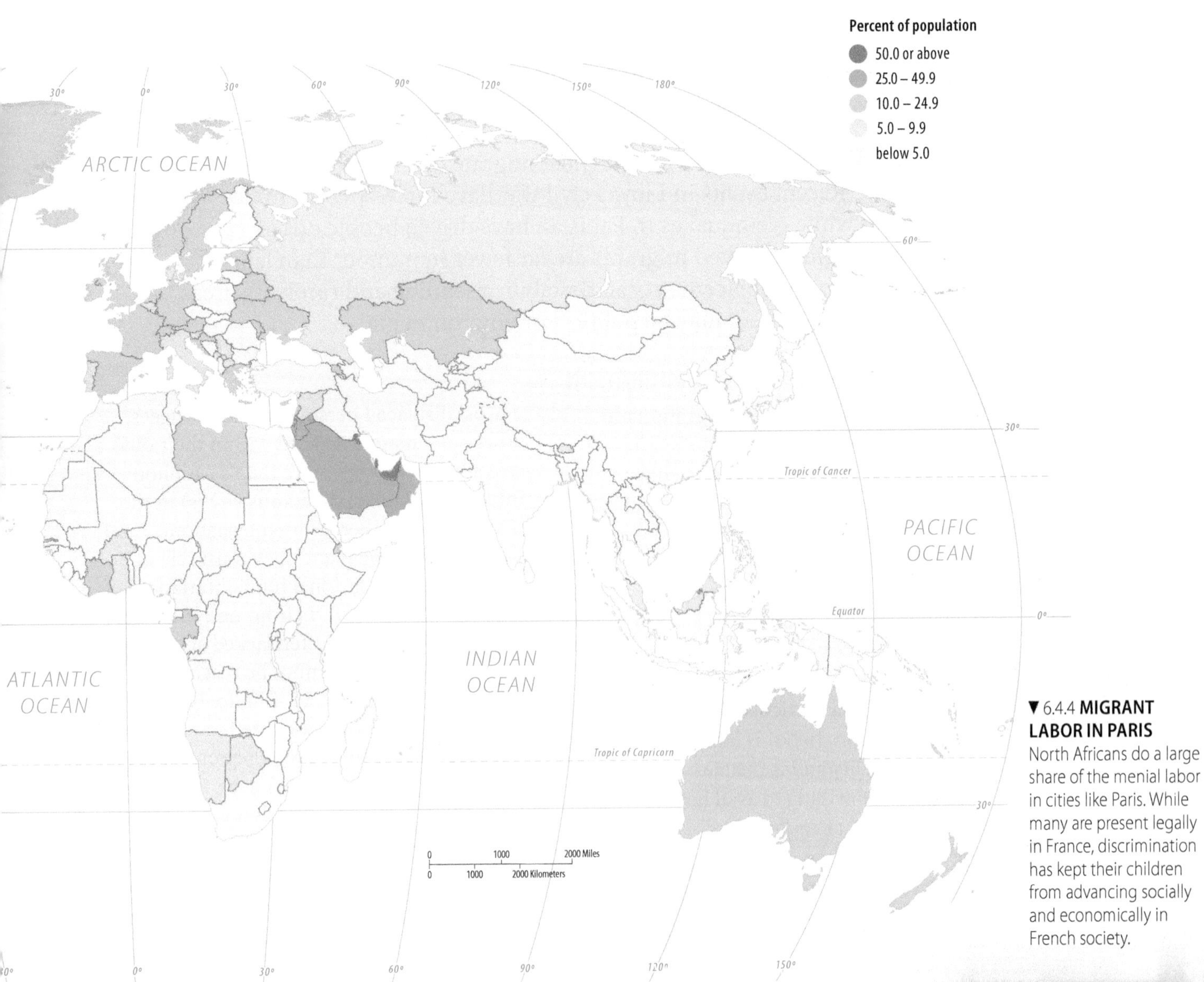

▼ 6.4.4 **MIGRANT LABOR IN PARIS**
North Africans do a large share of the menial labor in cities like Paris. While many are present legally in France, discrimination has kept their children from advancing socially and economically in French society.

DESTINATION: EUROPE

Europe has attracted economic migrants since the end of World War II. Germany, France, Italy, and other western European countries encouraged **guest workers** to immigrate and work in factories, helping to rebuild war-torn economies in western European. Although European countries today largely discourage immigration, the region attracts immigrants from Africa and the Middle East (Figure 6.4.4). These immigrants serve a useful role in Europe, because they take low-status and low-skilled jobs that local residents won't accept. In cities such as Berlin, Brussels, Paris, and Zurich, immigrants provide essential services, such as driving buses, collecting garbage, repairing streets, and washing dishes.

Although relatively low paid by European standards, immigrants earn far more than they would at home. By letting their people work elsewhere, poorer countries reduce their own unemployment problems. Immigrants also help their native countries by sending a large percentage of their earnings, called **remittances**, back home to their families. The injection of foreign currency then stimulates the local economy.

6.5 Forced Migration

- Some migrants are fleeing to avoid violence or disasters.
- Forced migration patterns tend to be local or regional.

Forced migration occurs when people must migrate or suffer terrible consequences. Recent events in Libya's civil war have forced people to flee combat zones while earthquakes in Pakistan have driven people out of mountainous regions. Forced migrants are far fewer in number than labor migrants, yet rapid **displacement** can destabilize sending and receiving areas. The cause of displacement may be political or environmental.

POLITICAL CAUSES OF FORCED MIGRATION

The political causes of forced migration are typically armed conflict or discrimination (Figure 6.5.1). The United Nations High Commission for Refugees (UNHCR) estimated there are 44 million forcibly displaced persons in the world in 2010. There were several categories of displaced persons. The 1951 Refugee Convention defines a **refugee** as a person with a

> "well-founded fear of being persecuted for reasons of race, religion, nationality, membership of a particular social group or political opinion, [who] is outside the country of his nationality and is unable or, owing to such fear, is unwilling to avail himself of the protection of that country."

There were more than 15 million refugees in the world in 2010.

Most displaced persons are not refugees, however, but are displaced within their own country. There were more than 27 million **internally displaced persons** (IDPs) in the world in 2010. There is no international legal protection for them and they must rely on themselves or occasional humanitarian relief to survive away from their homes. In Sudan's civil wars, IDPs have often lacked access to basic resources and cannot be reached by international aid agencies (Figure 6.5.2).

Displacement is concentrated in poor, unstable, and conflict-prone regions in the developing world. Most refugees, therefore, are seeking help from other poor or unstable countries. As a result, relatively stable countries such as Tanzania and Iran become a regional destination for refugees.

▼ 6.5.1 **SIGNIFICANT REFUGEE FLOWS, 2008**
The largest movements of refugees are within poor and conflict prone regions. Refugee migration tends to concentrate on neighboring states. Long distance refugee migrations tend to be invited and supported by host states as, for example, Somali refugees to the United States and Iraqi refugees to Sweden.

▼ 6.5.2 **FORCED MIGRATION IN SUDAN**

One of the largest countries in Africa, Sudan's unstable environmental and political conditions have driven millions from their homes. The creation of South Sudan in 2011 helped to resolve some displacement issues while creating new ones.

▲ 6.5.2a **REFUGEE CAMP, CHAD**

More than a quarter of a million refugees fled into Chad to escape the violence in Darfur. They sheltered in refugee camps along the border. Chad also hosts refugees from conflicts in the Central African Republic as well as its own IDP population.

▲ 6.5.2b **DESTROYED VILLAGE, DARFUR**

The brick walls of these huts are all that remain after Janjaweed militia raids killed and expelled the inhabitants. In 2011, there were an estimated two million IDPs from the Darfur conflict.

▲ 6.5.2c **HUNGER, NORTHERN KORDOFAN**

Regular droughts and poor economic policies contribute to hunger in many parts of Sudan, including Northern Kordofan. Famines in this region are frequent. Continued droughts have forced many to flee northern Sudan as environmentally displaced persons.

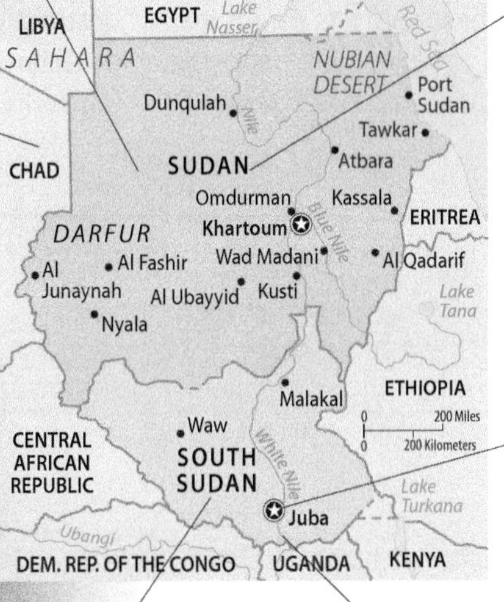

▲ 6.5.2d **SUDAN PEOPLE'S LIBERATION ARMY (SPLA)**

The SPLA is the military wing of the major southern Sudan political movement that sought independence from northern Sudan. They fought the government for more than 20 years in a conflict that contributed to more than one million civilian deaths. About 400,000 persons were displaced in South Sudan as of 2011.

► 6.5.2e **IDP CAMP, UGANDA**

Violence in northern Uganda involving the Lord's Resistance Army has filled camps with hundreds of thousands of IDPs, pushing thousands of refugees into South Sudan.

▲ 6.5.2f **UN TROOPS**

About 10,000 troops and advisors were sent by countries around the world to support the UN peacekeeping and humanitarian mission in southern Sudan. South Sudan became an independent country on July 9, 2011.

ENVIRONMENTAL CAUSES OF FORCED MIGRATION

Natural disasters such as flooding, tsunamis, earthquakes, landslides, and volcanic eruptions have always forced humans to relocate. Such migrants are termed **environmentally displaced persons** and are not protected by the 1951 convention. Recent environmental displacements include the evacuation of New Orleans after Hurricane Katrina and Haitians driven from their homes after the 2010 earthquake. In 2010, the UNHCR gave assistance to two million people in these situations, a fraction of those displaced by natural disasters.

6.6 Tourism Migration

- **Tourism is a form of temporary migration.**
- **Tourism has enormous economic and environmental effects on destinations.**

Whenever you vacation to the beach or spend a weekend in a different town, you are engaging in tourism. Tourism is by definition a form of temporary migration caused by an interest in visiting places and people other than where one resides. Tourists may travel within their own countries or internationally.

The number of international tourists has more than doubled during the last two decades. In 2010 there were an estimated 940 million international tourists measured as **arrivals** by foreigners to another country (Figure 6.6.1). If all tourists were citizens of one country, they would be the third largest country in the world. About half of these tourists are seeking leisure or recreation activities. These tourists might go the beach, enjoy the local food, or visit historical monuments (Figure 6.6.2). Another 25 percent of tourists visit family and friends, health providers, or religious activities. These tourists may be seeking medical treatment they cannot access at home or hiking pilgrimage routes between holy shrines.

Europe is the destination for half of the world's international tourists. China, Turkey, Malaysia, and Mexico are also among the most popular destinations. Most tourists come from European or Asian countries. Tourism brings cultures into contact (Figure 6.6.3).

▼6.6.1 **INTERNATIONAL TOURIST ARRIVALS** This map shows the count of each person arriving in a country as a tourist.

◄6.6.2 **CROWDED DESTINATIONS** Cities like Bangkok, Thailand have whole districts that swell with tourists, causing problems for city residents and officials.

THE TOURISM INDUSTRY

Tourism is also, by some accounts, the world's largest industry. International tourist receipts totaled $919 billion in 2010. Tourists spend large sums of money in their destinations on everything from hotels and meals to recreation, souvenirs, and taxes. In most countries, tourism is one of the fastest-growing sectors of the economy. Tourism is a significant amount of the income for many poor countries. Tourism receipts equal about 14 percent of Cambodia's GNI and about 7.5 of Egypt's GNI. Political turmoil in countries like these scares away tourists and stops the inflow of hard currency.

Many poor countries are pursuing economic development strategies that exploit areas of natural beauty. Unfortunately, these are often areas in which additional visitors may harm the natural environment. Tourists require hotels, transportation, and food that take up land and create pollution. This puts pressure on local natural areas. The problems are not limited to developing countries dependent on tourism. Resort areas in Europe's Mediterranean put an enormous strain on scarce fresh water supplies while dumping large amounts of wastewater into sensitive watersheds. **Ecotourism** refers to practices that focus upon lessening the environmental impact of visitors but its appeal remains limited (Figure 6.6.4). "Ecotourists" visiting areas of natural beauty and recreation might be more willing to use a bicycle and tent, for example, than conventional tourists.

▲ 6.6.3 **CLASH OF CULTURES**
Tourists from developed countries tend to have less restrictive attitudes about personal modesty and diet. Workers in many destinations are expected to put aside their cultural preferences to accommodate guests. In some Muslim countries, special secluded areas are created for male and female tourists to socialize, swim, or drink alcohol away from local staff.

▼ 6.6.4 **GORILLA WATCHING**
Traveling to view endangered wildlife means preserving sensitive habitats. Visitors to Rwanda's gorilla range must limit their impact on the forest.

6.7 Residential Mobility

- Relocation within countries is the most common form of permanent migration.
- Interregional and intraregional migrations are caused by different factors.

Internal migrations from part of a country to another are the most common form of permanent migration. In the United States, people move households an average of more than 10 times in their life, caused by such events as going away to school or relocating for a job. The ability for households to move within a country is termed **residential mobility.** Push and pull factors are often dominated by employment considerations but may also include strong pull factors to places with natural beauty, recreation, or lifestyle amenities.

INTERREGIONAL MIGRATION

Interregional migration is the relative long-distance movement from one region of a country to another (Figure 6.7.1). The most famous example of interregional migration is the opening of the American West. Two hundred years ago, the United States consisted of a collection of settlements concentrated on the Atlantic Coast. Through mass interregional migration, the rest of the continent was settled and developed.

INTRAREGIONAL MIGRATION

Movement constrained to within the same region is termed **intraregional migration.** Intraregional migration is much more common than interregional or international migration. Most intraregional migration has been from rural to urban areas or from cities to suburbs (Figure 6.7.2).

▲ 6.7.1 **CHINA'S INTERNAL MIGRATION (2000–2005)**
Rapid economic change in recent decades has fueled mass internal migration. Most migrants are bound for industrialized cities on China's east coast where export-oriented factories have multiplied. Rural poverty is the main push factor causing these migrants to leave towns and villages. They are pulled by the allure of better pay and the prospects of city life.

► 6.7.2 **RECESSION DRIVEN MOBILITY**

Open MapMaster North America in MasteringGEOGRAPHY

Select: *Impact of Recession* from *Economic* menu then *Cities* from the Political menu.

The recent recession is likely to make what cities more attractive to internal migrants?

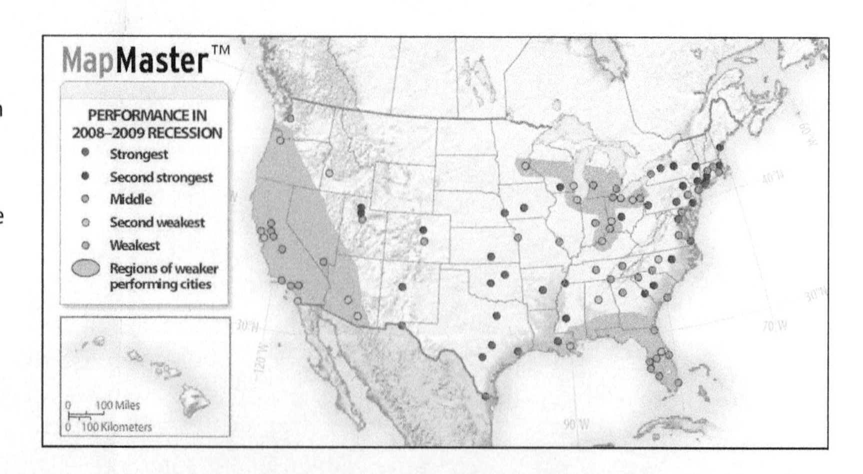

Rural to urban migration. Migration from rural to urban areas accelerated in the 1800s in Europe and North America as part of the Industrial Revolution. The percentage of people living in urban areas in the United States, for example, increased from 5 percent in 1800 to 50 percent in 1920. By some measures, nearly 90 percent of the population in the United States and other developed countries now live in urban areas. In recent years, large-scale rural to urban migration has occurred in developing countries of Asia, Latin America, and Africa. Worldwide, more than 20 million people are estimated to migrate each year from rural to urban areas. Like interregional migrants, most people who move from rural to urban areas seek economic advancement.

Migration from urban to suburban areas. Today most intraregional migration in developed countries is from central cities out to suburbs (Figure 6.7.3). The population of most cities in developed countries has declined since the mid-twentieth century, while suburbs have grown rapidly. Nearly twice as many Americans migrate from central cities to suburbs each year than migrate from suburbs to central cities. The suburbs of Provo, Utah grew by 59 percent during the 2000s while the central city grew by only 13 percent. Comparable patterns are found in Canada and Europe. The major reason for the large-scale migration to the suburbs is not related to employment, as is the case with other forms of migration. For most people, migration to suburbs does not coincide with changing jobs. Instead, people are pulled by a suburban lifestyle typified by detached houses rather than apartments, better schools, private yards, and access to outdoor or cultural recreation (Figure 6.7.4).

Migration from urban to rural areas. Developed countries witnessed a new migration trend during the late twentieth century. For the first time, more people immigrated into rural areas than emigrated out of them. In some places, there is net migration from urban to rural areas, called **counterurbanization**. Like suburbanization, people move from urban to rural areas for lifestyle reasons. Some are lured to rural areas to get away from high-paced urban life or to live on a farm where they can own horses or grow vegetables. Many still make their living by working in factories, shops, or even commuting to jobs in the city.

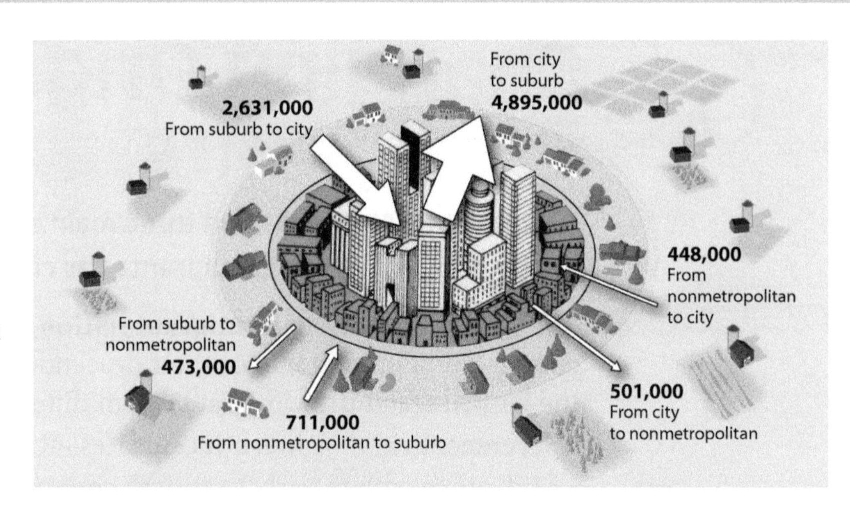

▲ 6.7.3
INTRAREGIONAL MIGRATION IN THE UNITED STATES (IN MILLIONS), 2007
Movements within any single metropolitan area may reveal different trends than the national totals shown here.

▼ 6.7.4
SUBURBANIZATION
The dramatic contrast of this Phoenix suburb and the desert marks the frontier of a rapidly growing city as it presses into the surrounding landscape.

6.8 Changing Origin of U.S. Immigrants

- ▶ **The United States has had three main eras of immigration.**
- ▶ **The principal source of migrants has changed in each era.**

The United States is an **immigrant nation**, meaning its population is almost entirely comprised of immigrants or their descendents. During three main eras of immigration, the United States drew migrants from different parts of the world (Figure 6.8.1):

- Seventeenth and eighteenth centuries—United Kingdom and Africa (Figure 6.8.2)
- Mid-nineteenth to early twentieth century—Europe (Figure 6.8.3)
- Late-twentieth to early twenty-first century—Latin America and Asia (Figure 6.8.4)

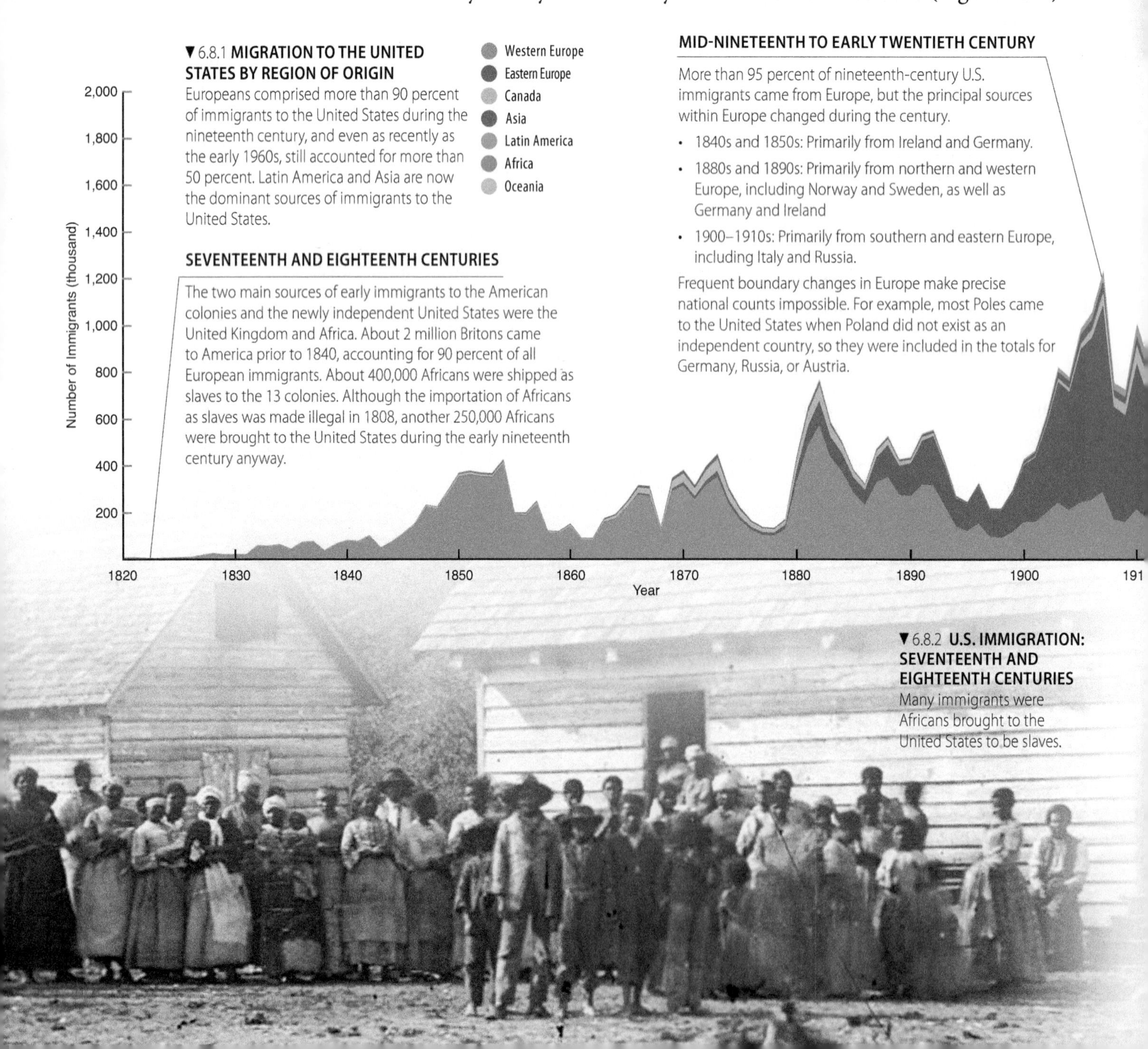

▼ 6.8.1 **MIGRATION TO THE UNITED STATES BY REGION OF ORIGIN**
Europeans comprised more than 90 percent of immigrants to the United States during the nineteenth century, and even as recently as the early 1960s, still accounted for more than 50 percent. Latin America and Asia are now the dominant sources of immigrants to the United States.

SEVENTEENTH AND EIGHTEENTH CENTURIES

The two main sources of early immigrants to the American colonies and the newly independent United States were the United Kingdom and Africa. About 2 million Britons came to America prior to 1840, accounting for 90 percent of all European immigrants. About 400,000 Africans were shipped as slaves to the 13 colonies. Although the importation of Africans as slaves was made illegal in 1808, another 250,000 Africans were brought to the United States during the early nineteenth century anyway.

MID-NINETEENTH TO EARLY TWENTIETH CENTURY

More than 95 percent of nineteenth-century U.S. immigrants came from Europe, but the principal sources within Europe changed during the century.

- 1840s and 1850s: Primarily from Ireland and Germany.
- 1880s and 1890s: Primarily from northern and western Europe, including Norway and Sweden, as well as Germany and Ireland
- 1900–1910s: Primarily from southern and eastern Europe, including Italy and Russia.

Frequent boundary changes in Europe make precise national counts impossible. For example, most Poles came to the United States when Poland did not exist as an independent country, so they were included in the totals for Germany, Russia, or Austria.

▼ 6.8.2 **U.S. IMMIGRATION: SEVENTEENTH AND EIGHTEENTH CENTURIES**
Many immigrants were Africans brought to the United States to be slaves.

▲ 6.8.3 **U.S. IMMIGRATION: MID-NINETEENTH TO EARLY TWENTIETH CENTURY**
Immigrants from southern and eastern Europe line up for entry into the United States.

LATE TWENTIETH TO EARLY TWENTY-FIRST CENTURY

The two leading sources of immigrants since the late twentieth century have been Latin America and Asia (right). About 13 million Latin Americans and 7 million Asians have migrated to the United States in the past half-century, compared to only 2 million and 1 million, respectively, in the two preceding centuries. Officially, Mexico passed Germany in 2006 as the country that has sent to the United States the most immigrants ever. The four leading sources of U.S. immigrants from Asia have been China (including Hong Kong), Philippines, India, and Vietnam.

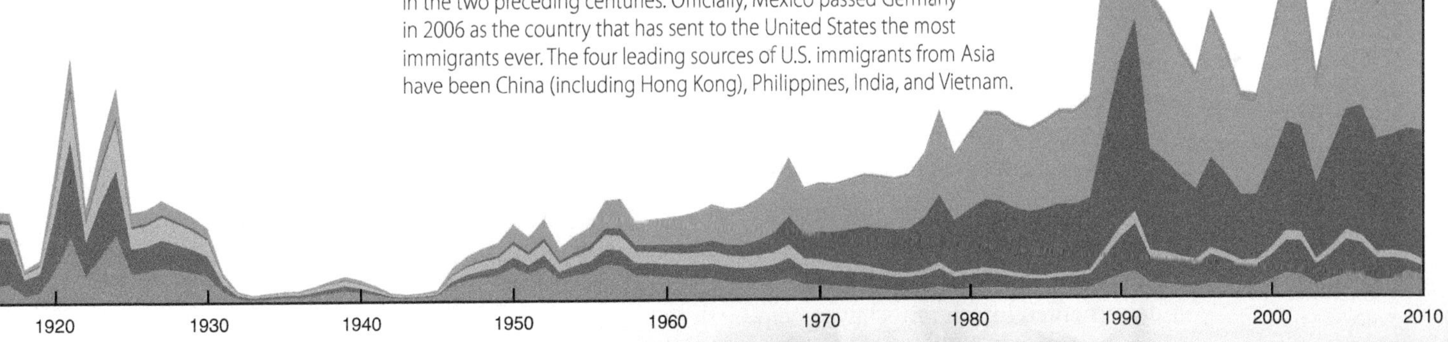

▲► 6.8.4 **U.S. IMMIGRATION: LATE TWENTIETH TO EARLY TWENTY-FIRST CENTURY**
(above) A family from Mexico City ready to leave. (right) A Chinese family in New York City's Chinatown.

6.9 Undocumented Migration

- ► **Some immigrants live and work in countries without permission.**
- ► **The reaction to undocumented migration differs among host countries.**

Governments want to control each immigrant's attempt to enter. Persons who enter a country without permission from the government are called unauthorized or **undocumented immigrants**. Yet governments often lack control over immigration given the sheer volume of migration and open borders.

Economically advanced countries, especially the United States, have record high numbers of people trying to enter to find work. An estimated 300,000 undocumented immigrants entered the United States in 2009, most along the nearly 2,000-mile-long U.S.-Mexican border guarded by about 17,000 federal agents.

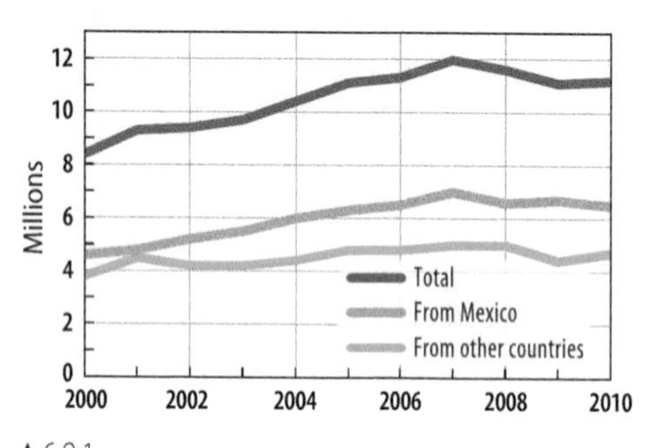

▲ 6.9.1
UNDOCUMENTED IMMIGRANTS IN THE UNITED STATES

The Pew Hispanic Center estimated that there were 11.2 million unauthorized immigrants living in the United States in 2010. The number increased rapidly during the first years of the twenty-first century (Figure 6.9.1). After hitting a peak in 2007, the figure declined because the severe recession starting in 2008 reduced job opportunities in the United States.

Other information about undocumented immigrants, according to Pew:

- **Source country.** Approximately 60 percent come from Mexico. The remainder are about evenly divided between other Latin American countries and other regions of the world.
- **Children.** The 11.2 million undocumented immigrants included 1 million children. In addition, while living in the United States undocumented immigrants have given birth to approximately 4.5 million babies, who are legal citizens of the United States.
- **Labor force.** Approximately 8 million undocumented immigrants are employed in the United States, accounting for around 5 percent of the total U.S. civilian labor force. Unauthorized immigrants are much more likely than the average American to be employed in construction and hospitality (food service and lodging) jobs and less likely to be in white-collar jobs such as education, health care, and finance.
- **Distribution.** California and Texas have the largest number of undocumented immigrants (Figure 6.9.2). Nevada has the largest percentage.

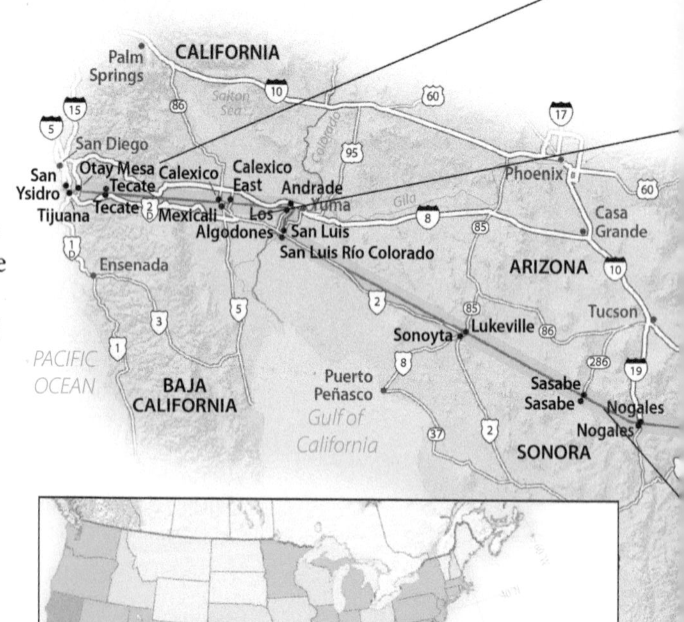

▲ 6.9.2 **UNDOCUMENTED IMMIGRANTS IN THE UNITED STATES**

Cities have distinctive economic bases.

Launch MapMaster North America in MasteringGEOGRAPHY

Select: *Political* then *Countries, States, and Provinces*

Select: *Population* then *Destination of Unauthorized Immigrants*. Adjust layer opacity to 50%.

Select: *Population* then *Distribution of African Americans*.

Deselect *African Americans* and select *Hispanic Americans*.

Deselect *Hispanic Americans* and select *Asian Americans*.

Which of the three groups matches most closely with the distribution of states that have the most undocumented immigrants?

▲ 6.9.3 **BORDER CROSSING BETWEEN SAN DIEGO AND TIJUANA**

CROSSING THE BORDER

The U.S.-Mexico border is 3,141 kilometers (1,951 miles) long. Guards heavily patrol border crossings in urban areas such as El Paso, Texas, and San Diego, California, or along highways (Figure 6.9.3). Rural areas are guarded by only a handful of agents (Figure 6.9.4). Crossing the border on foot legally is possible in several places (Figure 6.9.5). Elsewhere, the border runs mostly through sparsely inhabited regions (Figure 6.9.6). The United States has constructed a barrier covering approximately one-fourth of the border (Figure 6.9.7).

Actually finding the border is difficult in some remote areas. A joint U.S.–Mexican International Boundary and Water Commission is responsible for keeping official maps, on the basis of a series of nineteenth-century treaties. The commission is also responsible for marking the border by maintaining 276 six-foot-tall iron monuments erected in the late nineteenth century, as well as 440 fifteen-inch-tall markers added in the 1970s.

◄ 6.9.4 **BORDER CROSSING BETWEEN CALEXICO AND MEXICALI**
The border looks different in urban areas
Fly to *Mexicali, Mexico.*

1. **In which country are the green squares on the north side of the image?**
2. **What are the green squares?**

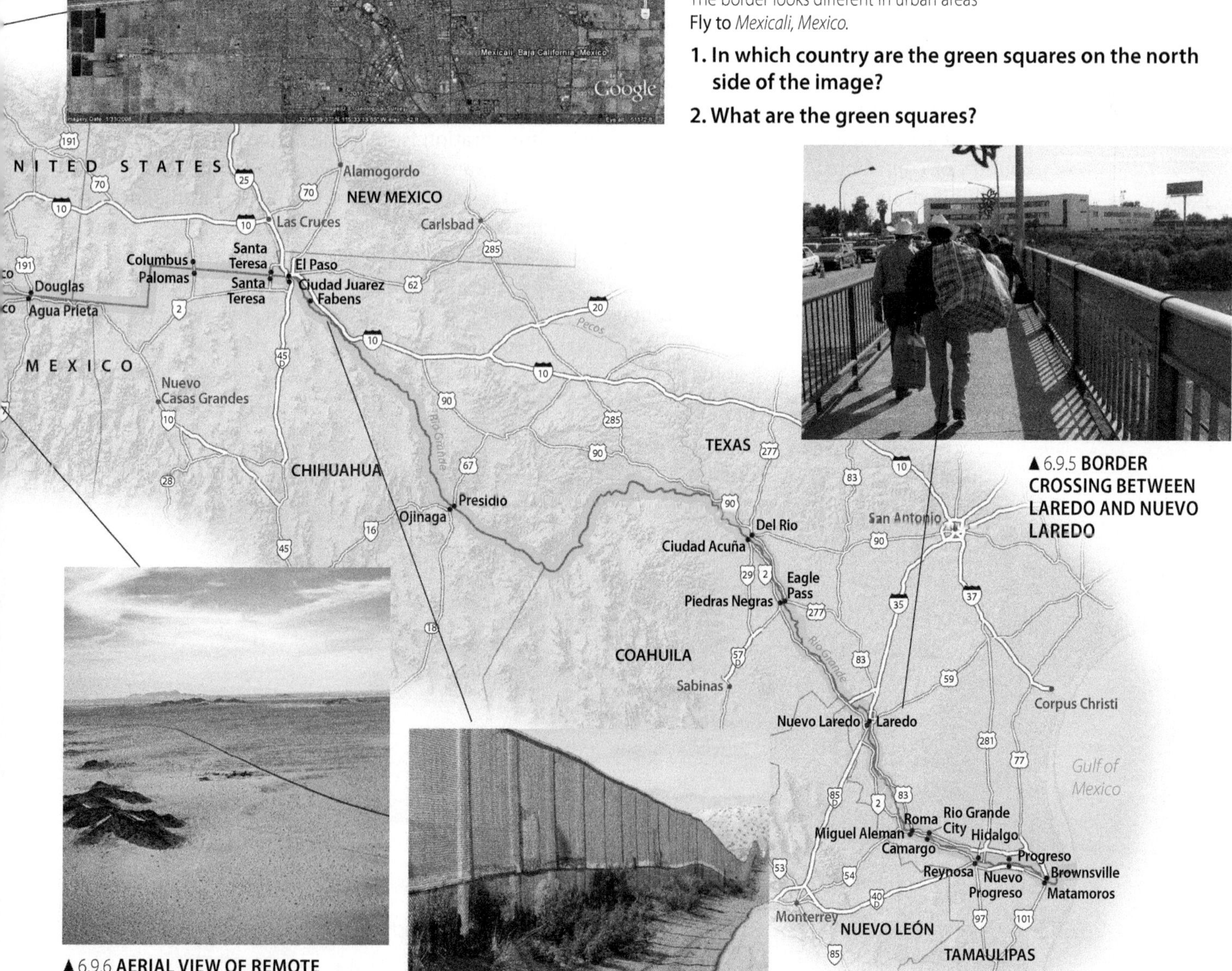

▲ 6.9.5 **BORDER CROSSING BETWEEN LAREDO AND NUEVO LAREDO**

▲ 6.9.6 **AERIAL VIEW OF REMOTE BORDER CROSSING BETWEEN RURAL ARIZONA AND SONORA**

▲ 6.9.7 **BORDER CROSSING BETWEEN EL PASO AND JUAREZ**

6.10 Controlling Migration

- Countries adopt a range of policies to promote or limit immigration.
- Many European countries have agreed to allow immigration between them.

Many countries view migrants suspiciously out of concern for undocumented immigration, security threats, or the movements of illegal materials. Some governments are concerned that immigrants will themselves become a burden on public resources, take away jobs from residents, or bring unwanted cultural change. While many of these fears are overblown, citizens expect their countries to control migration (Figure 6.10.1). Countries may adopt anti-immigration or **exclusionary policies** that punish migrants and those that hire them. Populist outrage and violence against immigrants may be considered an extension of such policies.

In contrast, **inclusionary policies** recognize that most immigration benefits the host economy and immigrants contribute as much as regular citizens by paying taxes, observing the law, and even serving in the militaries of their new country. Some immigrants bring professional skills that are in high demand in the destination country or contribute to economic growth (see "brain drain" in section 6.3). Others represent international business interests with considerable investments that appeal to the host country. In some places, immigrants are also encouraged to maintain their mother tongue and culture.

Most countries adopt **selective immigration policies** that excludes unwanted migrants and includes more desired ones. European countries that once maintained inclusionary policies to attract guestworkers from North Africa and the Middle East are today trying to limit immigration from these regions in favor of Europeans (Figure 6.10.2). The United States uses immigration caps to limit the number of persons coming from different countries. At the same time, the U.S. government provides special permission to highly desirable immigrants.

One way of imposing a selective policy is through government issued **visas** that require

▼ 6.10.1 **WEST BANK BORDER CROSSING** The border fence constructed by Israel around the West Bank is primarily a means to exclude Palestinians for fear of terrorist violence. Yet it also lets Israel control which Palestinians it will let in to work.

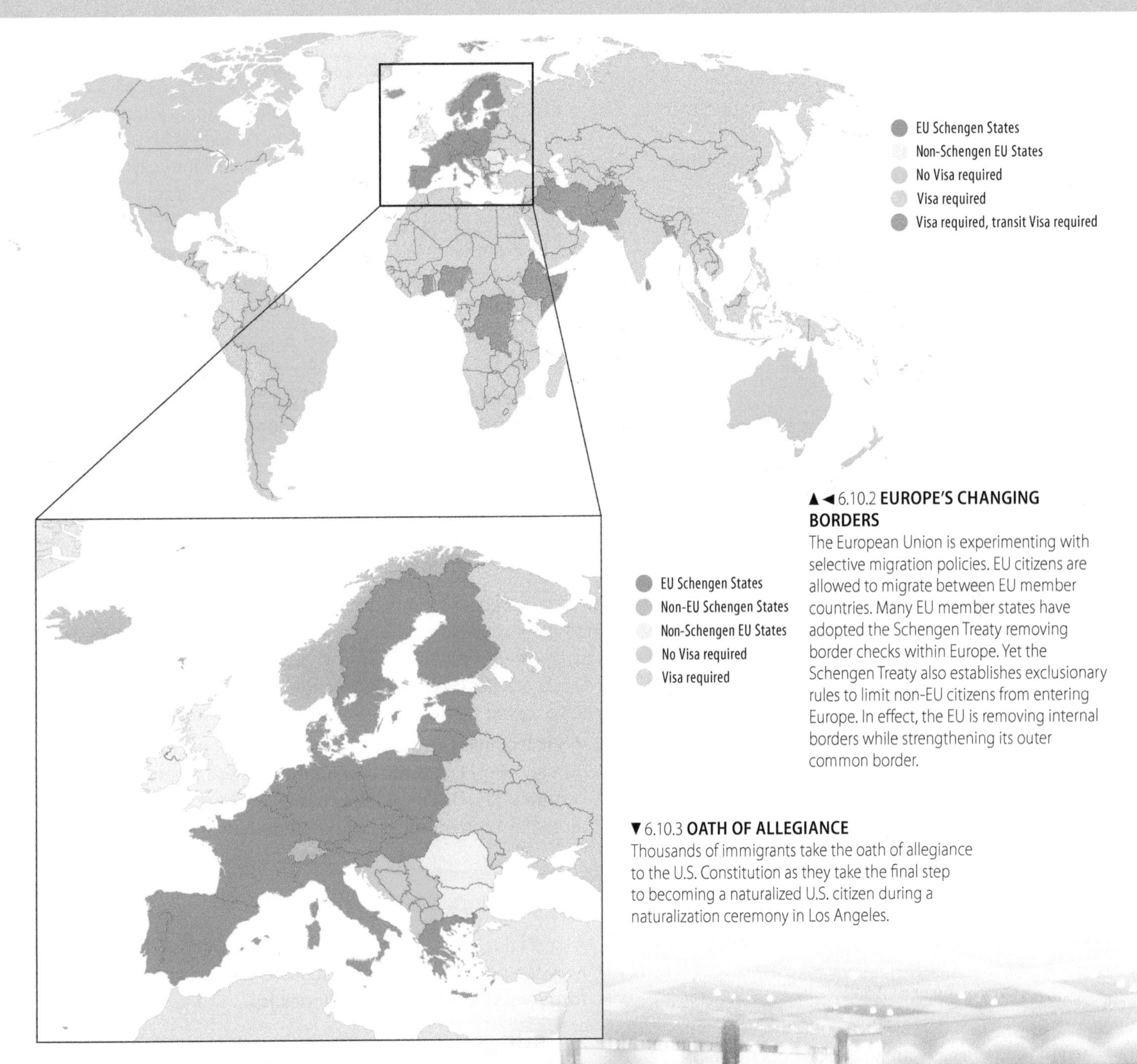

▲◄ 6.10.2 **EUROPE'S CHANGING BORDERS**
The European Union is experimenting with selective migration policies. EU citizens are allowed to migrate between EU member countries. Many EU member states have adopted the Schengen Treaty removing border checks within Europe. Yet the Schengen Treaty also establishes exclusionary rules to limit non-EU citizens from entering Europe. In effect, the EU is removing internal borders while strengthening its outer common border.

▼ 6.10.3 **OATH OF ALLEGIANCE**
Thousands of immigrants take the oath of allegiance to the U.S. Constitution as they take the final step to becoming a naturalized U.S. citizen during a naturalization ceremony in Los Angeles.

a migrant to gain permission to enter a country before arriving there. Special work or residency permits may also be used. Visas and permits typically differentiate between temporary migrants and long-term immigrants. Temporary migrants include tourists, commuters (who live outside the country but work within it), and those passing through the country. Those who are seeking to indefinitely or permanently immigrate must obtain a visa that will allow them to eventually **naturalize**, that is to become a legal permanent citizen of the host country (Figure 6.10.3). Each country has different rules for visas, permits, and naturalization.

Migration has always been a constant part of human life, whether to find food, jobs, or recreation. International migration reshaped the populations of many countries. Today internal and international migration continues to rework the distribution of humans on Earth.

Key Questions

What are the historical patterns of human migration?

- Prehistoric humans migrated out of Africa, eventually populating Earth.
- Mass migration between continents and countries in the past 500 years has dramatically changed the characteristics of local populations.
- The origins and destinations of modern mass migration have changed over the last five centuries.

Why do people migrate?

- Most people move in search of employment and to improve their economic situation. Even those who move for other reasons must support themselves economically in their new locations.
- Some migrate because political or environmental circumstances force them to flee their home areas.
- The most common form of international migration today is tourism, which is a temporary movement of people.

How does migration change population characteristics?

- The most common form of permanent migration is the relocation of households within countries. These migrants are typically leaving areas with contracting economies and moving to areas with expanding opportunities.
- The source countries of immigrants to the United States have changed in the last two centuries.

How do states deal with increasing immigration?

- Strong economies attract immigrants who may not have permission from the government to enter or work in that country.
- Governments adopt policies to promote or limit migration based on economic, political and cultural attitudes about migrants.

Thinking Geographically

Proponents of free trade argue that economic actors should be allowed to freely move goods across national borders. This is thought to benefit workers who produce goods for export. International labor migrants are economic actors yet their migration to a country is usually seen as a threat to workers there.

1. **Do the arguments in favor of free trade really exclude free migration? What economic and noneconomic factors do these debates take into consideration?**

The original interpretation of the 1951 Refugee Convention was that refugees should be granted permanent asylum in a host country. As refugee migration to Europe increased during the 1990s, many wealthy host countries granted only temporary status, requiring refugees to return to their home countries as soon as the immediate danger had passed.

2. **What are the reasons for and against refugees being granted permanent asylum? What are the reasons for and against their having to return to their home country?**

The current debate on immigration in the United States focuses on Hispanic migration, although much of this migration is legal. A common argument is that Hispanics will change the culture of the United States (Figure 6.CR.1). Some states have responded with measures that may discriminate against legal residents.

3. **What arguments were made against previous waves of immigration? Did these immigrants change the dominant culture of the U.S. or did they assimilate? Are current immigrant communities any different than historical ones?**

On the Internet

A list of basic terms, statistics, and descriptions of major migration issues are provided by the International Organization for Migration: **http://www.iom.int/jahia/jsp/index.jsp.**

The U.S. Census provides detailed statistics on U.S. migration and residential mobility: **http://www.census.gov/hhes/migration.**

► 6.CR.1 **THE MINUTEMAN PROJECT**
The group, which is opposed to undocumented immigration, rallies at the U.S. Capitol.

Interactive Mapping

TOURISM PATTERNS

Most tourists are seeking recreation. Examine the map in section 6.6.1

Launch Mapmaster World in MasteringGEOGRAPHY™

Select: *Climate* from the *Physical Environment* menu.

Examine the map in Figure 6.6.1 of *International Tourist Arrivals.*

Most international tourists travel to what types of climate(s)?

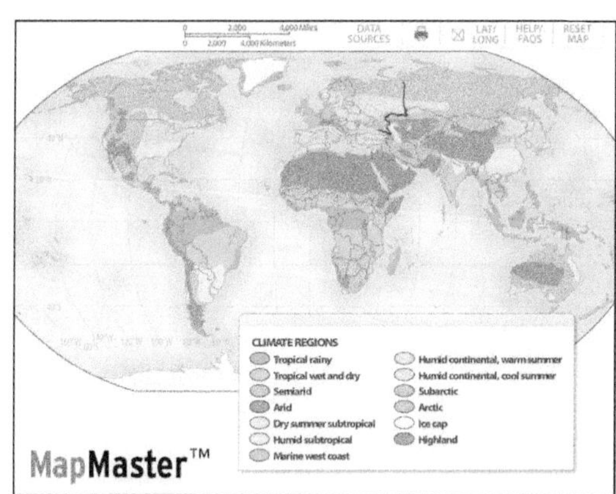

Explore

FORCED MIGRATION IN DARFUR

Use Google Earth to explore the destruction in Sudan's Darfur region, where government-backed militias have forced inhabitants to flee their villages.

Turn on *Layers>Global Awareness*

Select *USHMM Crisis in Darfur.*

Fly to: *Darfur, Sudan.*

Select: *Crisis in Darfur,* review images of destroyed villages.

1. **How have militias tried to revent displaced persons returning home?**
2. **Examine the local landscape for economic uses. What interest might the militias have in clearing Darfuris from their land?**

Key Terms

Arrivals
The entry of people into a country.

Brain drain
The loss of highly-trained professionals through emigration.

Counterurbanization
Residential relocation from urban and suburban places to rural ones.

Demic diffusion
The movement of people through space over time.

Diaspora
The widespread diffusion of a people from their region of origin.

Displacement
When people are compelled to move from one place to another.

Ecotourism
Tourism meant to lessen visitors' impact on the environment.

Emigration
Out-migration from one area to another area.

Environmentally displaced persons
Individuals compelled to flee natural disasters.

Exclusionary policies
Government rules to prevent immigration.

Forced migration
Any movement undertaken under coercive conditions, such as violence or disaster.

Guest workers
Labor immigrants admitted to meet demand for more workers.

Human origins
Where and when modern humans first appeared and how they peopled Earth.

Immigrant nation
A country whose population is primarily composed of immigrants and their descendants.

Immigration
In-migration to an area from another area.

Inclusionary policies
Government rules meant to accommodate or even encourage immigration.

Internally displaced persons (IDPs)
Persons compelled to migrate within their country of origin.

Interregional migration
Migration between two regions of the same country.

Intraregional migration
Migration within one region.

Migration stream
A sustained movement of people from the one source area to a common destination area.

Naturalize
The process of becoming a citizen of a country other than one's country of origin.

Net migration
The numerical difference between immigration and emigration.

Pull factors
A migration destination's features that attract in-migration.

Push factors
A migration origin's features that fuel out-migration.

Refugee
A person compelled to migrate outside their country of origin as defined by an international convention.

Remittances
Money migrants send to family or others in their place of origin.

Residential mobility
The movement of households from one place to another.

Seasonal migrants
Those who move in response to regular but temporary conditions, for example, to work jobs at harvest time or to escape cold winters.

Selective immigration policies
Government rules to include some migrants and exclude others.

Settler migration
Individuals and households that migrate to new colonies.

Temporary labor migrants
Migrants looking for work but who do not permanently migrate.

Undocumented immigrants
Migrants who enter a country without fulfilling a country's legal requirements to do so.

Visa
Permission to enter a country that is granted prior to or during arrival.

▶ LOOKING AHEAD

When people migrate they take their cultures with them. The age of global migration has helped to diffuse religions and languages to new places and, in the process, migrants' new experiences have changed their cultures.

7 Languages and Religions

Earth's languages and religions are important examples of cultural diversity. Geographers look for similarities and differences in the cultural features at different places, the reasons for their distribution, and the importance of these differences for world peace.

Geographers are not linguists or theologians, so they stay focused on those elements that are geographically significant. Some languages are spoken and some religions are practiced by people throughout the world, whereas other languages and religions are found in geographically limited areas.

Cultural values are like luggage: people carry them with them when they move from place to place. Migrants typically retain their religion, but learn the language of the new location.

Where are languages distributed?

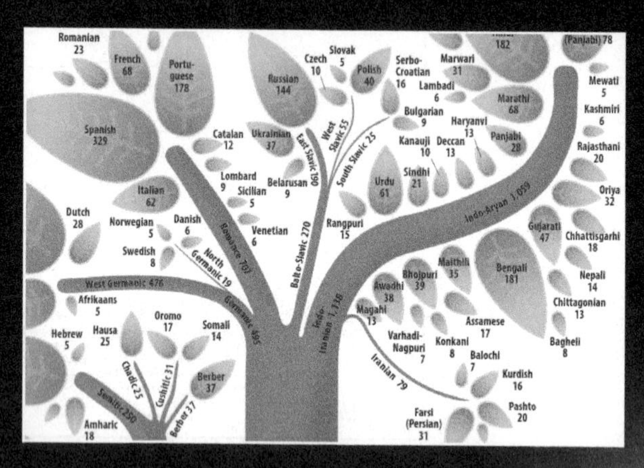

7.1 **Classifying Languages**

7.2 **Distribution of Languages**

7.3 **Origin and Diffusion of Languages**

TILES IN FRENCH AND VIETNAMESE IN THE BASILICA OF OUR LADY OF THE IMMACULATE CONCEPTION (NOTRE-DAME BASILICA), HO CHI MINH CITY, VIETNAM

How do languages share space?

7.4 **Dominant and Endangered Languages**

7.5 **French and Spanish in North America**

7.6 **Multilingual States**

Where are religions distributed?

SCAN FOR DETAILS ABOUT EVERY LANGUAGE OF THE WORLD

7.7 **Distribution of Religions**

7.8 **Geographic Branches of Religions**

7.9 **Origin of Religions**

7.10 **Diffusion of Universalizing Religions**

How do religions shape landscapes?

7.11 **Holy Places in Universalizing Religions**

7.12 **Ethnic Religions and the Landscape**

7.13 **Religious Conflicts in the Middle East**

7.1 Classifying Languages

► **The world's 6,000-plus languages can be classified into families, branches, and groups.**

► **Only around 100 of these languages are used by more than 5 million people each.**

Language is a system of communication through speech. It is a collection of sounds that a group of people understands to have the same meaning. An **official language** is one used by the government for conducting business and publishing documents. Many languages also have a **literary tradition**, or a system of written communication. Approximately 85 languages are spoken by at least 10 million people, and approximately 300 languages by between 1 million and 10 million people.

The world's languages can be organized into families, branches, and groups:

- **Language family:** a collection of languages related through a common ancestral language that existed long before recorded history (Figure 7.1.1).
- **Language branch:** a collection of languages within a family related through a common ancestral language that existed several thousand years ago; differences are not as extensive or as old as between language families, and archaeological evidence can confirm that the branches derived from the same family.
- **Language group:** a collection of languages within a branch that share a common origin in the relatively recent past and display many similarities in grammar and vocabulary.

Figure 7.1.2 attempts to depict relationships among language families, branches, and groups:

- Language families form the trunks of the trees.
- Some trunks divide into several branches, which logically represent language branches, as well as groups.
- Individual languages are displayed as leaves.

The larger the trunks and leaves are, the greater the number of speakers of those families and languages.

Numbers on the tree are in millions of **native speakers**. Native speakers are people for whom the language is their first language. The totals exclude those who use the languages as second languages.

Figure 7.1.2 displays each language family as a separate tree at ground level, because differences among families predate recorded history. Linguists speculate that language families were joined together as a handful of superfamilies tens of thousands of years ago. Superfamilies are shown as roots below the surface, because their existence is speculative.

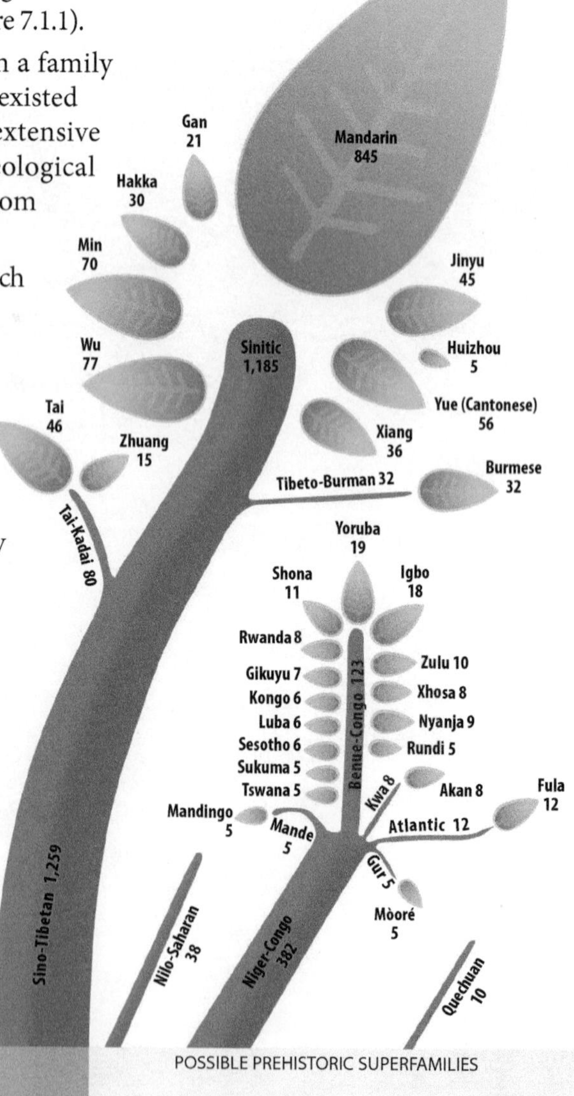

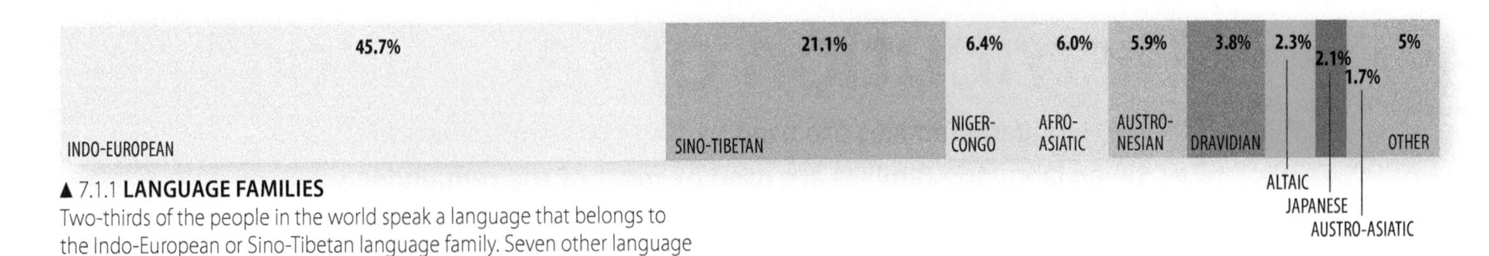

▲ 7.1.1 **LANGUAGE FAMILIES**
Two-thirds of the people in the world speak a language that belongs to the Indo-European or Sino-Tibetan language family. Seven other language families are used by between 2 and 6 percent of the world.

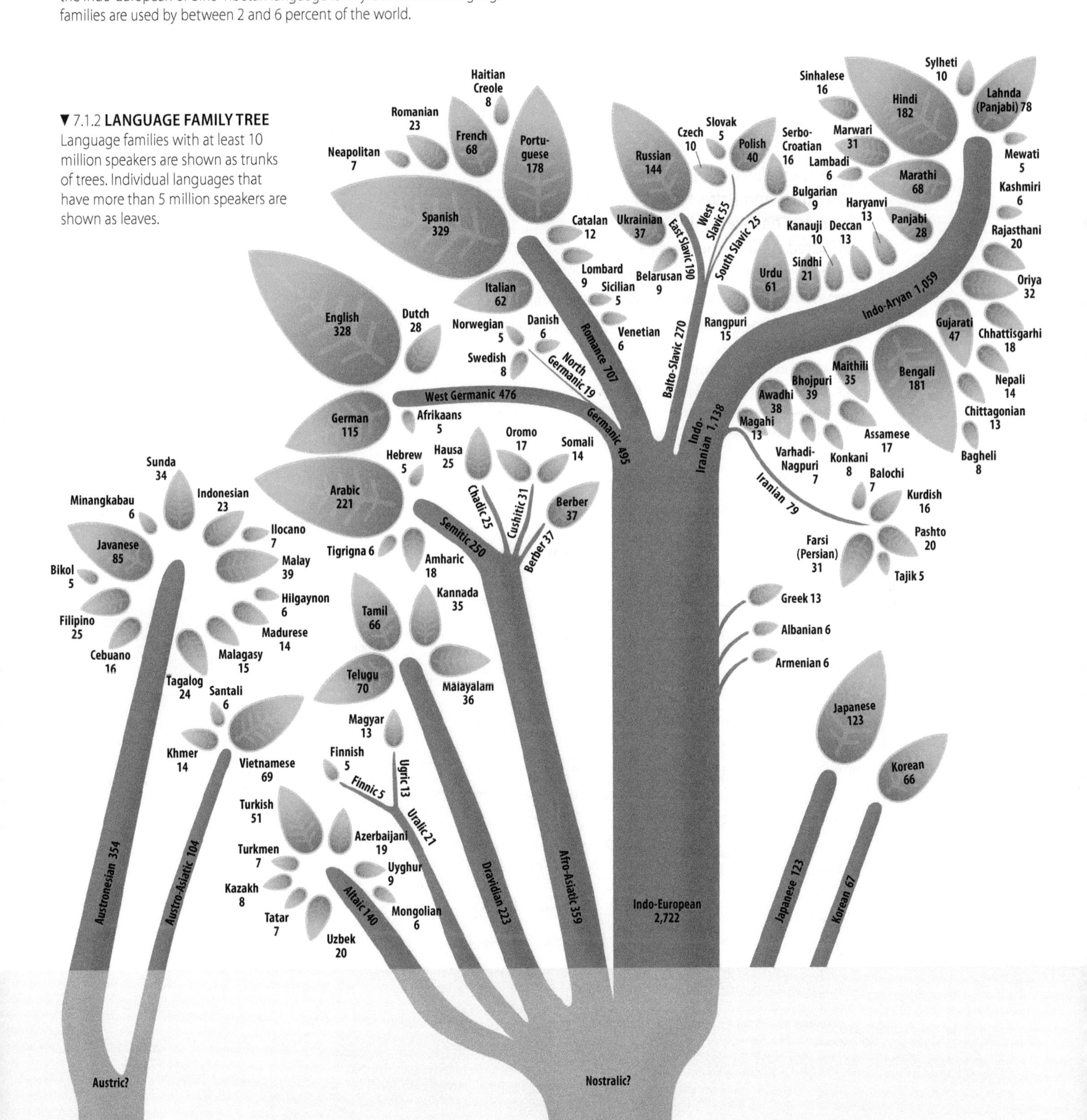

▼ 7.1.2 **LANGUAGE FAMILY TREE**
Language families with at least 10 million speakers are shown as trunks of trees. Individual languages that have more than 5 million speakers are shown as leaves.

7.2 Distribution of Languages

- Two language families are used by two-thirds of the world.
- Seven language families are used by most of the remainder.

Language families with at least 10 million native speakers are shown on Figure 7.2.1. Paragraphs placed around Figure 7.2.1 identify the nine families with at least 100 million native speakers.

The Indo-European family is divided into eight branches (Figure 7.2.2). The Indo-Aryan branch is the largest, with more than 1 billion speakers. Germanic, Romance, and Balto-Slavic branches have several hundred million speakers each. Four less extensively used Indo-European branches are Albanian, Armenian, Greek, and Celtic.

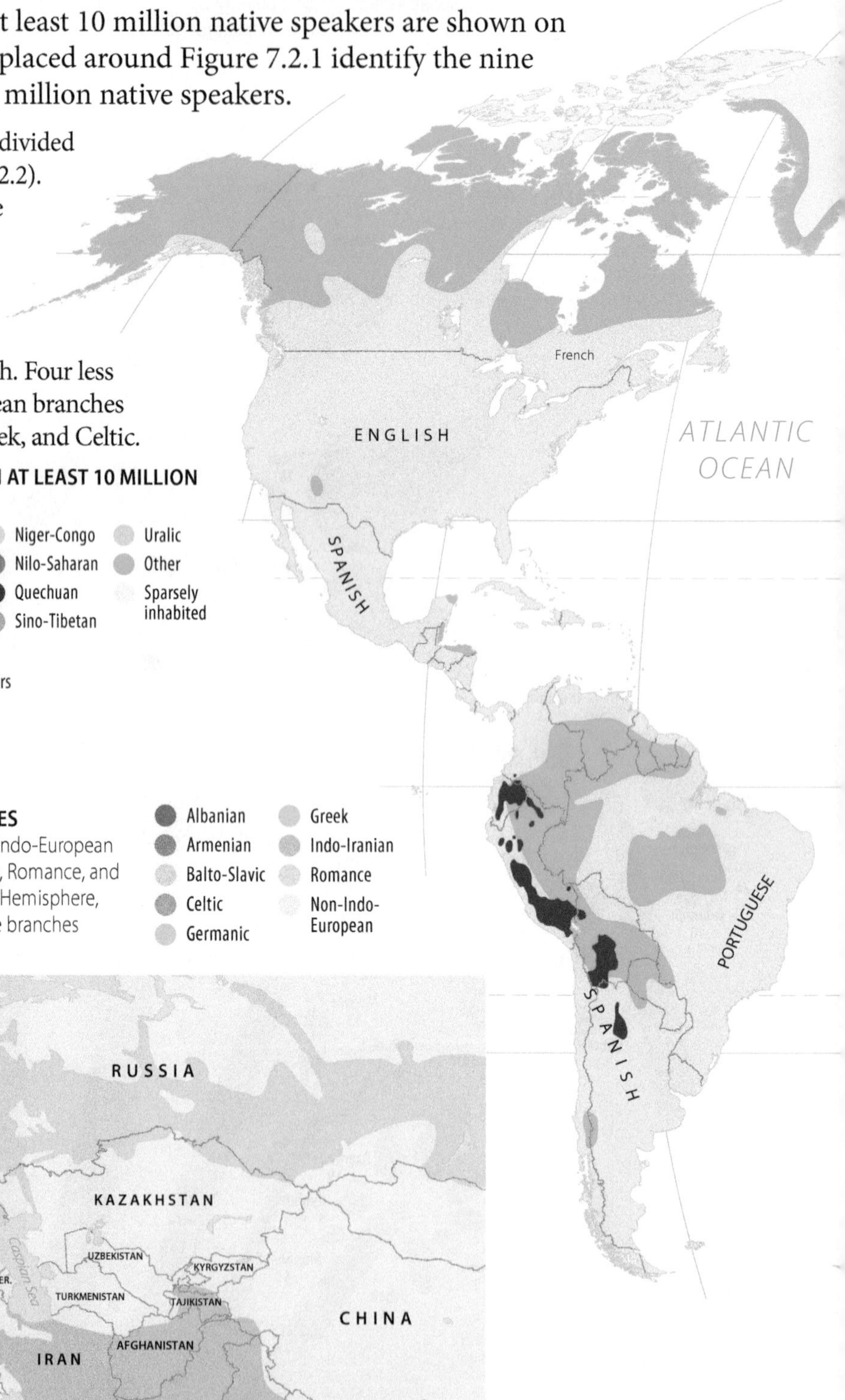

► 7.2.1 LANGUAGE FAMILIES WITH AT LEAST 10 MILLION SPEAKERS

Afro-Asiatic
Altaic
Austro-Asiatic
Austronesian
Dravidian
Indo-European
Japanese
Korean
Niger-Congo
Nilo-Saharan
Quechuan
Sino-Tibetan
Uralic
Other
Sparsely inhabited

SPANISH
Languages with more than 100 million speakers
French
Languages with 50–100 million speakers

French
ENGLISH
ATLANTIC OCEAN
SPANISH
PORTUGUESE
SPANISH

▼ 7.2.2 INDO-EUROPEAN BRANCHES

The most widely spoken branches of Indo-European are Indo-Iranian in Asia and Germanic, Romance, and Balto-Slavic in Europe. In the Western Hemisphere, the Germanic and Romance language branches predominate.

Albanian
Armenian
Balto-Slavic
Celtic
Germanic
Greek
Indo-Iranian
Romance
Non-Indo-European

ICELAND
ATLANTIC OCEAN
N
SWEDEN
FINLAND
NORWAY
RUSSIA
ESTONIA
LATVIA
DENMARK
LITHUANIA
UNITED KINGDOM
IRELAND
NETH.
BELARUS
GERMANY
POLAND
BELGIUM
CZECH REP.
UKRAINE
SLOVAKIA
KAZAKHSTAN
FRANCE
SWITZ.
AUSTRIA
HUNGARY
MOLDOVA
SLOVENIA
CROATIA
ROMANIA
BOS.& HERZ.
SERB.
Black Sea
ITALY
MONT.
BULGARIA
Caspian Sea
UZBEKISTAN
ALBANIA
PORTUGAL
SPAIN
MACEDONIA
GEORGIA
KYRGYZSTAN
ARMENIA
AZER.
GREECE
TURKEY
TURKMENISTAN
TAJIKISTAN
CHINA
SYRIA
Mediterranean Sea
AFGHANISTAN
IRAN
IRAQ
0 500 1,000 Miles
0 500 1,000 Kilometers
PAKISTAN
NEPAL
BHUTAN
INDIA
BANGLADESH
Arabian Sea
Bay of Bengal

INDO-EUROPEAN FAMILY

The world's most widely spoken family, shown in more detail in Figure 7.2.2. English belongs to the Indo-European language family.

ALTAIC FAMILY

Spoken across an 8,000-kilometer (5,000-mile) band of Asia between Turkey and China. Turkish, by far the most widely used Altaic language, was once written with Arabic letters. In 1928 the Turkish government, led by Kemal Ataturk, ordered that the language be written with the Roman alphabet as a symbol of modernization of the culture and economy.

SINO-TIBETAN FAMILY

Encompasses the languages of China. There is no single spoken Chinese language. Rather, the most important is Mandarin (or, as the Chinese call it, *pu tong hua*—common speech). Chinese languages are based on 420 one-syllable words, which in turn can be combined to form multi-syllable words.

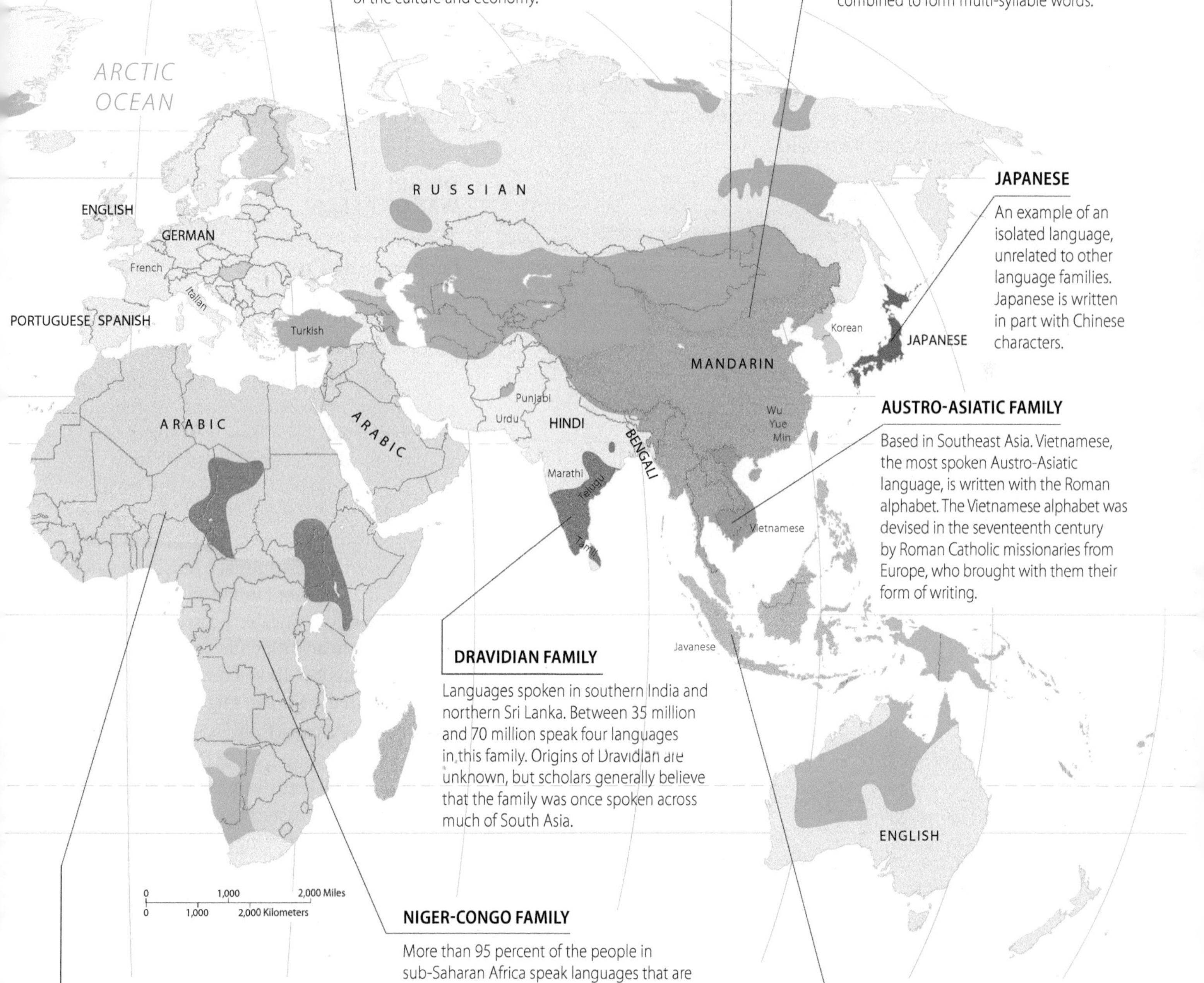

JAPANESE

An example of an isolated language, unrelated to other language families. Japanese is written in part with Chinese characters.

AUSTRO-ASIATIC FAMILY

Based in Southeast Asia. Vietnamese, the most spoken Austro-Asiatic language, is written with the Roman alphabet. The Vietnamese alphabet was devised in the seventeenth century by Roman Catholic missionaries from Europe, who brought with them their form of writing.

DRAVIDIAN FAMILY

Languages spoken in southern India and northern Sri Lanka. Between 35 million and 70 million speak four languages in this family. Origins of Dravidian are unknown, but scholars generally believe that the family was once spoken across much of South Asia.

AFRO-ASIATIC FAMILY

Includes Arabic and Hebrew. Arabic is the major Afro-Asiatic language, an official language in two dozen countries of the Middle East, and the language of Islam's holiest book the Quran. Hebrew, the language of much of the Jewish Bible and Christian Old Testament, is a rare case of an extinct language that was revived in the twentieth century as a modern language used in Israel (see Section 7.4).

NIGER-CONGO FAMILY

More than 95 percent of the people in sub-Saharan Africa speak languages that are generally classified as belonging to the Niger-Congo family. Most lack a written tradition and only five are spoken by more than 10 million people. More than 1,000 distinct languages have been documented in Africa, but no one knows the precise number, and scholars disagree on classifying those known into families.

AUSTRONESIAN FAMILY

Languages spoken mostly in Indonesia. The people of Madagascar speak Malagasy, which also belongs to the Austronesian family. This is evidence of migration to Madagascar from Indonesia, apparently in small boats 3,000 kilometers (1,900 miles) across the Indian Ocean, roughly 2,000 years ago.

7.3 Origin and Diffusion of Languages

- Languages diffuse from their place of origin through migration.
- Dialects within languages also emerge through migration and isolation.

A language originates at a particular place and traditionally diffuses to other locations through the migration of its speakers. The location of English-language speakers serves as a case study for understanding the process by which languages have been distributed around the world.

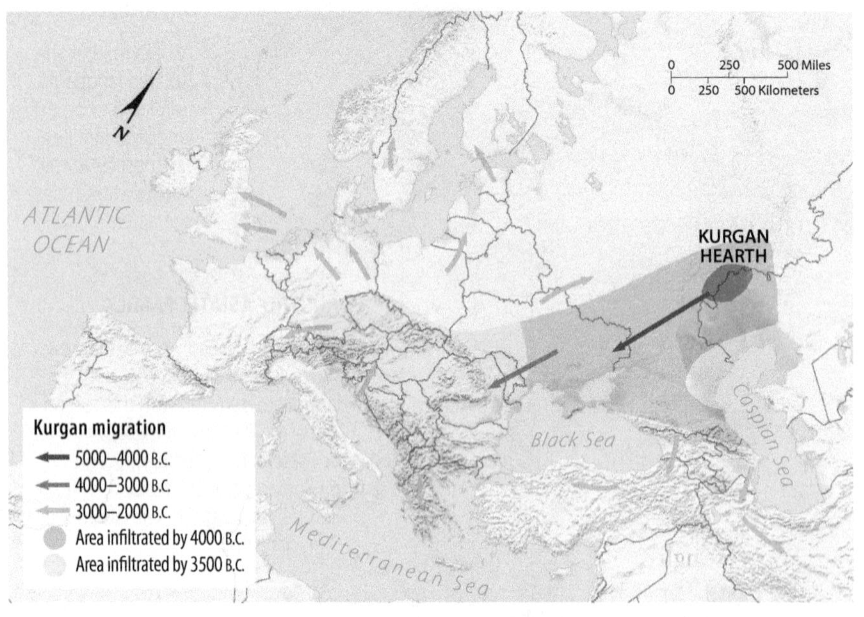

▲ 7.3.1 ORIGIN AND DIFFUSION OF INDO-EUROPEAN: NOMADIC WARRIOR HYPOTHESIS

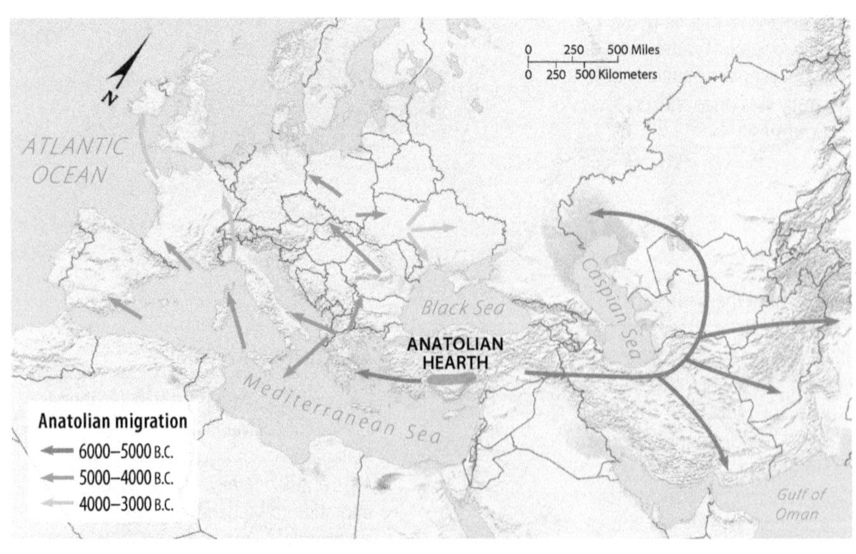

▲ 7.3.2 ORIGIN AND DIFFUSION OF INDO-EUROPEAN: SEDENTARY FARMER HYPOTHESIS

ORIGIN AND DIFFUSION: WAR OR PEACE?

The origin and early diffusion of language families predate recorded history. Linguists and anthropologists disagree on when and where Indo-European originated and the process and routes by which it diffused.

- **The "War" Hypothesis.** The first Indo-European speakers may have been the Kurgan people, who lived near the border of present-day Russia and Kazakhstan (Figure 7.3.1). The Kurgans were nomads, among the first to domesticate horses and cattle around 5,000 years ago. In search of grasslands for their animals, Kurgan warriors conquered much of Europe and South Asia, using their domesticated horses as weapons. According to an influential hypothesis by Marija Gimbutas, Indo-European language spread with the Kurgan migration.
- **The "Peace" Hypothesis.** Archaeologist Colin Renfrew argues that the first Indo-European speakers lived 2,000 years before the Kurgans, in eastern Anatolia, part of present-day Turkey (Figure 7.3.2). Renfrew believes that Indo-European speakers migrated into Europe and South Asia along with agricultural practices rather than by military conquest. The language triumphed because its speakers became more numerous and prosperous by growing their own food instead of relying on hunting.

Regardless of how Indo-European diffused, communication was poor among the people, whether warriors or farmers. After many generations of complete isolation, these migrants evolved into speaking increasingly distinct branches, groups, and individual languages.

ORIGIN AND DIFFUSION OF ENGLISH

English is the language of England because of migration to Britain from various parts of Europe (Figure 7.3.3):

- **Celtic tribes around 2000 B.C.** The Celts spoke languages classified as Celtic. We know nothing of earlier languages spoken in Britain.
- **Angles, Saxons, and Jutes around A.D. 450.** These tribes from northern Germany and southern Denmark pushed the Celtic tribes to remote northern and western parts of Britain, including Cornwall and the highlands of Scotland and Wales. The name England comes from Angles' land, and English people are often called Anglo-Saxons.
- **Vikings between 787 and 1171.** Vikings from present-day Norway landed on the northeast coast of England and raided several settlements there. Although unable to conquer Britain, Vikings remaining in the country contributed words from their language.
- **The Normans in 1066.** The Normans, from present-day Normandy in France, conquered England in 1066 and established French as the official language for the next 300 years. The British Parliament enacted the Statute of Pleading in 1362 to change the official language of court business from French to English, though Parliament itself continued to conduct business in French until 1489.

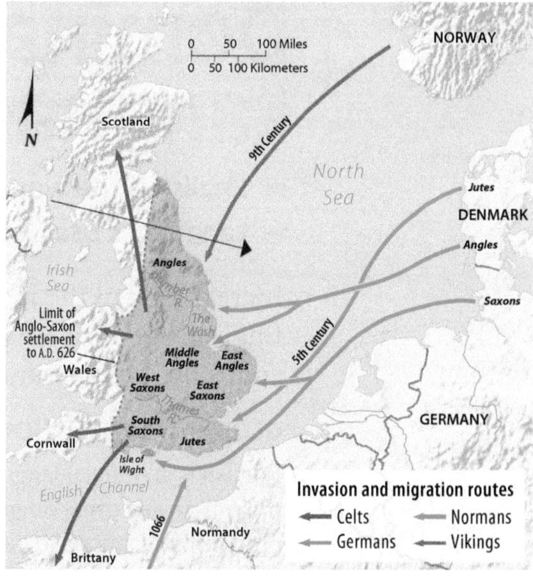

▲ 7.3.3 **INVASIONS OF ENGLAND**

DIALECTS

A **dialect** is a regional variation of a language distinguished by distinctive vocabulary, spelling, and pronunciation. Geographers are especially interested in differences in dialects, because they reflect distinctive features of the environments in which groups live. Dialects vary in three significant ways:

- Vocabulary • Spelling • Pronunciation

North Americans are well aware that they speak English differently than the British, not to mention people living in India, Pakistan, Australia, and other English-speaking countries. Differences result from centuries of isolation. Separated by the Atlantic Ocean, English in the United States and England evolved independently during the eighteenth and nineteenth centuries, with little influence on one another. Few residents of one country could visit the other, and the means to transmit the human voice over long distances would not become available until the twentieth century.

English also varies by regions within individual countries (Figure 7.3.4). In both the United States and England, northerners sound different from southerners (Figure 7.3.5).

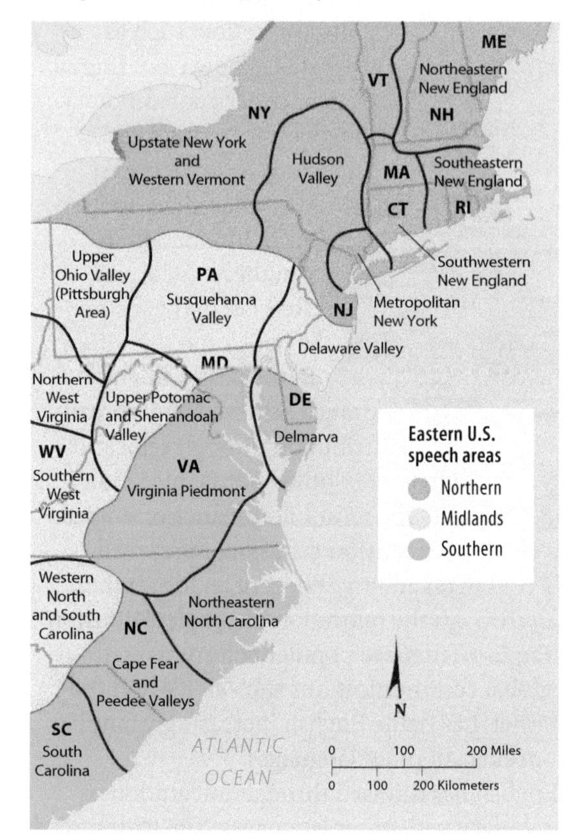

▲ 7.3.4 **U.S. DIALECTS**

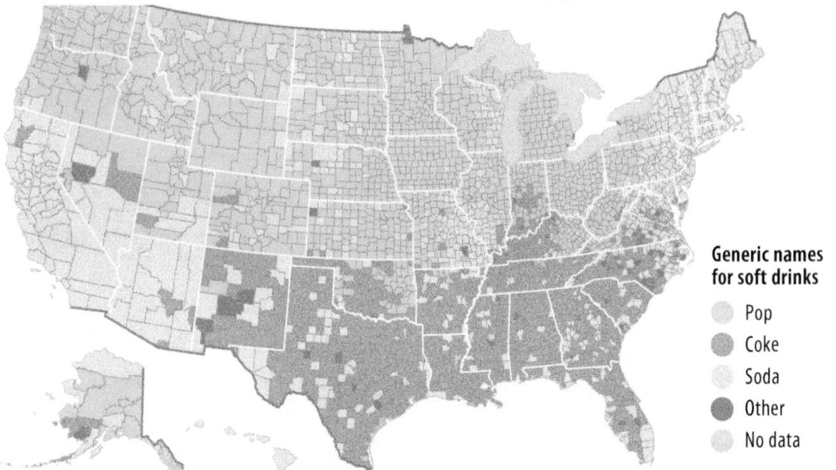

► 7.3.5 **SOFT-DRINK DIALECTS**
The most commonly used word for a soft drink–pop, cola, or soda–varies by region within the United States.

7.4 Dominant and Endangered Languages

- ▶ **English is the world's leading lingua franca.**
- ▶ **Languages used by only a few people may become extinct unless preserved.**

Increasingly in the modern world, the language of international communication is English. A Polish airline pilot who flies over France, for example, speaks to the traffic controller on the ground in English.

GLOBAL DISTRIBUTION OF ENGLISH

English is an official language in 55 countries, more than any other language, and is the predominant language in Australia, the United Kingdom, and the United States (Figure 7.4.1). Two billion people live in a country where English is an official language, even if they cannot speak it.

The contemporary distribution of English speakers around the world exists because the people of England migrated with their language when they established colonies during the past four centuries. English first diffused west from England to North American colonies in the seventeenth century. More recently, the United States has been responsible for diffusing English to several places.

▼ 7.4.1 ENGLISH-SPEAKING COUNTRIES

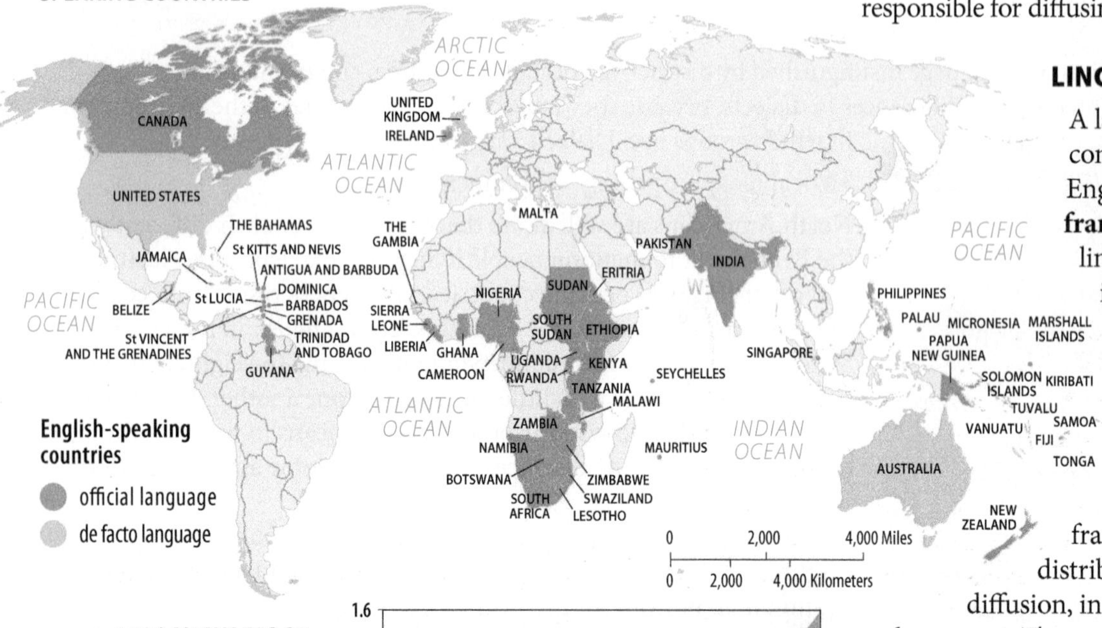

LINGUA FRANCA

A language of international communication, such as English, is known as a **lingua franca**. Other contemporary lingua franca languages include Swahili in East Africa, Hindi in South Asia, Indonesian in Southeast Asia, and Russian in the former Soviet Union.

In the past, a lingua franca achieved widespread distribution through relocation diffusion, in other words migration and conquest. The recent dominance of English is a result of expansion diffusion, the spread of a trait in an additive effect of an idea rather than through the relocation of people. Diffusion of English-language popular culture, as well as global communications such as TV and the Internet, has made English increasingly familiar to speakers of other languages.

English has diffused through integration of vocabulary with other languages. The widespread use of English in French is called **Franglais**, in Spanish **Spanglish**, and in German **Denglish**. English has also been the leading language of the Internet since its inception (Figure 7.4.2).

► 7.4.2 LANGUAGES OF ONLINE USERS

During the 1990s, three-fourths of the people online and three-fourths of websites used English. In recent years, other languages have been catching up to English, especially Chinese.

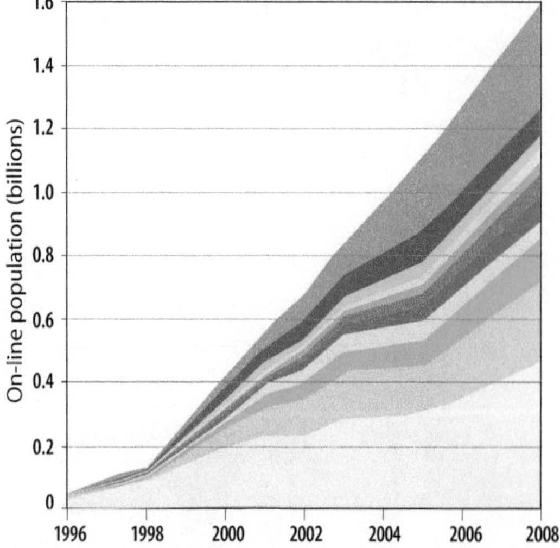

ISOLATED LANGUAGES

An **isolated language** is a language unrelated to any other and therefore not attached to any language family. The best example in Europe is Basque, apparently the only language currently spoken in Europe that survives from the period before the arrival of Indo-European speakers. Basque may have once been spoken over a wider area but was abandoned where its speakers came in contact with Indo-Europeans (Figure 7.4.3).

◀ 7.4.3 **BASQUE**
Basque is spoken by 600,000 people in the Pyrenees Mountains of Northern Spain and Southwestern France. Basque's lack of connection to other languages reflects the isolation of the Basque people in their mountainous homeland. The sign painted on the wall of a building in Bayonne, France, depicts the Basque flag and says in Basque "The people must live."

EXTINCT LANGUAGES

Thousands of languages are **extinct languages**, once in use—even in the recent past—but no longer spoken or read in daily activities by anyone in the world.

Hebrew is a rare case of an extinct language that has been revived. A language of daily activity in biblical times, Hebrew diminished in use in the fourth century B.C. and was thereafter retained only for Jewish religious services. At the time of Jesus, people in present-day Israel generally spoke Aramaic, which in turn was replaced by Arabic.

When Israel was established as an independent country in 1948, Hebrew became one of the new country's two official languages, along with Arabic. Hebrew was chosen because the Jewish population of Israel consisted of refugees and migrants from many countries who spoke many languages. Because Hebrew was still used in Jewish prayers, no other language could so symbolically unify the disparate cultural groups in the new country (Figure 7.4.4).

◀ 7.4.4 **HEBREW: A REVIVED LANGUAGE**
Grocery store, Jerusalem.

PRESERVING ENDANGERED LANGUAGES

Ethnologue considers approximately 500 languages as in danger of becoming extinct, but some are being preserved. The European Union has established the European Bureau for Lesser Used Languages (EBLUL), based in Dublin, Ireland, to provide financial support for the preservation of endangered languages, including several belonging to the Celtic family (Figure 7.4.5):

- **Irish Gaelic.** An official language of the Republic of Ireland, along with English; Ireland's government publications are in Irish as well as English.
- **Scottish Gaelic.** Most speakers live in remote highlands and islands of northern Scotland.
- **Welsh.** In Wales, teaching Welsh in schools is compulsory, road signs are bilingual, Welsh-language coins circulate, and a television and radio station broadcast in Welsh.
- **Cornish.** Became extinct in 1777, with the death of the language's last known native speaker; a standard written form of Cornish was established in 2008.
- **Breton.** Concentrated in France's Brittany region; Breton differs from the other Celtic languages in that it has more French words.

The survival of any language depends on the political and military strength of its speakers. The Celtic languages declined because the Celts lost most of the territory they once controlled to speakers of other languages.

▼ 7.4.5 **CELTIC**
Bilingual parking signs outside The Celtic Manor Resort where the 2010 Ryder Cup (golf) was held in Newport Gwent South East Wales

7.5 French and Spanish in North America

- French and Spanish are increasingly used in North America.
- Languages can mix to form new ones.

North America is dominated by English speakers. Yet other languages, especially French in Canada and Spanish in the United States, are becoming increasingly prominent. At the same time, French, Spanish, English, and other languages are mixing to form new languages.

▲ 7.5.1 FRENCH IN CANADA: "HELLO"

FRENCH IN CANADA

French is one of Canada's two official languages, along with English (Figure 7.5.1). French speakers comprise one-fourth of the country's population. Most are clustered in Québec, where they comprise more than three-fourths of the province's speakers (Figure 7.5.2).

Until recently, Québec was one of Canada's poorest and least developed provinces. Its economic and political activities were dominated by an English-speaking minority, and the province suffered from cultural isolation and a lack of French-speaking leaders.

The Québec government has made the use of French mandatory in many daily activities. Québec's Commission de Toponyme is renaming towns, rivers, and mountains that have names with English-language origins. The word *Stop* has been replaced by *Arrêt* on the red octagonal road signs, even though *Stop* is used throughout the world, even in France and other French-speaking countries. French must be the predominant language on all commercial signs, and the legislature passed a law banning non-French outdoor signs altogether (later ruled unconstitutional by the Canadian Supreme Court).

Many Québécois favored total separation of the province from Canada as the only way to preserve their cultural heritage. Voters in Québec have thus far rejected separation from Canada, but by a slim majority. Alarmed at these pro-French policies, many English speakers and major corporations moved from Montréal, Québec's largest city, to English-speaking Toronto, Ontario.

Confrontation during the 1970s and 1980s has been replaced in Québec by increased cooperation between French and English speakers. Montréal's neighborhoods, once highly segregated between French-speaking residents on the east and English-speaking residents on the west, have become more linguistically mixed.

Although French dominates over English, Québec faces a fresh challenge of integrating a large number of immigrants from Europe, Asia, and Latin America who don't speak French. Many immigrants would prefer to use English rather than French as their lingua franca but are strongly discouraged from doing so by the Québec government.

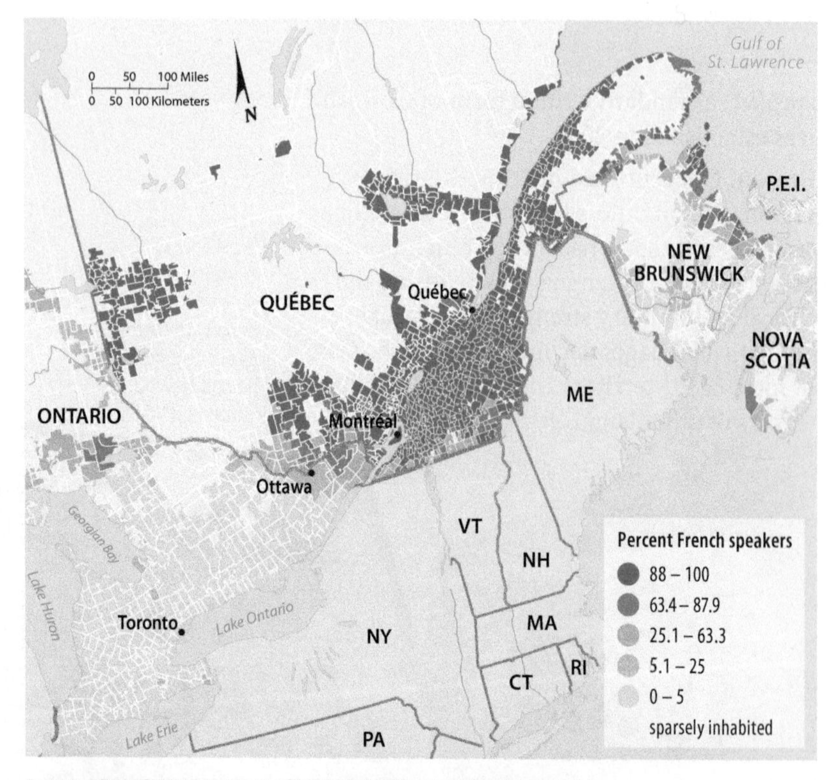

▲ 7.5.2 FRENCH/ENGLISH LANGUAGE BOUNDARY IN CANADA

SPANISH IN THE UNITED STATES

Spanish has become an increasingly important language in the United States because of large-scale immigration from Latin America. In some communities, government documents and advertisements are printed in Spanish. Several hundred Spanish-language newspapers and radio and TV stations operate in the United States, especially in southern Florida, the Southwest, and large northern cities (Figure 7.5.3).

▲ 7.5.3 **SPANISH IN THE UNITED STATES**
Little Havana, Miami, Florida.

Linguistic unity is an apparent feature of the United States, a nation of immigrants who learn English to become U.S. citizens. However, the diversity of languages in the United States is greater than it first appears. In 2008, a language other than English was spoken at home by 56 million Americans over age 5, 20 percent of the population. Spanish was spoken at home by 35 million people in the United States. More than 2 million spoke Chinese; at least 1 million each spoke French, German, Korean, Tagalog, and Vietnamese. In reaction against the increasing use of Spanish in the United States, 27 states and a number of localities have laws making English the official language.

Americans have debated whether schools should offer bilingual education. Some people want Spanish-speaking children to be educated in Spanish, because they think that children will learn more effectively if taught in their native language and that this will also preserve their own cultural heritage. Others argue that learning in Spanish creates a handicap for people in the United States when they look for jobs, virtually all of which require knowledge of English.

Promoting the use of English symbolizes that language is the chief cultural bond in the United States in an otherwise heterogeneous society. With the growing dominance of the English language in the global economy and culture, knowledge of English is important for people around the world, not just inside the United States.

CREOLIZED LANGUAGES

A **creole** or **creolized language** is defined as a language that results from the mixing of the colonizer's language with the indigenous language of the people being dominated (Figure 7.5.4). The word *creole* derives from a word in several Romance languages for a slave who is born in the master's house.

A creolized language forms when the colonized group adopts the language of the dominant group but makes some changes, such as simplifying the grammar and adding words from their former language. Creolized language examples include French Creole in Haiti, Papiamento (creolized Spanish) in Netherlands Antilles (West Indies), and Portuguese Creole in the Cape Verde Islands off the African coast.

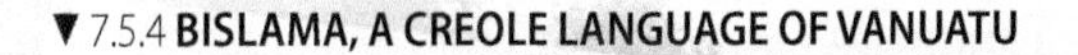

▼ 7.5.4 **BISLAMA, A CREOLE LANGUAGE OF VANUATU**
Public health campaign warning sign about AIDS.

7.6 Multilingual States

- Belgium and Switzerland are examples of multilingual states within Europe.
- Nigeria is an example of an African country with significant language diversity.

Difficulties can arise at the boundary between two languages. Note that the boundary between the Romance and Germanic branches of Indo-European runs through the middle of two small European countries, Belgium and Switzerland. Belgium has had more difficulty than Switzerland in reconciling the interests of the different language speakers.

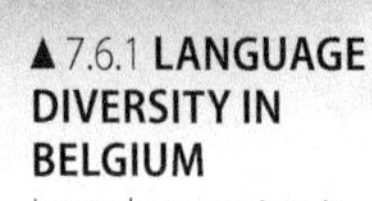

▲ 7.6.1 **LANGUAGE DIVERSITY IN BELGIUM**
Interchange sign in French (first) and Flemish.

BELGIUM

Motorists in Belgium see the language diversity on expressways (Figure 7.6.1). Belgium's language boundary sharply divides the country into two regions. Southern Belgians (known as Walloons) speak French, whereas northern Belgians (known as Flemings) speak a dialect of the Germanic language of Dutch, called Flemish (Figure 7.6.2).

Antagonism between the Flemings and Walloons is aggravated by economic and political differences. Historically, the Walloons dominated Belgium's economy and politics, and French was the official state language.

In response to pressure from Flemish speakers, Belgium was divided into two independent regions, Flanders and Wallonia. Each elects an assembly that controls cultural affairs, public health, road construction, and urban development in its region.

► 7.6.2 **LANGUAGES IN BELGIUM**
French is the principal language in Wallonia and Flemish (a dialect of Dutch) in Flanders.

SWITZERLAND

Switzerland peacefully exists with multiple languages (Figure 7.6.3). The key is a decentralized government, in which local authorities hold most of the power, and decisions are frequently made by voter referenda (direct voting). Switzerland has four official languages—German (used by 64 percent of the population), French (20 percent), Italian (7 percent), and Romansh (1 percent). Swiss voters made Romansh an official language in a 1978 referendum, despite the small percentage of people who use the language (Figure 7.6.4).

▲ 7.6.4 **LANGUAGE DIVERSITY IN SWITZERLAND**
Switzerland has four official languages, shown on the sign above: German (top left), French (top right), Italian (lower left), and Romansh (lower right). The sign prevents hikers, vehicles, and horses from entering the forest because of timber cutting.

▼ 7.6.3 **LANGUAGES IN SWITZERLAND**

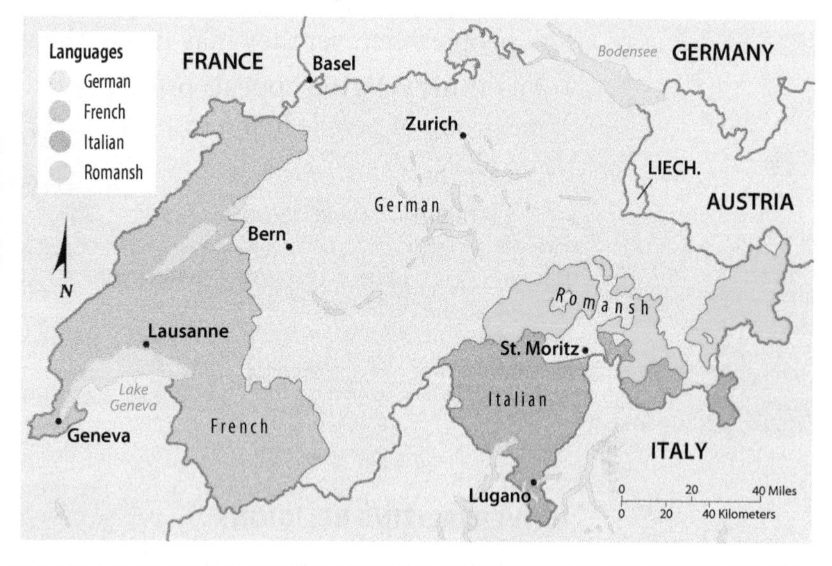

NIGERIA

Africa's most populous country, Nigeria, displays problems that can arise from the presence of many speakers of many languages. Nigeria has 493 distinct languages, according to *Ethnologue*, only three of which have widespread use. Hausa, Yoruba, and Igbo are spoken by approximately 15 percent each, and the remaining 55 percent of the population use one of the other 490 languages (Figure 7.6.5).

Groups living in different regions of Nigeria have often battled. The southern Igbos attempted to secede (withdraw) from Nigeria during the 1960s, and northerners have repeatedly claimed that the Yorubas discriminate against them. To reduce these regional tensions, the government has moved the capital from Lagos in the Yoruba-dominated southwest to Abuja in the center of Nigeria (Figure 7.6.6).

Nigeria reflects the problems that can arise when great cultural diversity—and therefore language diversity—is packed into a relatively small region. Nigeria also illustrates the importance of language in identifying distinct cultural groups at a local scale. Speakers of one language are unlikely to understand any of the others in the same language family, let alone languages from other families.

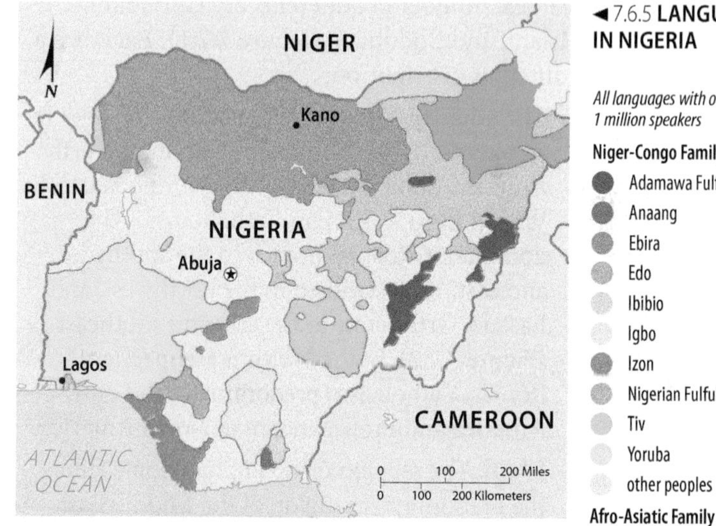

◄ 7.6.5 **LANGUAGES IN NIGERIA**

FARAWA

rko Allah ya ha-
kasa. 2 Ƙasa kwa
woñ kuma ; a kan
a sai dufu ; ruhun
otsi a bisa ruwaye.
che, Bari haske shi
kwa ya kasanche.
ga haske yana da
ya raba tsakanin
Allah ya che da
kwa ya che da shi
aiche akwai safiya
ke nan.
che, Bari sarari
kanin ruwaye, shi
ruwaye. 7 Allah
rin, ya raba ru-

kin nan biyu ; babban domin
mulkin yini, ƙaramin domin
mulkin dare : *ya yi* tamrari
17 Allah kuma ya sanya su
sararin sama domin su bada
bisa duniya, 18 su yi mulkin
dare, su raba tsakanin ha
dufu kuma : Allah kwa ya
da kyau. 19 Akwai maraich
safiya kuma, kwana na
nan.
20 Kuma Allah ya che, Bari
su yawaita haifan masu-
ɗanda ke da rai, tsuntsaye
tashi birbishin duniya chik
sama. 21 Allah kuma ya
katayen abubuwa na teku

▼ 7.6.6 **LANGUAGE DIVERSITY IN NIGERIA**
The Bible in Hausa.

7.7 Distribution of Religions

- ► Geographers distinguish between universalizing and ethnic religions.
- ► The two types of religions have different distributions.

Geographers distinguish between two types of religions:

- A **universalizing religion** attempts to be global, to appeal to all people wherever they may live in the world.
- An **ethnic religion** appeals primarily to one group of people living in one place.

► 7.7.1 DISTRIBUTION OF RELIGIONS

Universalizing Religions

Christianity
- Roman Catholic
- Protestant
- Eastern Orthodox
- Other

Islam
- Sunni
- Shiite

Other Universalizing Religions
- Buddhism
- Sikhism

Ethnic Religions
- Hinduism
- Judaism
- African
- Mixed with universalizing

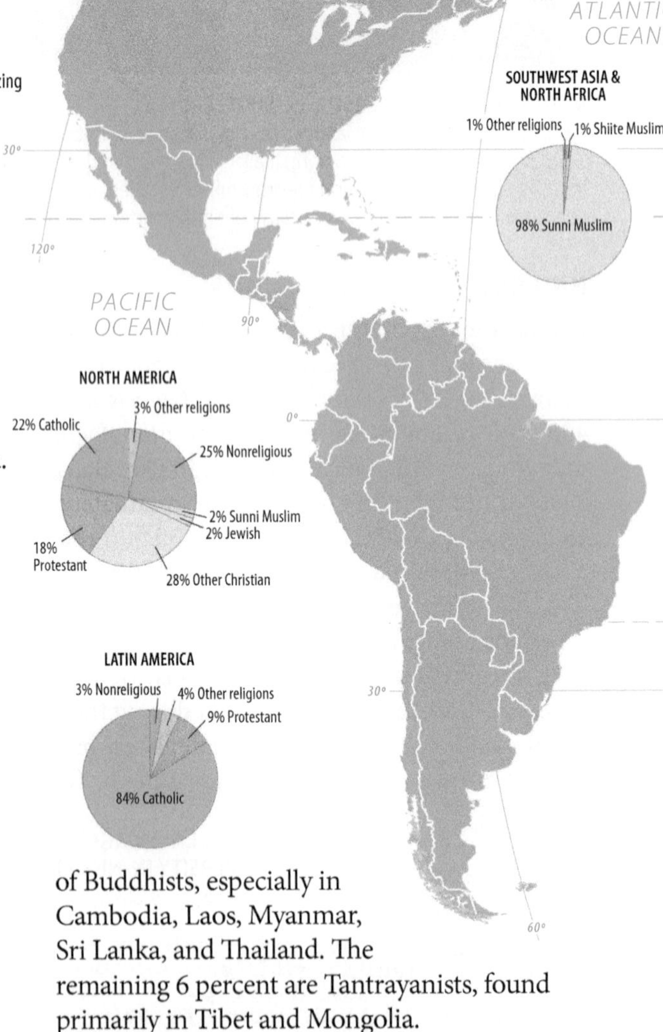

UNIVERSALIZING RELIGIONS

The three universalizing religions with the largest number of adherents are Christianity, Islam, and Buddhism (Figure 7.7.1). Each has a different distribution.

- **Christianity.** With more than 2 billion adherents, the predominant religion in North America, South America, Europe, and Australia. Within Europe, Roman Catholicism is the dominant Christian branch in the southwest and east, Protestantism in the northwest, and Eastern Orthodoxy in the east and southeast (Figure 7.7.2). In the Western Hemisphere, Roman Catholicism predominates in Latin America and Protestantism in North America.
- **Islam.** The religion of 1.3 billion people, and the predominant religion of the Middle East from North Africa to Central Asia. One-half of the world's Muslims (adherents of Islam) live outside the Middle East in Indonesia, Pakistan, Bangladesh, and India. The Sunni branch comprises 83 percent of Muslims and is the largest branch in most Muslim countries. The Shiite branch is clustered in Iran, Pakistan, and Iraq (Figure 7.7.3).
- **Buddhism.** With nearly 400 million adherents, mainly in China and Southeast Asia. Mahayanists account for about 56 percent of Buddhists, primarily in China, Japan, and Korea. Theravadists comprise about 38 percent of Buddhists, especially in Cambodia, Laos, Myanmar, Sri Lanka, and Thailand. The remaining 6 percent are Tantrayanists, found primarily in Tibet and Mongolia.
- **Other universalizing religions.** Sikhism and Bahá'í are the next two largest universalizing religions. All but 3 million of the world's 25 million Sikhs are clustered in the Punjab region of India. The 8 million Bahá'ís are dispersed among many countries, primarily in Africa and Asia.

▲ 7.7.2 CHRISTIANITY IN SWEDEN

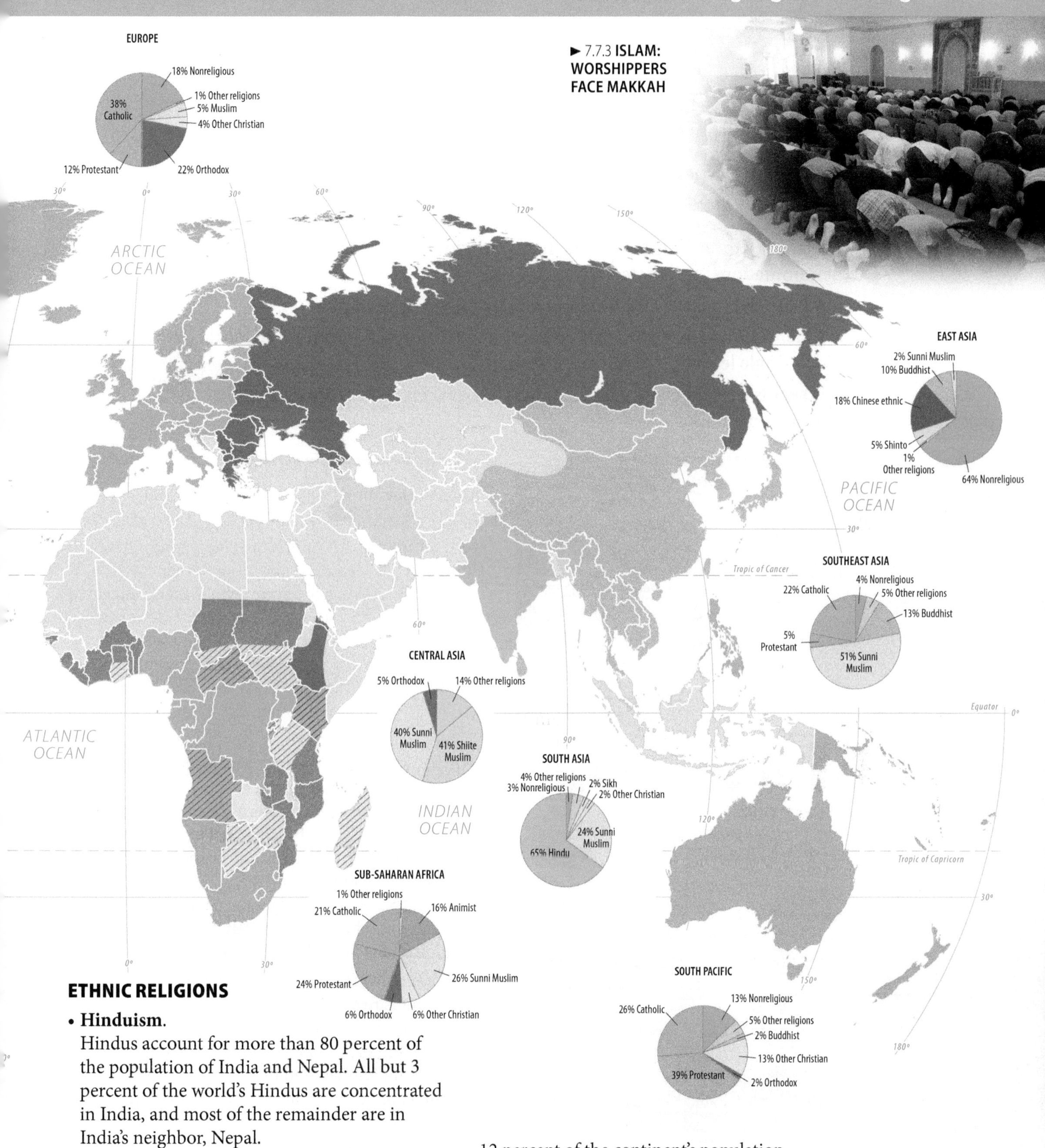

▶ 7.7.3 **ISLAM: WORSHIPPERS FACE MAKKAH**

ETHNIC RELIGIONS

- **Hinduism.**
  Hindus account for more than 80 percent of the population of India and Nepal. All but 3 percent of the world's Hindus are concentrated in India, and most of the remainder are in India's neighbor, Nepal.
- **Other ethnic religions.**
  Several hundred million people practice ethnic religions in East Asia, especially Confucianism and Daoism in China and Shintoism in Japan. Approximately 100 million Africans, 12 percent of the continent's population, follow traditional ethnic religions, sometimes called **animism**. Judaism has about 6 million adherents in the United States, 5 million in Israel, 2 million in Europe, and 1 million each in Asia and Latin America.

7.8 Geographic Branches of Religions

- **The three largest universalizing religions have different branches.**
- **Branches have distinctive regional distributions.**

Each of the three largest universalizing religions is subdivided into branches, denominations, and sects.

- A **branch** is a large and fundamental division within a religion.
- A **denomination** is a division of a branch that unites a number of local congregations in a single legal and administrative body.
- A **sect** is a relatively small group that has broken away from an established denomination.

▲▼ 7.8.1 **CHRISTIAN PLACES OF WORSHIP**
(top left) Protestant church in Edgartown, Massachusetts. (top right) Roman Catholic Cathedral in Pisa, Italy. (below) Eastern Orthodox church in Gifhorn, Germany.

BRANCHES OF CHRISTIANITY

Christianity has three major branches (Figure 7.8.1):

- **Roman Catholicism.** "Catholic," from the Greek word for *universal*, was first applied to the Christian Church in the second century. The Roman Catholic Church is headed by the Pope, who is also the Bishop of Rome. Bishops are considered the successors to Jesus's twelve original Apostles. Roman Catholics believe that the Pope possesses a universal primacy or authority, and that the Church is infallible in resolving theological disputes.
- **Eastern Orthodoxy.** A collection of 14 self-governing churches derive from the faith and practices in the Eastern part of the Roman Empire. The split between the Roman and Eastern churches dates to the fifth century and became final in 1054. The Russian Orthodox Church has more than 40 percent of all Eastern Orthodox Christians, the Romanian Church 20 percent, the Bulgarian, Greek, and Serbian Orthodox churches approximately 10 percent each, and nine others the remaining 10 percent.
- **Protestantism.** The Protestant Reformation movement is regarded as beginning when Martin Luther posted 95 theses on the door of the church at Wittenberg on October 31, 1517. According to Luther, individuals had primary responsibility for achieving personal salvation through direct communication with God. Grace is achieved through faith rather than through sacraments performed by the Church.

In the United States, roughly one-third each of the population are Roman Catholics and Protestants. The other one-third comprise other Christians, other religions, and nonreligious.

▲ 7.8.2 **BUDDHIST MONKS**
(left) Theravada Buddhist at Tooth Temple in Kandy, Sri Lanka. (right) Mahayana Buddhist at Great Buddha statue in Kamakura, Japan.

BRANCHES OF BUDDHISM

The two largest branches of Buddhism are Theravada and Mahayana (Figure 7.8.2).

- **Theravada,** which means "the way of the elders," emphasizes Buddha's life of wisdom, self-help, and solitary introspection.
- **Mahayana** ("the bigger ferry" or "raft"), which split from Theravada Buddhism about 2,000 years ago, emphasizes Buddha's life of teaching, compassion, and helping others.

DEITIES IN HINDUISM

Hinduism does not have a central authority or a single holy book, so each individual selects suitable rituals (Figure 7.8.4). The average Hindu has allegiance to a particular god or concept within a broad range of possibilities:

- The manifestation of God with the largest number of adherents—an estimated 68 percent—is Vaishnavism, which worships the god Vishnu, a loving god incarnated as Krishna.
- An estimated 27 percent adhere to Sivaism, dedicated to Siva, a protective and destructive god.
- Shaktism is a form of worship dedicated to the female consorts of Vishnu and Siva.

Some geographic concentration of support for these deities exists: Siva and Shakti in the north, Shakti and Vishnu in the east, Vishnu in the west, Siva and some Vishnu in the south. However, holy places for Siva and Vishnu are dispersed throughout India.

BRANCHES OF ISLAM

The word *Islam* in Arabic means "submission to the will of God," and it has a similar root to the Arabic word for *peace*. An adherent of the religion of Islam is known as a Muslim, which in Arabic means "one who surrenders to God." Islam is divided into two important branches:

- **Sunni.** From the Arabic word for "orthodox," Sunnis comprise two-thirds of Muslims and are the largest branch in most Muslim countries in the Middle East and Asia.
- **Shiite.** From the Arabic word for "sectarian," Shiites (sometimes written *Shia*), comprise nearly 90 percent of the population in Iran and a substantial share in neighboring countries.

Differences between the two main branches go back to the earliest days of Islam and reflect disagreement over the line of succession in Islamic leadership after the Prophet Muhammad, who had no surviving son, nor a follower of comparable leadership ability (Figure 7.8.3).

◄▲ 7.8.3 **MUSLIM PLACES OF WORSHIP**
(above) Sunni mosque in Manama, Bahrain. (left) Shiite mosque in Samarra, Iraq, which was destroyed in 2006.

▼ 7.8.4 **HINDUISM**
Bathing in the Ganges River at Varanasi, India.

7.9 Origins of Religions

- ► **Ethnic religions have unknown origins.**
- ► **Universalizing religions have precise places of origin.**

Universalizing and ethnic religions typically have different geographic origins:

- An ethnic religion, such as Hinduism, has unknown or unclear origins, not tied to single historical individuals.
- A universalizing religion, such as Christianity, Islam, and Buddhism, has a precise hearth, or place of origin, based on events in the life of a man. The hearths where the largest universalizing religions originated are all in Asia.

HINDUISM

Hinduism existed prior to recorded history (Figure 7.9.1). The earliest surviving Hindu documents were written around 1500 B.C. Aryan tribes from Central Asia invaded India about 1400 B.C. and brought with them Indo-European languages, as discussed earlier in this chapter. In addition to their language, the Aryans brought their religion. Archaeological explorations have unearthed Hindu objects relating to the religion from 2500 B.C. The word *Hinduism* originated in the sixth century B.C. to refer to people living in what is now India.

▲ 7.9.2 **ORIGIN OF CHRISTIANITY: CHURCH OF THE HOLY SEPULCHRE, JERUSALEM**
Many Christians believe that the church was constructed on the site of Jesus's crucifixion, burial, and Resurrection.

▼ 7.9.1 **HINDUISM'S UNKNOWN ORIGINS: MOUNT KAILĀS**
It is not known when or why people started making pilgrimages to the base of Mount Kailās. Because of its importance as a place of eternal bliss in Hinduism, as well as several other religions, no human in recorded history has ever climbed to the summit of Mount Kailās. Hindus believe that this mountain is home of Lord Siva, who is the destroyer of evil and sorrow.

CHRISTIANITY

Christianity was founded upon the teachings of Jesus, who was born in Bethlehem between 8 and 4 B.C. and died on a cross in Jerusalem about A.D. 30 (Figure 7.9.2). Raised as a Jew, Jesus gathered a small band of disciples and preached the coming of the Kingdom of God. He was referred to as *Christ,* from the Greek word for the Hebrew word *messiah,* which means "anointed."

In the third year of his mission, he was betrayed to the authorities by one of his companions, Judas Iscariot. After sharing the Last Supper (the Jewish Passover seder) with his disciples in Jerusalem, Jesus was arrested and put to death as an agitator. On the third day after his death, his tomb was found empty. Christians believe that Jesus died to atone for human sins, that he was raised from the dead by God, and that his Resurrection from death provides people with hope for salvation.

BUDDHISM

The founder of Buddhism, Siddharta Gautama, was born about 563 B.C. in Lumbinī, in present-day Nepal. The son of a lord, Gautama led a privileged life, with a beautiful wife, palaces, and servants.

According to Buddhist legend, Gautama's life changed after a series of four trips. He encountered a decrepit old man on the first trip, a disease-ridden man on the second trip, and a corpse on the third trip. After witnessing these scenes of pain and suffering, Gautama began to feel he could no longer enjoy his life of comfort and security.

On a fourth trip, Gautama saw a monk, who taught him about withdrawal from the world. Gautama lived under a bodhi (or bo) tree in a forest for seven weeks, thinking and experimenting with forms of meditation (Figure 7.9.3). He emerged as the Buddha, the "awakened or enlightened one."

► 7.9.3 **ORIGIN OF BUDDHISM: MAHABODHI TEMPLE, BODH GAYA**
Buddha reached perfect wisdom sitting under a bodhi tree. This temple has stood on the site since the third century B.C., and part of the current structure was built in the first century A.D.

◄ 7.9.4 **ORIGIN OF ISLAM: AL-MASJID AL-NABAWI (MOSQUE OF THE PROPHET), MADINAH, SAUDI ARABIA**
Muhammad is buried in this mosque, built on the site of his house. The mosque is the second holiest site in Islam and the second largest mosque in the world.

Fly to: *Al Masjid al Nabawi Mosque, Medina, Saudi Arabia.*

Use Google Earth's navigation tools and layer options to explore features of this area.

1. **What is the principal use of most of the largest buildings surrounding the mosque?**
2. **Why are these types of buildings located near the mosque?**

ISLAM

The Prophet Muhammad was born in Makkah about 570. Muhammad was a descendant of Ishmael, who was the son of Abraham and Hagar. Jews and Christians trace their history through Abraham's wife Sarah and their son Isaac. Sarah prevailed upon Abraham to banish Hagar and Ishmael, who wandered through the Arabian desert, eventually reaching Makkah.

Muslims believe that Muhammad received his first revelation from God, through the Angel Gabriel, at age 40 while he was engaged in a meditative retreat. The Quran, the holiest book in Islam, is a record of God's words, as revealed to the Prophet Muhammad through Gabriel.

As he began to preach the truth that God had revealed to him, Muhammad suffered persecution, and in 622 he was commanded by God to emigrate to the city of Yathrib (renamed Madinah, from the Arabic for "the City of the Prophet"), an event known as the Hijra (from the Arabic for immigration, sometimes spelled *hegira*). When he died in 632, Muhammad was buried in Madinah (Figure 7.9.4).

OTHER UNIVERSALIZING RELIGIONS

Other universalizing religions, with fewer adherents, also trace their origins to single individuals. For example:

- **Sikhism** was founded by Guru Nanak (1469–1539), who traveled widely through South Asia preaching his new faith. Many people became his Sikhs (a Hindi term for "disciples").
- **Bahá'í** was founded during the nineteenth century by Husayn 'Ali Nuri, known as Bahá'u'lláh (Arabic for "Glory of God"). Bahá'u'lláh was a disciple of Siyyid 'Alí Muhammad Shírází (1819–1850), known as the Bāb (Persian for "gateway"). As the prophet and messenger of God, Bahá'u'lláh sought to overcome the disunity of religions and establish a universal faith.

7.10 Diffusion of Universalizing Religions

- ► **Universalizing religions have diffused beyond their places of origin.**
- ► **Missionaries and military conquests have been important methods of diffusing universalizing religions.**

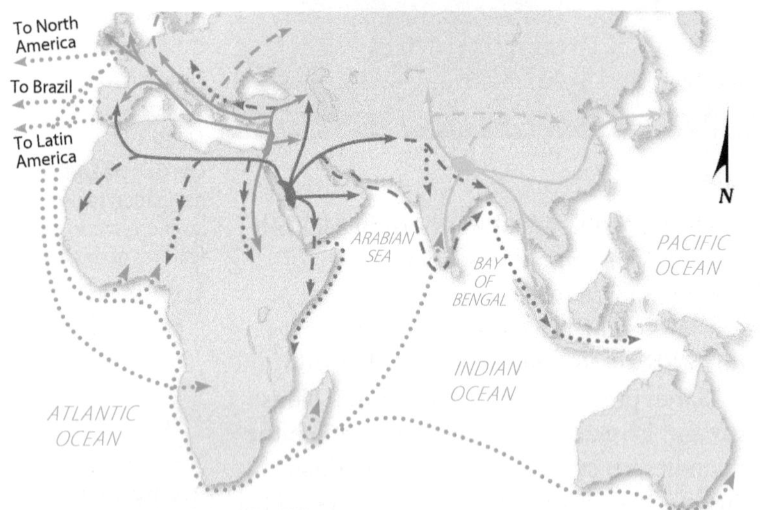

▲ 7.10.1 **DIFFUSION OF UNIVERSALIZING RELIGIONS**
Christianity diffused from present-day Israel primarily west towards Europe. Buddhism diffused from present-day Nepal primarily east towards East and Southeast Asia. Islam diffused from present-day Saudi Arabia primarily west across North Africa and east to Southwest and South Central Asia.

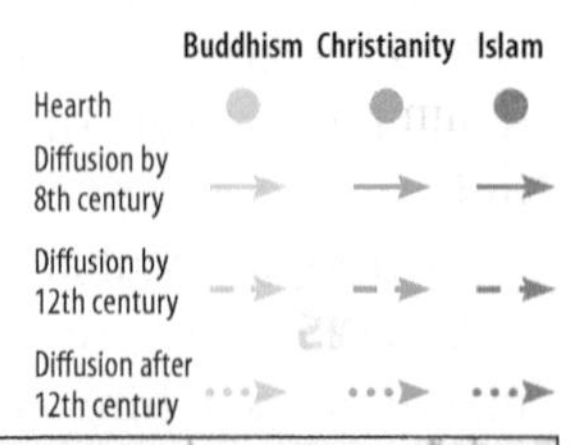

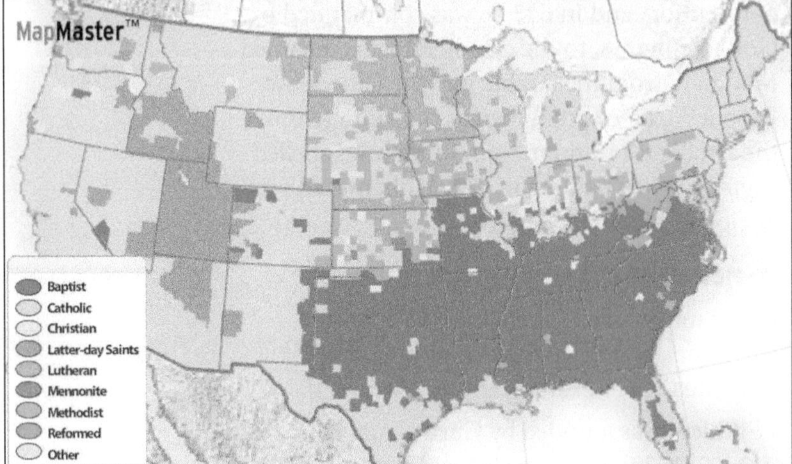

▲ 7.10.2 **DIFFUSION OF CHRISTIANITY IN NORTH AMERICA**
The distribution of Christian branches and Protestant denominations in North America is influenced by patterns of migration.

Launch MapMaster North America in MasteringGEOGRAPHY

Select: *Distribution of Christian Denominations* from the *Cultural* menu

1. **What is the predominant Christian branch in southwestern United States?**
2. **What pattern of immigration into the United States accounts for this distribution?**

The three main universalizing religions diffused from specific hearths or places of origin to other regions of the world. Followers transmitted the messages preached in the hearths to people elsewhere, diffusing them across Earth's surface along distinctive paths (Figure 7.10.1).

DIFFUSION OF CHRISTIANITY

Christianity's diffusion has been rather clearly recorded since Jesus first set forth its tenets in the Roman province of Palestine. In Chapter 1 we distinguished between relocation diffusion (through migration) and expansion diffusion (additive effect). Christianity diffused through a combination of both forms of diffusion.

Christianity first diffused from its hearth in Southwest Asia through migration. **Missionaries**—individuals who help to transmit a universalizing religion through relocation diffusion—carried the teachings of Jesus along the Roman Empire's protected sea routes and excellent road network to people in other locations. Migration and missionary activity by Europeans since the year 1500 has extended Christianity to other regions of the world (Figure 7.10.2).

Christianity spread widely within the Roman Empire through two forms of expansion diffusion:

- Contagious diffusion: daily contact between believers in the towns and nonbelievers in the surrounding countryside.
- Hierarchical diffusion: acceptance of the religion by the empire's key elite figure, the emperor; Emperor Constantine encouraged the spread of Christianity by embracing it in 313, and Emperor Theodosius proclaimed it the empire's official religion in 380.

▲ 7.10.3 **DIFFUSION OF BUDDHISM**
As Buddhism diffused across East and Southeast Asia, shrines were constructed in its hearth to mark key events in Buddha's life. Dhamek Stupa was built in Deer Park, Sarnath, India, around A.D. 500 to mark the spot where Buddha gave his first sermon.

DIFFUSION OF BUDDHISM

Buddhism did not diffuse rapidly from its point of origin in northeastern India (Figure 7.10.3). Most responsible for its diffusion was Asoka, emperor of the Magadhan Empire from about 273 to 232 B.C. Around 257 B.C., at the height of the Magadhan Empire's power, Asoka became a Buddhist and thereafter attempted to put into practice Buddha's social principles.

Emperor Asoka's son, Mahinda, led a mission to the island of Ceylon (now Sri Lanka), where the king and his subjects were converted to Buddhism. As a result, Sri Lanka is the country that claims the longest continuous tradition of practicing Buddhism. Missionaries were also sent in the third century B.C. to Kashmir, the Himalayas, Burma (Myanmar), and elsewhere in India.

In the first century A.D., merchants along the trading routes from northeastern India introduced Buddhism to China. Chinese rulers allowed their people to become Buddhist monks during the fourth century A.D., and in the following centuries Buddhism turned into a genuinely Chinese religion. Buddhism further diffused from China to Korea in the fourth century and from Korea to Japan two centuries later. During the same era, Buddhism lost its original base of support in India.

DIFFUSION OF ISLAM

Muhammad's successors organized followers into armies that extended the region of Muslim control over an extensive area of Africa, Asia, and Europe. Within a century of Muhammad's death, Muslim armies conquered Palestine, the Persian Empire, and much of India, resulting in the conversion of many non-Arabs to Islam, often through intermarriage.

To the west, Muslims captured North Africa, crossed the Strait of Gibraltar, and retained part of western Europe, particularly much of present-day Spain, until 1492 (Figure 7.10.4). During the same century during which the Christians regained all of western Europe, Muslims took control of much of southeastern Europe and Turkey.

As was the case with Christianity, Islam, as a universalizing religion, diffused well beyond its hearth in Southwest Asia through relocation diffusion of missionaries to portions of sub-Saharan Africa and Southeast Asia. Although it is spatially isolated from the Islamic core region in Southwest Asia, Indonesia, the world's fourth most populous country, is predominantly Muslim, because Arab traders brought the religion there in the thirteenth century.

DIFFUSION OF OTHER UNIVERSALIZING RELIGIONS

- **Bahá'í** diffused to other regions during the late nineteenth century, under the leadership of 'Abdu'l-Bahá, son of the prophet Bahá'u'lláh. During the twentieth century, Bahá'ís constructed a temple on every continent.
- **Sikhism** remained relatively clustered in South Asia, where the religion originated. When India and Pakistan became independent states in 1947, the Punjab region where most Sikhs lived was divided between the two countries.

▼ 7.10.4 **DIFFUSION OF ISLAM**
Islam diffused to Spain in 711. Shortly thereafter, a church in Cordoba, Spain, was modified to become the world's second largest mosque, known as the Mezquita de Córdoba. The mosque was captured by Christians in 1236 and reconsecrated as a cathedral.

7.11 Holy Places in Universalizing Religions

- **Universalizing religions honor holy places associated with the founder's life.**
- **Structures play distinctive roles in each of the universalizing religions.**

Religions elevate particular places to a holy position. A universalizing religion endows with holiness cities and sacred structures associated with the founder's life. Its holy places are not typically related to any particular feature of the physical environment.

CHRISTIAN CHURCHES

The church plays a more critical role in Christianity than buildings in other religions, because the structure is an expression of religious principles, an environment in the image of God. The word *church* derives from a Greek term meaning *lord, master*, and *power*. In many communities, the church is the largest and tallest building and has been placed at a prominent location.

Early churches were rectangular-shaped, modeled after Roman buildings for public assembly, known as *basilicas*. A raised altar, where the priest conducted the service, symbolized the hill of Calvary, where Jesus was crucified.

Since Christianity split into many branches and denominations, no single style of church construction has dominated (Figure 7.11.1). Eastern Orthodox churches follow an ornate architectural style that developed in the Byzantine Empire during the fifth century. Many Protestant churches in North America are austere, with little ornamentation, a reflection of the Protestant conception of a church as an assembly hall for the congregation.

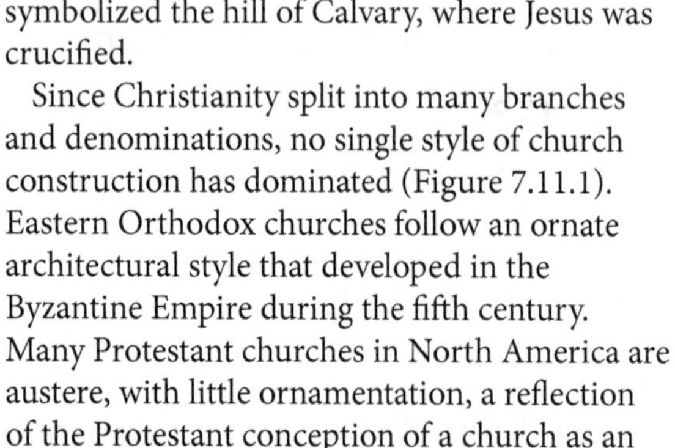

▼ 7.11.1 **HOLY PLACES IN CHRISTIANITY**
Basilica of St. Boniface, Munich, Germany.

▲ 7.11.2 **HOLY PLACES IN ISLAM**
Al-Masjid al-Harām (Sacred Mosque), Makkah, Saudi Arabia.

MUSLIM HOLY CITIES

The holiest places in Islam are in cities associated with the life of the Prophet Muhammad.
The holiest city for Muslims is Makkah, the birthplace of Muhammad. Every healthy Muslim who has adequate financial resources is expected to undertake a hajj to Makkah. A hajj is a **pilgrimage**, which is a journey to a place considered sacred for religious purposes.

The holiest object in the Islamic landscape, al-Ka'ba, a cubelike structure encased in silk, stands at the center of Makkah's Sacred Mosque, al-Masjid al-Harám (Figure 7.11.2). The second most holy geographic location is Madinah, where Muhammad received his first support and where he is buried (see Figure 7.9.4).

Muslims consider the mosque as a space for community assembly, but it is not a sanctified place like the Christian church. The mosque is organized around a central courtyard. The pulpit is placed at the end of the courtyard facing Makkah, the direction toward which all Muslims pray. A minaret or tower is where a man known as a *muzzan* summons people to worship.

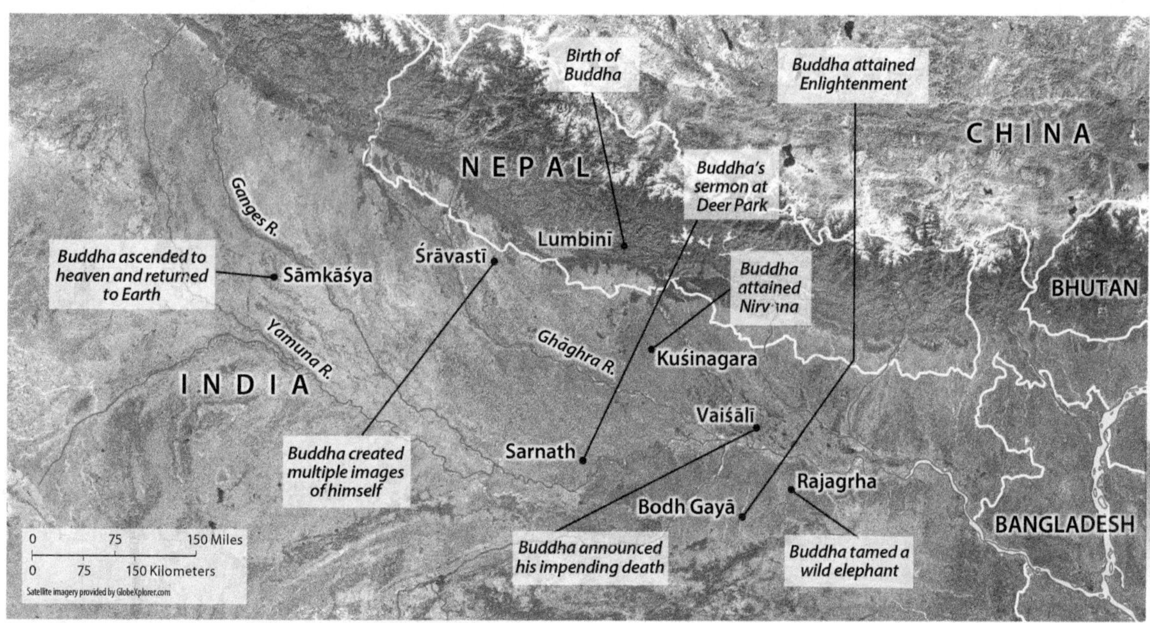

◄ 7.11.4 **HOLY PLACES IN BUDDHISM**

▲ 7.11.3 **HOLY PLACES IN BAHÁ'Í**
House of Worship, Delhi, India.

BAHÁ'Í HOLY PLACES

Bahá'ís have built Houses of Worship in every continent to dramatize that Bahá'ís is a universalizing religion with adherents all over the world. Sites include Wilmette, Illinois, in 1953; Sydney, Australia, and Kampala, Uganda, both in 1961; Lagenhain, near Frankfurt, Germany, in 1964; Panama City, Panama, in 1972; Tiapapata, near Apia, Samoa, in 1984; and New Delhi, India, in 1986 (Figure 7.11.3).

Additional Houses of Worship are planned in Tehran, Iran; Santiago, Chile; and Haifa, Israel. The first Bahá'ís House of Worship, built in 1908 in Ashgabat, Russia, now the capital of Turkmenistan, was turned into a museum by the Soviet Union and demolished in 1962 after a severe earthquake.

BUDDHIST HOLY PLACES

Eight places are holy to Buddhists because they were the locations of important events in Buddha's life. The four most important of the eight places are concentrated in a small area of northeastern India and southern Nepal (Figure 7.11.4).

The pagoda is a prominent and visually attractive element of the Buddhist landscape. Pagodas contain relics that Buddhists believe to be a portion of Buddha's body or clothing. Pagodas are not designed for congregational worship. Individual prayer or meditation is more likely to be undertaken at an adjacent temple, a remote monastery, or in a home.

SIKH HOLY PLACES

Sikhism's most holy structure, the Darbar Sahib (Golden Temple), was built at Amritsar during the seventh century (Figure 7.11.5). Sikhs seeking autonomy from India used the Golden Temple as a base to attack the Indian army. In 1984, the Indian army attacked a thousand Sikh separatists who sought sanctuary in the Temple. India's Prime Minister Indira Gandhi in turn was assassinated later that year by two of her guards, who were Sikhs.

▼ 7.11.5 **HOLY PLACES IN SIKHISM**
Darbar Sahib (Golden Temple) at Amritsar, India.

7.12 Ethnic Religions and the Landscape

- In ethnic religions, the calendar and beliefs in the origin of the universe are grounded in the physical environment.
- Ethnic religions are tied to the physical environment of a particular place.

Ethnic religions differ from universalizing religions in their understanding of relationships between human beings and nature. A variety of events in the physical environment are more likely to be incorporated into the principles of an ethnic religion.

THE CALENDAR IN JUDAISM

Calendars in ethnic religions are based upon the changing of the seasons because of the necessities of agricultural cycles. Prayers are offered in hope of favorable environmental conditions or to give thanks for past success.

Judaism is classified as an ethnic, rather than a universalizing, religion in part because its major holidays are based on events in the agricultural calendar of the religion's homeland in present-day Israel (Figure 7.12.1). The name *Judaism* derives from *Judah*, one of the patriarch Jacob's 12 sons; *Israel* is another biblical name for Jacob.

Israel—the only country where Jews are in the majority—uses a lunar rather than a solar calendar. The lunar month is only about 29 days long, so a lunar year of about 350 days quickly becomes out of step with the agricultural seasons. The Jewish calendar solves the problem by adding an extra month seven out of every 19 years, so that its principal holidays are celebrated in the same season every year.

Fundamental to Judaism is belief in one all-powerful God. It was the first recorded religion to espouse **monotheism**, and the belief that there is only one God. Judaism offered a sharp contrast to the **polytheism** practiced by neighboring people, who worshipped a collection of gods.

▼ 7.12.1 **JEWISH HOLIDAY OF SUKKOTH**
On the holiday of Sukkoth, Jews carry branches of date palm, myrtle, and willow to symbolize gratitude for the many agricultural bounties offered by God.

COSMOGONY IN CHINESE ETHNIC RELIGIONS

Cosmogony is a set of religious beliefs concerning the origin of the universe. The cosmogony underlying Chinese ethnic religions, such as Confucianism and Daoism, is that the universe is made up of two forces, yin and yang, which exist in everything. The force of yin (earth, darkness, female, cold, depth, passivity, and death) interacts with the force of yang (heaven, light, male, heat, height, activity, and life) to achieve balance and harmony. An imbalance results in disorder and chaos.

Confucianism, based on the sayings of the philosopher and teacher Confucius (551–479 B.C.), emphasizes the importance of the ancient Chinese tradition of *li*, which can be translated roughly as "propriety" or "correct behavior," such as following traditions, fulfilling obligations, and treating others with sympathy and respect (Figure 7.12.2).

Daoism, organized by a government administrator Lao-Zi (604–ca. 531 B.C.), emphasizes the mystical and magical aspects of life. Daoists seek *dao* (or *tao*), which means the "way" or "path." *Dao* cannot be comprehended by reason and knowledge, because not everything is subject to rational analysis, so myths and legends develop to explain events.

▲ 7.12.2 **CONFUCIUS TEMPLE, NANJING, CHINA**

SPIRITS IN INANIMATE OBJECTS

To animists, the powers of the universe are mystical, and only a few people on Earth can harness these powers for medical or other purposes (Figure 7.12.3). Spirits or gods can be placated, however, through prayer and sacrifice. Rather than attempting to transform the environment, animists accept environmental hazards as normal and unavoidable.

Animists believe that such inanimate objects as plants and stones, or such natural events as thunderstorms and earthquakes, are "animated," or have discrete spirits and conscious life. Many African animist religions are apparently based on monotheistic concepts, although below the supreme god there is a hierarchy of divinities. These divinities may be assistants to the supreme god or personifications of natural phenomena, such as trees or rivers.

As recently as 1980, some 200 million Africans—half the population of the region at the time—were classified as animists. Some atlases and textbooks persist in classifying Africa as predominantly animist, even though the actual percentage is small and declining. The rapid decline in animism, down to 100 million currently, has been caused by diffusion of the two largest universalizing religions, Christianity and Islam.

◄ 7.12.3 **AFRICAN ANIMIST RELIGIONS** The character and form of the Odo-Kuta have evolved from the animistic origins of this once hunter-gatherer people. The circular design of the mask represents a model of the world and the individual's place in it. The masks' power to enforce this model of order is considered absolute, since mask wisdom comes from beyond the human realm and renders their authority beyond the questioning of humans.

SACRED SPACE IN HINDUISM

Unlike universalizing religions, Hindus generally practice cremation rather than burial. The body is washed with water from the Ganges River and then burned with a slow fire on a funeral pyre (Figure 7.12.4). Burial is reserved for children, ascetics, and people with certain diseases. Cremation is considered an act of purification, although it tends to strain India's wood supply.

Motivation for cremation may have originated from unwillingness on the part of nomads to leave their dead behind, possibly because of fear that the body could be attacked by wild beasts or evil spirits, or even return to life. Cremation could also free the soul from the body for departure to the afterworld and provide warmth and comfort for the soul as it embarked on that journey.

▼ 7.12.4 **SACRED SPACE IN HINDUISM** The most common form of disposal of bodies in India is cremation. In middle-class families, bodies are more likely to be cremated in an electric oven at a crematorium. A poor person may be cremated in an open fire, such as this one on the banks of the River Ganges. High-ranking officials and strong believers in traditional religious practices may also be cremated on an outdoor fire.

7.13 Religious Conflicts in the Middle East

- ► **Jews, Muslims, and Christians have fought to control Israel/Palestine.**
- ► **Places holy to all three are clustered in Jersusalem.**

Jews, Christians, and Muslims have fought for 2,000 years to control a small strip of land in the Middle East.

- **Jews** consider the territory their Promised Land. The major events in the development of Judaism as an ethnic religion took place there, and the religion's customs and rituals acquire meaning from the agricultural life of the ancient Hebrew tribe.
- **Christians** consider Palestine the Holy Land and Jerusalem the Holy City because the major events in Jesus's life, death, and Resurrection were concentrated there.
- **Muslims** regard Jerusalem as their third holy city, after Makkah and Madinah, because it is the place from which Muhammad is thought to have ascended to heaven.

JEWISH PERSPECTIVES

Israel was created by the United Nations in 1947 as the only country in the world with a majority Jewish population (Figure 7.13.1). Opposed to having a predominantly Jewish country in their midst, neighboring Arab Muslim countries attacked Israel four times, in 1948, 1956, 1967, and 1973, without success.

During the 1967 Six-Day War, Israel captured territory from its neighbors, including the Old City portion of Jerusalem. Israel returned the

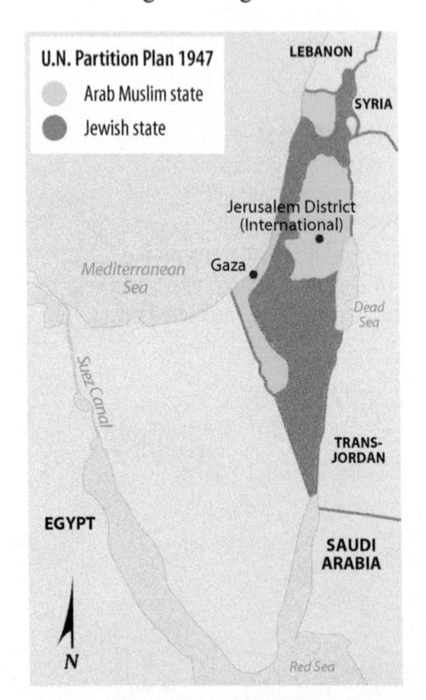

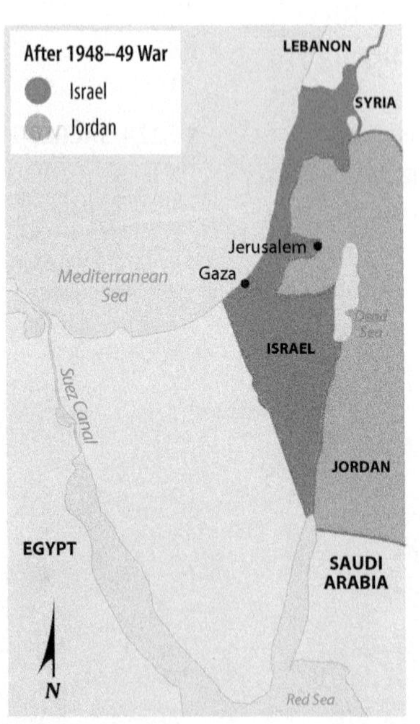

▲7.13.1. **BOUNDARY CHANGES IN PALESTINE/ISRAEL**

(left) The 1947 United Nations partition plan. Two countries were created, with the boundaries drawn to separate the predominantly Jewish areas from the predominantly Arab Muslim areas. Jerusalem was intended to be an international city, run by the UN.

(center) Israel after the 1948–1949 war. The day after Israel declared its independence, several neighboring states began a war, which ended in an armistice (agreement to cease fire). Israel's boundaries were extended beyond the UN partition to include the western suburbs of Jerusalem. Jordan gained control of the West Bank and East Jerusalem, including the Old City, where holy places are clustered.

(right) The Middle East since the 1967 war. Israel captured the Golan Heights from Syria, the West Bank and East Jerusalem from Jordan, and the Sinai Peninsula and Gaza Strip from Egypt. Israel returned Sinai to Egypt in 1979 and turned over Gaza and a portion of the West Bank to the Palestinians in 1994. Israel still controls the Golan Heights, most of the West Bank, and East Jerusalem.

Sinai Peninsula to Egypt in exchange for a peace treaty in 1979. The West Bank (formerly part of Jordan) and Gaza (formerly part of Egypt) have been joined to create an entity known as Palestine, with its own Arab Muslim government but with a strong continuing Israeli military presence.

Jerusalem is especially holy to Jews as the location of the Temple, their center of worship in ancient times. The Second Temple was destroyed by the Romans in A.D. 70, but its Western Wall survives as a site for daily prayers by observant Jews (Figure 7.13.2).

The most important Muslim structure in Jerusalem is the Dome of the Rock, built in A.D. 691. Muslims believe that the large rock beneath the building's dome is the place from which Muhammad ascended to heaven, as well as the altar on which Abraham prepared to sacrifice his son Isaac. Next to the Dome of the Rock is al-Aqsa Mosque, finished in A.D. 705.

The challenge facing Jews and Muslims is that al-Aqsa was built on the site of the ruins of the Jewish Second Temple. Through a complex arrangement of ramps, Muslims have free access to the Mosque without passing in front of the Wall. But with holy Muslim structures sitting literally on top of holy Jewish structures, the two cannot be logically divided by a line on a map.

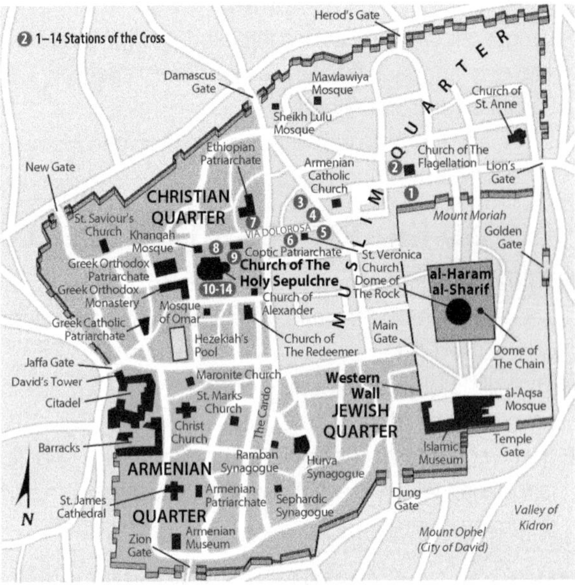

▼▲ 7.13.2 **JERUSALEM**
Less than 1/4 square mile, the Old City of Jerusalem contains religious structures important to Jews (the Western Wall), Muslims (Dome of the Rock and al-Aqsa Mosque), and Christians (Church of the Holy Sepulchre and Stations of the Cross). In the photo, the Dome of the Rock is in the top left and the Western Wall right foreground.

PALESTINIAN PERSPECTIVES

Palestinians emerged as Israel's principal opponent after the 1973 war. Egypt and Jordan renounced their claims to Gaza and the West Bank, respectively, and recognized the Palestinians as the legitimate rulers of these territories.

Five groups of people consider themselves Palestinians:

- People living in the territories captured by Israel in 1967.
- Muslim citizens of Israel.
- People who fled from Israel after Israel was created in 1948.
- People who fled from the occupied territories after the 1967 war.
- Citizens of other countries who identify themselves as Palestinians.

Palestinians see repeated efforts by Jewish settlers to increase the territory under their control.

Muslim hostility increased after the 1967 war, when Israel permitted some of its citizens to build settlements in some of the territories it had captured.

Some Palestinians are willing to recognize Israel with its Jewish majority in exchange for return of all territory taken in the 1967 war. Others still do not recognize the right of Israel to exist and want to continue fighting for control of the entire territory between the Jordan River and the Mediterranean Sea. For its part, to thwart attacks by Palestinians, Israel has constructed a barrier across the occupied territories.

The Middle East is one of many regions of the world with the potential for conflict resulting from cultural diversity. In the modern world of global economics and culture, the diversity of language and religions continues to play strong roles in people's lives.

Key Questions

Where are languages distributed?

- Languages can be classified into families, branches, and groups.
- The two language families with the most speakers are Indo-European and Sino-Tibetan.
- Language families originated before recorded history, and they have diffused through migration.

How do languages share space?

- Through migration and conquest, some languages have become more widespread, whereas others have become less widely used.
- Though English dominates North America, French and Spanish have become more widely used.
- Some countries face conflicts among speakers of different languages, whereas other countries peacefully embrace language diversity.

Where are religions distributed?

- A religion can be classified as universalizing or ethnic.
- Universalizing religions have more widespread distribution than do ethnic religions.
- Religions can be divided into branches, denominations, and sects.

How do religions shape landscapes?

- Universalizing religions revere places of importance in the lives of their founders.
- Ethnic religions are shaped by the physical geography and agriculture of its hearth.
- Adherents of different religions have fought to control the same space, especially in the Middle East.

▼ 7.CR.1 **DEMONSTRATION FOR INDEPENDENCE OF QUÉBEC**

Thinking Geographically

Thirty U.S. states have passed laws mandating English as the language of all government functions.

1. What are the benefits and the drawbacks for cultural integration and diversity resulting from this English-only mandate?

The province of Québec has debated declaring independence from Canada (Figure 7.CR.1).

2. What would be the impact of Québec's independence on the remainder of Canada, and on the United States?

Sharp demographic differences, such as NIR, CBR, and net migration, can be seen among Jews, Christians, and Muslims in the Middle East

3. How might demographic differences affect future relationships among the religious groups in the region?

On the Internet

Detailed information for every language of the world is provided at **www.ethnologue.com.** Areas of greatest diversity of languages appear in red on Ethnologue's world map.

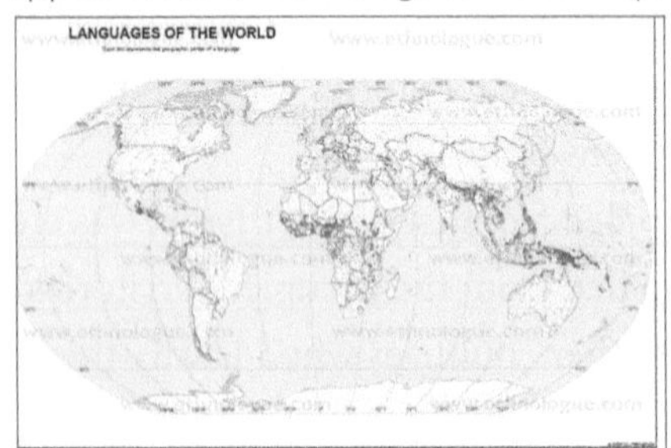

Statistics on the number of adherents to religions, branches, and denominations are at **www.adherents.com** or by scanning the QR on the opening page of this chapter.

Glenmary Research Center, which is affiliated with the Roman Catholic Church, provides maps of U.S. religions at **www.glenmary.org**. Glenmary has a map of Americans not affiliated with any religion (high percentage in red).

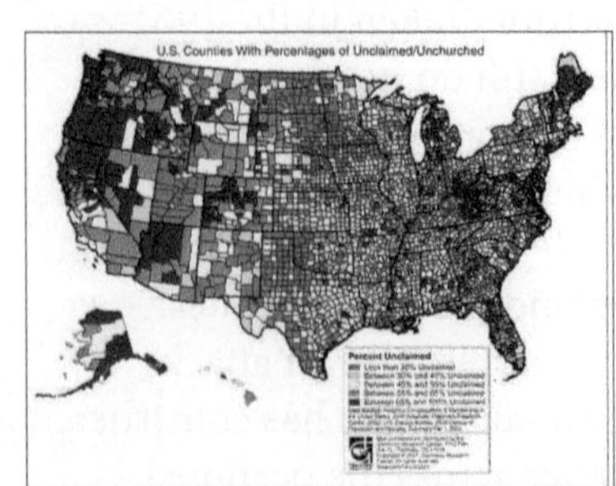

Interactive Mapping

FORCED MIGRATION IN SOUTH ASIA

Millions of people were forced to migrate after South Asia gained independence from the United Kingdom in 1947.

Launch Mapmaster South Asia in MasteringGEOGRAPHY

Select: *Religions* from the *Cultural* menu, then *Ethnic Division* from the *Population* menu.

What accounts for the migration pattern?

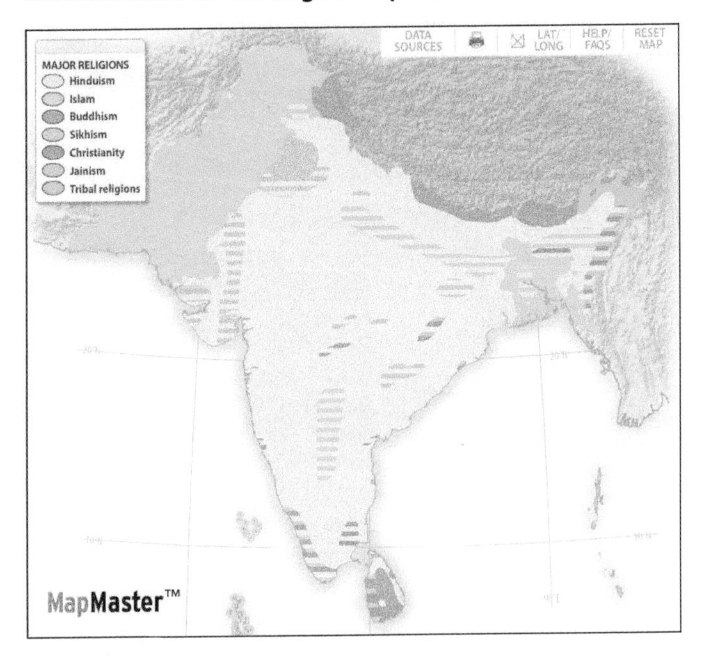

Explore

SAN FRANCISCO, CA, USA

Use Google Earth to explore language diversity in the United States.

Fly to: *400 Grant Ave, San Francisco, CA, USA*

Drag to enter *Street view*

Use mouse to continue north along Grant Ave.

1. **What languages other than English do you see on the business signs?**
2. **What name might be given to this section of San Francisco?**

Key Terms

Animism
Belief that objects, such as plants and stones, or natural events, like thunderstorms and earthquakes, have a discrete spirit and conscious life.

Branch (of a religion)
A large and fundamental division within a religion.

Cosmogony
A set of religious beliefs concerning the origin of the universe.

Creole or creolized language
A language that results from the mixing of a colonizer's language with the indigenous language of the people being dominated.

Denglish
Combination of German and English.

Denomination (of a religion)
A division of a branch that unites a number of local congregations in a single legal and administrative body.

Dialect
A regional variety of a language distinguished by vocabulary, spelling, and pronunciation.

Ethnic religion
A religion with a relatively concentrated spatial distribution whose principles are likely to be based on the physical characteristics of the particular location in which its adherents are concentrated.

Extinct language
A language that was once used by people in daily activities but is no longer used.

Franglais
A term used by the French for English words that have entered the French language; a combination of français and anglais, the French words for "French" and "English," respectively.

Isolated language
A language that is unrelated to any other languages and therefore not attached to any language family.

Language
A system of communication through the use of speech, a collection of sounds understood by a group of people to have the same meaning.

Language branch
A collection of languages related through a common ancestor that existed several thousand years ago. Differences are not as extensive or as old as with language families, and archaeological evidence can confirm that the branches derived from the same family.

Language family
A collection of languages related to each other through a common ancestor long before recorded history.

Language group
A collection of languages within a branch that share a common origin in the relatively recent past and display relatively few differences in grammar and vocabulary.

Lingua franca
A language mutually understood and commonly used in trade by people who have different native languages.

Literary tradition
A language that is written as well as spoken.

Missionary
An individual who helps to diffuse a universalizing religion.

Monotheism
The doctrine or belief of the existence of only one god.

Native speakers
People for whom a particular language is their first language.

Official language
The language adopted for use by the government for the conduct of business and publication of documents.

Pilgrimage
A journey to a place considered sacred for religious purposes.

Polytheism
Belief in or worship of more than one god.

Sect (of a religion)
A relatively small group that has broken away from an established denomination

Spanglish
Combination of Spanish and English, spoken by Hispanic Americans.

Universalizing religion
A religion that attempts to appeal to all people, not just those living in a particular location.

▸ LOOKING AHEAD

This chapter has displayed key elements of cultural diversity among the world's peoples. The next chapter looks at political problems that arise from this cultural diversity.

8 Political Geography

Viewed from space, Earth's continents show no natural signs of political units, no borders, and no conflict. Given a blank map, a newcomer to our world would never be able to guess how we have divided the Earth into political spaces, if they thought to divide it at all. Yet throughout history, humans have repeatedly subdivided Earth's lands and seas. Hunting territories, early civilizations, city-states, empires, and our current system of separate countries tell stories about how societies govern themselves according to changing ideas and circumstances.

We are also reminded that our world political map is the consequence of, and often the basis for, conflict within or between political units. Political geography is the study of how our world is politically divided, how political units spatially relate and interact, and the place-specific factors leading to conflict and peace.

U.S.-MEXICO BORDER NEAR NOGALES, ARIZONA

How is the world politically organized?

How are states organized internally?

8.5 **Governing States**

8.6 **Electoral Geography**

8.7 **Ethnicity and Nationality**

How does conflict vary by region?

8.8 **Conflicts in Western Asia**

8.9 **Ethnic Cleansing in the Balkans**

8.10 **Conflict and Genocide in Africa**

8.11 **Terrorism**

SCAN FOR DATA ON ALL COUNTRIES OF THE WORLD

8.1 A World of States

- **States are the basic political unit in the world today.**
- **States exist when other states recognize them.**

A **state** is an area organized into a political unit and ruled by an established government that has control over its internal and foreign affairs. A state occupies a defined territory on Earth's surface with a resident population. A state has **sovereignty**, which means control of its internal affairs without interference by other states. The term *country* is a synonym for *state*.

A map of the world shows that virtually all habitable land belongs to states, but for most of history, this was not so. As recently as the 1940s, the world contained only about 50 states, compared to 193 members of the United Nations as of 2011 (Figure 8.1.1). As of the same year, South Sudan is the world's newest state. Historically, existing states had to grant **recognition** or formally acknowledge each new state's claim to sovereign independence for it to be considered valid. Since World War II, membership in the UN has the same effect as recognition (Figure 8.1.2).

The entire area of a state is supposed to be managed by its national government, laws, army, and leaders. Sometimes other forces challenge this control. Some states have areas that are occupied by different cultural groups or by political movements seeking a separate state for themselves (Figure 8.1.3). Most of the new states in the twentieth century were achieved by anticolonial movements seeking independence from their European colonial masters. Territories that could be new countries are sometimes claimed by one or more other states, blocking attempts to form a new state (Figures 8.1.4 and 8.1.5). The Kashmir region is an example of a territory claimed in part by China, India, and Pakistan yet it could also be its own country.

▼ 8.1.1 **MEMBERS OF THE UNITED NATIONS**

▼ 8.1.2 **THE UNITED NATIONS BUILDING**
New York City was selected after World War II as the site of the United Nations headquarters. Diplomats of the world's countries meet here to discuss international peace and security.

▼ 8.1.3 **KOSOVO: UNRECOGNIZED SOVEREIGNTY**
The Republic of Kosovo declared its independence from Serbia in 2008, following ethnic cleansing and war crimes by some Serb leaders. Serbia's ally Russia has blocked Kosovo's membership in the United Nations so Kosovo's sovereignty must be instead recognized by a majority of the world's countries. The United States and most European countries recognize Kosovo as an independent sovereign state, but Serbia, Russia, and most countries of Africa and Asia do not.

► 8.1.4 TAIWAN: A SOVEREIGN STATE?

The governments of most other states consider China (officially, the People's Republic of China) and Taiwan (officially, the Republic of China) as separate and sovereign states. According to China's government, Taiwan is not sovereign, but a part of China. This confusing situation arose from a civil war in China during the late 1940s between the Nationalists and the Communists. After losing, Nationalist leaders in 1949 fled to Taiwan, 200 kilometers (120 miles) off the Chinese coast. The Nationalists proclaimed that they were still the legitimate rulers of the entire country of China. Until some future occasion when they could defeat the Communists and recapture all of China, the Nationalists argued, at least they could continue to govern one island of the country. The United Nations transferred China's seat from the Republic of China to the People's Republic of China in 1971, and the United States transferred diplomatic recognition to the People's Republic in 1979.

◄ 8.1.5 WESTERN SAHARA: NOT A SOVEREIGN STATE

Once a Spanish colony, the sparsely populated desert of Western Sahara remains the last major African territory without its own government. In 1975, the Spanish left administration of Western Sahara to Morocco and Mauritania. When Mauritania left in 1979, Morocco sent settlers to colonize Western Sahara. Morocco's claims have been challenged by Western Saharans seeking an independent country. The UN has repeatedly asked for a referendum on independence but without success, mainly because Morocco would like to hold on to the mineral resources in the territory.

8.2 State Space

- States have territories that include land, air, and sea.
- A state's shape affects its internal administration.

The physical space claimed by a sovereign state comprises its **territory**. A state's territory includes land, subsoil, internal bodies of water, and the airspace over these features. States with a sea border also have territorial waters near their shoreline, as well as special rights that extend far into the oceans (Figure 8.2.1).

The shape of a state may aid or hinder a government's administration of its territory. Any state may be classified as one of five basic shapes—prorupted, compact, elongated, fragmented, and perforated (Figure 8.2.2).

▼ 8.2.1 **STATE TERRITORY**
A 1967 international treaty prohibits states from claiming outer space or celestial objects as part of their sovereign territory.

There is no clear boundary between outer space and airspace. Current technological limits leave a wide gap between airspace and outer space. Regular airplanes do not fly above 32 kilometers (20 miles). Minimum heights for orbiting spacecraft vary from 161–362 kilometers (100–225 miles) above the Earth.

State's have the right to control the airspace over their territory.

State territory includes its land, resources, and internal waters, such as lakes and rivers, as well as subsoil resources such as oil and minerals.

INTERNAL WATERS

BASELINE

International law recognizes a state's baseline as the low-tide elevation of the line running along the general path of the shore including islands near shore. Bays, marshes, and waters behind islands are part of a state's internal waters.

A state's **territorial waters** include the area from 0 to 12 nautical miles or n.m. from the state's baseline. States have full sovereign authority over this territory but they must allow the "right of innocent passage" to ships passing through these waters peacefully.

TERRITORIAL WATERS
0–12 n.m. from baseline.

INTERNATIONAL WATERS
These include contiguous waters, exclusive economic zone, and high seas.

CONTIGUOUS WATERS
12–24 n.m. from baseline.

A state has limited authority over the **contiguous waters** between 12 and 24 n.m. from its baseline. In this zone a state may prevent and punish violations of its laws on customs, taxes, immigration, and sanitation.

EXCLUSIVE ECONOMIC ZONE
0–200 n.m. from baseline.

The **Exclusive Economic Zone (EEZ)** extends 200 n.m. from a state's baseline. A state has the right to control all resources within this zone, from fish to oil and gas. Some states with wide continental shelves claim these resource rights up to 350 n.m. from their baselines.

HIGH SEAS
The waters that lie beyond the EEZ are excluded from sovereign claims.

▼ 8.2.2 STATE SHAPES

► 8.2.2a **PRORUPTED STATE**

An otherwise compact state with a large, protruding extension is a **prorupted state**. Proruptions are the cause of boundary-drawing to provide a state access to water or another resource, as is the case for Congo's 500-kilometer (300-mile) extension along the Congo River to the Atlantic Ocean. The Caprivi Strip is the prorupted portion of northwestern Namibia that gave the country's colonial masters access to the Zambezi River and also cut connections between rival colonial territories.

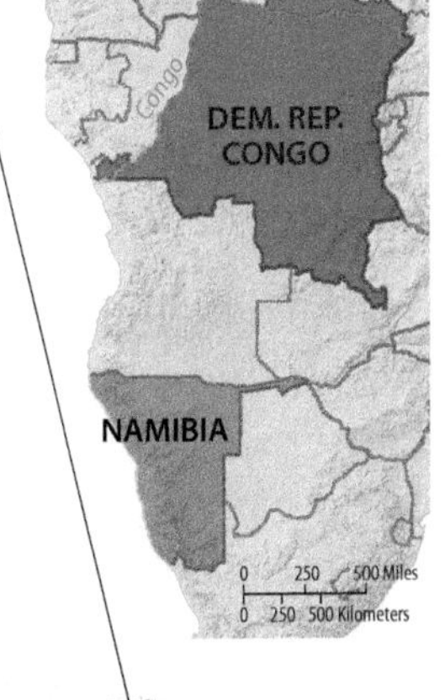

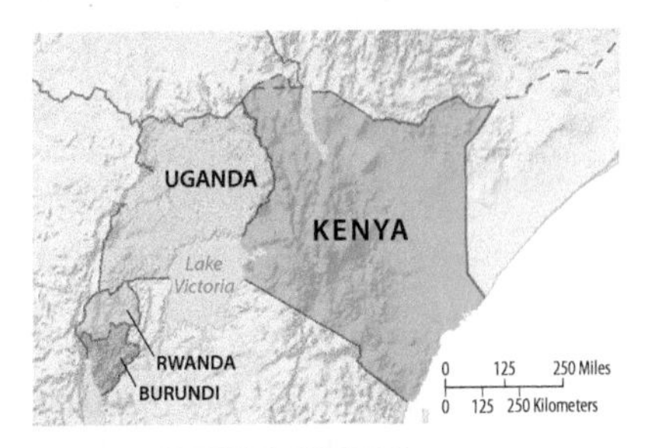

▲ 8.2.2b **COMPACT STATES**

Compact states have shapes that begin to approximate a circle, lessening the distance between any two points. This facilitates connections between any two places within the country. In Africa, Burundi, Rwanda, Kenya, and Uganda have fairly compact shapes yet each has also experienced political unrest.

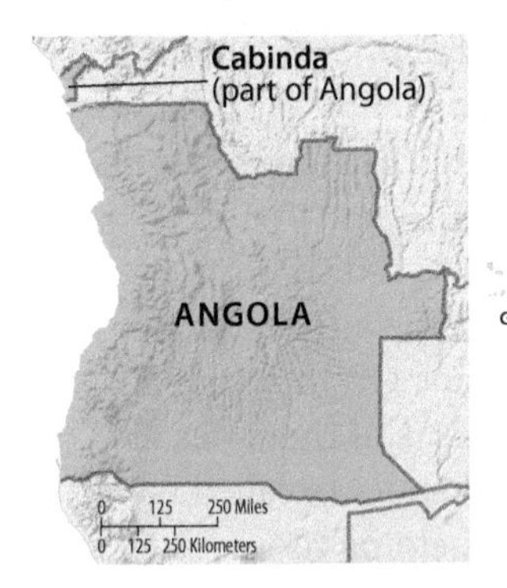

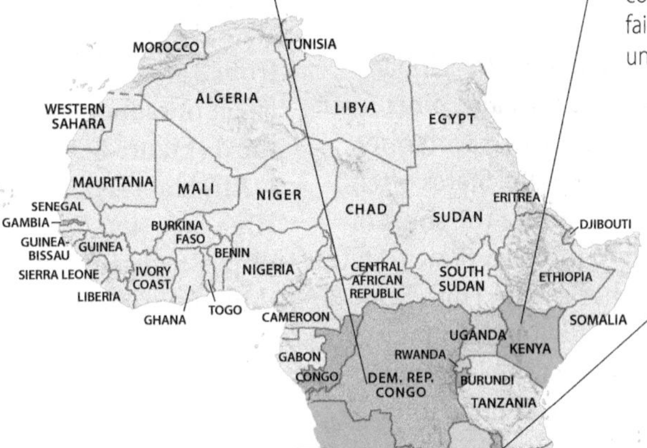

▼ 8.2.2c **ELONGATED STATE**

Elongated states have a long and narrow shape. Malawi, which measures about 850 kilometers (530 miles) north–south but only 100 kilometers (60 miles) east–west, is a good example. The Gambia, Italy, and Chile are other examples of elongated states. This shape creates long distances and high transportation costs that may economically and socially isolate remote territories.

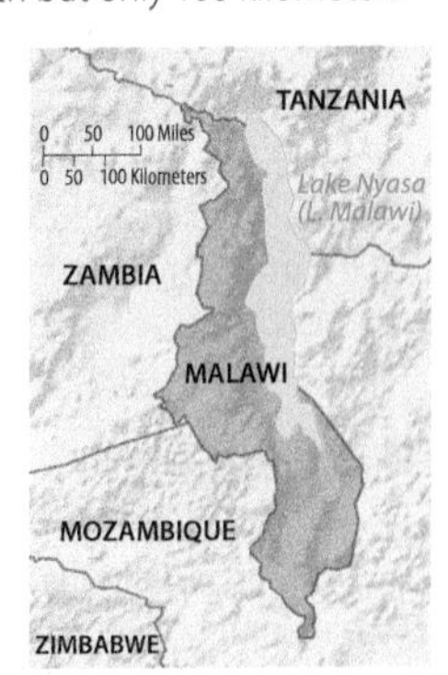

▲▼ 8.2.2d **FRAGMENTED STATE**

A **fragmented state** has two or more disconnected pieces of territory. Troubles arise when states are fragmented by another state's territory. The minor fragment is called an **exclave** of the country's mainland. Movement between the mainland and exclave territory requires permission of other states, which may not be granted. In some cases, exclaves are home to minority groups that reject mainland rule, such as the Cabinda exclave of Angola. Another example is Kaliningrad, an important exclave of Russia.

Fragmented states may also be divided by water, like Tanzania and Zanzibar islands. Fragmentation by water applies to many countries, including the United States, and makes settlement and administration more costly.

▼ 8.2.2e **PERFORATED STATE**

Perforated states have other state territories within them. The perforation may be around another whole state, as South Africa is to Lesotho. This is also the case with Vatican City and San Marino in Italy. The perforation may be caused by exclaves belonging to other states. This is the case with small territories along the border between India and Bangladesh, between Spain and France, between Belgium and the Netherlands, among others. The surrounded territory is termed an **enclave**.

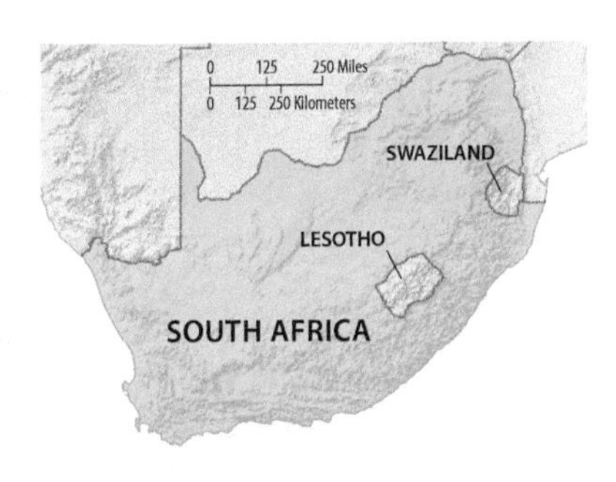

FROM LUANDA TO CABINDA

Use Google Earth to find a path for moving large freight from the town of Cabinda in Angola's exclave to the capital, Luanda. What modes of transportation are available and what routes would be necessary?

Fly to: *Cabinda, Angola.*

Zoom in to find major transportation nodes and pan to follow routes to Luanda, located 379 km (235 mi) south of Cabinda on the Atlantic Coast.

1. **What mode of transportation requires the longest route?**
2. **What mode allows the shortest route?**
3. **What mode is likely least costly?**
4. **How close are other major towns to Cabinda?**
5. **What countries are they in?**

8.3 Non-state Spaces

- Parts of our globe belong to no individual state.
- States seek agreements on how to manage common spaces.

Large parts of the Earth are not claimed by individual states. Agreements between states, called **international treaties**, forbid states from extending their sovereign claims over the oceans, Antarctica, and outer space. These treaties also provide the means for states to act together or individually to protect or manage these spaces.

THE HIGH SEAS

The problem of how to manage the open waters of the world oceans is centuries old. As European states began to compete for global colonies between the seventeenth and nineteenth centuries, their navies often clashed with one another to block passage or trade. States often hired mercenary sailors to capture the ships of other states, giving birth to modern piracy. States eventually realized that sharing the High Seas were in their own interest and fought piracy to secure open shipping lanes. Yet the legal basis for managing a common space and its limited resources remained unclear until recently.

The United Nations 1982 Convention on the Law of the Sea defines the High Seas as those parts of the oceans or seas beyond the territorial waters and Exclusive Economic Zone (EEZ) claimed by coastal states (refer to section 8.2). The convention set the extent of each coastal state's EEZ as 200 nautical miles (230 miles) from a state's coastline. This had the immediate effect of reducing the High Seas by about one-third (Figure 8.3.1).

Any state is permitted to sail ships on the High Seas and fish its waters as long as the ship is registered to a state. On the High Seas, ships are subject to the laws of the states to which they are registered but are called upon to sustain the tradition of helping those lost at sea or shipwrecked. So, too, are states expected to combat and prosecute piracy under a principle called **universal jurisdiction**, meaning any state can prosecute the offenders. The convention also stipulates that the High Seas be reserved for peaceful purposes, although naval powers at war would likely ignore this requirement.

▼ 8.3.1 **SHRINKING THE HIGH SEAS**
The United Nations 1982 Convention on the Law of the Sea allows coastal states, including island nations, to claim up to 200 n.m. (230 miles) of the ocean off their shores as an Exclusive Economic Zone. These EEZs are shown here in light blue. Some states can claim even more of the oceans if they have long continental shelves.

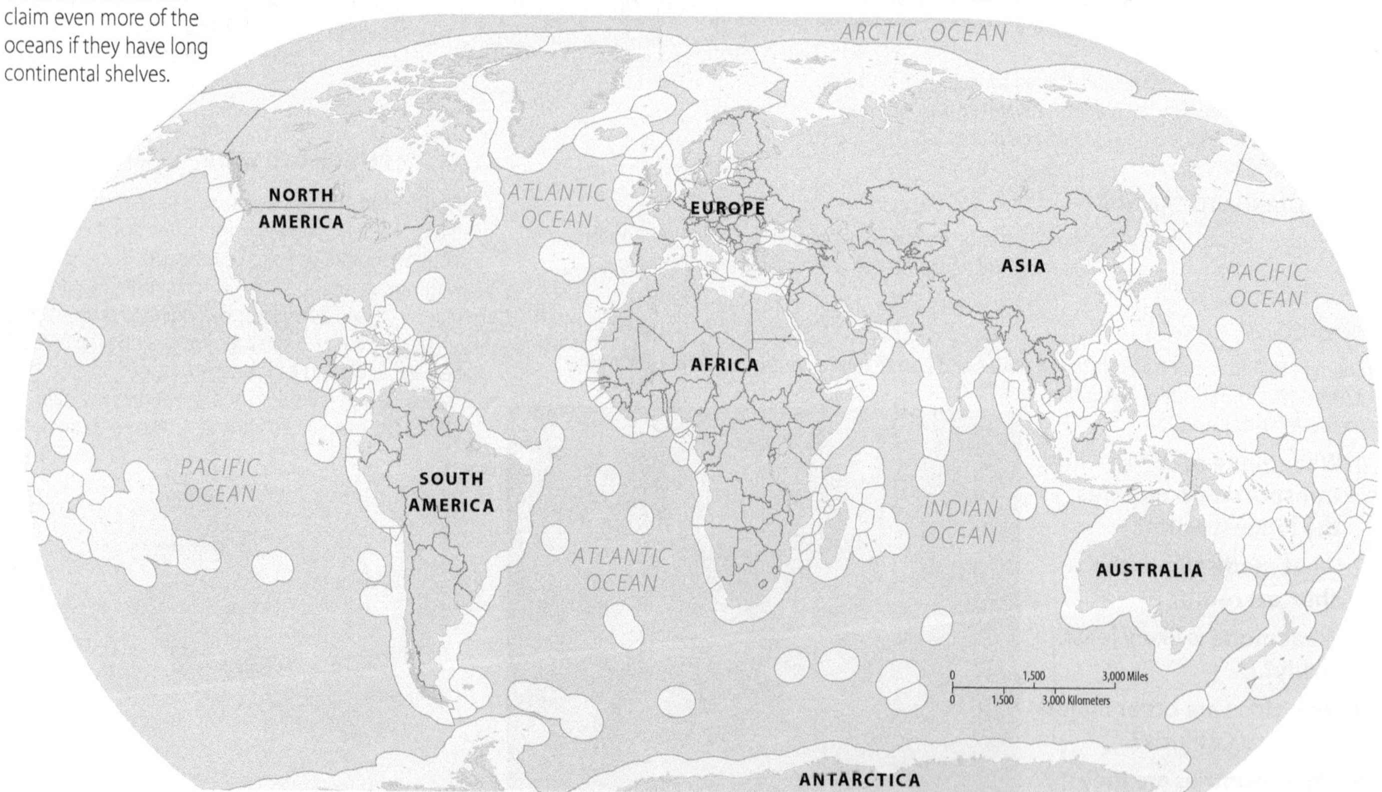

ANTARCTICA

The treaties creating non-state spaces often cite the importance of peaceful uses, scientific access, and enjoyment by all of humanity. These principles are enshrined in the 1959 Antarctic Treaty, which sets aside the territorial claims already made on the continent up to that point (Figure 8.3.2). It also bars states from laying new claims to sovereign territory in Antarctica. There is no native population and the only dwellings are scientific stations staffed by researchers who are not permanent residents. Experts have suggested that the treaty might be abandoned if valuable resources were located under the melting ice.

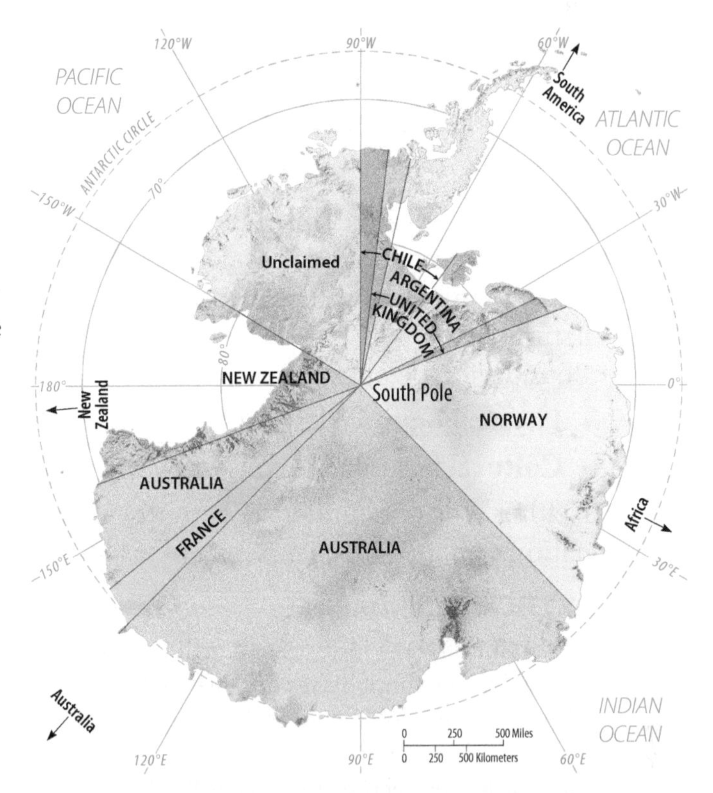

► 8.3.2 **ANTARCTICA**
Antarctica is the only large landmass in the world that is not part of a sovereign state. It comprises 14 million square kilometers (5.4 million square miles), which makes it 50 percent larger than Canada. Portions are claimed by Argentina, Australia, Chile, France, New Zealand, Norway, and the United Kingdom; claims by Argentina, Chile, and the United Kingdom are conflicting.

OUTER SPACE

Like the principles underlying the Law of the Sea and the Antarctic Treaty, the 1967 Outer Space Treaty limits states' exclusive use of outer space and celestial objects, including the moon. Besides barring claims of sovereignty, the treaty bans nuclear weapons in outer space, although conventional weapons are allowed. It shares with the Antarctic Treaty an emphasis on scientific exploration for the good of humanity. The extensive use of Earth's orbital zone for military and commercial use has led to problems of space debris that can damage other spacecraft (Figure 8.3.3). The treaty holds states responsible for damage done by their spacecraft.

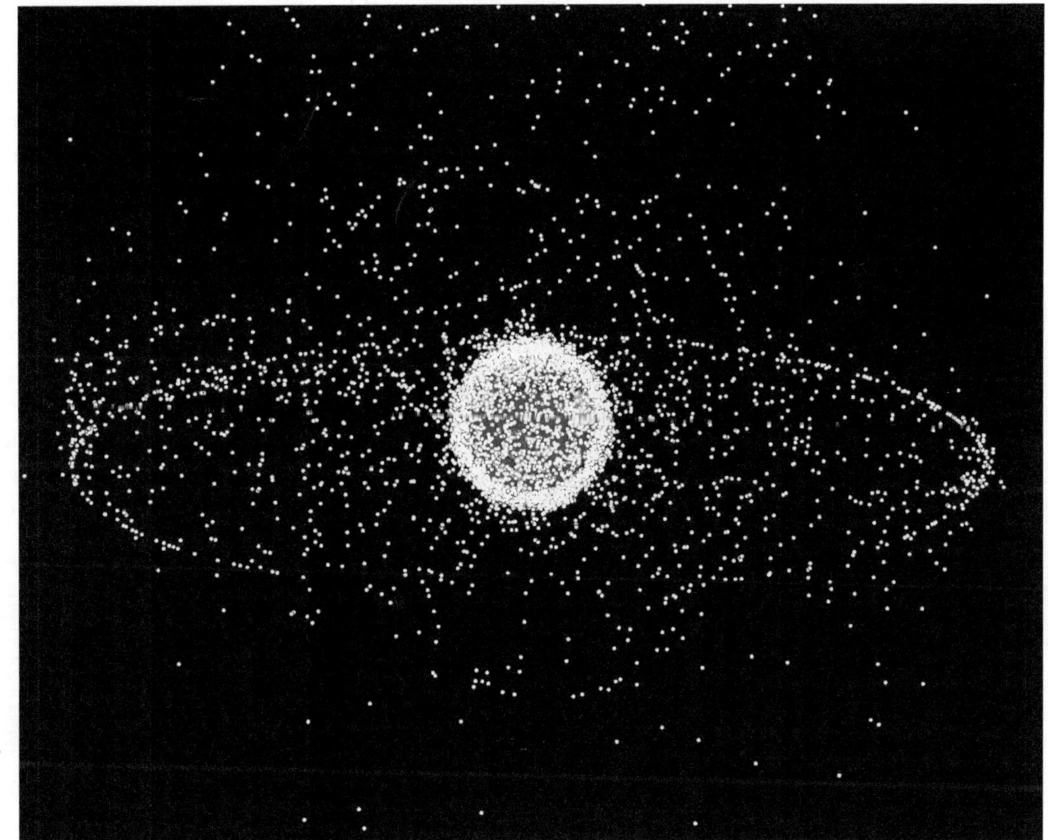

▲ 8.3.3 **OUTER SPACE**
Far from being empty, Earth's orbital space is filled with active satellites, spacecraft, and debris left by earlier space missions. This NASA generated map is a snapshot of the location of approximately 19,000 pieces of orbiting material 10 centimeters or larger (not to scale). Objects in geosynchronous orbit form a band of debris cast off from satellites that move over a fixed spot on the Earth's rotating surface. NASA tracks these materials to help spacecraft avoid collision with damaging objects.

8.4 Boundaries

- ► Borders separate political spaces.
- ► Borders can be defined in different ways.

When looking at satellite images of Earth, we see physical features like mountains and oceans, but not boundaries between countries. Boundary lines are not painted on Earth, but they might as well be, because for many people they are more meaningful than natural features.

Boundaries are of two types:

- **Physical boundaries** coincide with significant features of the natural landscape.
- **Cultural boundaries** follow the distribution of cultural characteristics.

Neither type of boundary is better or more "natural," and many boundaries are a combination of both types.

PHYSICAL BOUNDARIES

Important physical features on Earth's surface can make good boundaries because they are easily seen, both on a map and on the ground. Three types of physical elements serve as boundaries between states (Figure 8.4.1):

- **Desert boundaries.** Deserts make effective boundaries because they are hard to cross and sparsely inhabited. In North Africa, the Sahara has generally proved to be a stable boundary separating Algeria, Libya, and Egypt on the north from Mauritania, Mali, Niger, Chad, and the Sudan on the south.
- **Mountain boundaries.** Mountains can be effective boundaries if they are difficult to cross. Contact between nationalities living on opposite sides may be limited, or completely impossible if passes are closed by winter storms. Mountains are also useful boundaries because they are rather permanent and are usually sparsely inhabited.
- **Water boundaries.** Rivers, lakes, and oceans are commonly used as boundaries, because they are readily visible on maps and aerial imagery. Historically, water boundaries offered good protection against attack from another state, because an invading state had to transport its troops by ship and secure a landing spot in the country being attacked. The state being invaded could concentrate its defense at the landing point.

▲► 8.4.1 **PHYSICAL BOUNDARIES**
(above) Desert boundary between Libya and Chad. (right) Mountain boundary between Argentina and Chile. (far right) Water boundary between Germany and France.

CULTURAL BOUNDARIES

Two types of cultural boundaries are common—geometric and ethnic. Geometric boundaries are simply straight lines drawn on a map. Ethnic boundaries between states coincide with differences in ethnicity, especially language and religion.

- **Geometric boundaries.** Part of the northern U.S. boundary with Canada is a 2,100-kilometer (1,300-mile) straight line (more precisely, an arc) along 49° north latitude, running from Lake of the Woods between Minnesota and Manitoba to the Strait of Georgia between Washington State and British Columbia (Figure 8.4.2). This boundary was established in 1846 by a treaty between the United States and the United Kingdom, which then controlled Canada. The two countries share an additional 1,100-kilometer (700-mile) geometric boundary between Alaska and the Yukon Territory along the north–south arc of 141° west longitude.
- **Ethnic boundaries.** Boundaries between countries have been placed where possible to separate ethnic groups (Figure 8.4.3). Language is an important cultural characteristic for drawing boundaries, especially in Europe. Religious differences often coincide with boundaries between states, but in only a few cases has religion been used to select the actual boundary line.

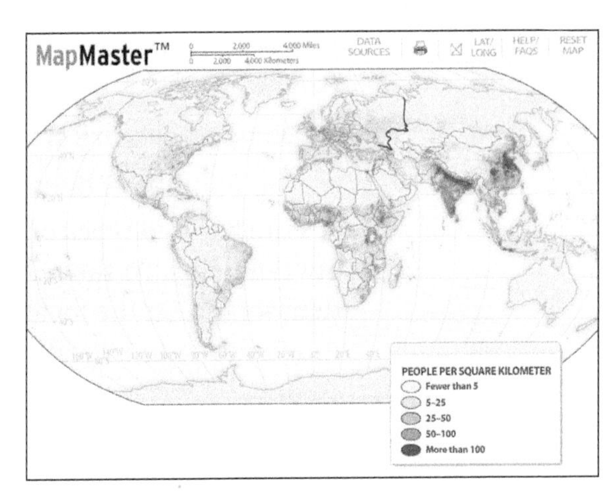

▲8.4.2 **GEOMETRIC BORDERS AND POPULATIONS**
Diplomats drawing boundaries in the colonial period often drew straight lines through areas unknown to them, especially in the Sahara Desert and along the U.S.–Canadian border.

Open MapMaster World Layered Thematic

Select: *Continents and Country Borders* from the Political menu, and *Population Density* from the Population menu.

Do geometric borders cut through highly populated places?

◀8.4.3 **ETHNIC BOUNDARY: GREEK AND TURKISH CYPRUS**
Cyprus, the third-largest island in the Mediterranean Sea, contains two nationalities—Greek and Turkish. Several Greek Cypriot military officers who favored unification of Cyprus with Greece seized control of the government in 1974. Shortly after, Turkey invaded Cyprus to protect the Turkish Cypriot minority, and the portion of the island controlled by Turkey declared itself the independent Turkish Republic of Northern Cyprus in 1983.

FRONTIERS

A **frontier** is a zone where no state exercises complete political control. A frontier is an area often many kilometers wide that is either uninhabited or sparsely settled. Historically, frontiers rather than boundaries separated many states (Figure 8.4.4). Almost universally, frontiers between states have been replaced by boundaries. Modern communications systems permit countries to monitor and guard boundaries effectively, even in previously inaccessible locations.

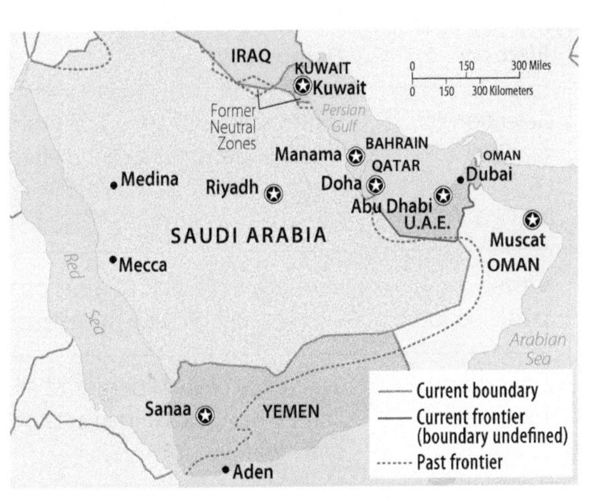

◀8.4.4 **FRONTIERS: ARABIAN PENINSULA**

8.5 Governing States

- ▶ Democracies are states ruled by their people.
- ▶ The number of democracies in the world has increased since 1800.

A state has two types of government—a national government and local governments. At the national scale, a government can be more or less democratic. At the local scale, the national government can determine how much power to allocate to local governments.

NATIONAL GOVERNMENT REGIMES

National governments can be classified as democratic, autocratic, or anocratic (Figure 8.5.1). According to the Center for Systemic Peace, a **democracy** and an autocracy differ in three essential elements (Figure 8.5.2). An **autocracy** is a country that is run according to the interests of the ruler rather than the people. An **anocracy** is a country that is not fully democratic or fully autocratic, but rather displays a mix of the two types. Anocracies are vulnerable to instability.

▶ 8.5.1 **REGIME TYPE, 2010**

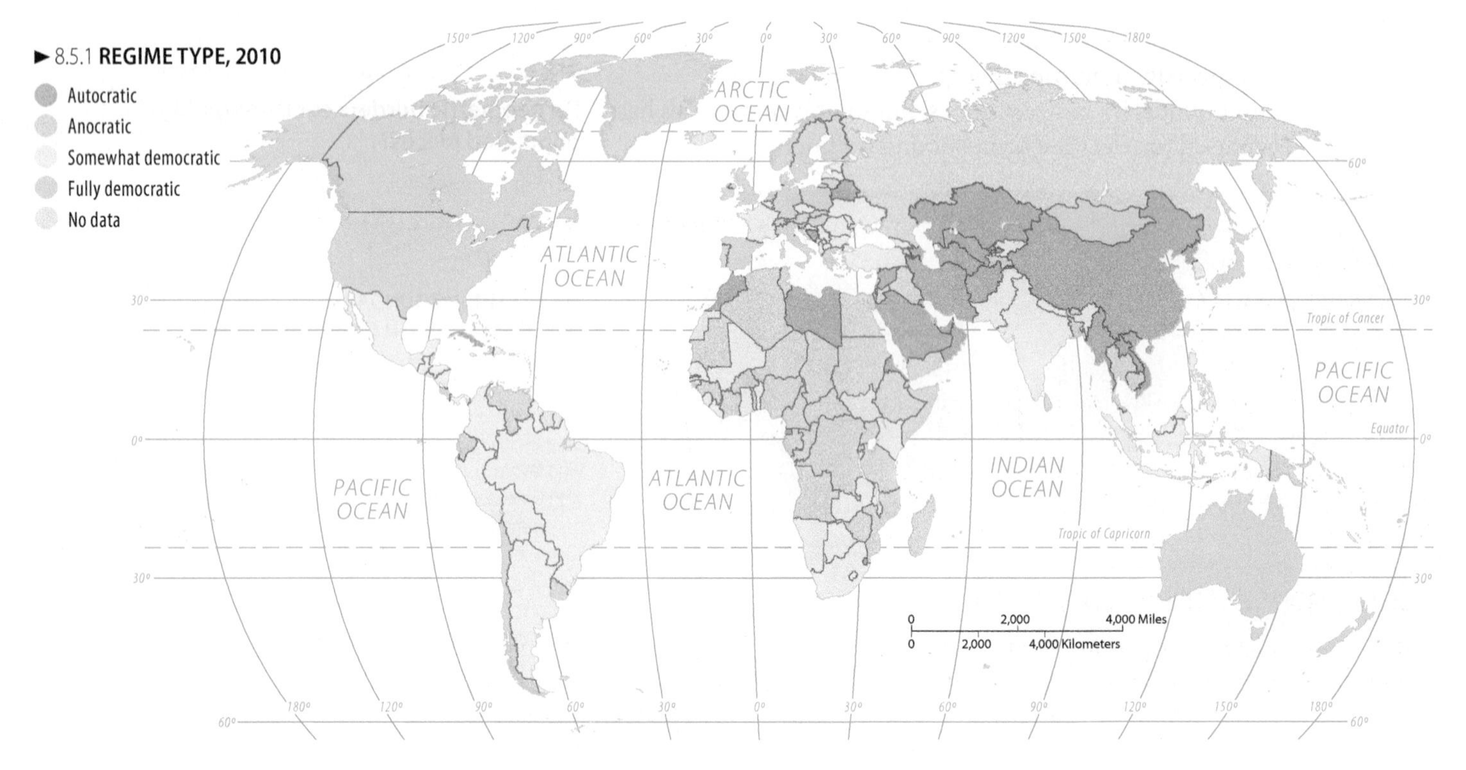

▶ 8.5.2 **DIFFERENCES BETWEEN DEMOCRACY AND AUTOCRACY**

Element	Democracy	Autocracy
Selection of leaders	Institutions and procedures through which citizens can express effective preferences about alternative policies and leaders.	Leaders are selected according to clearly defined (usually hereditary) rules of succession from within the established political elite.
Citizen participation	Institutionalized constraints on the exercise of power by the executive.	Citizens' participation is sharply restricted or suppressed.
Checks and balances	Guarantee of civil liberties to all citizens in their daily lives and in acts of political participation.	Leaders exercise power with no meaningful checks from legislative, judicial, or civil society institutions.

▲ 8.5.3 **ARAB SPRING**
Beginning in late 2010, mass protests and revolutionary movements swept across the Arab world. Autocratic regimes were removed in Tunisia and Egypt. In Libya, the uprising led to armed rebellion that overthrew the government. In Syria the government responded violently against protestors.

TREND TOWARDS DEMOCRACY

Overall, the world has become more democratic over the last 200 years. The Center for Systemic Peace cites three reasons:

- The replacement of autocracies, especially monarchies, with elected governments that are more responsive to citizens' needs and more respectful of their rights (Figure 8.5.3).
- Increased citizen participation in government by voting and running for public office.
- The diffusion of democratic government structures created in Europe and North America to other regions of the world (Figure 8.5.4).

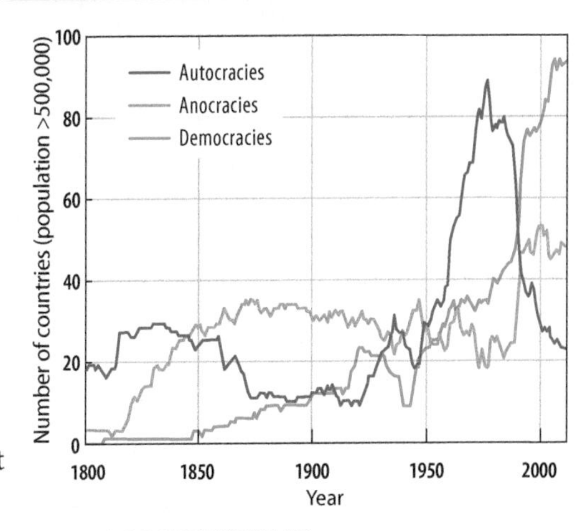

▲ 8.5 4 **DEMOCRACY TREND**

LOCAL GOVERNMENT: UNITARY STATE

The governments of states are organized according to one of two approaches: unitary and federal. The **unitary state** allocates most power to the national government, and local governments have relatively few powers. In principle, the unitary government system works best in nation-states characterized by few internal cultural differences and a strong sense of national unity. Because the unitary system requires effective communications with all regions of the country, smaller states are more likely to adopt it. Unitary states are especially common in Europe (Figure 8.5.5). Some multinational states, like Kenya and Rwanda, have unitary systems, so that the values of the dominant nationality can be imposed on other groups.

► 8.5.5 **UNITARY STATE**
Monaco.

LOCAL GOVERNMENT: FEDERAL STATE

In a **federal state**, strong power is allocated to units of local government within the country. In a federal state, such as the United States, local governments possess more authority to adopt their own laws. Multinational states may adopt a federal system of government to empower different nationalities, especially if they live in separate regions of the country (Figure 8.5.6). Under a federal system, local government boundaries can be drawn to correspond with regions inhabited by different ethnicities.

The federal system is also more suitable for very large states because the national capital may be too remote to provide effective control over isolated regions. Most of the world's largest states are federal, including Russia, Canada, the United States, Brazil, and India. However, the size of the state is not always an accurate predictor of the form of government: tiny Belgium is a federal state (to accommodate the two main cultural groups, the Flemish and the Waloons, as discussed in Chapter 5), whereas China is a unitary state (to promote Communist values).

In recent years there has been a strong global trend toward federal government. Unitary systems have been sharply curtailed in a number of countries and scrapped altogether in others.

► 8.5.6 **GOVERNMENT IN GERMANY**
Germany has a federal government with 16 Länder, or states. Each Länder has its own parliamentary government that establishes and enforces local laws and policies.

8.6 Electoral Geography

- ► **Changing electoral unit boundaries can change election outcomes.**
- ► **Gerrymandering is the drawing of electoral districts to favor the party in power.**

The boundaries separating legislative districts within the United States and other countries are redrawn periodically to ensure that each district has approximately the same population. Boundaries must be redrawn principally because migration inevitably results in some districts gaining population, whereas others are losing. The 435 districts of the U.S. House of Representatives are redrawn every 10 years following the release of official population figures by the Census Bureau.

▲ 8.6.2 **"WASTED VOTE" GERRYMANDERING**
"Wasted vote" spreads opposition supporters across many districts as a minority. If the Blue Party controls the redistricting process, it could do a "wasted vote" gerrymander by creating four districts with a slender majority of Blue Party voters and one district (#1) with a strong majority of Red Party voters.

The process of redrawing legislative boundaries for the purpose of benefiting the party in power is called **gerrymandering**. The term gerrymandering was named for Elbridge Gerry (1744–1814), governor of Massachusetts (1810–12) and vice president of the United States (1813–14). As governor, Gerry signed a bill that redistricted the state to benefit his party. An opponent observed that an oddly shaped new district looked like a "salamander," whereupon another opponent responded that it was a "gerrymander." A newspaper subsequently printed an editorial cartoon of a monster named "gerrymander" with a body shaped like the district (Figure 8.6.1).

▼ 8.6.1 **THE ORIGINAL GERRYMANDER CARTOON**
It was drawn in 1812 by Elkanah Tinsdale.

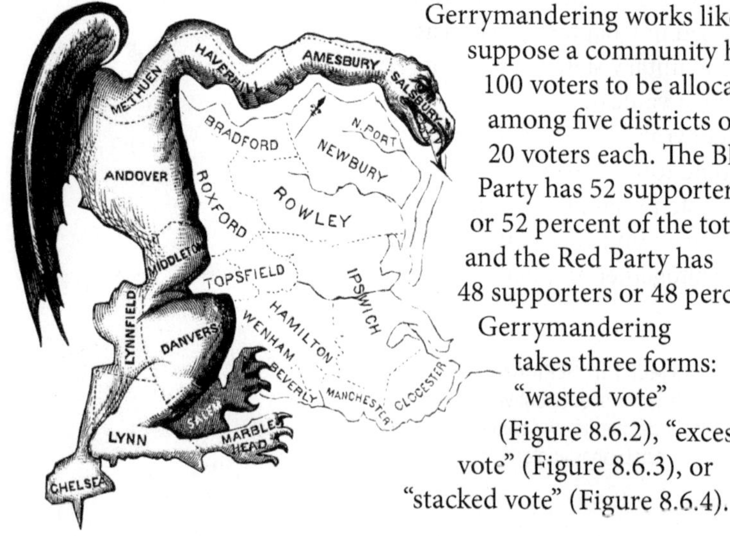

Gerrymandering works like this: suppose a community has 100 voters to be allocated among five districts of 20 voters each. The Blue Party has 52 supporters or 52 percent of the total, and the Red Party has 48 supporters or 48 percent. Gerrymandering takes three forms: "wasted vote" (Figure 8.6.2), "excess vote" (Figure 8.6.3), or "stacked vote" (Figure 8.6.4).

The job of redrawing boundaries in most European countries is entrusted to independent commissions. Commissions typically try to create compact homogeneous districts without regard for voting preferences or incumbents. A couple of U.S. states, including Iowa and Washington, also use independent or bipartisan commissions (Figure 8.6.5), but in most U.S. states the job of redrawing boundaries is entrusted to the state legislature. The political party in control of the state legislature naturally attempts to redraw boundaries to improve the chances of its supporters to win seats.

The U.S. Supreme Court ruled gerrymandering illegal in 1985 but did not require dismantling of existing oddly shaped districts, and a 2001 ruling allowed North Carolina to add another oddly shaped district that ensured the election of an African American Democrat. Through gerrymandering, only about one-tenth of Congressional seats are competitive, making a shift of more than a few seats unlikely from one election to another in the United States except in unusual circumstances.

Boundaries must be redrawn every 10 years after release of census data to assure that the population is the same in each district. Political parties may offer competing plans designed to favor their candidates (Figure 8.6.6).

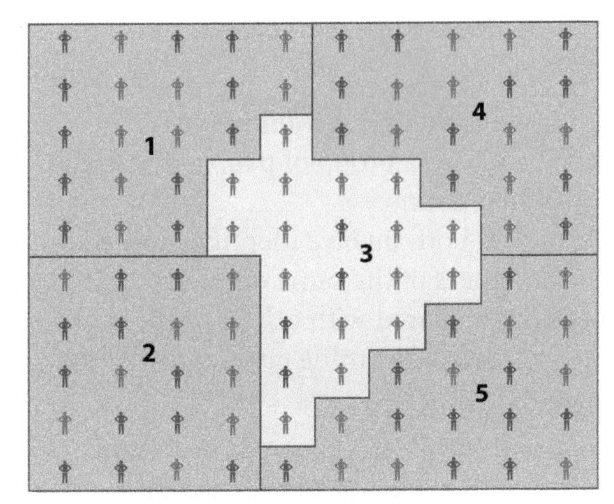

▲ 8.6.3 **"EXCESS VOTE" GERRYMANDERING**
"Excess vote" concentrates opposition supporters into a few districts. If the Red Party controls the redistricting process, it could do an "excess vote" gerrymander by creating four districts with a slender majority Red Party voters and one district (#3) with an overwhelming majority of Blue Party voters.

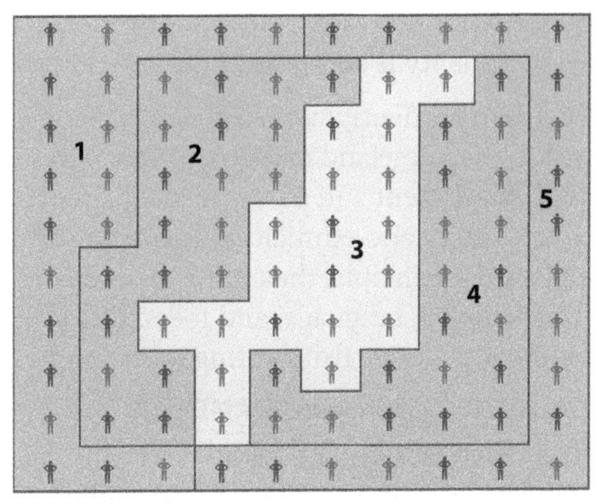

▲ 8.6.4 **"STACKED VOTE" GERRYMANDERING**
A "stacked vote" links distant areas of like-minded voters through oddly shaped boundaries. In this example, Red Party controls redistricting and creates five oddly shaped districts, four with a slender majority Red Party voters and one (#3) with an overwhelming majority of Blue Party voters.

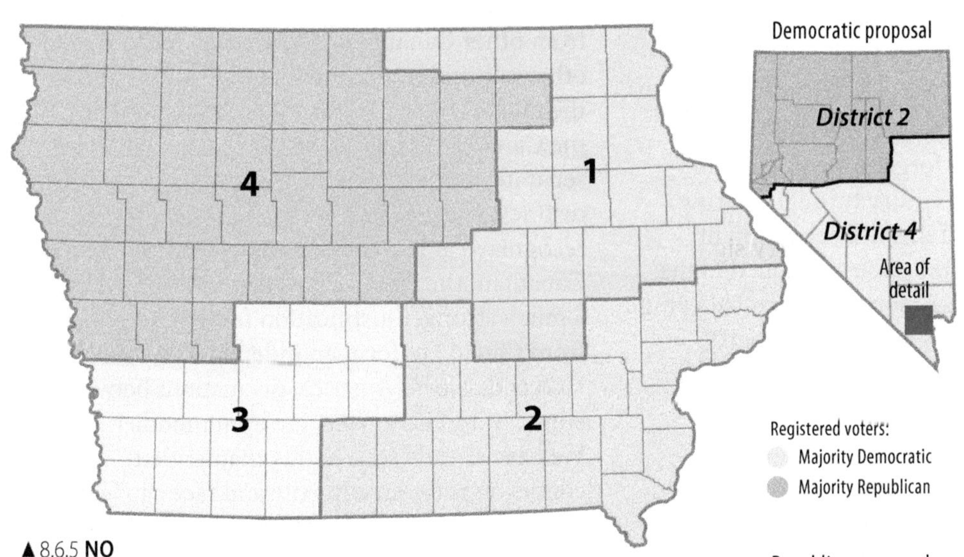

▲ 8.6.5 **NO GERRYMANDERING: IOWA**
Iowa does not have gerrymandered congressional districts. Each district is relatively compact, and boundaries coincide with county boundaries.

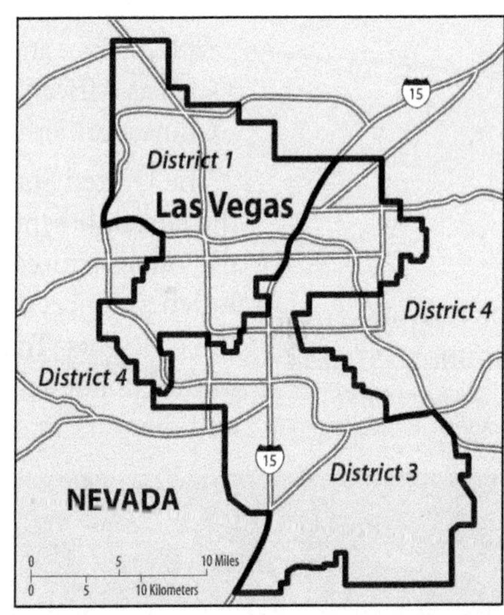

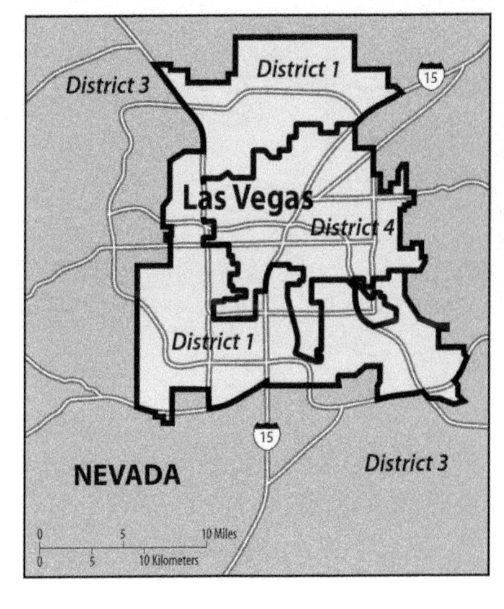

► 8.6.6 **GERRYMANDERING NEVADA: TWO PROPOSALS**
Competing proposals to draw boundaries for Nevada's four congressional districts illustrate all three forms of gerrymandering.
"Wasted vote" gerrymander (top right). Although Nevada as a whole has slightly more registered Democrats than Republicans (43 percent to 37 percent), the Democratic plan made Democrats more numerous than Republicans in three of the four districts.
"Excess vote" gerrymander (bottom right). By clustering a large share of the state's registered Democrats in District 4, the Republican plan gave Republicans the majority of registered voters in two of the four districts.
"Stacked vote" gerrymander (both plans). In the Republican plan (bottom), District 4 has a majority Hispanic population, and is surrounded by a "C" shaped District 1. The Democratic plan (top) created a long, narrow District 3.

8.7 Ethnicity and Nationality

- **Ethnicity and nationality are different ways people identify with each other.**
- **Nationality is sometimes defined in ethnic terms.**

Ethnicity is identity with a group of people who share the cultural traditions of a particular homeland or hearth. **Nationality** is identity with a group of people who share legal attachment and personal allegiance to a particular country.

Nationality and ethnicity are similar concepts, in that both involve identification with a place. In principle, the cultural values shared with others of the same ethnicity derive from religion, language, and folk culture, whereas those shared with others of the same nationality derive from voting, obtaining a passport, and performing civic duties.

NATIONALITIES IN NORTH AMERICA

In the United States, nationality is generally kept reasonably distinct from ethnicity in common usage:

- The American ***nationality*** identifies citizens of the United States of America, including those born in the country and those who immigrated and became citizens.
- ***Ethnicity*** in the United States identifies groups with distinct ancestry and cultural traditions, such as African Americans, Hispanic Americans, Chinese Americans, and Polish Americans.

The United States forged a nationality in the late eighteenth century through sharing the values expressed in the Declaration of Independence, the U.S. Constitution, and the Bill of Rights. To be American meant believing in the "unalienable rights" of "life, liberty, and the pursuit of happiness," and electing a president rather than submitting to a hereditary monarch (Figure 8.7.1). Initially, the last part was only true if you were a white male: African Americans weren't considered full citizens until the nineteenth century, and women weren't allowed to vote for president until the twentieth century.

In Canada, the Québécois are clearly distinct from other Canadians in language, religion, and other cultural traditions (Figure 8.7.2). But do the Québécois form a distinct ethnicity within the Canadian nationality or a second nationality separate altogether from Anglo-Canadian? The distinction is critical, because if Québécois is recognized as a separate nationality from Anglo-Canadian, the Québec government would have a much stronger justification for breaking away from Canada to form an independent country.

Outside North America, distinctions between ethnicity and nationality are even muddier. We have already seen in this chapter that confusion between ethnicity and race can lead to discrimination and segregation. Confusion between ethnicity and nationality can lead to violent conflicts.

▼ 8.7.1 **NATIONALITY IN THE UNITED STATES**
Independence Day parade in Rhode Island.

▼ 8.7.2 **NATIONALITY IN CANADA**
Canada Day (July 1) celebration in Ottawa.

NATIONALISM

Nationalism is loyalty and devotion to a nationality. Nationalism typically promotes a sense of national consciousness that exalts one nationality above all others and emphasizes its culture and interests as opposed to those of other nationalities. People display nationalism by supporting a state that preserves and enhances the culture and attitudes of their nationality.

Nationalism is an important example of a **centripetal force,** which is an attitude that tends to unify people and enhance support for a state. (The word centripetal means "directed toward the center"; it is the opposite of centrifugal, which means "to spread out from the center.") Most states find that the best way to achieve citizen support is to emphasize shared attitudes that unify the people.

States foster nationalism by promoting symbols such as flags and songs. The symbol of the hammer and sickle or the color red was long synonymous with the beliefs of communism (Figure 8.7.3). After the fall of communism, one of the first acts in those countries was to redesign a flag without the hammer and sickle (Figure 8.7.4).

Nationalism can be a negative force, especially when a nation is defined in ethnic terms. The sense of unity within ethnic nationalism is often achieved by creating negative images of other nations. German nationalism openly denigrated minority groups, such as Jews and Slavs, who became victims of the Nazi regime in World War II. Another example is Yugoslavia, which was a multinational country that prospered during the twentieth century. As the socialist era came to an end in the late 1980s, nationalist movements blamed each other for the country's economic and political crises. Elections put in power nationalists who then launched wars of "ethnic cleansing" against civilians from other national groups, causing the country to break apart (Figure 8.7.5, refer also to section 8.9).

◄ 8.7.3 **NATIONALISM IN UKRAINE UNDER COMMUNISM**
Under communism, Ukraine observed Revolution Day (November 7), when the Communists declared victory in 1917. As part of the Soviet Union, Ukraine's flag included symbols of communism, including the hammer and sickle and a partially red field.

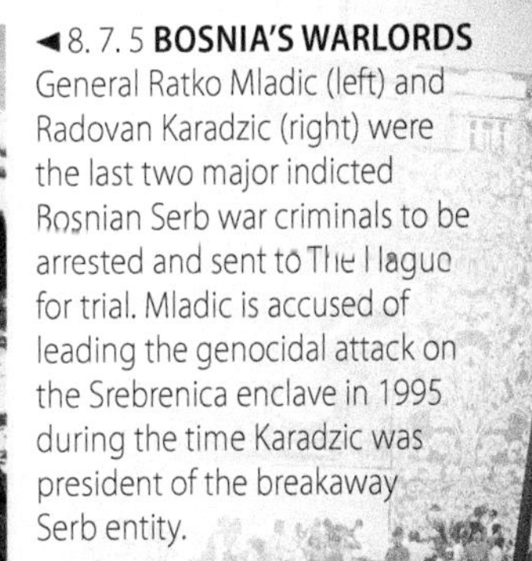

▼ 8.7.4 **NATIONALISM IN UKRAINE AFTER COMMUNISM**
After the fall of communism, Ukraine changed its day of national celebration to August 24, when it became independent of the Soviet Union. Ukraine also adopted its pre-Soviet flag.

◄ 8. 7. 5 **BOSNIA'S WARLORDS**
General Ratko Mladic (left) and Radovan Karadzic (right) were the last two major indicted Bosnian Serb war criminals to be arrested and sent to The Hague for trial. Mladic is accused of leading the genocidal attack on the Srebrenica enclave in 1995 during the time Karadzic was president of the breakaway Serb entity.

8.8 Conflicts in Western Asia

- **Western Asia is a complex area of nationalities and ethnicities.**
- **Conflict in Western Asia has resulted in part from a mismatch between ethnicities and nationalities.**

States in Western Asia have territories that rarely correspond to ethnicity. As a result, states trying to foster national unity have frequently imposed the dominant ethnic culture on other ethnic groups. This has led to conflict. Dozens of ethnicities inhabit the region, which is divided into seven nationalities (Figure 8.8.1):

- Iraqi nationality. Most numerous ethnicity is Arab. The major ethnicities are divided into numerous tribes and clans. Most Iraqis actually have stronger loyalty to a tribe or clan than to the nationality or a major ethnicity (Figure 8.8.2).
- Armenian nationality. Armenian is both an ethnicity and a nationality (Figure 8.8.3).
- Azerbaijani nationality. Most numerous is Azeri, but Armenians represent an important minority (refer to Figure 8.8.3).
- Georgian nationality. Most numerous is Georgian (refer to Figure 8.8.3).
- Afghan nationality. Most numerous ethnicities are Pashtun, Tajik, and Hazara (Figure 8.8.4).
- Iranian nationality. Most numerous is Persian, but Azeri and Baluchi represent important minorities (Figure 8.8.5).
- Pakistani nationality. Most numerous are Punjabi, but the border area with Afghanistan is principally Baluchi and Pashtun (Figure 8.8.6).

▼ 8.8.1 **ETHNIC DIVERSITY IN WESTERN ASIA**
Azeri refugees from Nagorno-Karabakh living in abandoned housing, Baku, Azerbaijan.

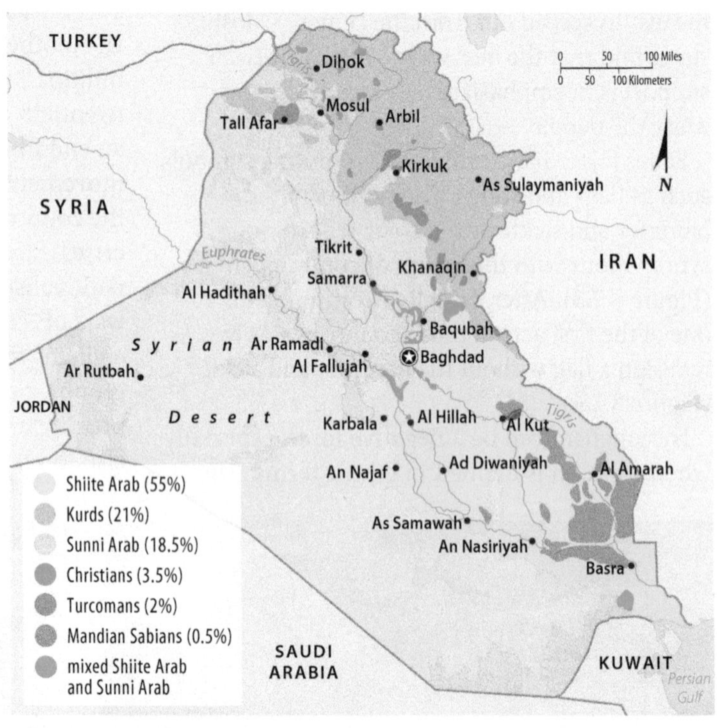

▲ 8.8.2 **ETHNICITIES IN IRAQ**
The United States led an attack against Iraq in 2003 in order to depose the country's longtime President Saddam Hussein. U.S. officials justified removing Hussein, because he ran a brutal dictatorship, created weapons of mass destruction, and allegedly had close links with terrorists. Having invaded Iraq and removed Hussein from power, the United States expected an enthusiastic welcome from the Iraqi nation. Instead, the United States became embroiled in a complex and violent struggle among ethnic groups.

- Kurds welcomed the United States because they gained more security and autonomy than they had under Hussein.
- Sunni Muslim Arabs opposed the U.S.-led attack, because they feared loss of power and privilege given to them by Hussein, who was a Sunni.
- Shiite Muslim Arabs also opposed the U.S. presence. Although they had been treated poorly by Hussein and controlled Iraq's post-Hussein government, Shiites shared a long-standing hostility toward the United States with their neighbors in Shiite-controlled Iran.

Most Iraqis have stronger loyalty to a tribe or clan than to the nationality or major ethnicity. A tribe ('ashira) is divided into several clans (fukhdhs), which in turn encompass several houses (beit), which in turn include several extended families (khams). Tribes are grouped into more than a dozen federations (qabila).

8.8.3 ETHNICITIES IN ARMENIA, AZERBAIJAN, AND GEORGIA

These nationalities became independent of the Soviet Union in 1991.

- Armenians once controlled an extensive empire, but more than a million were massacred by the Turks during the late nineteenth and early twentieth centuries.
- Azeris trace their roots to Turkish invaders who migrated from Central Asia in the eighth and ninth centuries and merged with the existing Persian population.
- Armenians and Azeris have clashed several times over control of Nagorno-Karabakh, an enclave within Azerbaijan inhabited primarily by Armenians.
- Georgia is a more diverse country. Ossetians and Abkhazians have both attempted to leave Georgia and set up independent countries with support from nearby Russia.

8.8.4 ETHNICITIES IN AFGHANISTAN

The current unrest among Afghanistan's ethnicities dates from 1979, with the start of a rebellion by several ethnic groups against the government, which was being defended by more than 100,000 troops from the Soviet Union. Unable to subdue the rebellion, the Soviet Union withdrew its troops in 1989, and the Soviet-installed government in Afghanistan collapsed in 1992. After several years of infighting among ethnicities, a faction of the Pashtun called the Taliban (which means "religious students") gained control over most of the country in 1995. The Taliban imposed very harsh, strict laws on Afghanistan, according to Islamic values as the Taliban interpreted them. The United States invaded Afghanistan in 2001 and overthrew the Taliban-led government, because it was harboring terrorists. Removal of the Taliban unleashed a new struggle for control of Afghanistan among the country's many ethnic groups, including the Taliban.

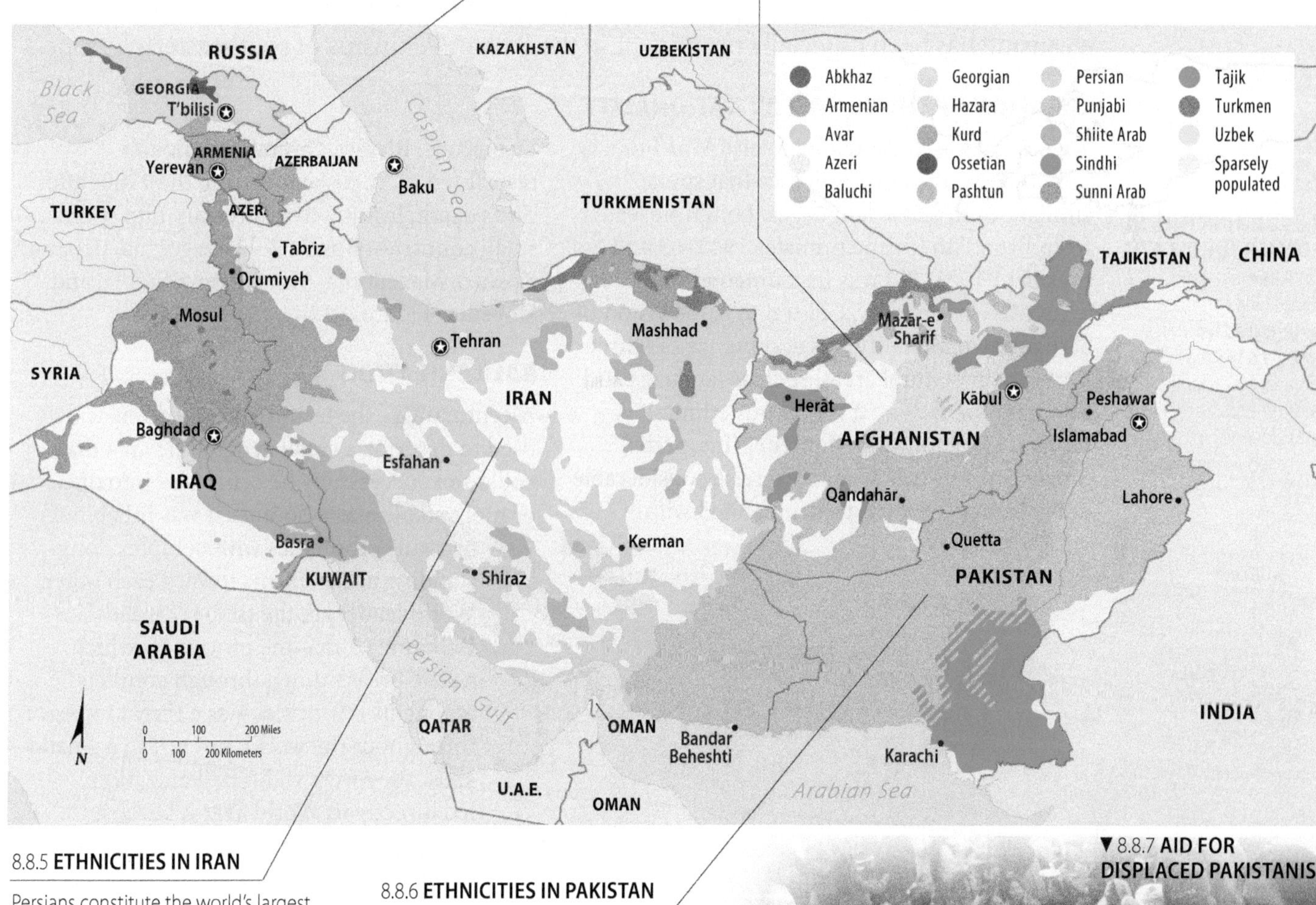

8.8.5 ETHNICITIES IN IRAN

Persians constitute the world's largest ethnic group that adheres to Shiite Islam. Persians are believed to be descendents of the Indo-Europeans tribes that began migrating from Central Asia into what is now Iran several thousand years ago. The Persian Empire extended from present-day Iran west as far as Egypt during the fifth and fourth centuries B.C. After the Muslim army conquered Persia in the seventh century, most Persians converted to Sunni Islam. The conversion to Shiite Islam came primarily in the fifteenth century.

8.8.6 ETHNICITIES IN PAKISTAN

The Punjabi have been the most numerous ethnicity since ancient times in what is now Pakistan. As with the neighboring Pashtun, the Punjabi converted to Islam after they were conquered by the Muslim army in the seventh century. The Punjabi remained Sunni Muslims rather than convert to Shiite Islam like their neighbors the Pashtun, who comprise Pakistan's second largest ethnicity, especially along the border with Afghanistan. (right) Fighting between Pakistan's army and supporters of the Taliban forced Pakistanis to leave their homes and move into camps, where they were fed by international relief organizations.

▼ 8.8.7 AID FOR DISPLACED PAKISTANIS

8.9 Ethnic Cleansing in the Balkans

- **Ethnic cleansing is the forcible removal of an ethnic group by a more powerful one.**
- **Southeastern Europe has suffered from ethnic cleansing in recent years.**

Ethnic cleansing is a process in which a more powerful ethnic group forcibly removes a less powerful one in order to create an ethnically homogeneous region. Ethnic cleansing is undertaken to rid an area of an entire ethnicity so that the surviving ethnic group can be the sole inhabitants. Rather than a clash between armies of male soldiers, ethnic cleansing involves the removal of every member of the less powerful ethnicity—women as well as men, children as well as adults, the frail elderly as well as the strong youth. Ethnic cleansing has been especially prominent in the Balkan Peninsula of southeastern Europe.

CREATION OF THE YUGOSLAV NATIONALITY

Yugoslavia was created after World War I to unite several Balkan ethnicities that spoke similar South Slavic languages. Longtime leader Josip Broz Tito (prime minister 1943–63 and president 1953–80) was instrumental in forging a Yugoslav nationality. Central to Tito's vision of a Yugoslav nationality was acceptance of ethnic diversity in cultural areas, such as language and religion. The five most numerous ethnicities—Croats, Macedonians, Montenegrins, Serbs, Slovenes—were allowed to exercise considerable control over the areas they inhabited within Yugoslavia. Rivalries among ethnicities resurfaced in Yugoslavia during the 1980s after Tito's death, leading to its breakup into seven small countries: Bosnia & Herzegovina, Croatia, Kosovo, Macedonia, Montenegro, Serbia, and Slovenia (Figure 8.9.1).

▼ 8.9.1 **ETHNICITIES IN FORMER YUGOSLAVIA**
Until its breakup in 1992, Yugoslavia comprised six so-called republics (Bosnia & Herzegovina, Croatia, Macedonia, Montenegro, Serbia, and Slovenia), plus two so-called autonomous regions (Kosovo and Vojvodina).

BALKANIZATION

A century ago, the term Balkanized was widely used to describe a small geographic area that could not successfully be organized into one or more stable states because it was inhabited by many ethnicities with complex, long-standing antagonisms toward each other. World leaders at the time regarded **Balkanization**—the process by which a state breaks down through conflicts among its ethnicities—as a threat to peace throughout the world, not just in a small area. They were right: Balkanization led directly to World War I, because the various nationalities in the Balkans dragged into the war the larger powers with which they had alliances.

After two world wars and the rise and fall of communism during the twentieth century, the Balkans has once again become Balkanized in the twenty-first century. Peace has come to the Balkans in a tragic way, through the "success" of ethnic cleansing. Millions of people were rounded up and killed or forced to migrate because they constituted ethnic minorities.

◄ 8.9.2 **ETHNIC CLEANSING IN BOSNIA & HERZEGOVINA**
The Stari Most (old bridge), built by the Turks in 1566 across the Neretva River, was an important symbol and tourist attraction in the city of Mostar. (left) The bridge was blown up by Croat forces in 1993 in an attempt to demoralize Bosnian Muslims as part of ethnic cleansing. (right) With the end of the war in Bosnia & Herzegovina, the bridge was rebuilt in 2004.

BOSNIA & HERZEGOVINA

At the time of Yugoslavia's breakup, Bosnia & Herzegovina was a mix of three ethnicities: 48 percent Bosnian Muslim, 37 percent Serb, and 14 percent Croat. Rather than live in an independent country with a Muslim plurality, Bosnia & Herzegovina's Serbs and Croats fought to unite the portions of the republic that they inhabited with neighboring Serbia and Croatia, respectively.

To strengthen their cases for breaking away from Bosnia & Herzegovina, Serbs and Croats engaged in the ethnic cleansing of Bosnian Muslims. Ethnic cleansing by Bosnian Serb forces against Bosnian Muslims was especially severe, because much of the territory inhabited by Bosnian Serbs was separated from Serbia by areas with Bosnian Muslim majorities (Figure 8.9.2). By ethnically cleansing Bosnian Muslims from intervening areas, Bosnian Serbs created one continuous area of Bosnian Serb domination rather than several discontinuous ones (Figure 8.9.3).

Accords reached in Dayton, Ohio, in 1996 by leaders of the three ethnicities divided Bosnia & Herzegovina into three regions, each one dominated, respectively, by the Bosnian Croats, Muslims, and Serbs. In recognition of the success of their ethnic cleansing, Bosnian Serbs and Croats received more land than their share of the population warranted.

BOSNIA & HERZEGOVINA

Predominantly Croat
Predominantly Bosnian
Predominantly Serb
Bosnian-Croat mix

► 8.9.3 **BOSNIA AFTER ETHNIC CLEANSING**
Compare the current distribution of ethnicities to Figure 8.9.1, which shows the distribution before ethnic cleansing.

KOSOVO

The population of Kosovo is more than 90 percent ethnic Albanian. At the same time, Serbs consider Kosovo an essential place in the formation of the Serb ethnicity, because they fought an important—though losing—battle against the Ottoman Empire there in 1389.

As part of Yugoslavia, Kosovo had been an autonomous province. With the breakup of Yugoslavia, Serbia took direct control of Kosovo and launched a campaign of ethnic cleansing of the Albanian majority. At its peak in 1999, Serb ethnic cleansing had forced 750,000 of Kosovo's 2 million ethnic Albanian residents from their homes, mostly to camps in Albania (Figure 8.9.4).

Outraged by the ethnic cleansing, the United States and western European countries, operating through the North Atlantic Treaty Organization (NATO), launched an air attack against Serbia. The bombing campaign ended when Serbia agreed to withdraw all of its soldiers and police from Kosovo.

Kosovo declared independence from Serbia in 2008. The United States and most European countries have recognized the independence, but countries allied with Serbia, including China and Russia, oppose it.

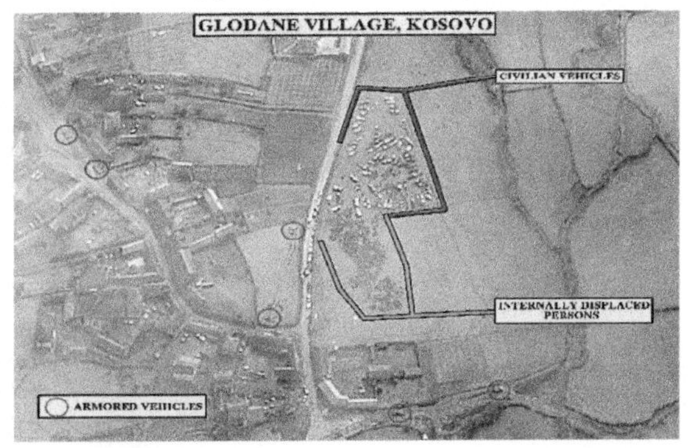

◄ 8.9.4 **ETHNIC CLEANSING IN KOSOVO**
In this photo taken by NATO air reconnaissance in 1999, the village of Glodane is on the west (left) side of the road. The villagers and their vehicles have been rounded up and placed in the field east of the road. The red circles show the locations of Serb armored vehicles.

8.10 Conflict and Genocide in Africa

- **Genocide is the mass killing of a population by another group.**
- **Genocide has been practiced in several areas of sub-Saharan Africa.**

Ethnicities compete in many places to dominate territory and control the defining of a nationality. This competition can lead to war and ethnic cleansing, as discussed in previous sections. In a handful of the most extreme cases, competition can lead to the most extreme action, which is genocide.

Genocide is the mass killing of a group of people in an attempt to eliminate the entire group out of existence. Among other places, including former Yugoslavia, sub-Saharan Africa has been plagued by conflicts among ethnic groups that have resulted in genocide in recent years, especially in Sudan and in central Africa.

ETHNIC COMPETITION AND GENOCIDE IN SUDAN

Sudan, a country of 42 million inhabitants, has had several civil wars in recent years (Figure 8.10.1).

- **South Sudan** Black Christian and animist ethnicities in the south resisted attempts by the Arab–Muslim-dominated government forces in the north to impose a unified nationality based on fundamentalist Muslim principles, including harsh repression of women. A north–south war between 1983 and 2005 resulted in the death of an estimated 1.9 million Sudanese, mostly civilians. The war ended with the establishment of South Sudan as an independent state in 2011. However, fighting resumed as the governments of Sudan and South Sudan could not agree on boundaries between the two countries.
- **Darfur** As Sudan's religion-based civil war was winding down, an ethnic war erupted in Sudan's westernmost Darfur region. Resenting discrimination and neglect by the national government, Darfur's black Africans launched a rebellion in 2003. Marauding Arab nomads, known as Janjaweed, with the support of the Sudanese government, crushed Darfur's black population, made up mainly of settled farmers. 480,000 have been killed and another 2.8 million have been living in dire conditions in refugee camps in the harsh desert environment of Darfur. Actions of Sudan's government troops, including mass murders and rape of civilians, have been termed genocide by many other countries, and charges of war crimes have been filed against Sudan's leaders (Figure 8.10.2).
- **Eastern Front** Ethnicities in the east fought Sudanese government forces between 2004 and 2006 with the support of neighboring Eritrea. At issue was disbursement of profits from oil.

▼ 8.10.1 SUDAN AND SOUTH SUDAN

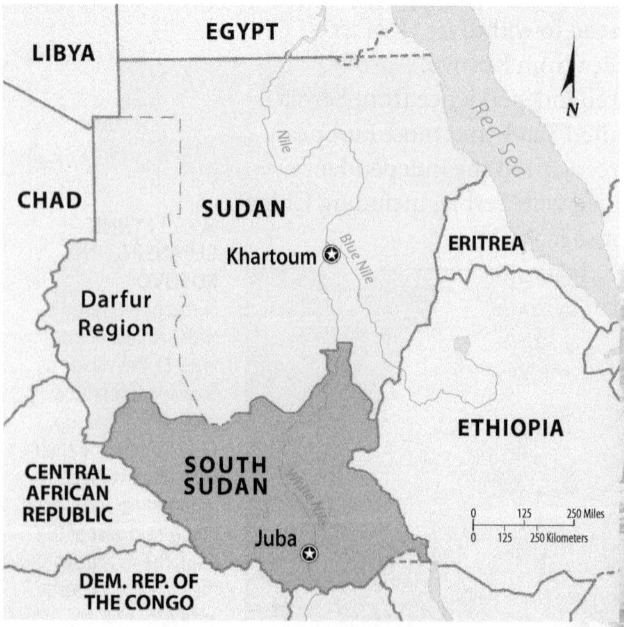

▼ 8.10.2 DARFUR
Sudanese police stand guard in the 4 sq km Abu Shouk refugee camp in Al Fasher, North Darfur, Sudan.

GENOCIDE IN CENTRAL AFRICA

Long-standing conflicts between two ethnic groups, the Hutus and Tutsis, lie at the heart of genocide in central Africa.

- Hutus were settled farmers, growing crops in the fertile hills and valleys of present-day Rwanda and Burundi, known as the Great Lakes region of central Africa.
- Tutsis were cattle herders who migrated to present-day Rwanda and Burundi from the Rift Valley of western Kenya beginning 400 years ago.

Relations between settled farmers and herders are often uneasy—this is also an element of the ethnic cleansing in Darfur discussed earlier. Genocide has been most severe in Rwanda and the Congo:

- **Rwanda** Genocide in Rwanda in 1994 involved Hutus murdering hundreds of thousands of Tutsis (as well as Hutus sympathetic to the Tutsis). Hutus constituted a majority of the population of Rwanda historically, but Tutsis controlled the kingdom of Rwanda for several hundred years and had turned the Hutu into their serfs. As a colony of Germany and then Belgium during the first half of the twentieth century, Tutsis were given more privileges than the Hutus. Shortly before Rwanda gained its independence in 1962, Hutus killed or ethnically cleansed many Tutsis out of fear that the Tutsis would seize control of the newly independent country. After an airplane carrying the presidents of Rwanda and Burundi was shot down in 1994, probably by a Tutsi, the genocide began (Figure 8.10.3).
- **Democratic Republic of the Congo** The conflict between Hutus and Tutsis spilled into neighboring countries, especially the Democratic Republic of the Congo. Several million have died in the Congo in a war that began in 1998 and is the world's deadliest since World War II. The war started after Tutsis helped to overthrow the Congo's longtime president, Joseph Mobutu. Mobutu had amassed a several-billion-dollar personal fortune from the sale of minerals while impoverishing the rest of the country. After succeeding Mobutu as president, Laurent Kabila relied heavily on Tutsis and permitted them to kill some of the Hutus who had been responsible for atrocities against Tutsis back in the early 1990s. But Kabila soon split with the Tutsis, and the Tutsis once again found themselves offering support to rebels seeking to overthrow the Congo's government. Kabila turned for support to Hutus, and armies from neighboring countries came to Kabila's aid. Kabila was assassinated in 2001 and succeeded by his son, who negotiated an accord with rebels the following year.

▼ 8.10.3 **RWANDA**
Thousands of Rwandan Hutu refugees from Gisenyi cross the border into Goma, Democratic Republic of the Congo.

8.11 Terrorism

- Terrorists use violence to coerce governments to change their policies.
- Extremist Islamist groups have targeted many governments around the world.

Terrorism is the systematic use of violence by a group in order to intimidate a population or coerce a government into granting its demands. Terrorists attempt to achieve their objectives through organized acts that spread fear and anxiety among the population, such as bombing, kidnapping, hijacking, taking of hostages, and assassination. They consider violence necessary as a means of bringing widespread publicity to goals and grievances that are not being addressed through peaceful means. Belief in their cause is so strong that terrorists do not hesitate to strike despite knowing they will probably die in the act.

Distinguishing terrorism from other acts of political violence can be difficult. For example, if a Palestinian suicide bomber kills several dozen Israeli teenagers in a Jerusalem restaurant, is that an act of terrorism or wartime retaliation against Israeli government policies and army actions? Competing arguments are made: Israel's sympathizers denounce the act as a terrorist threat to the country's existence, whereas advocates of the Palestinian cause argue that long-standing injustices and Israeli army attacks on Palestinian civilians provoked the act.

▲ 8.11.1 **SEPTEMBER 11, 2001, ATTACKS**

February 26, 1993: A car bomb parked in the underground garage damaged New York's World Trade Center, killing 6 and injuring about 1,000.

April 19, 1995: A truck bomb killed 168 people in the Alfred P. Murrah Federal Building in Oklahoma City.

December 21, 1988: A terrorist bomb destroyed Pan Am Flight 103 over Lockerbie, Scotland, killing all 259 aboard, plus 11 on the ground.

June 25, 1996: A truck bomb blew up an apartment complex in Dhahran, Saudi Arabia, killing 19 U.S. soldiers who lived there and injuring more than 100 people.

September 11, 2001: Airplanes crash into the World Trade Center in New York, the Pentagon near Washington, and Shanksville, PA, killing 3,000.

August 7, 1998: U.S. embassies in Kenya and Tanzania were bombed, killing 190 and wounding nearly 5,000.

October 12, 2000: The USS *Cole* was bombed while in the port of Aden, Yemen, killing 17 U.S. service personnel.

► 8.11.2 **TERRORISM AGAINST AMERICANS**
Major attacks, 1988–2001

TERRORISM AGAINST AMERICANS

The most dramatic terrorist attack against the United States came on September 11, 2001. Two of the tallest buildings in the United States, the 110-story twin towers of the World Trade Center in New York City were destroyed, and the Pentagon in Washington, D.C., was damaged (Figure 8.11.1). The attacks resulted in nearly 3,000 fatalities.

Prior to the 9/11 attacks, the United States had suffered several terrorist attacks during the late twentieth century (Figure 8.11.2). Some were by American citizens operating alone or with a handful of others.

- Theodore J. Kaczynski, known as the Unabomber, was convicted of killing three people and injuring 23 others by sending bombs through the mail during a 17-year period. His targets were mainly academics in technological disciplines and executives in businesses whose actions he considered to be adversely affecting the environment.
- Timothy J. McVeigh was convicted and executed for the Oklahoma City bombing, and for assisting him Terry I. Nichols was convicted of conspiracy and involuntary manslaughter. McVeigh claimed he had been provoked by U.S. government actions, including the FBI's 51-day siege of the Branch Davidian religious compound near Waco, Texas, culminating with an attack on April 19, 1993, that resulted in 80 deaths.

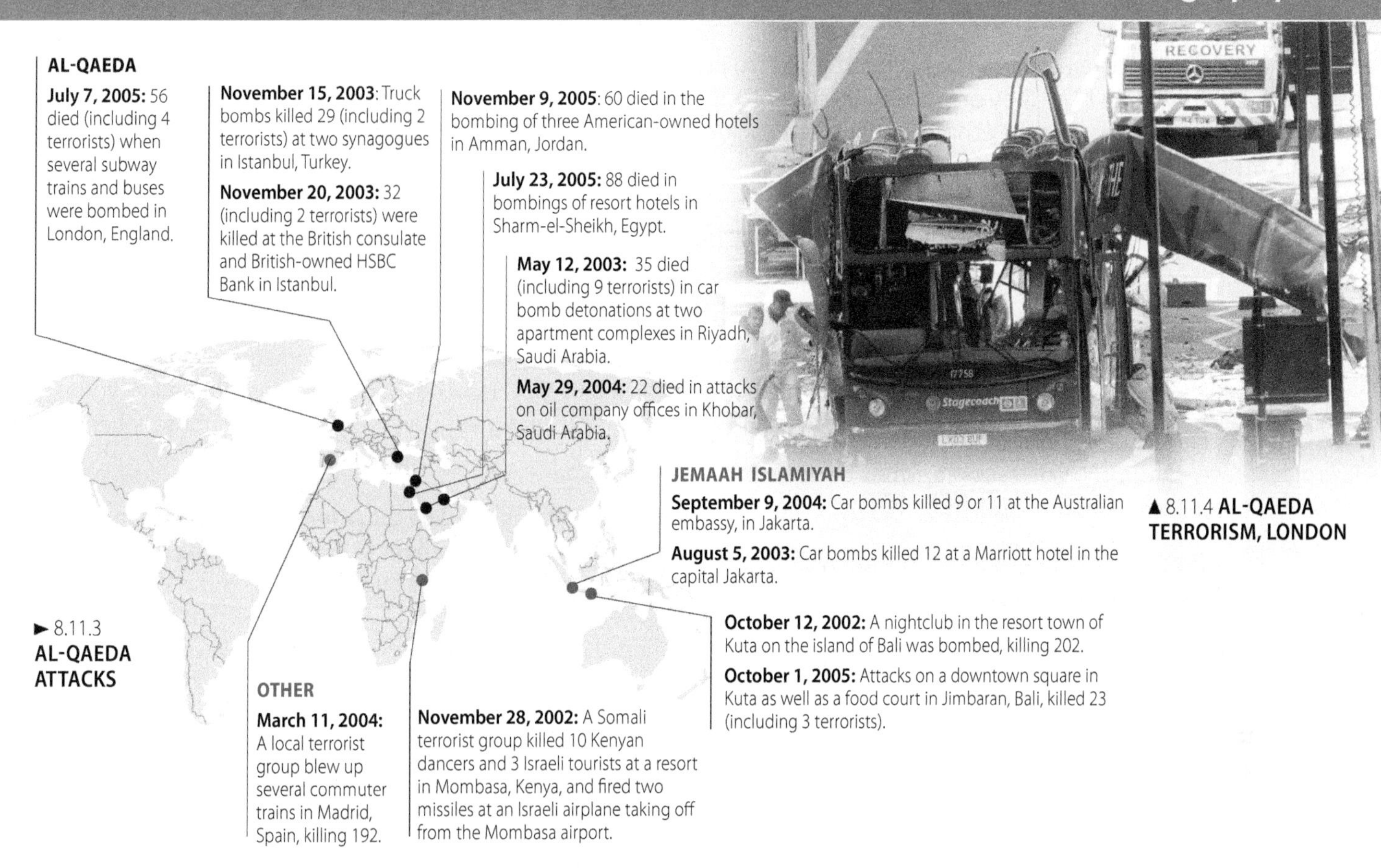

► 8.11.3 **AL-QAEDA ATTACKS**

▲ 8.11.4 **AL-QAEDA TERRORISM, LONDON**

AL-QAEDA

Responsible or implicated in most of the anti-U.S. terrorism in Figure 8.11.2, including the September 11, 2001, attack, was the al-Qaeda network (Figure 8.11.3). Al-Qaeda (an Arabic word meaning "the foundation" or "the base") has been implicated in several bombings since 9/11 (Figure 8.11.4).

Al-Qaeda's founder Osama bin Laden (1957–2011) issued a declaration of war against the United States in 1996 because of U.S. support for Saudi Arabia and Israel. In a 1998 fatwa ("religious decree"), bin Laden argued that Muslims had a duty to wage a holy war against U.S. citizens because the United States was responsible for maintaining the Saud royal family as rulers of Saudi Arabia and a state of Israel dominated by Jews. Destruction of the Saudi monarchy and the Jewish state of Israel would liberate from their control Islam's three holiest sites of Makkah (Mecca), Madinah, and Jerusalem.

In some respects, al-Qaeda operates like a business. A leadership council sets policy and oversees committees that specialize in such areas as finance, military, media, and religious policy. The organization keeps records and reimburses its members for expenses, such as purchasing bomb-making equipment. After U.S. Navy SEALS killed bin Laden in 2011, the al-Qaeda council replaced him as commander with his deputy Ayman al-Zawahiri.

Al-Qaeda is not a single unified organization. In addition to the original organization responsible for the World Trade Center attack, al-Qaeda also encompasses local franchises concerned with country-specific issues, as well as imitators and emulators ideologically aligned with al-Qaeda but not financially tied to it. For example, Jemaah Islamiyah, an al-Qaeda franchise trying to create fundamentalist Islamic governments in Southeast Asia, launches attacks in the world's most populous Muslim country, Indonesia.

Al-Qaeda's use of religion to justify attacks has posed challenges to Muslims and non-Muslims alike. For many Muslims, the challenge has been to express disagreement with the policies of governments in the United States and Europe yet disavow the use of terrorism. For many Americans and Europeans, the challenge has been to distinguish between the peaceful but unfamiliar principles and practices of the world's 1.3 billion Muslims and the misuse and abuse of Islam by a handful of terrorists.

Humans have divided their globe into political spaces to administer and manage people and resources. States vary in how they are internally organized and ruled. Armed conflicts may erupt within or between states. Recent ethnic conflicts, genocides, and terrorist acts have targeted civilians rather than just soldiers.

Key Questions

How is the world politically organized?

- Our globe has been divided politically for the purpose of creating separate and independent states.
- Most large areas, such as the oceans and Antarctica, as well as outer space, remain undivided and unclaimed by states.
- Boundaries separate states but vary in how they are defined and enforced.

How are states organized internally?

- States' internal organization varies widely from autocracies to democracies, as well as the degree of power held by central versus local government.
- Redistricting affects the outcomes of elections.
- States rarely have just one nationality and often include multiple nationalities or ethnic groups.

How does conflict vary by region?

- Armed conflicts within and between states emerge when competing interests cannot be peacefully settled.
- Many recent wars have featured violence against civilians to change populations and boundaries.
- Recent Islamist terrorism has targeted countries around the world in an effort to change their policies.

▼ 8.CR.1 **BORDER BETWEEN PAKISTAN AND INDIA**
UN peacekeeping troops, contributed by other countries, have monitored the ceasefire line between Pakistan and India since 1949.

Thinking Geographically

Rapidly increasing flows of goods, finances, and migrants across international boundaries have been interpreted as a challenge to state sovereignty. Governments are increasingly pressured to allow in goods and investments but many voters want to stop immigration and the outflow of jobs to other countries.

1. How can states manage these forces? Are the means of controlling the movement of goods, finances, and people still located at a state's boundary?

In most U.S. states, redistricting is conducted by whatever political party happens to be in power after each census. Parties use redistricting to deepen their hold on power by wasting some votes. Many have argued that this is contrary to a democracy in which each voter should have an equal say to effect a change in government ("one person, one vote").

2. Can you think of methods for redrawing districts that would help ensure greater voter rights? Can you think of a system for selecting U.S. representatives that would abandon districts all together? How might these changes affect the balance of federal and local power?

There have been many bloody civil wars since the 1990s. Some think that this is due to a mismatch between state territories and the homelands idealized by ethnic or national groups (8.CR.1). This view, in other words, sees ethnic and national groups as politically incompatible.

3. Identify states where multiple ethnic or national groups live together peacefully. What geographical factors, such as territory or the role of local government, contribute to the stability of these states?

On the Internet

The United Nations provides maps of its peacekeeping missions that are useful for keeping up with current events **(http://www.un.org/depts/Cartographic/english/htmain.htm)**.

The U.S. Census Bureau makes public the data used for congressional redistricting **(http://www.census.gov/rdo/)**.

Interactive Mapping

NATIONALISM AND ETHNICITY IN LATIN AMERICA

Select MapMaster Latin America in MasteringGEOGRAPHY

Select *Political* then *Countries.*

Select *Geopolitical* then *Shifting Political Boundaries – 1830.*

Select *Cultural* then *Dominant Ethnic Group.*

1. Which countries in Latin America have boundaries today that closely match those at the time of independence?

2. Which of these countries appears to have one ethnicity occupying all or nearly all of the territory of the state?

Explore

ARCTIC MARITIME BORDERS

Use Google Earth to explore the the emerging conflict over Arctic oil and gas resources.

Fly to: *Lomonosov Ridge,* a long underwater elevation running from Greenland to Russia under the North Pole.

1. **What states have Exclusive Economic Zones in the Arctic Ocean? Use the ruler tool to measure 200 nautical miles from various coastlines to get a sense of how far their claims to energy sources might reach.**
2. **Russia claims that the Lomonosov Ridge is an extension of its continental shelf that would push its EEZ deep into the Arctic. If this claim were upheld in international court, which states might dispute each other EEZs?**
3. **What other states might object to the Russian claim and why?**

Key Terms

Anocracy
Type of government with mix of democratic and autocratic features.

Autocracy
Type of government in which leaders are selected from the established elite and have few limits on their powers.

Balkanization
Breakdown of a state because of internal ethnic differences.

Centripetal force
Political attitude that supports the state.

Compact state
A state whose shape approximates a circle.

Contiguous waters
Area of the ocean from 12 to 24 nautical miles from a state's shore over which it has limited sovereignty.

Cultural boundaries
Areal limit of states that follow the distribution of cultural characteristics.

Democracy
Type of limited government in which citizens can elect leaders and stand for office.

Elongated state
A state whose shape is long and narrow.

Enclave
A territory completely surrounded by another state.

Ethnic cleansing
Process in which a more powerful ethnic group forcibly removes a less powerful one in order to create an ethnically homogeneous region.

Ethnicity
Identity with a group of people who share the cultural traditions of a particular homeland or hearth.

Exclave
A piece of territory separated from the rest of the country by other countries.

Exclusive Economic Zone (EEZ)
The area up to 200 nautical miles from a state's shore in which it has sole authority to exploit natural resources.

Federal state
Internal organization of states giving some powers to local government.

Fragmented state
A state with disconnected pieces of territory.

Frontier
A zone where no state exercises complete political control.

Genocide
Murderous campaign to eliminate an ethnic, racial, religious, or national group.

Gerrymandering
Redrawing election districts to favor one party over another.

International treaties
Agreements between states by which they agree to limit their sovereign rights or to cooperate on shared problems.

Nationalism
Loyalty and devotion to a nationality.

Nationality
Identity with a group of people who share legal attachment and personal allegiance to a particular country.

Perforated state
A country with other state territories enclosed within it.

Physical boundaries
Areal limit of states that coincide with significant features of the natural landscape.

Prorupted state
A state whose shape is compact with a protruding extension.

Recognition
Formal acknowledgement by existing states of a new state's claim to sovereign independence.

Sovereignty
A state's right to independently rule itself without interference from other states.

State
An area organized into a political unit and ruled by an established government with a resident population.

Territorial waters
Area of the ocean within 12 nautical miles of a state's shore over which it has full sovereign authority.

Territory
The physical space of a political unit. For states, this includes land, subsoil, waters, and airspace

Terrorism
Systematic use of violence by a group in order to intimidate a population or coerce a government into granting its demands.

Unitary state
Internal organization of states keeping most powers for central government.

Universal jurisdiction
A legal principle that gives all states the right to prosecute crimes in non-state spaces, for example piracy on the High Seas.

▸ LOOKING AHEAD

The second half of the book concentrates on economic elements of geography, beginning with the division of the world into more and less developed regions.

9 Development

The world is divided into developed countries and developing countries. The one-fifth of the world's people living in developed countries consume five-sixths of the world's goods, whereas the 14 percent of the world's people who live in Africa consume about 1 percent.

The United Nations recently contrasted spending between developed and developing countries in picturesque terms: Americans spend more per year on cosmetics ($8 billion) than the cost of providing schools for the 2 billion people in the world in need of them ($6 billion). Europeans spend more on ice cream ($11 billion) than the cost of providing a working toilet to the 2 billion people currently without one at home ($9 billion).

To reduce disparities between rich and poor countries, developing countries must develop more rapidly. This means increasing wealth and using that wealth to make more rapid improvements in people's health and well-being.

How does development vary among regions?

BUILDING A NEW ROAD, MOZAMBIQUE

How can countries promote development?

What are future challenges for development?

SCAN TO ACCESS THE UN's HUMAN DEVELOPMENT REPORT

9.1 Human Development Index

- ► **Countries are classified as developed or developing.**
- ► **The Human Development Index (HDI) measures a country's level of development.**

Earth's nearly 200 countries can be classified according to their level of **development**, which is the process of improving the material conditions of people through diffusion of knowledge and technology. The development process is continuous, involving never-ending actions to constantly improve the health and prosperity of the people. Every place lies at some point along a continuum of development.

The United Nations classifies countries as developed or developing:

- A **developed country**, also known as a **more developed country** (MDC) or a **relatively developed country,** has progressed further along the development continuum. The UN considers these countries to have very high development.
- A **developing country**, also frequently called a **less developed country (LDC),** has made some progress towards development though less than developed countries. Recognizing that progress has varied widely among developing countries, the UN divides them into high, medium, and low development.

To measure the level of development of every country, the UN created the **Human Development Index (HDI)**. The UN has computed HDIs for countries every year since 1990, although it has occasionally modified the method of computation. The HDI considers development to be a function of three factors:

- A decent standard of living.
- Access to knowledge.
- A long and healthy life.

Each country gets a score for each of these three factors, which are then combined into an overall HDI (Figure 9.1.1). The highest HDI possible is 1.0, or 100 percent. These factors are discussed in more detail in sections 9.2, 9.3, and 9.4.

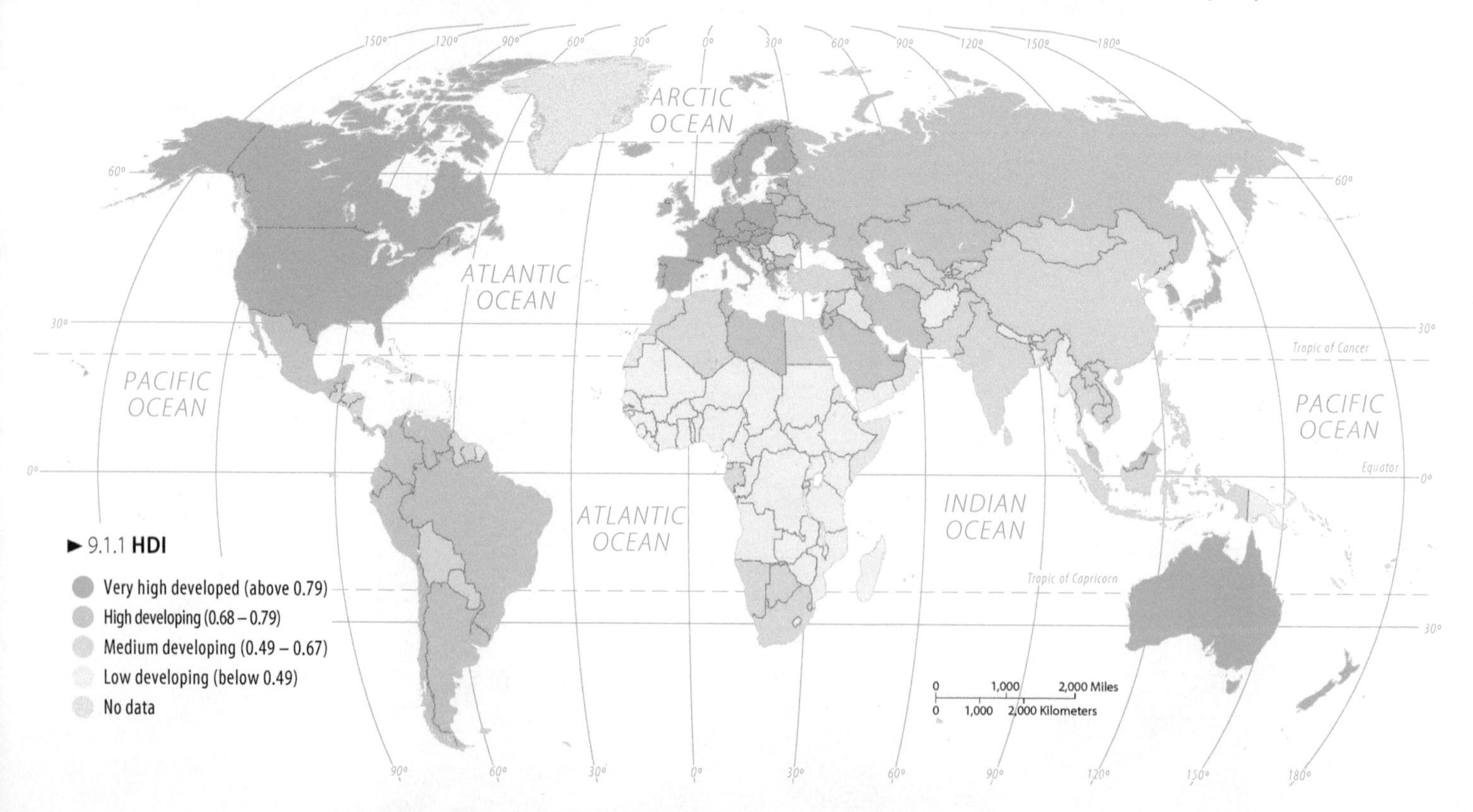

► 9.1.1 **HDI**

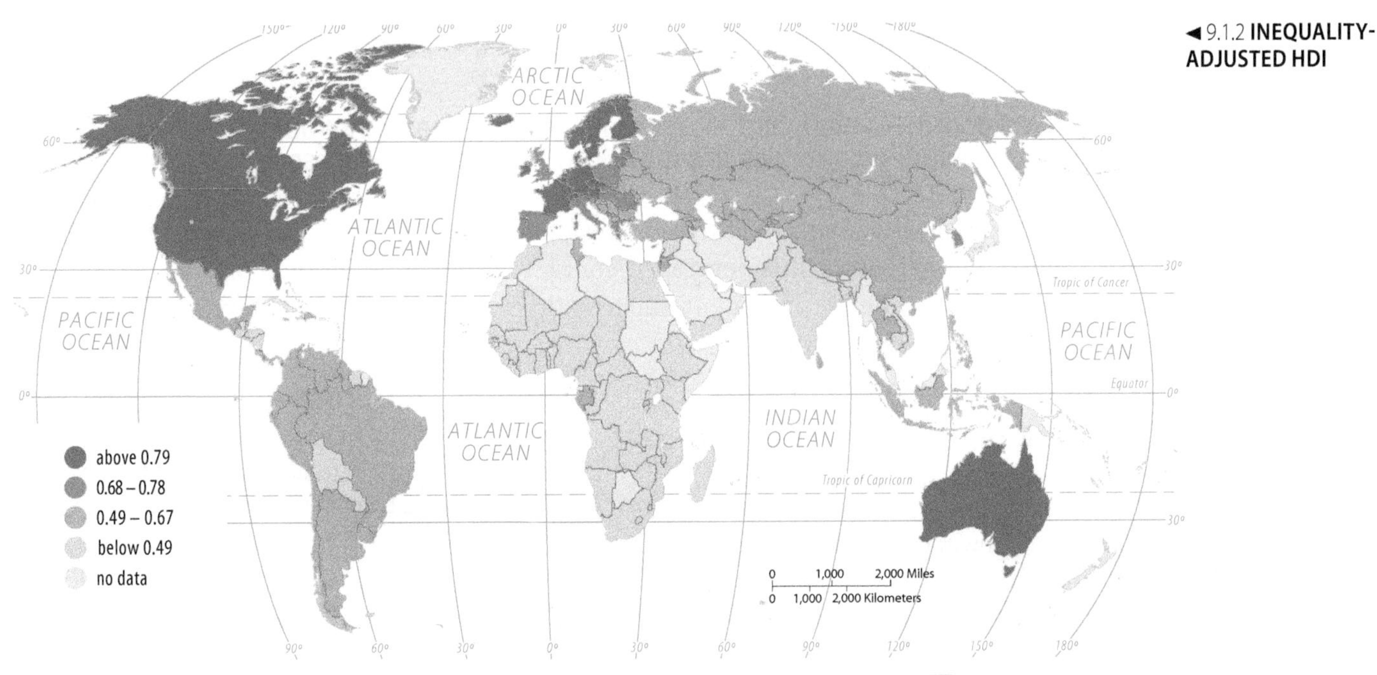

◄ 9.1.2 **INEQUALITY-ADJUSTED HDI**

INEQUALITY-ADJUSTED HDI

The United Nations believes that every person should have access to decent standards of living, knowledge, and health. The **Inequality-adjusted HDI (IHDI)** modifies the HDI to account for inequality (Figure 9.1.2).

Under perfect equality the HDI and the IHDI are the same. If the IHDI is lower than the HDI, the country has some inequality; the greater the difference in the two measures, the greater the inequality. A country where only a few people have high incomes, college degrees, and good health care would have a lower IHDI than a country where differences in income, level of education, and access to health care are minimal.

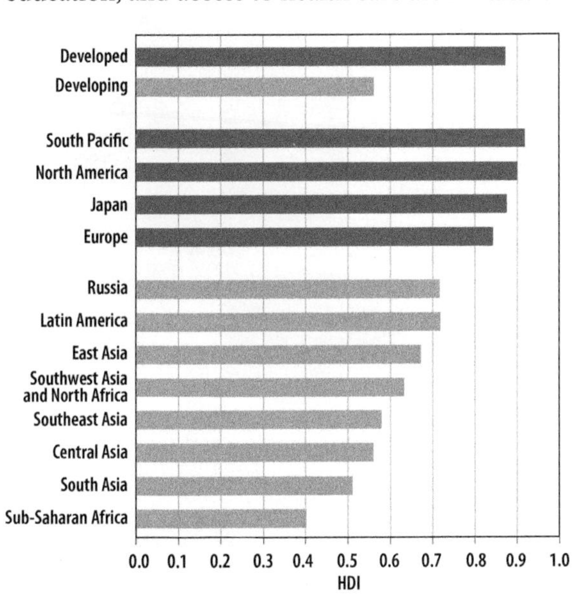

▲ 9.1.4 **HDI BY REGION**

ARCTIC OCEAN
PACIFIC OCEAN
ATLANTIC OCEAN
PACIFIC OCEAN
INDIAN OCEAN
Europe
North America
Latin America
Southwest Asia and North Africa
Sub-Saharan Africa
Central Asia
East Asia
South Asia
Southeast Asia
Japan
Russia
South Pacific
0 1,000 2,000 Miles
0 1,000 2,000 Kilometers

▲ 9.1.3 **NINE WORLD REGIONS**

FOCUS ON WORLD REGIONS

Geographers divide the world into nine regions according to physical, cultural, and economic features (Figure 9.1.3). Two of the nine regions—North America and Europe—are considered developed (Figure 9.1.4). The other seven regions—Latin America, East Asia, Southwest Asia & North Africa, Southeast Asia, Central Asia, South Asia, and sub-Saharan Africa—are considered developing. In addition to these nine regions, three other distinctive areas can be identified—Japan, Russia, and South Pacific. Japan and South Pacific are grouped with the developed regions. Because of limited progress in development both under and since communism, Russia is now classified as a developing country by the United Nations. In each of the remaining nine sections of this chapter, one of the nine regions is highlighted in relation to the topic of the section.

9.2 Standard of Living

► Developed countries have higher average incomes than developing countries.
► People in developed countries are more productive and possess more goods.

Key to development is enough wealth for a decent standard of living. The average individual earns a much higher income in a developed country than in a developing one. Geographers observe that people generate and spend their wealth in different ways in developed countries than in developing countries.

INCOME

The United Nations measures the average income in countries through a complex index called annual gross national income per capita at purchasing power parity. The figure is approximately $40,000 in developed countries compared to approximately $5,000 in developing countries (Figure 9.2.1).

Gross national income (GNI) is the value of the output of goods and services produced in a country in a year, including money that leaves and enters the country. Dividing GNI by total population measures the contribution made by the average individual towards generating a country's wealth in a year. Older studies refer to **gross domestic product**, which is also the value of the output of goods and services produced in a country in a year, but it does not account for money that leaves and enters the country.

Purchasing power parity (PPP) is an adjustment made to the GNI to account for differences among countries in the cost of goods. For example, if a resident of country A has the same income as a resident in country B but must pay more for a Big Mac or a Starbucks latte, the resident of country B is better off.

ECONOMIC STRUCTURE

Average per capita income is higher in developed countries because people typically earn their living by different means than in developing countries. Jobs fall into three categories:

- **Primary sector** (including agriculture).
- **Secondary sector** (including manufacturing).
- **Tertiary sector** (including services).

Developing countries have a higher share of primary and secondary sector workers and a smaller share of tertiary sector workers than developed countries (Figure 9.2.2). The relatively low percentage of primary-sector workers in developed countries indicates that a handful of farmers produce enough food for the rest of society. Freed from the task of growing their own food, most people in a developed country can contribute to an increase in the national wealth by working in the secondary and tertiary sectors (Figure 9.2.3).

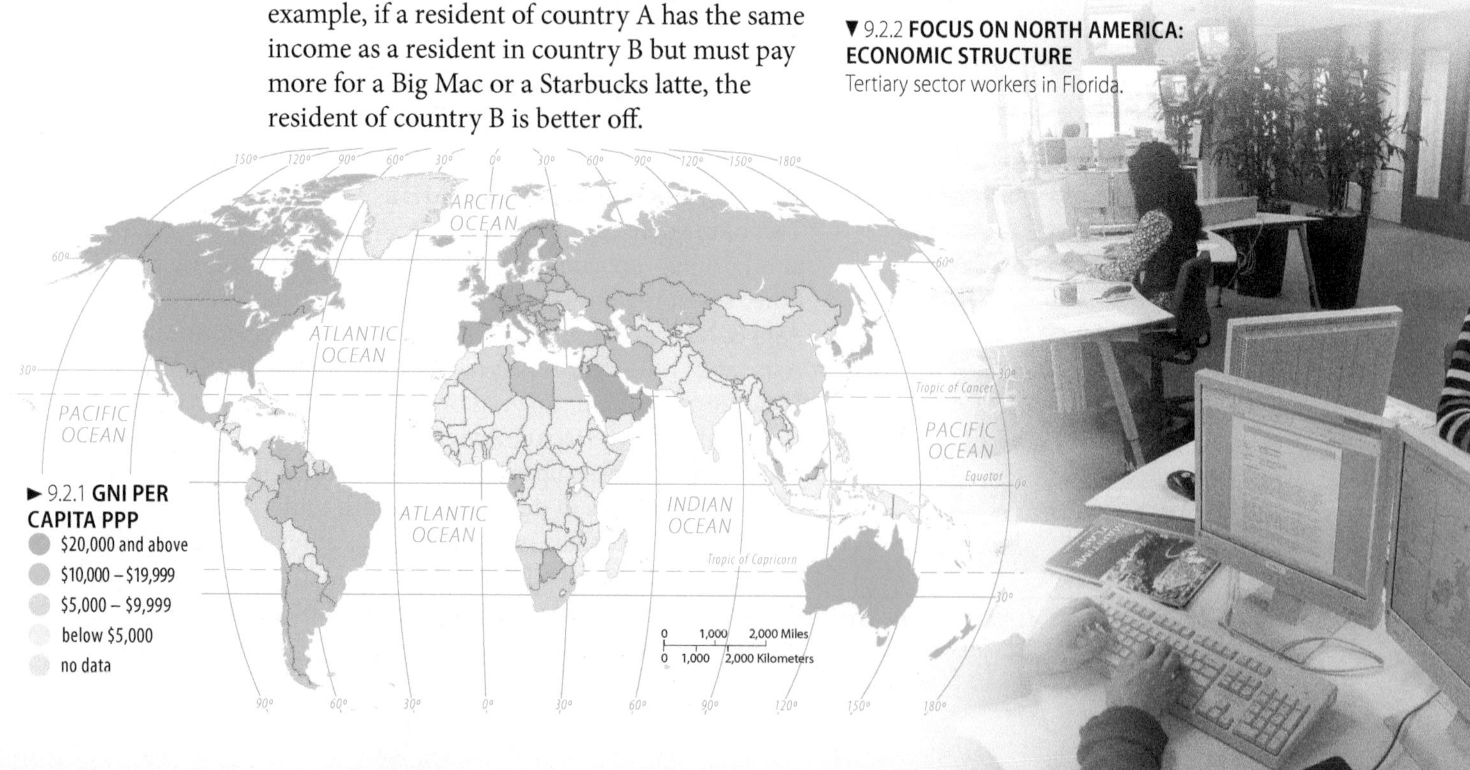

▼ 9.2.2 **FOCUS ON NORTH AMERICA: ECONOMIC STRUCTURE**
Tertiary sector workers in Florida.

► 9.2.1 **GNI PER CAPITA PPP**

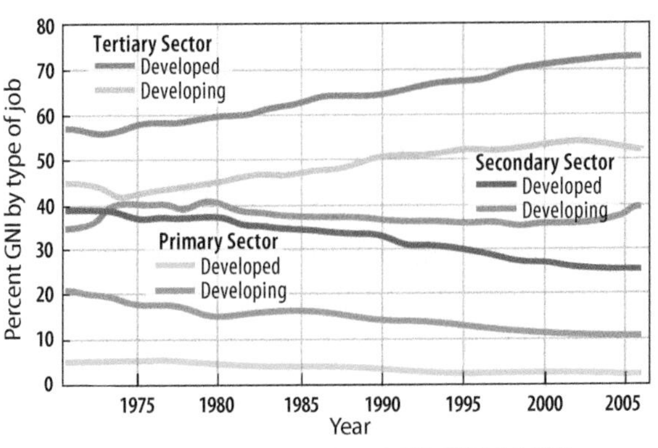

▲ 9.2.3 PERCENT GNI CONTRIBUTED BY TYPE OF JOB

▲ 9.2.4 FOCUS ON NORTH AMERICA: PRODUCTIVITY
Manufacturing computers in California.

PRODUCTIVITY

Workers in developed countries are more productive than those in developing ones. **Productivity** is the value of a particular product compared to the amount of labor needed to make it. Productivity can be measured by the value added per worker. The **value added** in manufacturing is the gross value of the product minus the costs of raw materials and energy. Workers in developed countries produce more with less effort because they have access to more machines, tools, and equipment to perform much of the work (Figure 9.2.4).

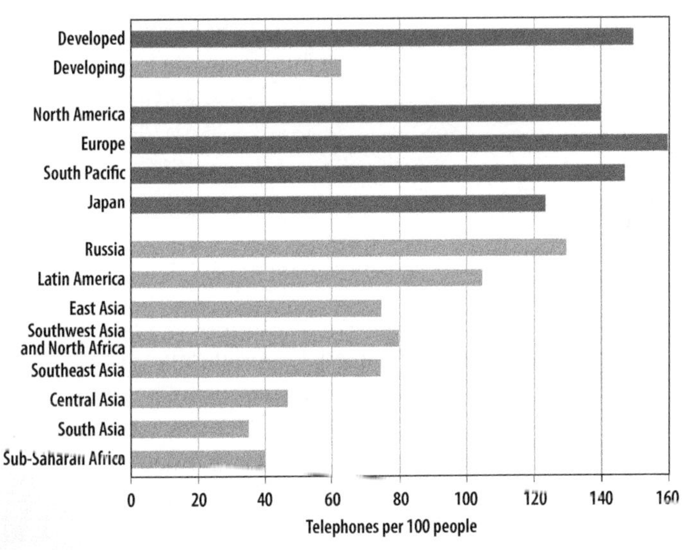

▲ 9.2.5 TELEPHONES PER 100 PEOPLE

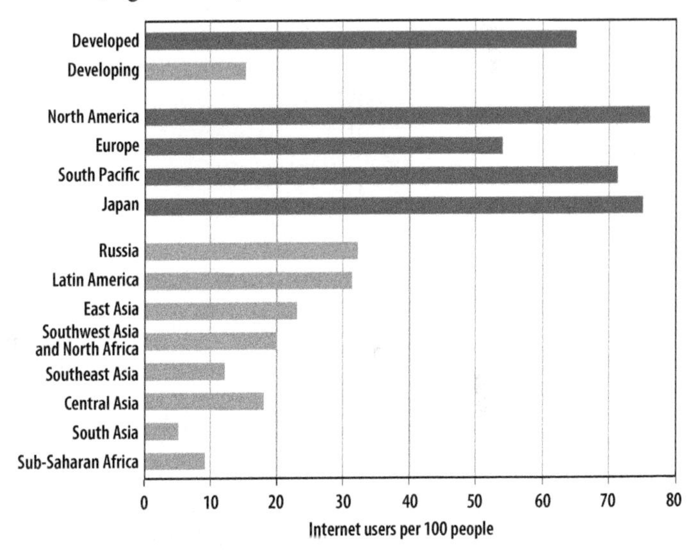

▲ 9.2.6 INTERNET USERS PER 100 PEOPLE

CONSUMER GOODS

Part of the wealth generated in developed countries is used to purchase goods and services. Especially vital to the economy's functioning and growth are goods and services related to communications, such as telephones and computers. Computers and telephones are not essential to people who live in the same village as their friends and relatives and work all day growing food in nearby fields.

Telephones enhance interaction with providers of raw materials and customers for goods and services (Figure 9.2.5). Computers facilitate the sharing of information with other buyers and suppliers (Figure 9.2.6). Developed countries average 150 telephones and 65 Internet users per 100 persons, compared to 60 telephones and 15 Internet users per 100 in developing countries.

FOCUS ON NORTH AMERICA

North America is the region with the world's highest per capita income. North America was once the world's major manufacturer of steel, motor vehicles, and other goods, but since the late twentieth century other regions have taken the lead. Now the region has the world's highest percentage of tertiary-sector employment, especially health care, leisure, and financial services. North Americans remain the leading consumers and the world's largest market for many products. The wealth generated in the United States and Canada enables the residents of those countries to purchase more consumer goods than in other regions.

9.3 Access to Knowledge

- ► People in developed countries complete more years of school.
- ► Developed countries have lower pupil/teacher ratios and higher literacy.

Development is about more than possession of wealth. The United Nations believes that access to knowledge is essential for people to have the possibility of leading lives of value. In general, the higher the level of development, the greater are both the quantity and the quality of a country's education. For many in developing countries, education is the ticket to better jobs and higher social status.

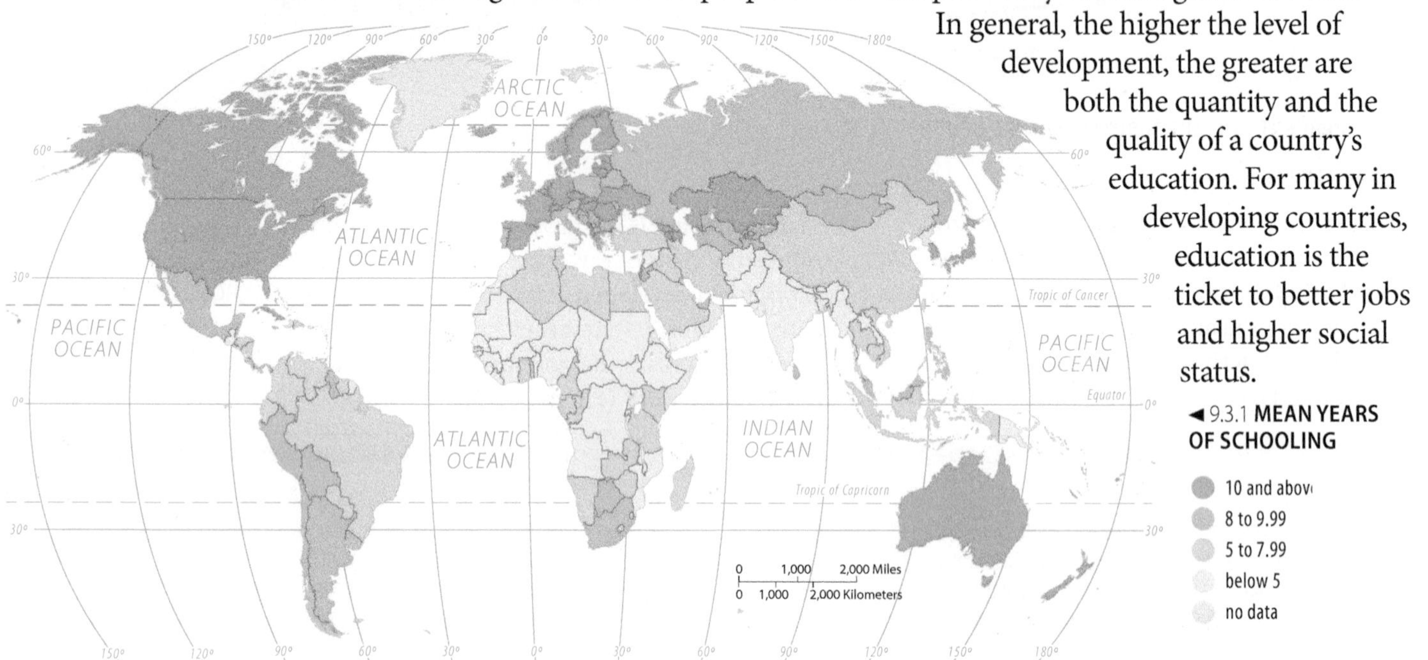

◄ 9.3.1 **MEAN YEARS OF SCHOOLING**

QUANTITY OF SCHOOLING

The United Nations considers years of schooling to be the most critical measure of the ability of an individual to gain access to knowledge needed for development. The assumption is that no matter how poor the school, the longer the pupils attend, the more likely they are to learn something.

To form the access to knowledge component of HDI, the United Nations combines two measures of quantity of schooling:

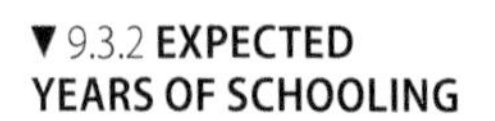

▼ 9.3.2 **EXPECTED YEARS OF SCHOOLING**

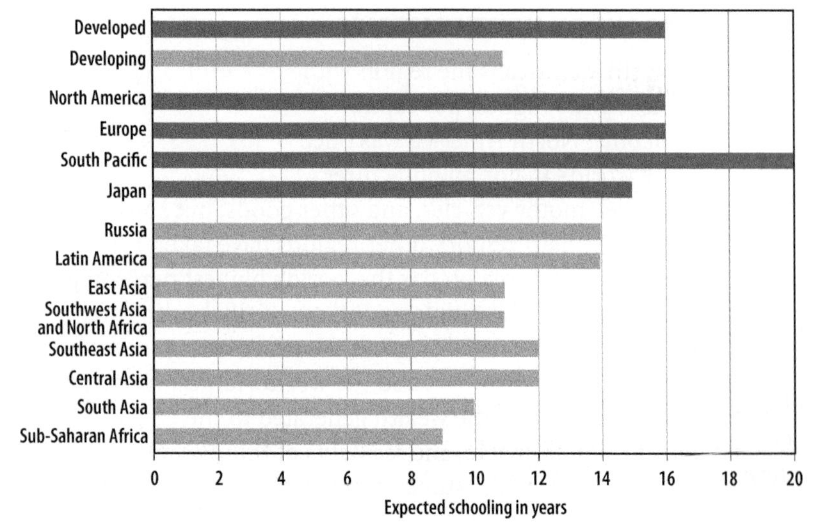

- ***Years of schooling.*** The number of years that the average person aged 25 or older in a country has spent in school. The average pupil has attended school for approximately 11 years in developed countries, compared to approximately 6 years in developing countries (Figure 9.3.1).
- ***Expected years of schooling.*** The number of years that an average 5 year old child is expected to spend with his or her education in the future. The United Nations expects that today's 5-year-old will attend an average of 16 years of school in developed countries and 11 years in developing ones (Figure 9.3.2). Sub-Saharan Africa and South Asia are expected to lag in schooling compared to other regions.

Thus, the United Nations expects children around the world to receive an average of five years more education in the future, but the gap in education between developed and developing regions will remain high. Otherwise stated, the United Nations expects that roughly half of today's 5-year-olds will graduate from college in developed countries, whereas less than half will graduate from high school in developing ones.

QUALITY OF SCHOOLING

Two measures of quality of education include:

- ***Pupil/teacher ratio.*** The fewer pupils a teacher has, the more likely that each student will receive instruction. The pupil/teacher ratio is twice as high in developing countries—approximately 30 pupils per teacher—compared to only 15 in developed countries (Figure 9.3.3). Pupil/teacher ratio exceeds 40 in sub-Saharan Africa and South Asia.
- ***Literacy rate.*** A higher percentage of people in developed countries are able to attend school and as a result learn to read and write. The **literacy rate** is the percentage of a country's people who can read and write. It exceeds 99 percent in developed countries (Figure 9.3.4). Among developing regions, the literacy rate exceeds 90 percent in East Asia and Latin America, but is less than 70 percent in sub-Saharan Africa and South Asia.

Most books, newspapers, and magazines are published in developed countries, in part because more of their citizens read and write. Developed countries dominate scientific and nonfiction publishing worldwide—this textbook is an example. Students in developing countries must learn technical information from books that usually are not in their native language but are printed in English, German, Russian, or French.

Improved education is a major goal of many developing countries, but funds are scarce. Education may receive a higher percentage of the GNI in developing countries, but their GNI is far lower to begin with, so they spend far less per pupil than do developed countries.

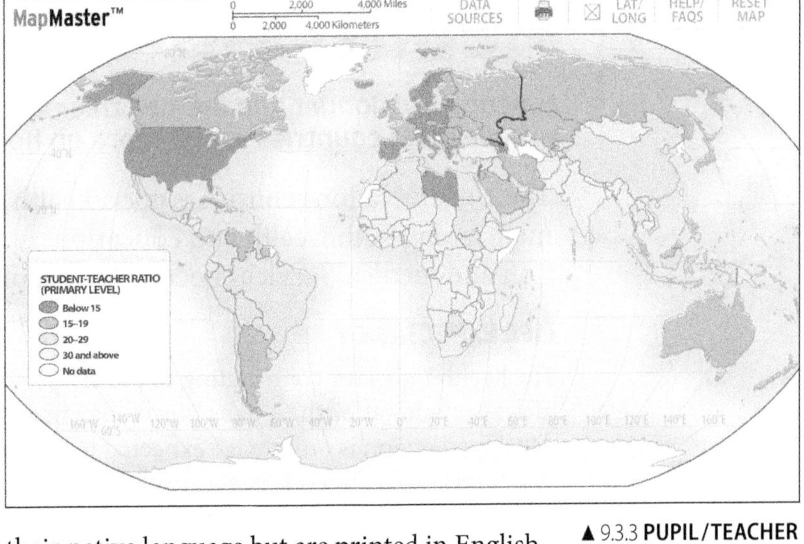

▲ 9.3.3 **PUPIL/TEACHER RATIO**

Open MapMaster World in MasteringGEOGRAPHY

Select: *Cultural* then *Students per teacher in primary school.*
Select: *Popul ation* then *Percentage of population under age 15.*

Are class sizes larger or smaller in countries that have a high percentage of population under age 15?

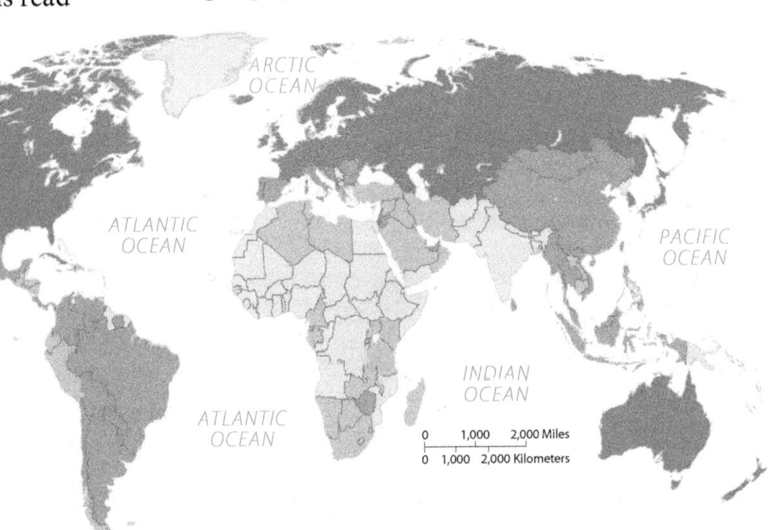

► 9.3.4 **LITERACY RATE**

◄ 9.3.5 **FOCUS ON EUROPE: SCHOOLING**
Spain has one of the world's most favorable pupil/teacher ratios.

FOCUS ON EUROPE

Within Europe, the HDI is the world's highest in a core area that extends from southern Scandinavia to western Germany. These countries have especially high levels of schooling, favorable pupil/teacher ratios, and universal literacy (Figure 9.3.5). Europe's overall development indicators are somewhat lower because of inclusion of Eastern European countries that developed under communist rule for much of the twentieth century. Europe must import food, energy, and minerals, but can maintain its high level of development by providing high value goods and services, such as insurance, banking, and luxury motor vehicles.

9.4 Health Indicators

- People live longer and are healthier in developed countries.
- Developed countries spend more on health care.

The United Nations considers good health to be a third important measure of development, along with wealth and education. A goal of development is to provide the nutrition and medical services needed for people to lead long and healthy lives.

LIFE EXPECTANCY

The health indicator contributing to the HDI is life expectancy at birth. A baby born today in a developed region is on average expected to live ten years longer than one born in a developing region (Figure 9.4.1, and refer to Figure 5.4.3 for world map). Variation among developing regions is especially wide; life expectancy in East Asia and Latin America is comparable to the level in developed countries, but it is much lower in sub-Saharan Africa.

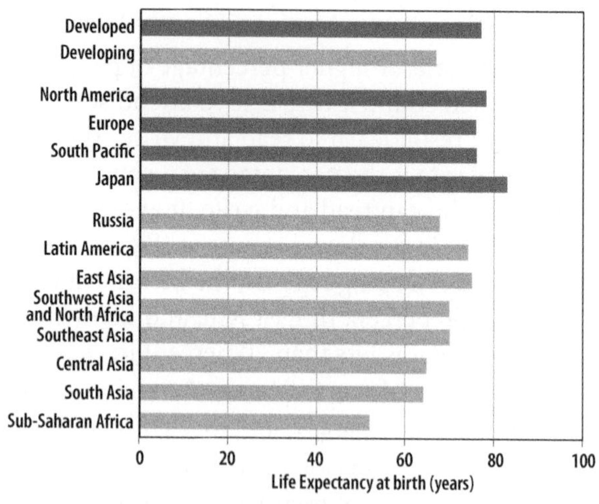

▲ 9.4.1 LIFE EXPECTANCY BY REGION

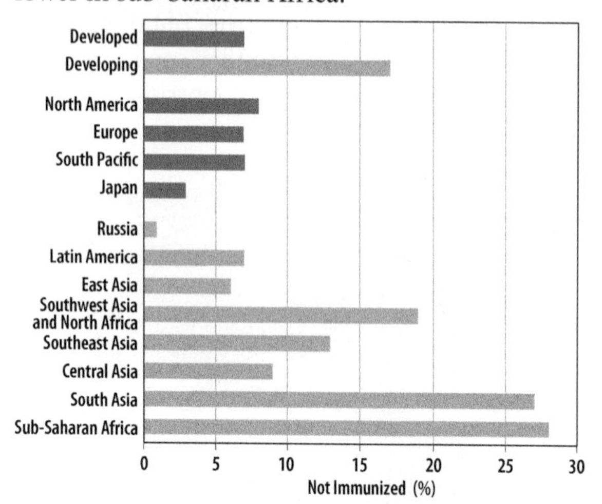

► 9.4.2 CHILDREN LACKING MEASLES IMMUNIZATION

HEALTH CARE ACCESS

People live longer and are healthier in developed countries than in developing ones because of better access to health care. The greater wealth that is generated in developed countries is used in part to obtain health care. A healthier population in turn can be more economically productive. For example, 17 percent of children in developing countries are not immunized against measles, compared to 7 percent in developed ones. More than one-fourth of children lack measles immunization in South Asia and sub-Saharan Africa (Figure 9.4.2).

When people get sick, developed countries possess the resources to care for them. For example, developed countries on average have 50 hospital beds per 10,000 population compared to only 20 in developing countries (Figures 9.4.3 and 9.4.4).

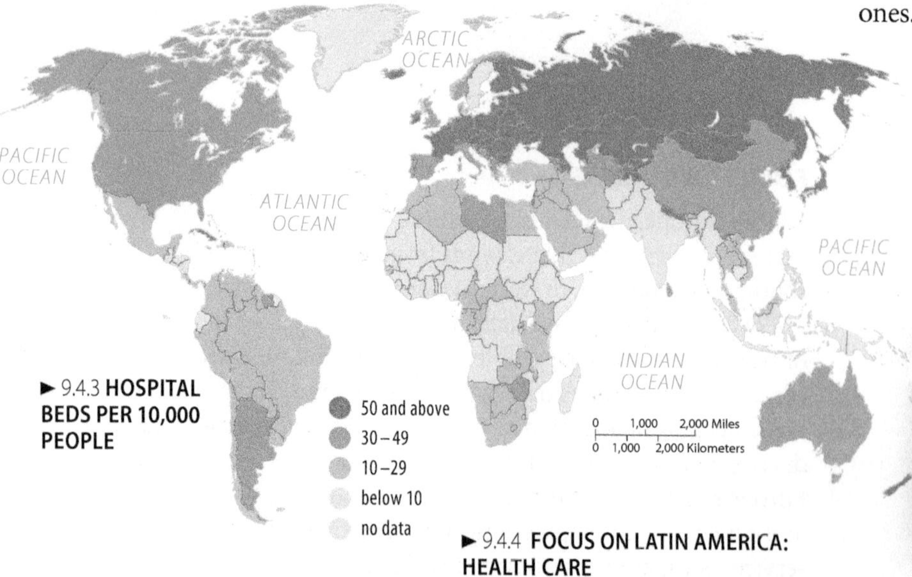

► 9.4.3 HOSPITAL BEDS PER 10,000 PEOPLE

► 9.4.4 FOCUS ON LATIN AMERICA: HEALTH CARE
Clinic in Haiti run by American missionaries.

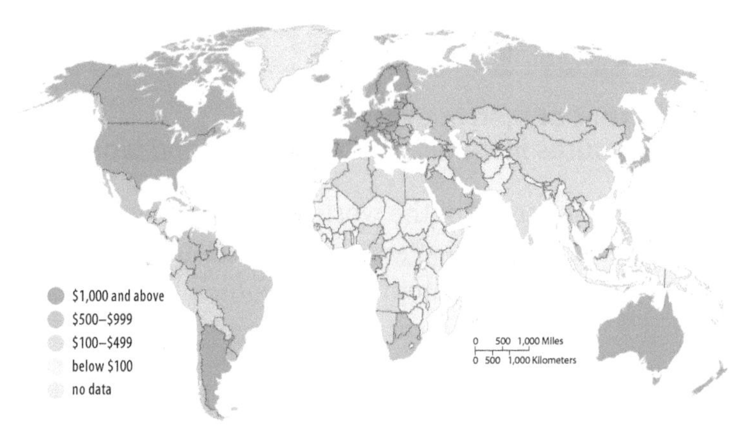

▲ 9.4.5 **HEALTH CARE EXPENDITURE PER CAPITA**

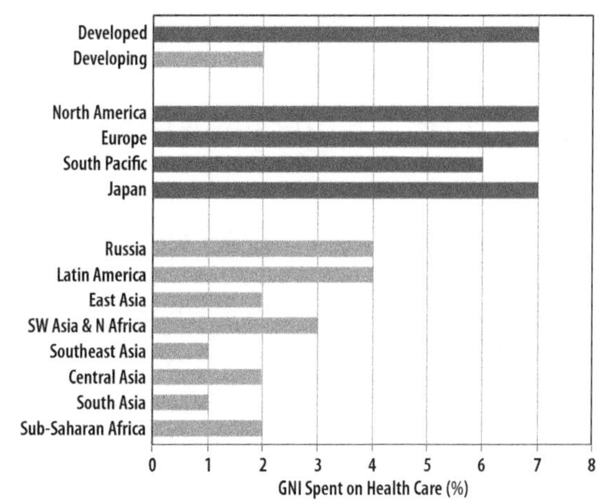

▲ 9.4.6 **HEALTH CARE EXPENDITURE AS PERCENTAGE OF GNI**

HEALTH CARE EXPENDITURES

The gap between developed and developing countries is especially high in expenditures on health care. Developed countries spend more than $4,000 per person annually on health care, compared to approximately $200 per person in developing countries (Figure 9.4.5). Hospitals, medicines, doctors—spending is much higher in developed countries.

Total expenditures on health care exceed 7 percent of GNI in developed countries, compared to 2 percent in developing ones. So not only do developed countries have much higher GNI per capita than developing countries, they spend a higher percentage of that GNI on health care (Figure 9.4.6).

In most developed countries, health care is a public service that is available at little or no cost. The government programs pay more than 70 percent of health care costs in most European countries, and private individuals pay less than 30 percent. In comparison, private individuals must pay more than half of the cost of health care in developing countries. An exception is the United States, where private individuals are required to pay 55 percent of health care, more closely resembling the pattern in developing countries.

Developed countries also use part of their wealth to protect people who, for various reasons, are unable to work. In these countries some public assistance is offered to those who are sick, elderly, poor, disabled, orphaned, veterans of wars, widows, unemployed, or single parents. European countries such as Denmark, Norway, and Sweden typically provide the highest level of public-assistance payments.

Developed countries are hard-pressed to maintain their current levels of public assistance. In the past, rapid economic growth permitted these states to finance generous programs with little hardship. But in recent years economic growth has slowed, whereas the percentage of people needing public assistance has increased. Governments have faced a choice between reducing benefits or increasing taxes to pay for them.

▼ 9.4.7 **FOCUS ON LATIN AMERICA: HEALTH CARE**
Clinic in Colombia for displaced people.

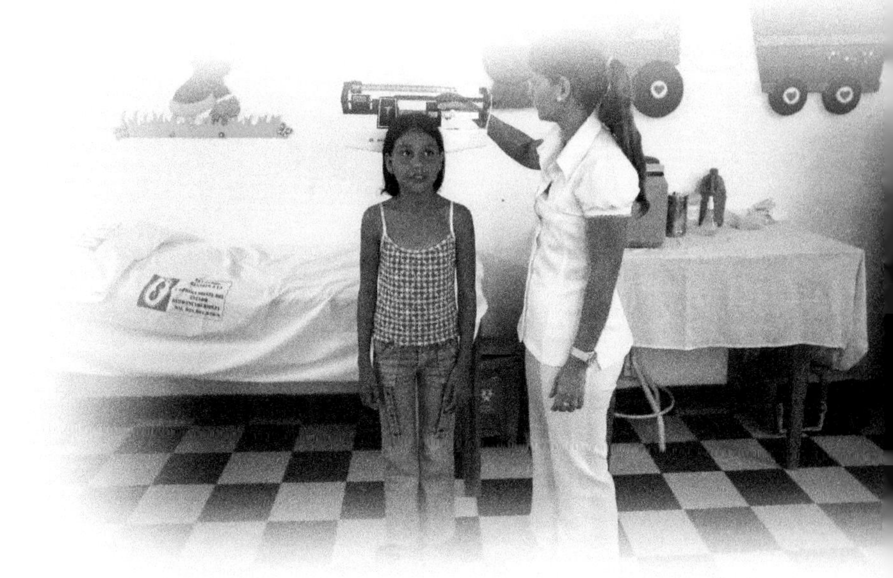

FOCUS ON LATIN AMERICA

The level of development varies sharply within Latin America. Neighborhoods within some large cities along the South Atlantic Coast enjoy a level of development comparable to that of developed countries. The coastal area as a whole has a relatively high GNI per capita. Outside the coastal area, development is lower. Among developing regions, Latin America—along with East Asia—has relatively high life expectancy, high immunization rates, more hospital beds per capita, and more money spent on health care. The levels lag, though, compared with developed regions.

9.5 Gender-Related Development

- The status of women is lower than that of men in every country.
- The Gender Inequality Index (GII) measures inequality between men and women.

The United Nations has not found a single country in the world where women are treated as well as men. At best women have achieved near equality with men in some countries, whereas in other countries the level of development of women lags far behind the level for men.

To measure the extent of each country's gender inequality, the United Nations has created the **Gender Inequality Index (GII)**. The higher the score the greater is the inequality between men and women (Figure 9.5.1). As with the other indices, the GII combines multiple measures, in this case reproductive health, empowerment, and labor.

EMPOWERMENT

The empowerment dimension is measured by two indicators:

- The percentage of seats held by women in the national legislature (Figures 9.5.2 and 9.5.3).
- The percentage of women who have completed high school.

Both measures are lower in developing regions than in developed ones.

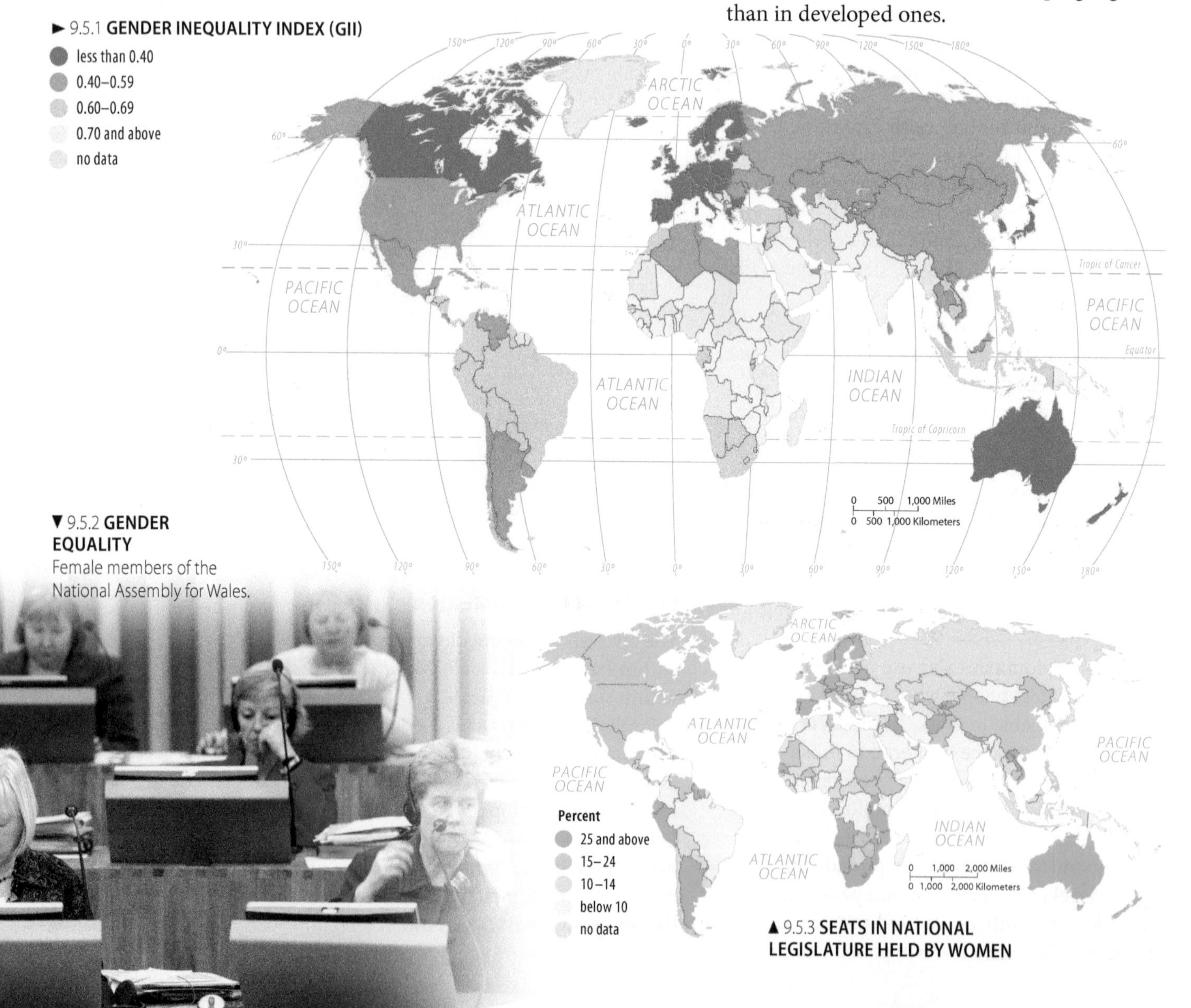

► 9.5.1 **GENDER INEQUALITY INDEX (GII)**

▼ 9.5.2 **GENDER EQUALITY** Female members of the National Assembly for Wales.

▲ 9.5.3 **SEATS IN NATIONAL LEGISLATURE HELD BY WOMEN**

LABOR

The labor force participation rate is the percent of women holding full-time jobs outside the home. Women in developing countries are less likely than women in developed countries to hold full-time jobs outside the home (Figure 9.5.4).

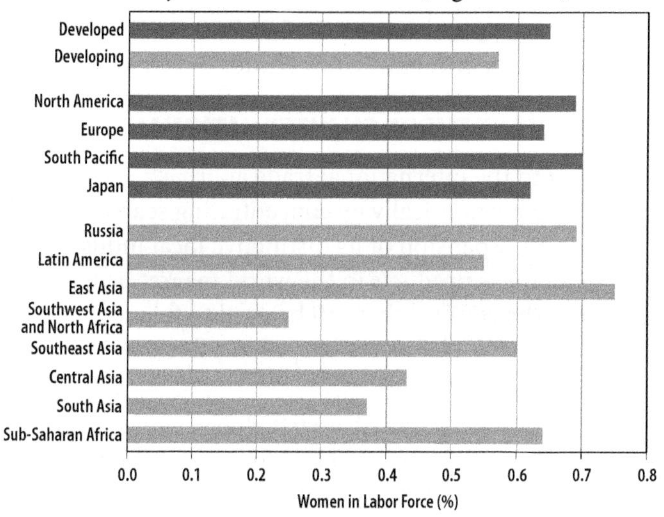

▲ 9.5.4 **WOMEN IN LABOR FORCE**
(above left) Percent of women in the labor force by world region.
(above right) Female workers in optical fiber factory, Guangzhou, China.

▼ 9.5.5 **ADOLESCENT FERTILITY RATE (right) TEENAGE MOTHER IN OHIO (below)**

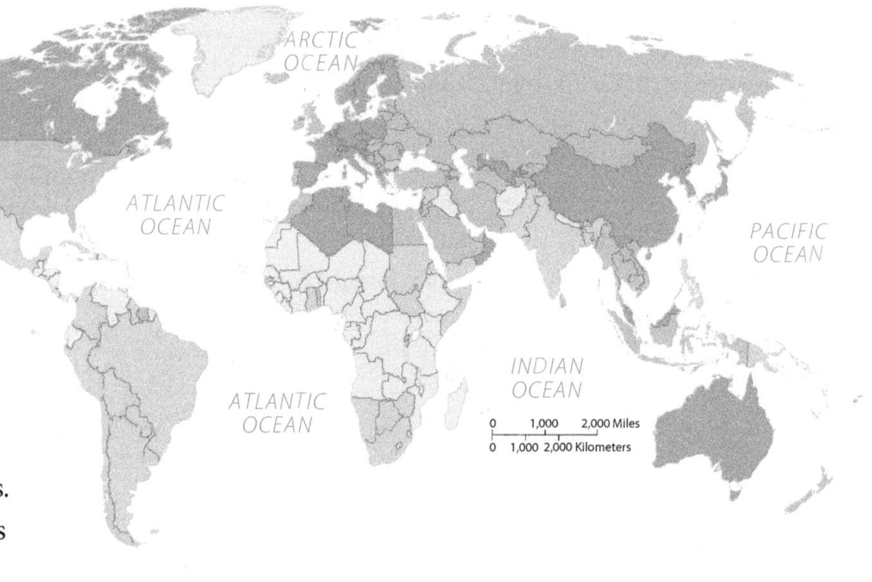

REPRODUCTIVE HEALTH

The health dimension is also measured by two indicators:

- **Maternal mortality ratio** is the number of women who die giving birth per 100,000 births.
- **Adolescent fertility rate** is the number of births per 1,000 women age 15–19.

Women in developing regions are more likely than women in developed regions to die in childbirth and to give birth as teenagers (Figure 9.5.5).

In general, the GII is higher in developing regions than in developed ones. Sub-Saharan Africa, South Asia, Central Asia, and Southwest Asia are the developing regions with the highest levels of gender inequality. Reproductive health is the largest contributor to gender inequality in these regions. South and Southwest Asia also have relatively poor female empowerment scores. The United States ranks especially poor in the percentage of teenagers who give birth and in the percentage of women serving in Congress.

FOCUS ON EAST ASIA

The Gender Inequality Index in East Asia is comparable to that of developed regions. Compared to other developing countries, China has high female education levels and participation in the labor force and low maternal mortality and teenage fertility rates. Now the world's second largest economy, behind only the United States, China accounts for one-third of total world economic growth, and GNI per capita has risen faster there than in any other country. Under communism, the government took strong control of most components of development.

9.6 Two Paths to Development

- The self-sufficiency development path erects barriers to trade.
- The international trade path allocates scarce resources to a few activities.

To promote development, developing countries typical follow one of two development models. One emphasizes self-sufficiency, the other international trade.

DEVELOPMENT THROUGH SELF-SUFFICIENCY

Self-sufficiency, or balanced growth, was the more popular of the development alternatives for most of the twentieth century. According to the self-sufficiency approach:

- Investment is spread as equally as possible across all sectors of a country's economy and in all regions.
- The pace of development may be modest, but the system is fair because residents and enterprises throughout the country share the benefits of development.
- Reducing poverty takes precedence over encouraging a few people to become wealthy consumers.
- Fledgling businesses are isolated from competition with large international corporations.
- The import of goods from other places is limited by barriers such as tariffs, quotas, and licenses.

SELF-SUFFICIENCY EXAMPLE: INDIA

India once followed the self-sufficiency model (Figure 9.6.1). India's barriers to trade included:

- To import goods into India, most foreign companies had to secure a license that had to be approved by several dozen government agencies.
- An importer with a license was severely restricted in the quantity it could sell in India.
- Heavy taxes on imported goods doubled or tripled the price to consumers.
- Indian money could not be converted to other currencies.
- Businesses required government permission to sell a new product, modernize a factory, expand production, set prices, hire or fire workers, and change the job classification of existing workers.

▼ 9.6.1 **SELF-SUFFICIENCY EXAMPLE: INDIA**
Basmati rice on sale at a market in Haryana.

DEVELOPMENT THROUGH INTERNATIONAL TRADE

According to the international trade approach, a country can develop economically by concentrating scarce resources on expansion of its distinctive local industries. The sale of these products in the world market brings funds into the country that can be used to finance other development. W. W. Rostow proposed a five-stage model of development in 1960.

- **The traditional society.** A very high percentage of people engaged in agriculture and a high percentage of national wealth allocated to what Rostow called "nonproductive" activities, such as the military and religion.
- **The preconditions for takeoff.** An elite group of well-educated leaders initiates investment in technology and infrastructure, such as water supplies and transportation systems, designed to increase productivity.
- **The takeoff.** Rapid growth is generated in a limited number of economic activities, such as textiles or food products.
- **The drive to maturity.** Modern technology, previously confined to a few takeoff industries, diffuses to a wide variety of industries.
- **The age of mass consumption.** The economy shifts from production of heavy industry, such as steel and energy, to consumer goods, such as motor vehicles and refrigerators.

INTERNATIONAL TRADE EXAMPLES

Among the first countries to adopt the international trade alternative during the twentieth century:

- **The "Four Dragons."** South Korea, Singapore, Taiwan, and the then-British colony of Hong Kong (also known as the "four little tigers" and "the gang of four") developed by producing a handful of manufactured goods, especially clothing and electronics, that depended on low labor costs.
- **Petroleum-rich Arabian Peninsula countries.** Once among the world's least developed countries, they were transformed overnight into some of the wealthiest thanks to escalating petroleum prices during the 1970s (Figure 9.6.2).

SELF-SUFFICIENCY SHORTCOMINGS

The experience of India and other developing countries revealed two major problems with self-sufficiency:

- **Self-sufficiency protected inefficient industries.** Businesses could sell all they made, at high government-controlled prices, to customers culled from long waiting lists. So they had little incentive to improve quality, lower production costs, reduce prices, or increase production. Nor did they keep abreast of rapid technological changes elsewhere.
- **A large bureaucracy was needed to administer the controls.** A complex administrative system encouraged abuse and corruption. Aspiring entrepreneurs found that struggling to produce goods or offer services was less rewarding financially than advising others how to get around the complex regulations.

▲ 9.6.2 **INTERNATIONAL TRADE EXAMPLE: UNITED ARAB EMIRATES** Development in Dubai, United Arab Emirates.

INTERNATIONAL TRADE SHORTCOMINGS

Three factors have hindered countries outside the four Asian dragons and the Arabian Peninsula from developing through the international trade:

- **Local hardships.** Building up a handful of takeoff industries has forced some developing countries to cut back on production of food, clothing, and other necessities for their own people.
- **Slow market growth.** Developing countries trying to take advantage of their low-cost labor find that markets in developed countries are growing more slowly than when the "four dragons" used this strategy a generation ago.
- **Low commodity prices.** Some developing countries have raw materials sought by manufacturers and producers in developed countries. The sale of these raw materials could generate funds for developing countries to promote development. International trade worked in the Arabian Peninsula because the price of petroleum has escalated so rapidly, but other developing countries have not been so fortunate because of low prices for their commodities.

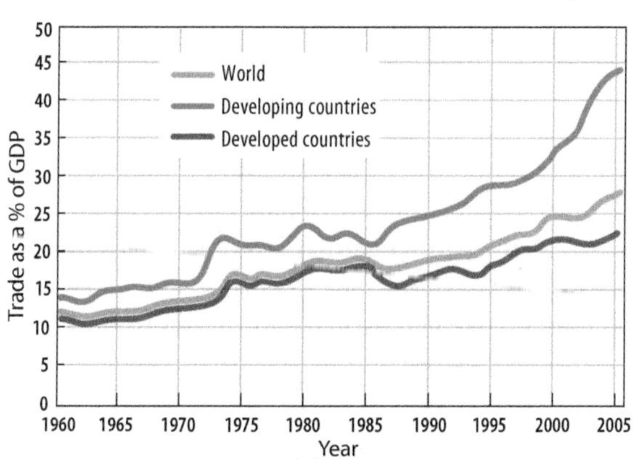

▲ 9.6.3 **WORLD TRADE AS PERCENT OF INCOME**

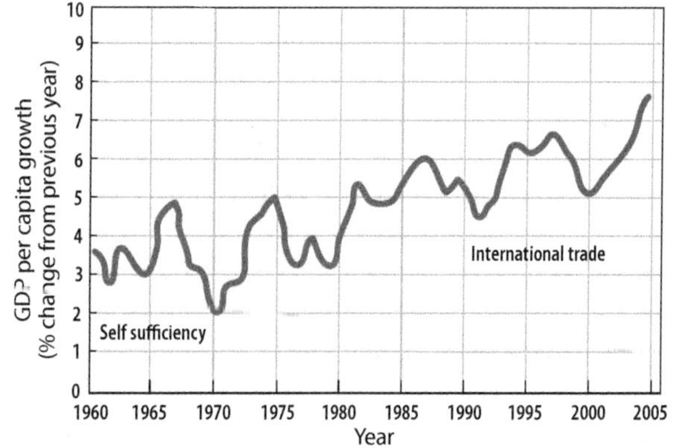

▲ 9.6.4 **GDP PER CAPITA CHANGE IN INDIA**

INTERNATIONAL TRADE TRIUMPHS

Countries have converted from self-sufficiency to international trade (Figure 9.6.3). For example, India has:

- Reduced taxes and restrictions on imports and exports.
- Eliminated many monopolies.
- Encouraged improvement of the quality of products.

India's per capita income has increased more rapidly since conversion to international trade (Figure 9.6.4).

FOCUS ON SOUTHWEST ASIA AND NORTH AFRICA

Countries in Southwest Asia and North Africa that are oil-rich have used petroleum revenues to finance large-scale projects, such as housing, highways, airports, universities, and telecommunications networks. Imported consumer goods are readily available. However, some business practices typical of international trade are difficult to reconcile with Islamic religious principles. Women are excluded from holding many jobs and visiting some public places. All business halts several times a day when Muslims are called to prayer.

9.7 World Trade

- ► **The World Trade Organization has facilitated adoption of international trade.**
- ► **Transnational corporations are a major source of development funds.**

To promote the international trade development model, most countries have joined the World Trade Organization (WTO). Private corporations are especially eager to promote international trade.

► 9.7.1 **WORLD TRADE ORGANIZATION**
Open MapMaster World in Mastering GEOGRAPHY

Select: *Economic* then *Gross National Income Per Capita at Purchasing Power Parity*.
Select: *Geopolitical* then *World Trade Organization Members*.

In what three of the nine major world regions (excluding Russia) are most of the countries not members of the WTO?

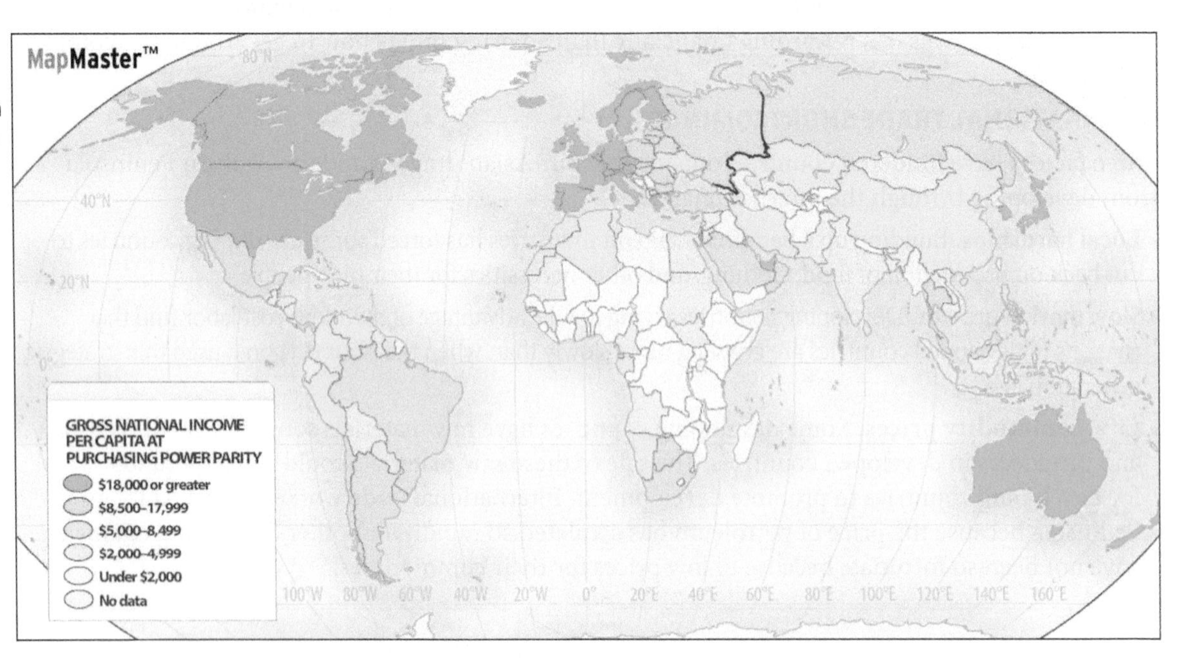

WORLD TRADE ORGANIZATION

To promote the international trade development model, countries representing 97 percent of world trade established the WTO in 1995. Russia is the largest economy that has not joined the WTO (Figure 9.7.1). The WTO works to reduce barriers to trade in three principal ways:

1. Reduce or eliminate restrictions:
   - On trade of manufactured goods, such as government subsidies of exports, quotas, and tariffs.
   - On international movement of money by banks, corporations, and wealthy individuals.
2. Enforce agreements:
   - By ruling on whether a country has violated WTO agreements.
   - By ordering remedies when one country has been found to have violated the agreements.
3. Protect intellectual property:
   - By hearing charges from an individual or corporation concerning copyright and patent violations in other countries.
   - By ordering illegal copyright or patent activities to stop.

The WTO has been sharply attacked by critics (Figure 9.7.2). Protesters routinely gather in the streets outside high-level meetings of the WTO:

- Progressive critics charge that the WTO is antidemocratic, because decisions made behind closed doors promote the interest of large corporations rather than the poor.
- Conservative critics charge that the WTO compromises the power and sovereignty of individual countries because it can order changes in taxes and laws that it considers unfair trading practices.

◄ 9.7.2 **THE WORLD TRADE ORGANIZATION GENERATES STRONG SUPPORT AND OPPOSITION**

FOREIGN DIRECT INVESTMENT

International trade requires corporations based in a particular country to invest in other countries (Figure 9.7.3). Investment made by a foreign company in the economy of another country is known as **foreign direct investment (FDI)**. FDI in developing countries has grown from $2 trillion in 1990 to $7 trillion in 2000 and $17 trillion in 2009 (Figure 9.7.4).

FDI does not flow equally around the world. Only 30 percent of FDI in 2009 went from a developed to a developing country, whereas 70 percent moved between two developed countries. Among developing regions, more than one-fourth each was directed to East Asia and Latin America (Figure 9.7.5).

The major sources of FDI are transnational corporations (TNCs). A **transnational corporation** invests and operates in countries other than the one in which its headquarters are located. Of the 100 largest TNCs in 2009, 61 had headquarters in Europe, 19 in the United States, 10 in Japan, 3 in other developed countries, and only 7 in developing countries.

▲ 9.7.3 **FOREIGN DIRECT INVESTMENT**
Japanese carmakers have built several assembly plants in Thailand.

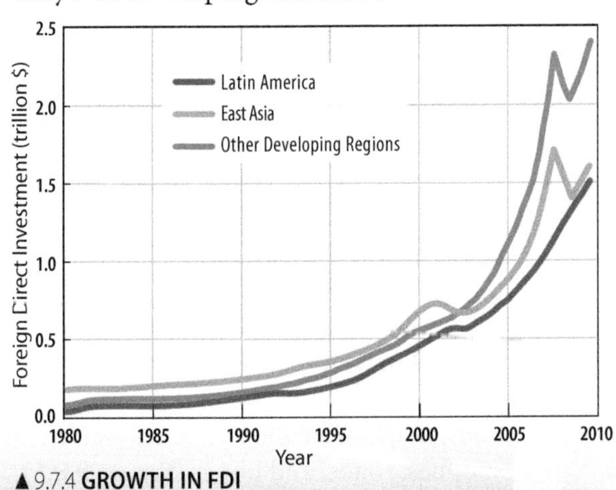

▲ 9.7.4 **GROWTH IN FDI**

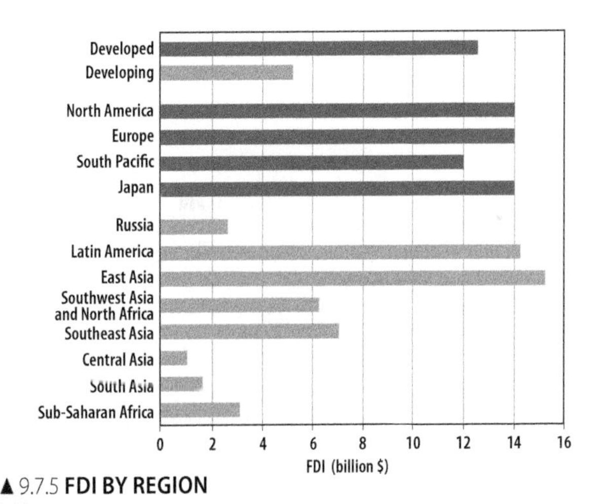

▲ 9.7.5 **FDI BY REGION**

▲ 9.7.6 **FOCUS ON SOUTHEAST ASIA: INTERNATIONAL TRADE**
Female labor in clothing factory.

FOCUS ON SOUTHEAST ASIA

Southeast Asia has become a major manufacturer of textiles and clothing, taking advantage of cheap labor. Thailand has become the region's center for the manufacturing of automobiles and other consumer goods. Indonesia, the world's fourth most populous country, is a major producer of petroleum (Figure 9.7.6). Development has slowed because of painful reforms to restore confidence among international investors shaken by unwise and corrupt investments made possible by lax regulations and excessively close cooperation among manufacturers, financial institutions, and government agencies.

9.8 Financing Development

- ▶ **Developing countries finance some development through foreign aid and loans.**
- ▶ **To qualify for loans, a country may need to enact economic reforms.**

Developing countries lack the money needed to finance development. So they obtain grants and loans from governments, banks, and international organizations based in developed countries.

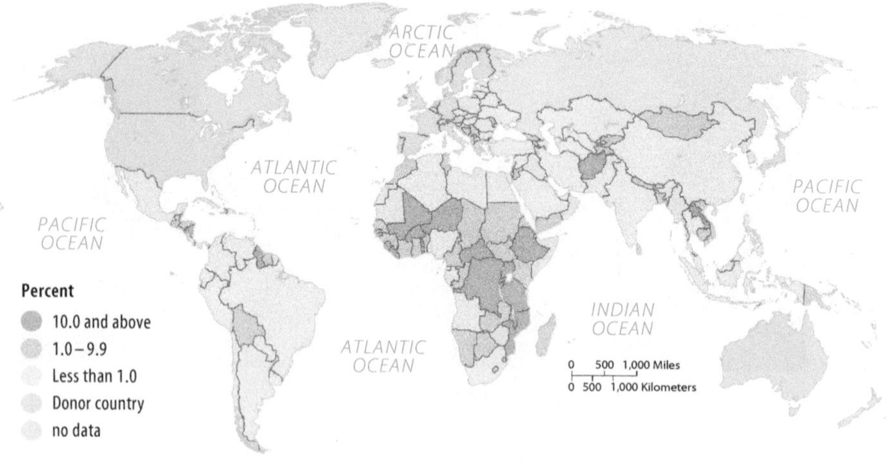

▲ 9.8.1 **FOREIGN AID AS PERCENT OF GNI**

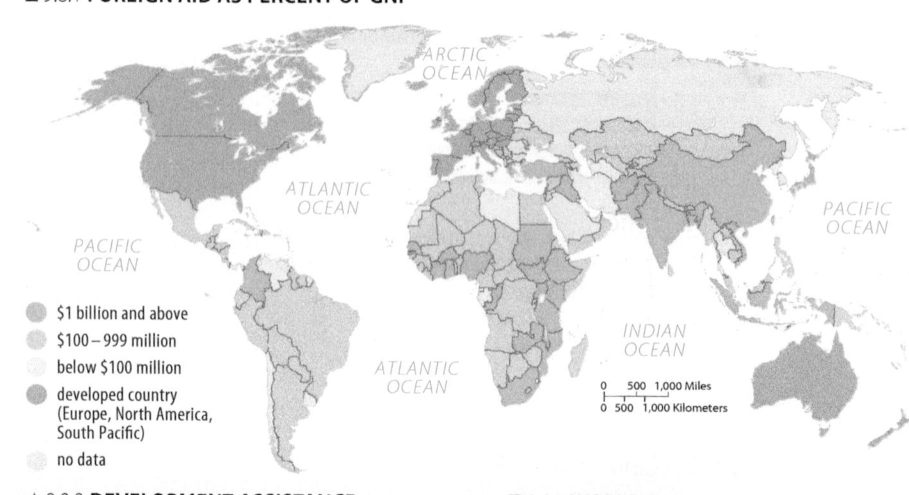

▲ 9.8.2 **DEVELOPMENT ASSISTANCE**

▼ 9.8.3 **WORLD BANK INVESTMENT: THE PHILIPPINES**
Wind farm in Bangui Bay, the Philippines, financed with a World Bank loan.

FOREIGN AID

Most developing countries also receive aid directly from governments of developed countries. The U.S. government allocates approximately 0.2 percent (1/5 of 1%) of its GNI to foreign aid. European countries average a good bit more, approximately 0.5 percent (Figure 9.8.1).

LOANS

The two major international lending organizations are the World Bank and the International Monetary Fund (IMF). The World Bank and IMF were conceived in 1944 to promote development after the devastation of World War II and to avoid a repetition of the disastrous economic policies contributing to the Great Depression of the 1930s. The IMF and World Bank became specialized agencies of the United Nations when it was established in 1945. Twenty-six countries received at least $1 billion in 2009 (Figure 9.8.2).

Developing countries borrow money to build new infrastructure, such as hydroelectric dams, electric transmission lines, flood-protection systems, water supplies, roads, and hotels (Figure 9.8.3). The theory is that the new infrastructure attracts businesses, which in turn pays taxes used to repay the loans and to improve people's living conditions.

In reality, the World Bank itself judges half of the projects it has funded in Africa to be failures. Common reasons include:

- Projects do not function as intended because of faulty engineering.
- Aid is squandered, stolen, or spent on armaments by recipient nations.
- New infrastructure does not attract other investment.

STRUCTURAL ADJUSTMENT PROGRAMS

Some developing countries have had difficulty repaying their loans. The IMF, World Bank, and banks in developed countries fear that granting, canceling, or refinancing debts without strings attached would perpetuate bad habits in developing countries. Therefore before getting debt relief, a developing country is required to prepare a Policy Framework Paper (PFP) outlining a structural adjustment program.

A **structural adjustment program** includes economic "reforms" or "adjustments." Requirements placed on a developing country typically include:

- Spend only what it can afford.
- Direct benefits to the poor not just the elite.
- Divert investment from military to health and education spending.
- Invest scarce resources where they would have the most impact.
- Encourage a more productive private sector.
- Reform the government, including a more efficient civil service, more accountable fiscal management, more predictable rules and regulations, and more dissemination of information to the public.

Critics charge that poverty worsens under structural adjustment programs. By placing priority on reducing government spending and inflation, structural adjustment programs may result in:

- Cuts in health, education, and social services that benefit the poor.
- Higher unemployment.
- Loss of jobs in state enterprises and the civil service.
- Less support for those most in need, such as poor pregnant women, nursing mothers, young children, and elderly people.

In short, structural reforms allegedly punish Earth's poorest people for actions they did not commit—waste, corruption, misappropriation, military build-ups.

International organizations respond that the poor suffer more when a country does not undertake reforms. Economic growth is what benefits the poor the most in the long run. Nevertheless, in response to criticisms, the IMF and World Bank now encourage innovative programs to reduce poverty and corruption, and consult more with average citizens. A safety net must be included to ease short-term pain experienced by poor people. Meanwhile, in the twenty-first century, it is the developed countries that have piled up the most debt, especially in the wake of the severe recession of 2007-09 (Figure 9.8.4).

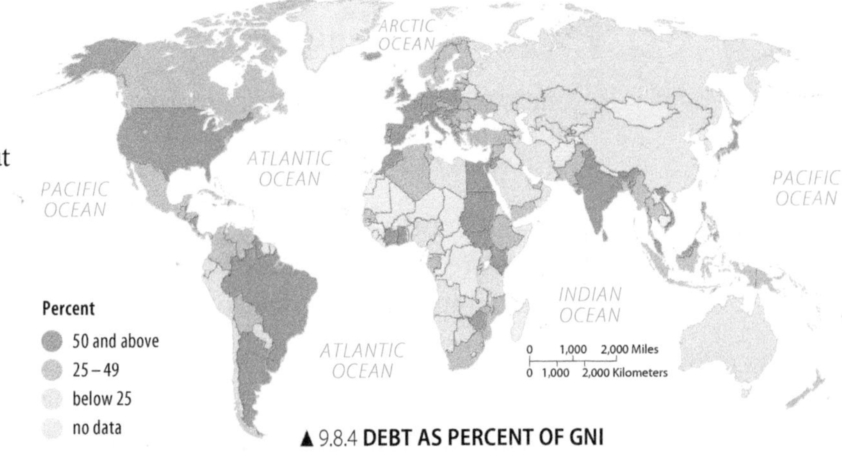

▲ 9.8.4 **DEBT AS PERCENT OF GNI**

◄ 9.8.5 **WORLD BANK INVESTMENT IN AFGHANISTAN**

Use Google Earth™ to explore development aid in Afghanistan.

Fly to: *Kabul Airport, Afghanistan.*

Drag to enter *Street view.*

Exit: *Ground level view.*

Zoom out until airport is visible.

The World Bank paid for the long straight dark narrow strip.

1. **What is it?**
2. **Does it appear to be in good condition or in poor condition?**

FOCUS ON CENTRAL ASIA

Within Central Asia, the level of development is relatively high in Kazakhstan and Iran. Not by coincidence, these two countries are the region's leading producers of petroleum. In Kazakhstan, rising oil revenues are being used to finance a carefully managed improvement in overall development. In Iran, a large share of the rising oil revenues has been used to maintain low consumer prices rather than to promote development.

Since coming to power in a 1979 revolution, Iran's Shiite leaders have also used oil revenues to promote revolutions elsewhere in the region and to sweep away elements of development and social customs they perceive to be influenced by Europe or North America. War-torn Afghanistan has received more development assistance than any other country in recent years (Figure 9.8.5).

9.9 Fair Trade

- **Fair trade is a model of development that is meant to protect small businesses and workers.**
- **With fair trade, a higher percentage of the sales price goes back to the producers.**

A variation of the international trade model of development is **fair trade**, in which products are made and traded following practices and standards that protect workers and small businesses in developing countries.

Two sets of standards distinguish fair trade:

- Fairtrade Labelling Organizations International (FLO) sets international standards for fair trade (Figure 9.9.1).
- Standards applied to workers on farms and in factories.

► 9.9.1 **FAIR TRADE CLOTHING LABEL**

FAIR TRADE PRODUCER PRACTICES

Many farmers and artisans in developing countries are unable to borrow from banks the money they need to invest in their businesses. By banding together in fair trade cooperatives, they can get credit, reduce their raw material costs, and maintain higher and fairer prices for their products (Figure 9.9.2).

Cooperatives are managed democratically, so farmers and artisans learn leadership and organizational skills. The people who grow or make the products have a say in how local resources are utilized and sold. Safe and healthy working conditions can be protected. Cooperatives thus benefit the local farmers and artisans who are members, rather than absentee corporate owners interested only in maximizing profits.

For fair trade coffee, consumers pay prices comparable to those charged by gourmet brands. However, fair trade coffee producers receive a significantly higher price per pound than traditional coffee producers: around $1.20 compared to around $0.80 per pound. Through bypassing exploitative middlemen and working directly with producers, fair trade organizations are able to cut costs and return a greater percentage of the retail price to the producers.

In North America, fair trade products have been primarily craft products such as decorative home accessories, jewelry, textiles, and ceramics. Ten Thousand Villages is the largest fair trade organization in North America specializing in handicrafts. In Europe, most fair trade sales are in food, including coffee, tea, bananas, chocolate, cocoa, juice, sugar, and honey products. TransFair USA certifies the products sold in the United States that are fair trade.

◄ 9.9.2 **FAIR TRADE FOOD**
Fair trade rice for export in Dehradun, India.

FAIR TRADE WORKER STANDARDS

Fair trade requires employers to:

- Pay workers fair wages (at least the country's minimum wage).
- Permit union organizing.
- Comply with minimum environmental and safety standards.

In contrast, protection of workers' rights is a low priority in the international trade development path, according to its critics:

- People in developing countries allegedly work long hours in poor conditions for low pay with minimal oversight by governments and international lending agencies.
- The workforce may include children or forced labor.
- Poor sanitation and safety may result in health problems and injuries.
- Injured, ill, or laid-off workers are not compensated.

Fair trade returns on average one-third of the price back to the producer in the developing country. The rest goes to the wholesaler who imports the item and for the retailer's rent, wages, and other expenses. On the other hand, only a tiny percentage of the price a consumer pays for a good reaches the individual in the developing country responsible for making or growing it, charge critics of international trade. A Haitian sewing clothing for the U.S. market, for example, earns less than 1 percent of the retail price, according to the National Labor Committee.

▼ 9.9.3 **FOCUS ON SOUTH ASIA: GRAMEEN BANK**

FOCUS ON SOUTH ASIA

Many would-be entrepreneurs in developing countries are too poor to qualify for regular bank loans. An alternative source of loans for development, the Grameen Bank, based in Bangladesh, has made several hundred thousand loans to women in South Asia. Only 1 percent of the borrowers have failed to make their weekly loan repayments, an extraordinarily low percentage for a bank (Figure 9.9.3). Several million loans have also been provided to women by the Bangladesh Rural Advancement Committee. For founding the bank, Muhammad Yunus was awarded the Nobel peace prize in 2006.

9.10 Millennium Development Goals

- By most development measures, the gap between developing and developed countries has narrowed.
- The United Nations has set eight goals to further reduce the gap in development.

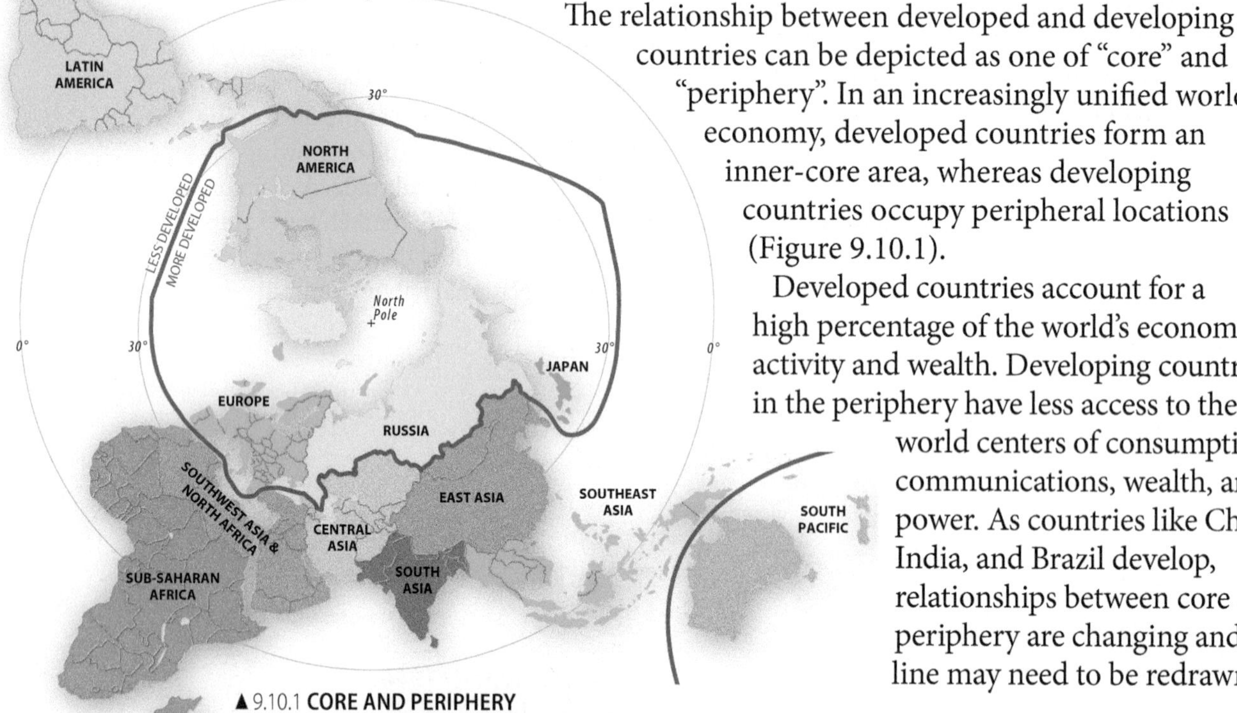

▲ 9.10.1 **CORE AND PERIPHERY**
This unorthodox world map projection emphasizes the central role played by developed countries at the core of the world economy and the secondary role of developing countries at the periphery.

The relationship between developed and developing countries can be depicted as one of "core" and "periphery". In an increasingly unified world economy, developed countries form an inner-core area, whereas developing countries occupy peripheral locations (Figure 9.10.1).

Developed countries account for a high percentage of the world's economic activity and wealth. Developing countries in the periphery have less access to the world centers of consumption, communications, wealth, and power. As countries like China, India, and Brazil develop, relationships between core and periphery are changing and the line may need to be redrawn.

CLOSING THE GAP

Since the United Nations began measuring HDI in 1980, all but three countries have had improved HDI scores (Figure 9.10.2). The exceptions are in sub-Saharan Africa—Democratic Republic of the Congo, Zambia, Zimbabwe—and the region as a whole has improved only from 0.29 to 0.39. In contrast, East Asia's HDI has improved especially rapidly, from 0.39 to 0.65, and South Asia's has improved from 0.32 to 0.52 (Figure 9.10.3).

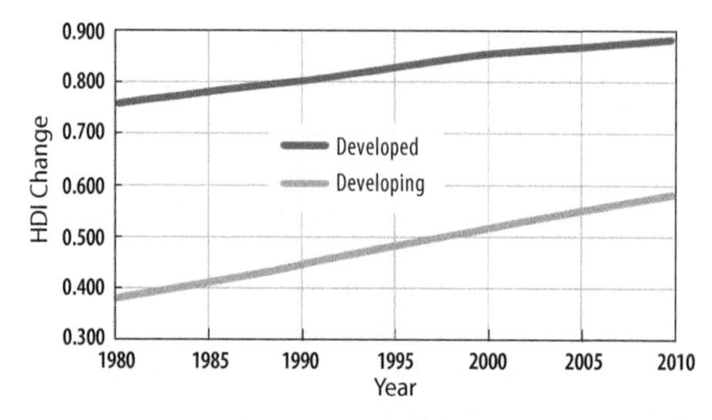

▲ 9.10.2 **HDI CHANGE, 1980–2010**

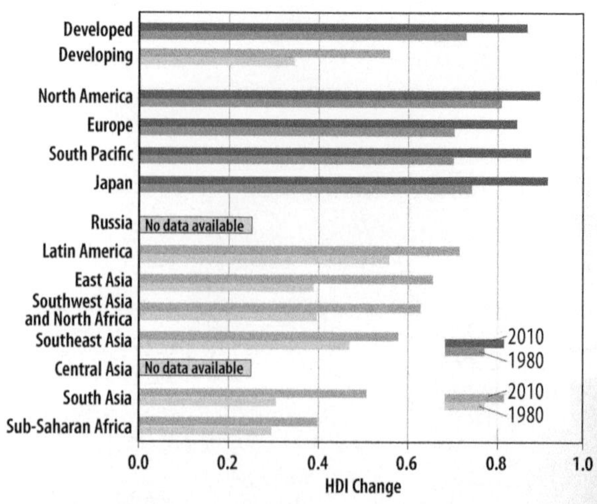

▲ 910.3 **HDI CHANGE BY REGION**

The gap between developed and developing countries is narrowing in health and education. For example, during the 1950s people lived on average more than two decades longer in developed countries than in developing ones. In the twenty-first century, the gap is less than ten years (Figure 9.10.4). On the other hand, the gap in wealth between developed and developing countries has widened (Figure 9.10.5).

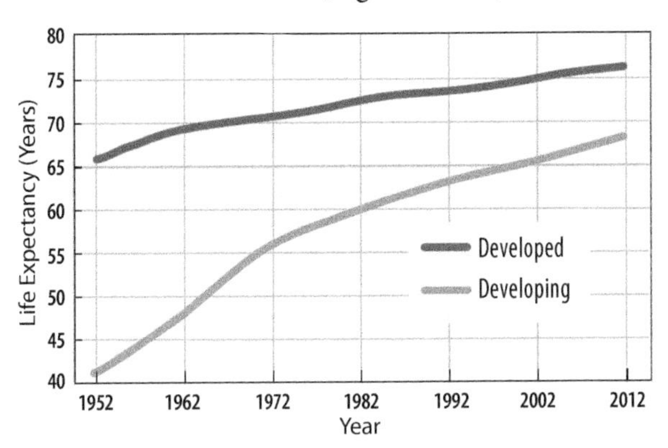

▲ 910.4 **CHANGE IN LIFE EXPECTANCY**

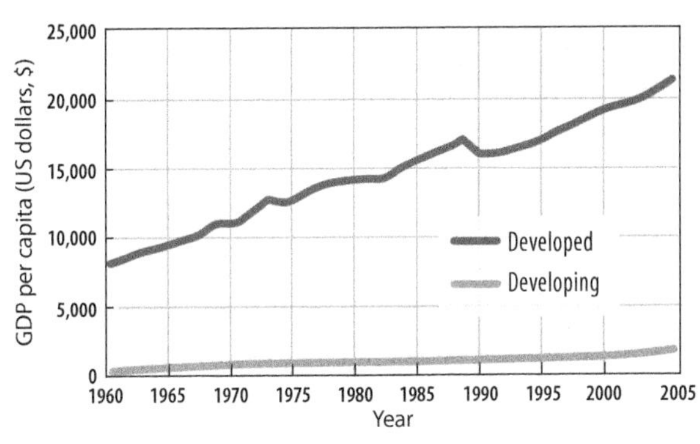

▲ 910.5 **CHANGE IN GDP PER CAPITA**

▼ 910.6 **FOCUS ON SUB-SAHARAN AFRICA: EDUCATION**
School in Kenya.

EIGHT GOALS

To reduce disparities between developed and developing countries, the United Nations has set eight Millennium Development Goals:

Goal 1: End poverty and hunger
Progress: Extreme poverty has been cut substantially in the world, primarily because of success in Asia, but it has not declined in sub-Saharan Africa.

Goal 2: Achieve universal primary (elementary school) education
Progress: The percentage of children not enrolled in school remains relatively high in South Asia and sub-Saharan Africa.

Goal 3: Promote gender equality and empower women.
Progress: Gender disparities remain in all regions, as discussed in Section 9.5

Goal 4: Reduce child mortality
Progress: Infant mortality rates have declined in most regions, except sub-Saharan Africa.

Goal 5: Improve maternal health
Progress: One-half million women die from complications during pregnancy; 99 percent of these women live in developing countries.

Goal 6: Combat HIV/AIDS, malaria, and other diseases
Progress: The number of people living with HIV remains high, especially in sub-Saharan Africa, as discussed in Chapter 5.

Goal 7: Ensure environmental sustainability
Progress: Water scarcity and quality, deforestation, and overfishing are still especially critical environmental issues, according to the United Nations.

Goal 8: Develop a global partnership for development
Progress: Aid from developed to developing countries has instead been declining.

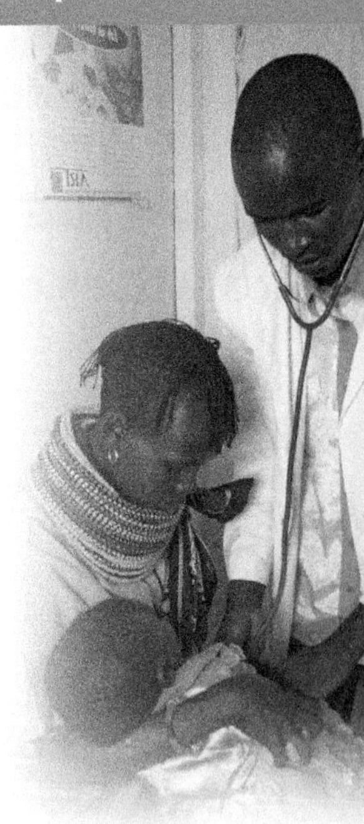

▲ 9.10.7 **FOCUS ON SUB-SAHARAN AFRICA: HEALTH**
Clinic in Kenya.

FOCUS ON SUB-SAHARAN AFRICA

Sub-Saharan Africa has the least favorable prospect for development. The region has the world's highest percentage of people living in poverty and suffering from poor health and low education levels (Figures 9.10.6 and 9.10.7). And conditions are getting worse: the average African consumes less today than a quarter-century ago. The fundamental problem in many countries of sub-Saharan Africa is a dramatic imbalance between the number of inhabitants and the capacity of the land to feed the population.

The world is divided into developed countries and developing ones. Developed and developing countries can be compared according to a number of indicators.

Key Questions

How does development vary among regions?

- The United Nations has created the Human Development Index to measure the level of development of every country.
- Gross National Income measures the standard of living in a country.
- Developed countries display higher levels of education and literacy.
- People in developed countries have a longer life expectancy.
- A Gender Inequality Index compares the level of development of women and men in every country.

How can countries promote development?

- The two principal paths to development are self-sufficiency and international trade.
- Self-sufficiency was the most commonly used path in the past, but most countries now follow international trade.
- Developing countries finance trade through loans, but may required to undertake economic reforms.

What are future challenges for development?

- Fair trade is an alternative approach to development through trade that provides greater benefits to the producers in developing countries.
- The United Nations has set Millennium Development Goals for countries to enhance their level of development.

Thinking Geographically

Review the major economic, social, and demographic characteristics that contribute to a country's level of development.

1. Which indicators can vary significantly by gender within countries and between countries at various levels of development? Why?

Some geographers have been attracted to the concepts of Immanuel Wallerstein, who argued that the modern world consists of a single entity, the capitalist world economy that is divided into three regions: the core, semi-periphery, and periphery (refer to Figure 9.10.1).

2. How have the boundaries among these three regions changed?

Opposition to international trade , as well as the severe recession of the early twenty-first century, has encouraged some countries to switch from international trade back to self-sufficiency (Figure 9.CR.1).

3. What are the advantages and challenges of returning to self-sufficiency in poor economic conditions?

On the Internet

Each year's United Nations Human Development Index Report, including numerous indicators for every country, can be accessed at **http://hdr.undp.org,** or scan QR on first page of this chapter.

Indicators cited in this chapter that are not part of the HDI can be found through the Earth Trends portion of the World Resources Institute (WRI) web site at **http://earthtrends.wri.org/.**

Several data sources, including the United Nations and the CIA, are brought together at **www.NationMaster.com**.

◄ 9.CR.1 **WTO PROTESTS**
Protesters at WTO meeting in Seattle, 1999.

Interactive Mapping

INTERNAL VARIATIONS IN DEVELOPMENT

The level of development varies within Latin America's two most populous countries, Brazil and Mexico.

Open MapMaster Latin America in MasteringGEOGRAPHY

Select: *Economic* then *Mapping Poverty and Prosperity.*
Select: *Population* then *Population Density.*

Use the slider tool to adjust the layer opacity.

Are high population concentration within Brazil and Mexico found primarily in the poorer or the wealthier regions of the two countries?

Explore

BRASILIA

A number of countries have built or are considering constructing new cities to promote development in poorer regions. One example is Brasilia, which was started in the 1950s and became the capital of Brazil in 1960.

Fly to: *National Congress of Brazil, Brasilia, Brazil*
Click on box in middle of screen.

What is the predominant style of housing constructed for the residents?

Key Terms

Adolescent fertility rate
The number of births per 1,000 women age 15-19.

Developed country (more developed country or MDC)
A country that has progressed relatively far along a continuum of development.

Developing country (less developed country or LDC)
A country that is at a relatively early stage in the process of economic development.

Development
A process of improvement in the material conditions of people through diffusion of knowledge and technology.

Fair trade
Alternative to international trade that emphasizes small businesses and worker-owned and democratically run cooperatives and requires employers to pay workers fair wages, permit union organizing, and comply with minimum environmental and safety standards.

Foreign direct investment
Investment made by a foreign company in the economy of another country.

Gender Inequality Index (GII)
Indicator constructed by the United Nations to measure the extent of each country's gender inequality.

Gross domestic product (GDP)
The value of the total output of goods and services produced in a country in a year, not accounting for money that leaves and enters the country.

Gross national income (GNI)
The value of the output of goods and services produced in a country in a year, including money that leaves and enters the country.

Human Development Index (HDI)
Indicator of level of development for each country, constructed by United Nations, combining income, literacy, education, and life expectancy.

Inequality-adjusted HDI (IHDI)
Indicator of level of development for each country that modifies the HDI to account for inequality.

Literacy rate
The percentage of a country's people who can read and write.

Maternal mortality rate
The number of women who die giving birth per 100,000 births.

Primary sector
The portion of the economy concerned with the direct extraction of materials from Earth's surface, generally through agriculture, although sometimes by mining, fishing, and forestry.

Productivity
The value of a particular product compared to the amount of labor needed to make it.

Secondary sector
The portion of the economy concerned with manufacturing useful products through processing, transforming, and assembling raw materials.

Structural adjustment program
Economic policies imposed on less developed countries by international agencies to create conditions encouraging international trade, such as raising taxes, reducing government spending, controlling inflation, selling publicly owned utilities to private corporations, and charging citizens more for services.

Tertiary sector
The portion of the economy concerned with transportation, communications, and utilities, sometimes extended to the provision of all goods and services to people in exchange for payment.

Value added
The gross value of the product minus the costs of raw materials and energy.

▸ LOOKING AHEAD

One of the most fundamental differences between developed and developing countries is the predominant methods of agriculture.

10 Food and Agriculture

When you buy food in the supermarket, are you reminded of a farm? Not likely. The meat is carved into pieces that no longer resemble an animal and is wrapped in paper or plastic film. Often the vegetables are canned or frozen. The milk and eggs are in cartons.

Providing food in the United States and Canada is a vast industry. Only a few people are full-time farmers, and they may be more familiar with the operation of computers and advanced machinery than the typical factory or office worker is.

The mechanized, highly productive American or Canadian farm contrasts with the subsistence farm found in much of the world. In China and India, more than half of the people are farmers who grow enough food for themselves and their families to survive, with little surplus. This sharp contrast in agricultural practices constitutes one of the most fundamental differences between the world's developed countries and developing countries.

What do people eat?

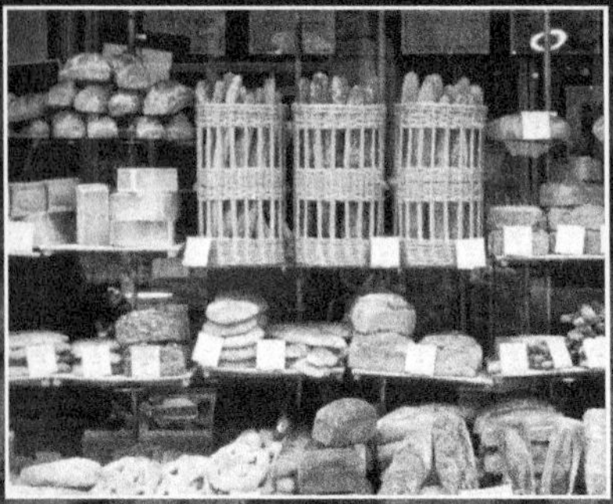

10.1 **Origin of Agriculture**

10.2 **Diet**

10.3 **Food Preferences**

10.4 **Nutrition and Hunger**

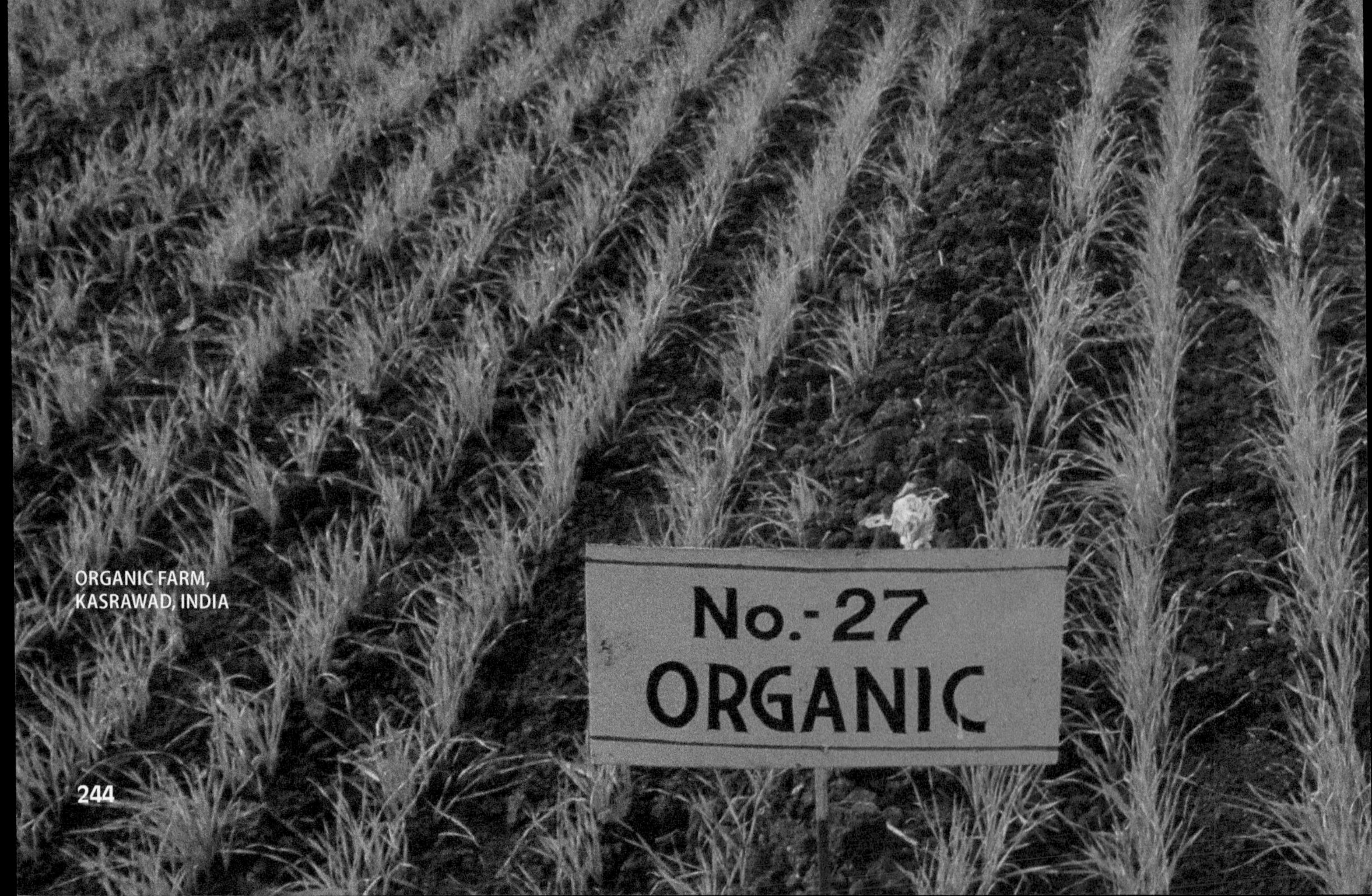

ORGANIC FARM, KASRAWAD, INDIA

How is agriculture distributed?
What challenges does agriculture face?
SCAN TO ACCESS THE UN's FOOD AND AGRICULTURE DATA
IR64680-81-2-2-1-3
BASAL 24DAT PI
45 45 45
Kg N/ha

10.1 Origin of Agriculture

- Early humans obtained food through hunting and gathering.
- Agriculture originated in multiple hearths and diffused in many directions.

Agriculture is the deliberate modification of Earth's surface through cultivation of plants and rearing of animals to obtain sustenance or economic gain. Agriculture originated when humans domesticated plants and animals for their use. The word cultivate means "to care for," and a **crop** is any plant cultivated by people.

HUNTERS AND GATHERERS

Before the invention of agriculture, all humans probably obtained the food they needed for survival through hunting for animals, fishing, or gathering plants (including berries, nuts, fruits, and roots). Hunters and gatherers lived in small groups, with usually fewer than 50 persons, because a larger number would quickly exhaust the available resources within walking distance.

Typically, the men hunted game or fished, and the women collected berries, nuts, and roots. This division of labor sounds like a stereotype but is based on evidence from archaeology and anthropology, although exceptions to this pattern have been documented. They collected food often, perhaps daily. The food search might take only a short time or much of the day, depending on local conditions.

The group traveled frequently, establishing new home bases or camps. The direction and frequency of migration depended on the movement of game and the seasonal growth of plants at various locations. We can assume that groups communicated with each other concerning hunting rights, intermarriage, and other specific subjects. For the most part, they kept the peace by steering clear of each other's territory.

Today perhaps a quarter-million people still survive by hunting and gathering rather than by agriculture. Examples include the Spinifex (also known as Pila Nguru) people, who live in Australia's Great Victorian Desert; the Sentinelese people, who live in India's Andaman Islands; and the Bushmen, who live in Botswana and Namibia (Figure 10.1.1). Contemporary hunting and gathering societies are isolated groups living on the periphery of world settlement, but they provide insight into human customs that prevailed in prehistoric times, before the invention of agriculture.

▲ 10.1.1 **HUNTERS AND GATHERERS**
Botswana.

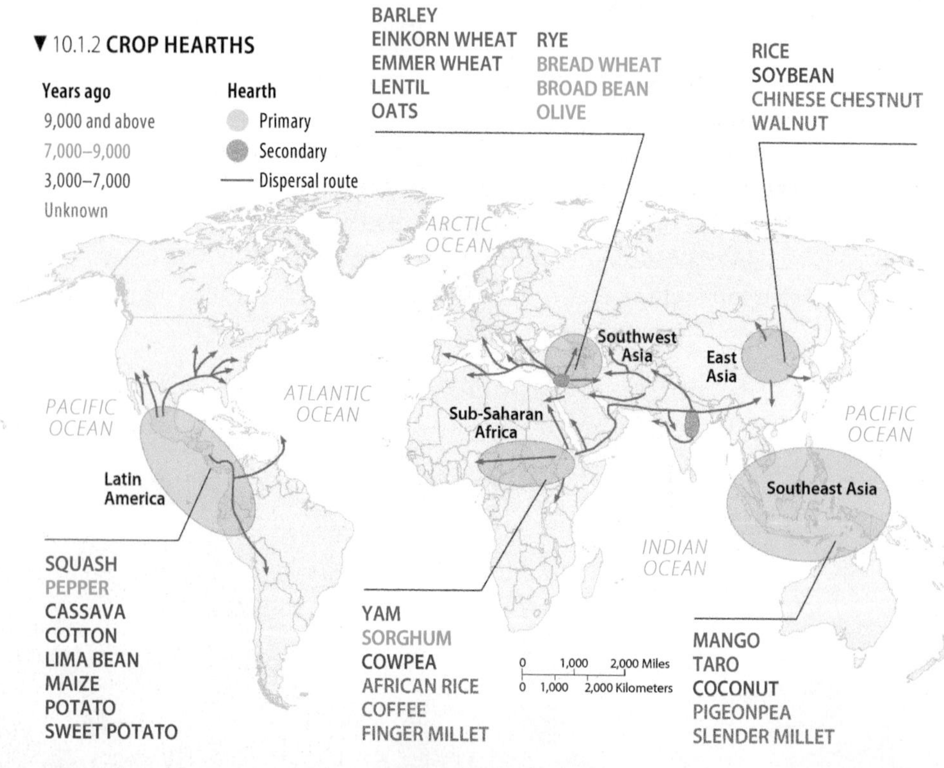

▼ 10.1.2 **CROP HEARTHS**

CROP HEARTHS

Why did most nomadic groups convert from hunting, gathering, and fishing to agriculture? Geographers and other scientists agree that agriculture originated in multiple hearths around the world. They do not agree on when agriculture originated and diffused, or why. Early centers of crop domestication include Southwest Asia, sub-Saharan Africa, Latin America, East Asia, and Southeast Asia (Figure 10.1.2). Crop cultivation diffused from these multiple hearths:

- From Southwest Asia: west to Europe and east to Central Asia.
- From sub-Saharan Africa: south to southern Africa.
- From Latin America: north to North America and south to tropical South America.

ANIMAL HEARTHS

Animals were also domesticated in multiple hearths at various dates. Southwest Asia is thought to be the hearth for the domestication of the largest number of animals that would prove to be most important for agriculture. Animals thought to be domesticated in Southwest Asia between 8,000 and 9,000 years ago include cattle, goats, pigs, and sheep. (Figure 10.1.3). The turkey is thought to have been domesticated in the Western Hemisphere (Figure 10.1.4).

Inhabitants of Southwest Asia may have been the first to integrate cultivation of crops with domestication of herd animals such as cattle, sheep, and goats. These animals were used to plow the land before planting seeds and, in turn, were fed part of the harvested crop. Other animal products, such as milk, meat, and skins, may have been exploited at a later date. This integration of plants and animals is a fundamental element of modern agriculture.

Domestication of the dog is thought to date from around 12,000 years ago, also in Southwest Asia. The horse is considered to have been domesticated in Central Asia; diffusion of the domesticated horse may have been associated with the diffusion of the Indo-European language, as discussed in Chapter 7.

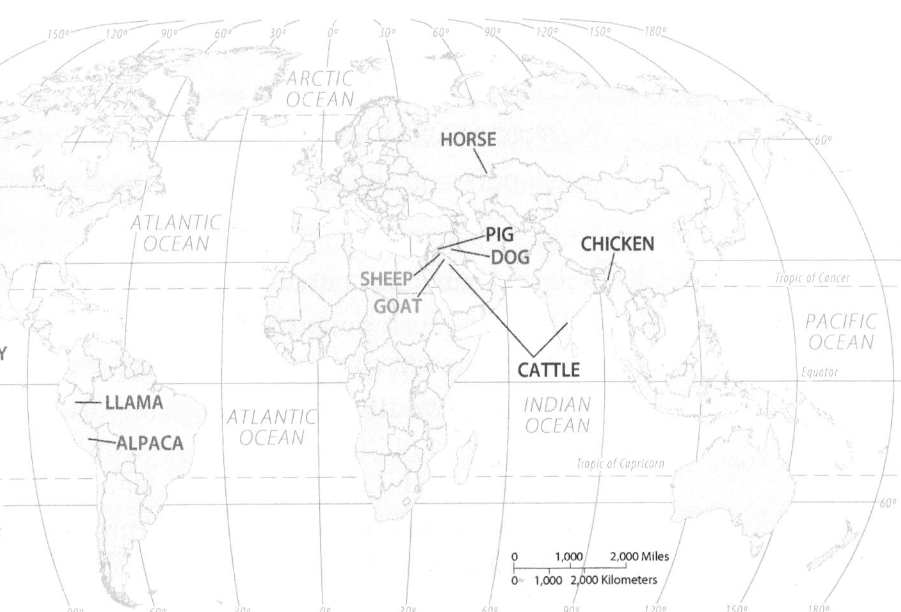

▲ 10.1.3 **ANIMAL HEARTHS**
Years ago
12,000
9,000
8,000
6,000
Unknown

WHY AGRICULTURE ORIGINATED

Scientists do not agree on whether agriculture originated primarily because of environmental factors or cultural factors. Probably a combination of both factors contributed.

- **Environmental factors.** The first domestication of crops and animals around 10,000 years ago coincides with climate change. This marked the end of the last ice age, when permanent ice cover receded from Earth's midlatitudes to polar regions, resulting in a massive redistribution of humans, other animals, and plants at that time.
- **Cultural factors.** Preference for living in a fixed place rather than as nomads may have led hunters and gatherers to build permanent settlements and to store surplus vegetation there.

In gathering wild vegetation, people inevitably cut plants and dropped berries, fruits, and seeds. These hunters probably observed that, over time, damaged or discarded food produced new plants. They may have deliberately cut plants or dropped berries on the ground to see if they would produce new plants.

Subsequent generations learned to pour water over the site and to introduce manure and other soil improvements. Over thousands of years, plant cultivation apparently evolved from a combination of accident and deliberate experiment.

That agriculture had multiple origins means that, from earliest times, people have produced food in distinctive ways in different regions. This diversity derives from a unique legacy of wild plants, climatic conditions, and cultural preferences in each region.

▼ 10.1.4 **WILD TURKEY**
California.

10.2 Diet

- ▶ **Most people derive most of their food energy from cereals.**
- ▶ **Climate and the level of development influence choice of food.**

Everyone needs food to survive. Consumption of food varies around the world, both in total amount and source of nutrients. The variation results from a combination of:

- **Level of development.** People in developed countries tend to consume more food and from different sources than do people in developing countries.
- **Physical conditions.** Climate is important in influencing what can be most easily grown and therefore consumed in developing countries. In developed countries, though, food is shipped long distances to locations with different climates.
- **Cultural preferences.** Some food preferences and avoidances are expressed without regard for physical and economic factors, as discussed in the next section.

TOTAL CONSUMPTION OF FOOD

Dietary energy consumption is the amount of food that an individual consumes. The unit of measurement of dietary energy is a kilocalorie (kcal), or Calorie in the United States. One gram (or ounce) of each food source delivers a kcal level that nutritionists can measure.

Humans derive most of their kilocalories through consumption of **cereal grain,** or simply cereal, which is a grass that yields grain for food. The three leading cereal grains—maize (corn in North America), wheat, and rice—together account for nearly 90 percent of all grain production and more than 40 percent of all dietary energy consumed worldwide.

Wheat is the principal cereal grain consumed in the developed regions of Europe and North America (Figure 10.2.1). Wheat is consumed in the form of bread, pasta, cake, and many other forms (Figure 10.2.2). It is also the most consumed **grain** in the developing regions of Central and Southwest Asia, where relatively dry conditions are more suitable for growing wheat than other grains.

▲ 10.2.2 **WHEAT CONSUMPTION**
Lille, France.

Rice is the principal cereal grain consumed in the developing regions of East, South, and Southeast Asia (Figure 10.2.3). It is the most suitable cereal crop for production in tropical climates.

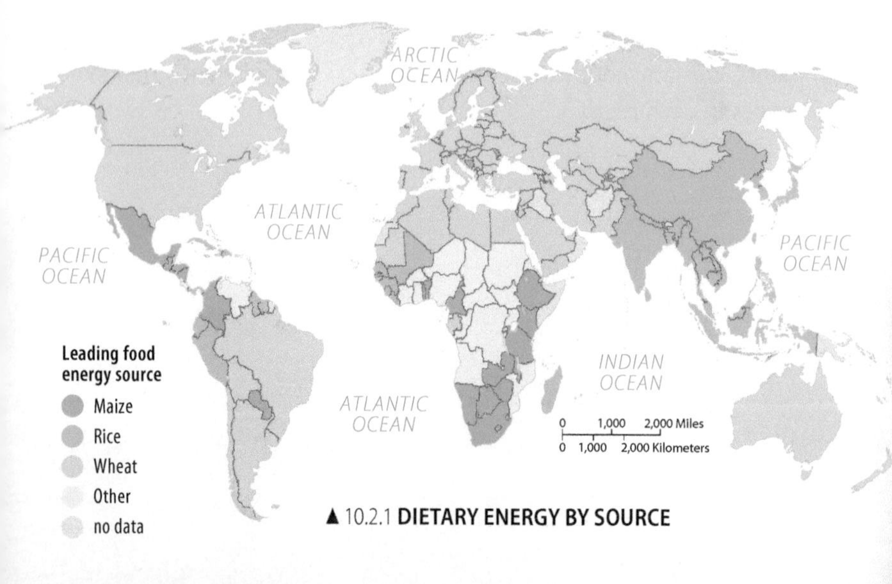

▲ 10.2.1 **DIETARY ENERGY BY SOURCE**

▼ 10.2.3 **RICE CONSUMPTION**
Ho Chi Minh City, Vietnam.

Maize is the leading crop in the world, though much of it is grown for purposes other than direct human consumption, especially as animal feed. It is the leading crop in some countries of sub-Saharan Africa (Figure 10.2.4).

A handful of countries obtain the largest share of dietary energy from other crops, especially in sub-Saharan Africa. These include cassava, sorghum, millet, plantains, sweet potatoes, and yams (Figure 10.2.5). Sugar is the leading source of dietary energy in several Latin American countries.

▲ 10.2.4 **MAIZE CONSUMPTION**
Weekly market, Madagascar.

▲ 10.2.5 **YAM CONSUMPTION**
Ghana.

SOURCE OF NUTRIENTS

Protein is a nutrient needed for growth and maintenance of the human body. Many food sources provide protein of varying quantity and quality. One of the most fundamental differences between developed and developing regions is the primary source of protein (Figure 10.2.6). In developed countries, the leading source of protein is meat products, including beef, pork, and poultry (Figure 10.2.7). Meat accounts for approximately one-third of all protein intake in developed countries, compared to approximately one-tenth in developing ones (Figure 102.8). In most developing countries, cereal grains provide the largest share of protein.

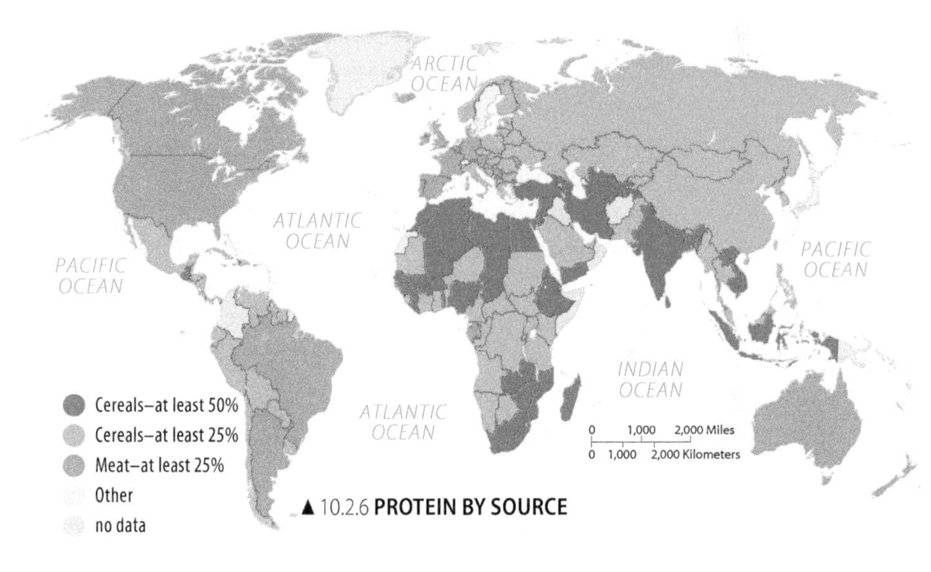

▲ 10.2.6 **PROTEIN BY SOURCE**

▲ 10.2.7 **MEAT CONSUMPTION**
Dublin, Ireland.

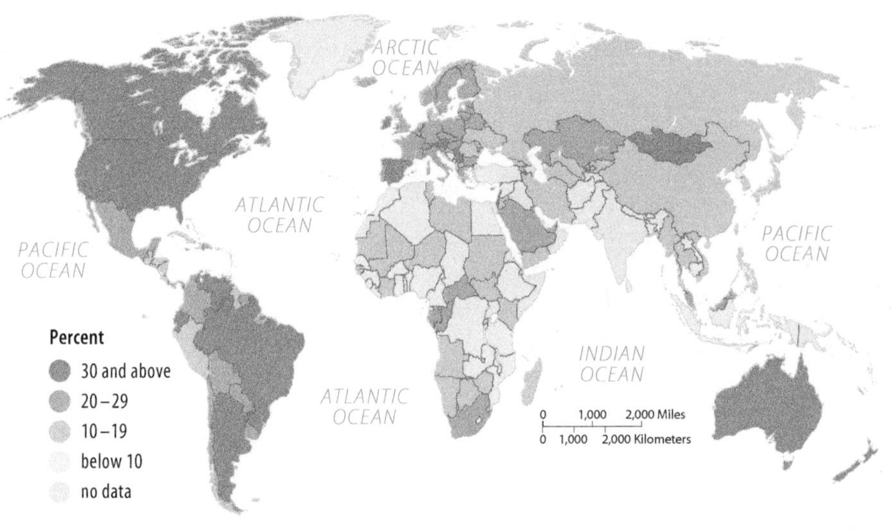

▲ 10.2.8 **PROTEIN FROM MEAT**

10.3 Food Preferences

- People embrace or avoid specific foods for cultural reasons.
- Food preferences are influenced in part by environmental factors.

Cultural preferences and environmental features influence choice of foods to consume in both developing and developed regions.

FOOD TABOOS

A restriction on behavior imposed by social custom is a **taboo**. Taboos are especially strong in the area of food. Relatively well-known taboos against consumption of certain foods can be found in the Bible. These Biblical taboos were developed through oral tradition and by rabbis into the kosher laws observed today by some Jews.

The Biblical food taboos were established in part to set the Hebrew people apart from others. That Christians ignore the Biblical food injunctions reflects their desire to distinguish themselves from Jews beginning 2,000 years ago. Furthermore, as a universalizing religion, Christianity was less tied to taboos that originated in the Middle East (see Chapter 7).

Among the Biblical taboos is prohibition against consuming animals that do not chew their cud or that have cloven feet, such as pigs, and seafood lacking fins or scales, such as lobsters (Figure 10.3.1). Muslims share the taboo against consuming pork.

As a result of taboos against consuming pork, the number of pigs raised in different regions of the world varies sharply. Pigs are especially scarce in predominantly Muslim regions, such as Southwest Asia and North Africa (Figure 10.3.2). On the other hand, China, where consumption of pork is embraced, has nearly one-half of the world's pig stock.

▼ 10.3.1 **KOSHER RESTAURANT**
Paris, France.

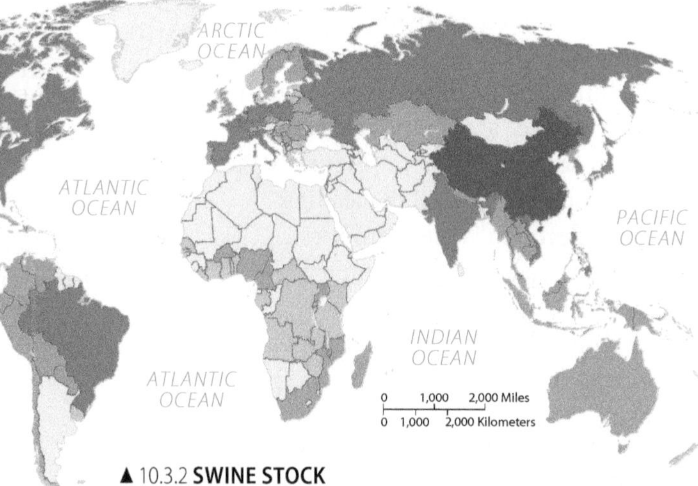

▲ 10.3.2 **SWINE STOCK**

ENVIRONMENTAL INFLUENCES

Food preferences are embedded especially strongly in the physical environment. Humans eat mostly plants and animals—living things that spring from the soil and water of a region. Inhabitants of a region must consider the soil, climate, terrain, vegetation, and other environmental features in deciding to produce particular foods.

People refuse to eat particular plants or animals that are thought to be strongly linked to negative forces in the environment.

- Biblical taboos arose partially from concern for the environment by the Hebrews, who lived as pastoral nomads in lands bordering the eastern Mediterranean. The pig, for example, is prohibited in part because it is more suited to sedentary farming than pastoral nomadism and in part because its meat spoils relatively quickly in hot climates, such as the Mediterranean.

- Muslims, like Jews, embrace the taboo against pork. Pigs would compete with humans for food and water without offering compensating benefits, such as being able to pull a plow, carry loads, or provide milk and wool. Widespread raising of pigs would be an ecological disaster in the deserts of Southwest Asia and North Africa.
- In India, Hindu sanctions against consuming cows are explained in part by the need to maintain a large supply of oxen (castrated male cows), the traditional choice for pulling plows as well as carts. A large supply of oxen must be maintained in India, because every field has to be plowed at approximately the same time—when the monsoon rains arrive.

Environmental features also influence food preferences as well as avoidances:

- In Asia, soybeans are widely grown, but raw they are toxic and indigestible. Lengthy cooking renders them edible, but in Asia fuel is scarce. Asians have adapted to this environmental dilemma by deriving foods from soybeans that do not require extensive cooking. These include bean sprouts (germinated seeds), soy sauce (fermented soybeans), and bean curd (steamed soybeans).
- In Europe, traditional preferences for quick-frying foods resulted in part from fuel shortages in Italy. In northern Europe, an abundant wood supply encouraged the slow stewing and roasting of foods over fires, which also provided home heat in the colder climate.

FOOD AND PLACE: THE CONCEPT OF TERROIR

The environment not only influences food preferences, but also contributes to the characteristics of foods produced in a particular area. The contribution of a location's distinctive physical features to the way food tastes is known by the French term **terroir**. The word comes from the same root as terre (French for land or earth), but terroir does not translate precisely into English; it has a similar meaning to the English expressions "grounded" or "sense of place." Terroir is the sum of the effects of the local environment on a particular food item.

Terroir is frequently used to refer to the combination of soil, climate, and other physical features that contribute to the distinctive taste of a wine:

- Climate. Vineyards are best cultivated in temperate climates of moderately cold, rainy winters and fairly long, hot summers (Figure 10.3.3). Hot, sunny weather is necessary in the summer for the fruit to mature properly, whereas winter is the preferred season for rain, because plant diseases that cause the fruit to rot are more active in hot, humid weather.
- Landforms. Vineyards are planted on hillsides, if possible, to maximize exposure to sunlight and to facilitate drainage. A site near a lake or river is also desirable because water can temper extremes of temperature.
- Soil. Grapes can be grown in a variety of soils, but the best wine tends to be produced from grapes grown in soil that is coarse and well drained—a soil not necessarily fertile for other crops. For example, the soil is generally sandy and gravelly in the Burgundy wine region, chalky in Champagne country, and of a slate composition in the Moselle Valley (Figure 10.3.4).

The distinctive character of each region's wine is especially influenced by the unique combination of trace elements, such as boron, manganese, and zinc, in the rock or soil. In large quantities these elements could destroy the plants, but in small quantities they lend a unique taste to the grapes.

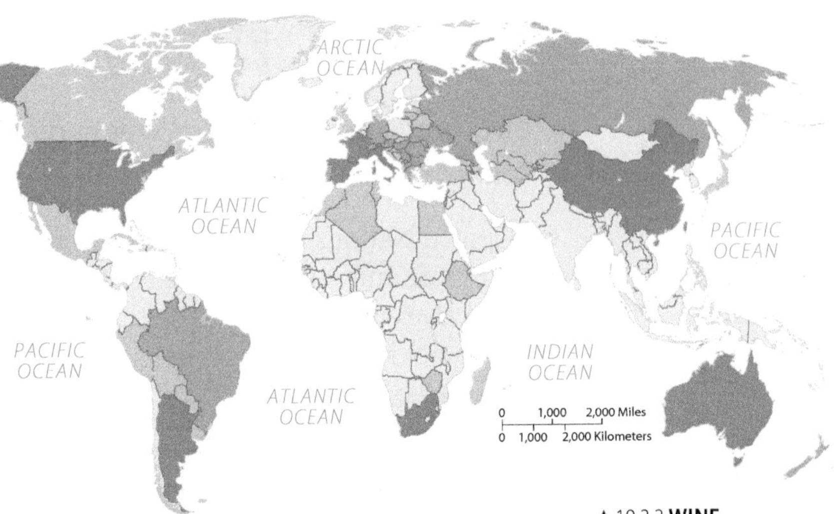

▲ 10.3.3 **WINE PRODUCTION**

▼ 10.3.4 **VINEYARDS** Burgundy, France.

10.4 Nutrition and Hunger

- On average, the world produces enough food to meet dietary needs.
- Some developing countries lack food security and are undernourished.

The United Nations defines **food security** as physical, social, and economic access at all times to safe and nutritious food sufficient to meet dietary needs and food preferences for an active and healthy life. By this definition, roughly one-eighth of the world's inhabitants do not have food security.

DIETARY ENERGY NEEDS

To maintain a moderate level of physical activity, according to the United Nations Food and Agricultural Organization, an average individual needs to consume on a daily basis at least 1,800 kcal.

Average consumption worldwide is 2,780 kcal per day, or roughly 50 percent more than the recommended minimum. Thus, most people get enough food to survive. People in developed countries are consuming on average nearly twice the recommended minimum, 3,470 kcal per day (Figure 10.4.1). The United States has the world's highest consumption, 3,800 kcal per day per person. The consumption of so much food is one reason that obesity rather than hunger is more prevalent in the United States, as well as other developed countries (Figure 10.4.2).

In developing regions, average daily consumption is 2,630 kcal, still above the recommended minimum. However, the average in sub-Saharan Africa is only 2,290, an indication that a large percentage of Africans are not getting enough to eat. Diets are more likely to be deficient in countries where people have to spend a high percentage of their income to obtain food (Figure 10.4.3).

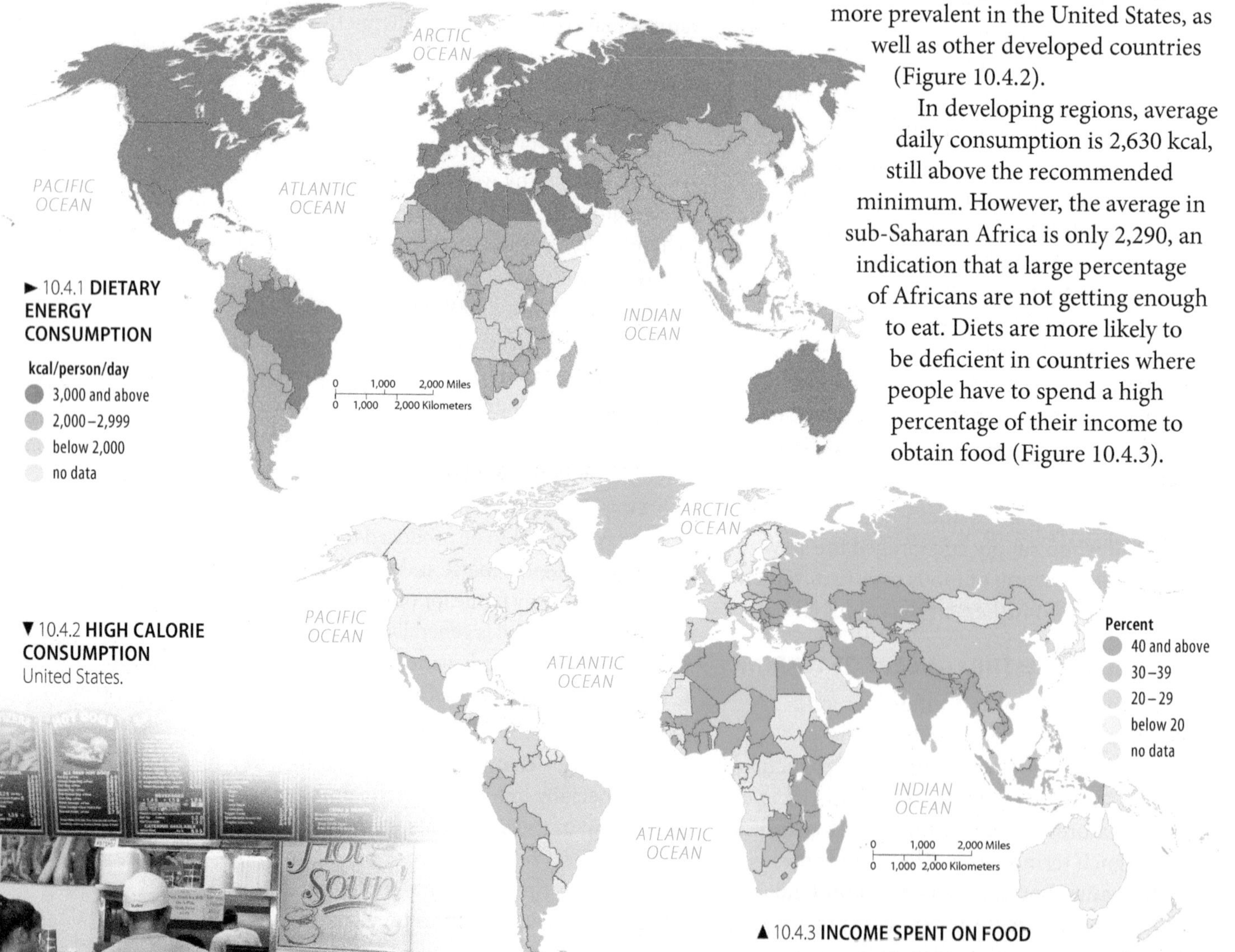

► 10.4.1 **DIETARY ENERGY CONSUMPTION**

▼ 10.4.2 **HIGH CALORIE CONSUMPTION**
United States.

▲ 10.4.3 **INCOME SPENT ON FOOD**

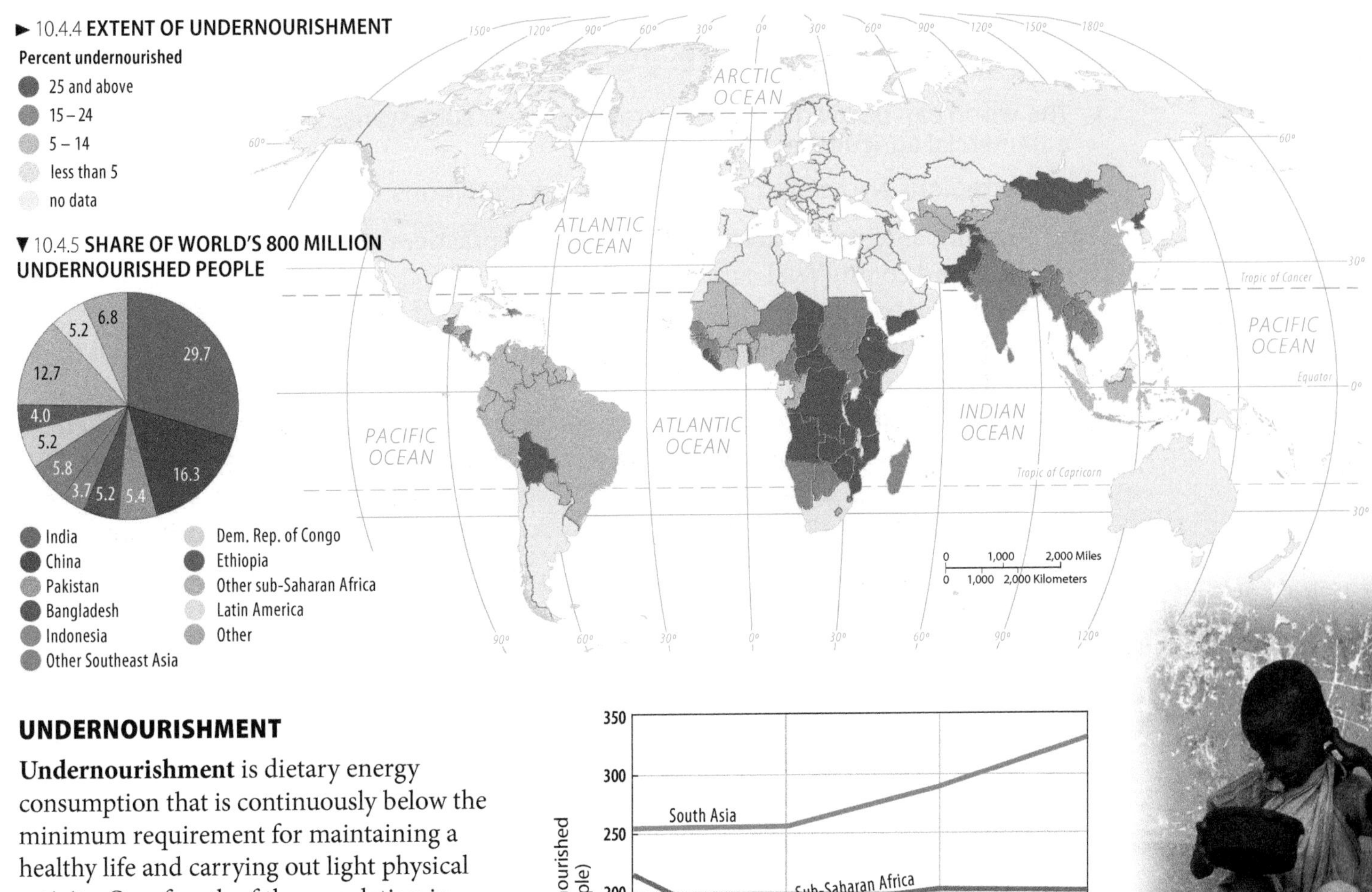

► 10.4.4 **EXTENT OF UNDERNOURISHMENT**

▼ 10.4.5 **SHARE OF WORLD'S 800 MILLION UNDERNOURISHED PEOPLE**

UNDERNOURISHMENT

Undernourishment is dietary energy consumption that is continuously below the minimum requirement for maintaining a healthy life and carrying out light physical activity. One-fourth of the population in sub-Saharan Africa, one-fifth in South Asia, and one-sixth in all developing countries are classified as undernourished (Figure 10.4.4).

India has by far the largest share of the world's 800 million undernourished people, 238 million, followed by China with 130 million (Figure 10.4.5). Worldwide, the total number of undernourished people has not changed much in several decades (Figure 10.4.6).

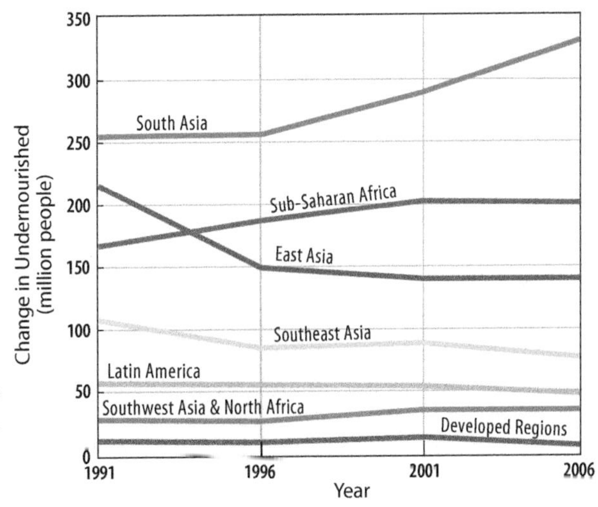

▲ 10.4.6 **CHANGE IN NUMBER UNDERNOURISHED**

▲ 10.4.7 **UNDERNOURISHMENT** Somalia.

AFRICA'S FOOD-SUPPLY STRUGGLE

Sub-Saharan Africa is struggling to keep food production ahead of population growth (Figure 10.4.7). Since 1961, food production has increased substantially in sub-Saharan Africa, but so has population (Figure 10.4.8). As a result, food production per capita has changed little in a half-century.

The threat of famine is particularly severe in the Sahel. Traditionally, this region supported limited agriculture. With rapid population growth, farmers overplanted, and herd size increased beyond the capacity of the land to support the animals. Animals overgrazed the limited vegetation and clustered at scarce water sources.

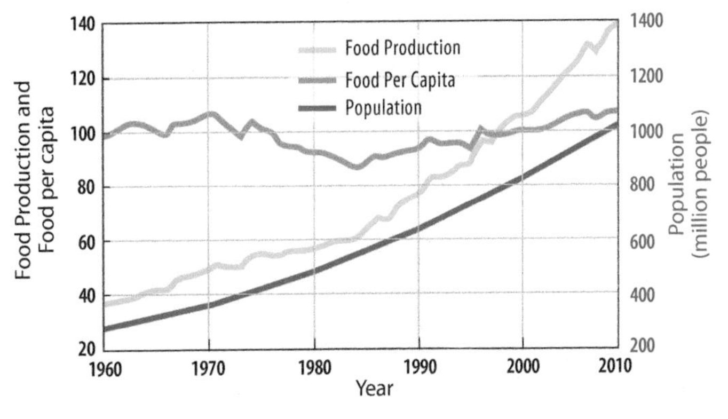

◄ 10.4.8 **POPULATION AND FOOD PRODUCTION IN AFRICA**

10.5 Agricultural Regions

- **The world can be divided into several regions of subsistence agriculture and commercial agriculture.**
- **These regions are related in part to climate conditions.**

The most fundamental differences in agricultural practices are between subsistence agriculture and commercial agriculture.

- **Subsistence agriculture** is generally practiced in developing countries (Figure 10.5.1). It is designed primarily to provide food for direct consumption by the farmer and the farmer's family (Figure 10.5.2).
- **Commercial agriculture,** generally practiced in developed countries, is undertaken primarily to generate products for sale off the farm to food-processing companies (Figure 10.5.3).

The most widely used map of world agricultural regions was prepared by geographer Derwent Whittlesey in 1936. Climate regions played an important role in determining agricultural regions, such as pastoral nomadism (Figure 10.5.4) and mixed crop and livestock (Figure 10.5.5).

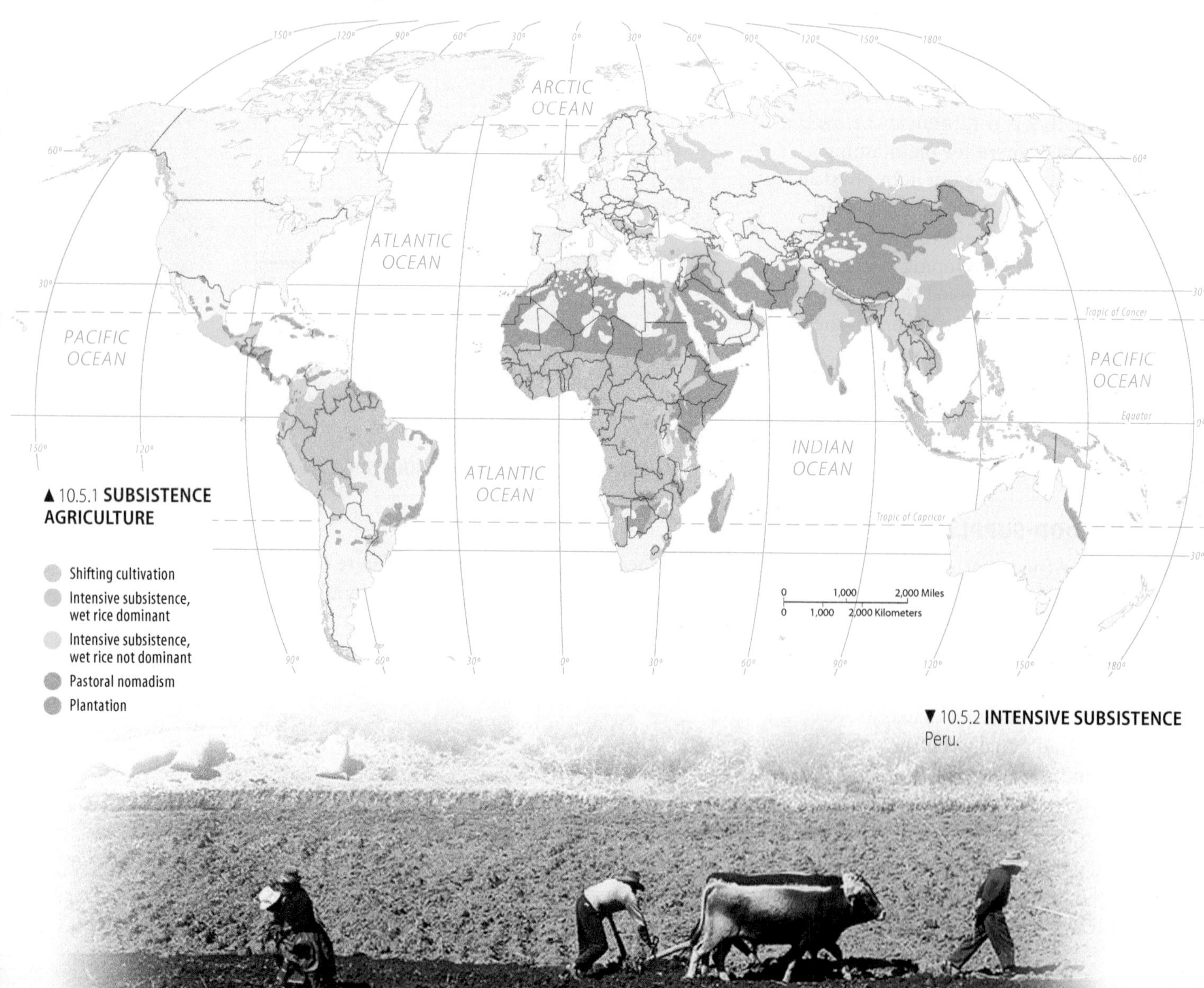

▲ 10.5.1 **SUBSISTENCE AGRICULTURE**

▼ 10.5.2 **INTENSIVE SUBSISTENCE**
Peru.

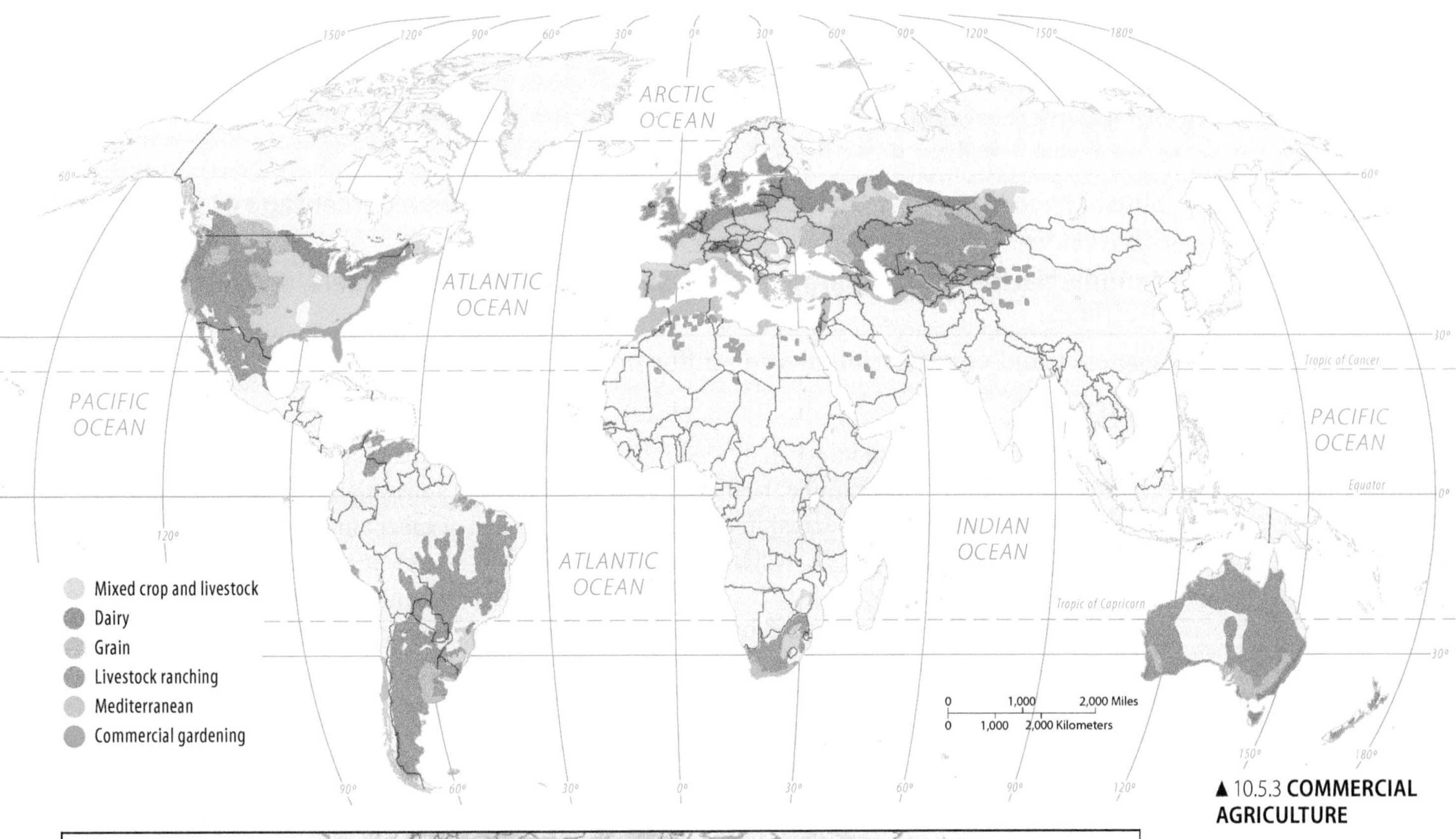

▲ 10.5.3 **COMMERCIAL AGRICULTURE**

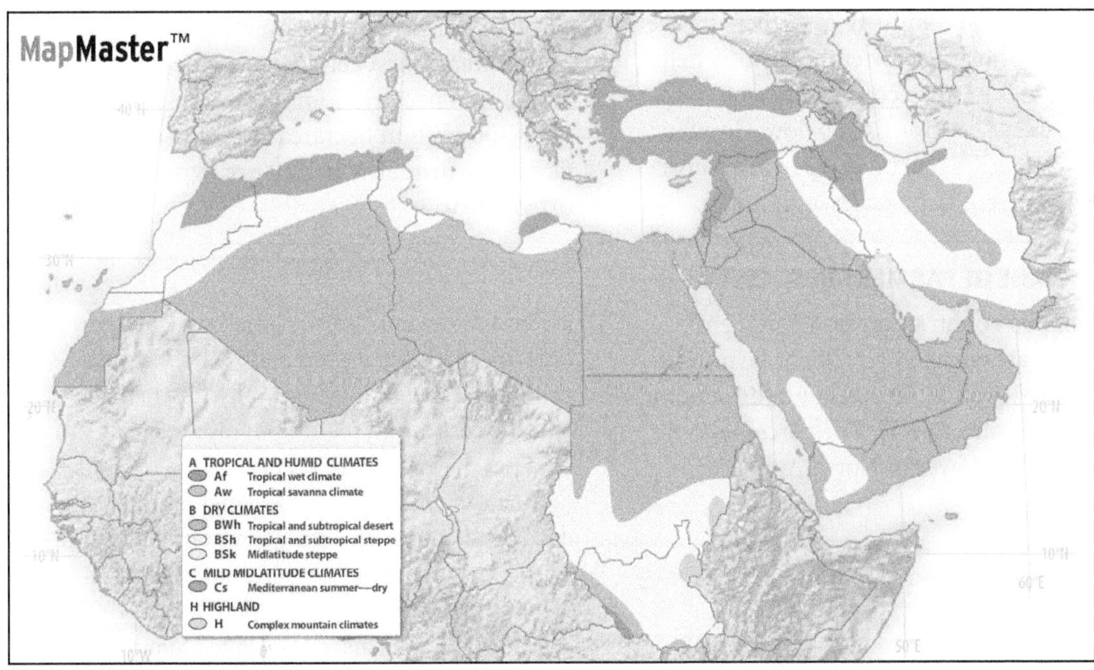

▲ 10.5.4 **CLIMATE REGIONS**

Climate influences the crop that is grown, or whether animals are raised instead of growing any crop.

Launch MapMaster Southwest Asia and North Africa in MasteringGEOGRAPHY

Select: *Climate* from the *Physical Environment* menu, then *Agricultural Regions* from the *Economic* menu.

What climate region is correlated with pastoral nomadism?

▼ 10.5.5 **MIXED CROP AND LIVESTOCK**

France.

10.6 Comparing Subsistence and Commercial Agriculture

- **Subsistence farming is characterized by small farms, a high percentage of farmers, and few machines.**
- **Commercial farming has large farms, a small percentage of farmers, and many machines.**

Subsistence and commercial farming differ in several key ways.

FARM SIZE

The average farm size is much larger in commercial agriculture. For example, farms average about 161 hectares (418 acres) in the United States, compared to about 1 hectare in China.

Commercial agriculture is dominated by a handful of large farms. In the United States, the largest 5 percent of farms produce 75 percent of the country's total agriculture. Despite their size, most commercial farms in developed countries are family owned and operated—90 percent in the United States. Commercial farmers frequently expand their holdings by renting nearby fields.

Large size is partly a consequence of mechanization, as discussed below. Combines, pickers, and other machinery perform most efficiently at very large scales, and their considerable expense cannot be justified on a small farm. As a result of the large size and the high level of mechanization, commercial agriculture is an expensive business.

Farmers spend hundreds of thousands of dollars to buy or rent land and machinery before beginning operations. This money is frequently borrowed from a bank and repaid after the output is sold.

The United States had 13 percent more farmland in 2000 than in 1900, primarily through irrigation and reclamation. However, in the twenty-first century it has been losing 1.2 million hectares (3 million acres) per year of its 400 million hectares (1 billion acres) of farmland, primarily because of expansion of urban areas.

PERCENTAGE OF FARMERS IN SOCIETY

In developed countries around 5 percent of workers are engaged directly in farming, compared to around 50 percent in developing countries (Figure 10.6.1). The percentage of farmers is even lower in North America, only around 2 percent. Yet the small percentage of farmers in the United States and Canada produces enough food not only for themselves and the rest of the region but also a surplus to feed people elsewhere.

The number of farmers declined dramatically in developed countries during the twentieth century. The United States had 60 percent fewer farms and 85 percent fewer farmers in 2000 than in 1900. The number of farms in the United States declined from about 6 million farms in 1940 to 4 million in 1960 and 2 million in 1980. Both push and pull migration factors have been responsible for the decline: people were pushed away from farms by lack of opportunity to earn a decent income, and at the same time they were pulled to higher-paying jobs in urban areas. The number of U.S. farmers has stabilized since 1980 at around 2 million.

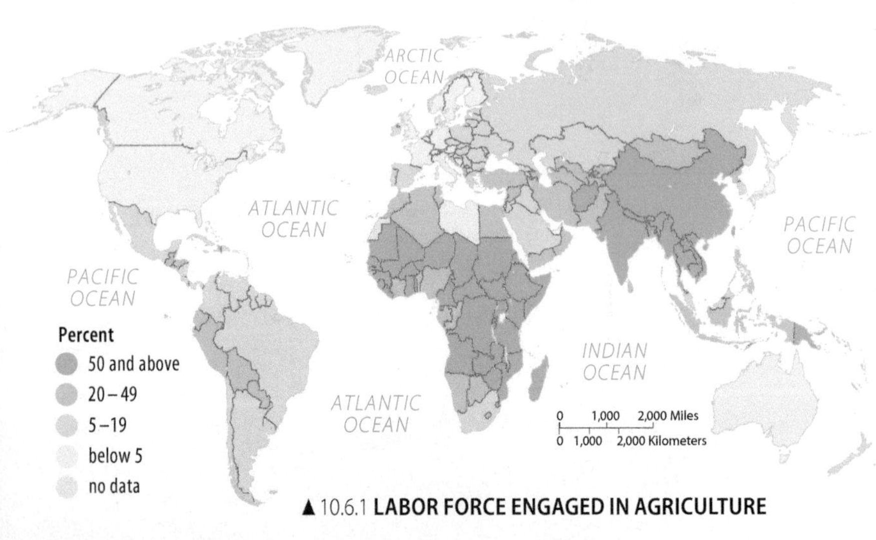

▲ 10.6.1 **LABOR FORCE ENGAGED IN AGRICULTURE**

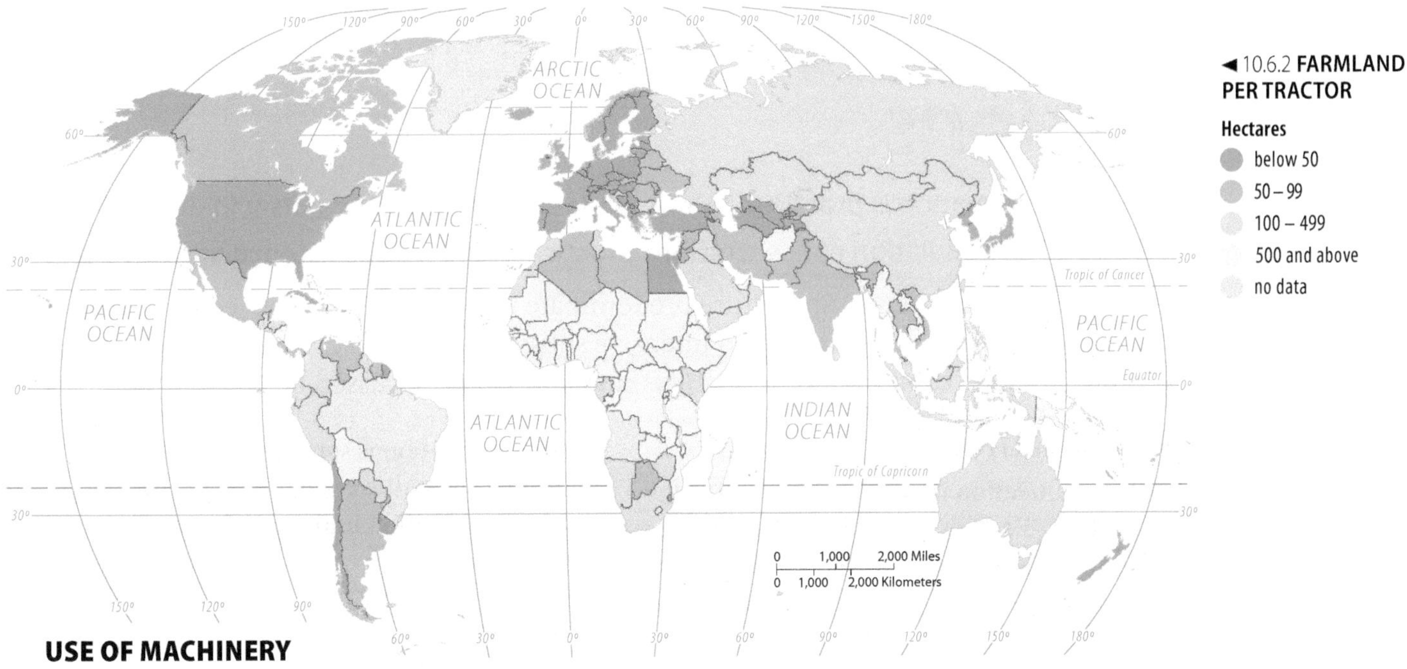

◀ 10.6.2 **FARMLAND PER TRACTOR**

Hectares
- below 50
- 50 – 99
- 100 – 499
- 500 and above
- no data

USE OF MACHINERY

In developed countries, a small number of farmers can feed many people because they rely on machinery to perform work, rather than relying on people or animals (Figure 10.6.2). In developing countries, farmers do much of the work with hand tools and animal power.

Traditionally, the farmer or local craftspeople made equipment from wood, but beginning in the late eighteenth century, factories produced farm machinery. The first all-iron plow was made in the 1770s and was followed in the nineteenth and twentieth centuries by inventions that made farming less dependent on human or animal power. Tractors, combines, corn pickers, planters, and other factory-made farm machines have replaced or supplemented manual labor (Figure 10.6.3).

Transportation improvements also aid commercial farmers. The building of railroads in the nineteenth century, and highways and trucks in the twentieth century, have enabled farmers to transport crops and livestock farther and faster. Cattle arrive at market heavier and in better condition when transported by truck or train than when driven on hoof. Crops reach markets without spoiling.

Commercial farmers use scientific advances to increase productivity. Experiments conducted in university laboratories, industry, and research organizations generate new fertilizers, herbicides, hybrid plants, animal breeds, and farming practices, which produce higher crop yields and healthier animals. Access to other scientific information has enabled farmers to make more intelligent decisions concerning proper agricultural practices. Some farmers conduct their own on-farm research.

Electronics also aid commercial farmers. Global positioning system (GPS) units determine the precise coordinates for spreading different types and amounts of fertilizers. On large ranches, GPS is also used to monitor the location of cattle. Satellite imagery monitors crop progress. Yield monitors attached to combines determine the precise number of bushels being harvested.

▼ 10.6.3 **USE OF MACHINERY**
Combines harvest wheat, Colarado.

10.7 Subsistence Agriculture Regions

- **Shifting cultivation is practiced in wet lands and pastoral nomadism in dry lands.**
- **Asia's large population concentrations practice intensive subsistence agriculture.**

Three types of subsistence agriculture predominate in developing countries: shifting cultivation, pastoral nomadism, and intensive subsistence. Plantation, a form of commercial agriculture found in developing countries, is also discussed here.

SHIFTING CULTIVATION

Shifting cultivation is practiced in much of the world's humid tropics, which have relatively high temperatures and abundant rainfall. Each year villagers designate for planting an area near the settlement. Before planting, they remove the dense vegetation using axes and machetes.

On a windless day the debris is burned under carefully controlled conditions; consequently, shifting cultivation is sometimes called **slash-and-burn agriculture** (Figure 10.7.1). The rains wash the fresh ashes into the soil, providing needed nutrients.

The cleared area is known by a variety of names, including *swidden, ladang, milpa, chena,* and *kaingin*. The **swidden** can support crops only briefly, usually 3 years or less, before soil nutrients are depleted. Villagers then identify a new site and begin clearing it, leaving the old swidden uncropped for many years, so that it is again overrun by natural vegetation.

Shifting cultivation is being replaced by logging, cattle ranching, and cultivation of cash crops. Selling timber to builders or raising beef cattle for fast-food restaurants is a more effective development strategy than maintaining shifting cultivation. Defenders of shifting cultivation consider it a more environmentally sound approach for tropical agriculture.

▼ 10.7.1 **SHIFTING CULTIVATION**
Venezuela.

PASTORAL NOMADISM

Pastoral nomadism is a form of subsistence agriculture based on the herding of domesticated animals. It is adapted to dry climates, where planting crops is impossible. Pastoral nomads live primarily in the large belt of arid and semiarid land that includes most of North Africa and Southwest Asia, and parts of Central Asia. The Bedouins of Saudi Arabia and North Africa and the Maasai of East Africa are examples of nomadic groups (Figure 10.7.2).

Pastoral nomads depend primarily on animals rather than crops for survival. The animals provide milk, and their skins and hair are used for clothing and tents. Like other subsistence farmers, though, pastoral nomads consume mostly grain rather than meat. Their animals are usually not slaughtered, although dead ones may be consumed. To nomads, the size of their herd is both an important measure of power and prestige and their main security during adverse environmental conditions.

Only about 15 million people are pastoral nomads, but they sparsely occupy about 20 percent of Earth's land area. Nomads used to be the most powerful inhabitants of the dry lands. Today, national governments control the nomadic population, using force, if necessary. Governments force groups to give up pastoral nomadism because they want the land for other uses.

▼ 10.7.2 **PASTORAL NOMADISM**
Sahara Desert, Africa.

INTENSIVE SUBSISTENCE

In densely populated East, South, and Southeast Asia, most farmers practice **intensive subsistence agriculture.** Because the agricultural density—the ratio of farmers to arable land—is so high in parts of East and South Asia, families must produce enough food for their survival from a very small area of land.

Most of the work is done by hand or with animals rather than with machines, in part due to abundant labor, but largely from lack of funds to buy equipment.

The intensive agriculture region of Asia can be divided between areas where **wet rice** dominates and areas where it does not. The term **wet rice** refers to the practice of planting rice on dry land in a nursery and then moving the seedlings to a flooded field to promote growth. Wet rice is most easily grown on flat land, because the plants are submerged in water much of the time.

The pressure of population growth in parts of East Asia has forced expansion of areas under rice cultivation (Figure 10.7.3). One method of developing additional land suitable for growing rice is to terrace the hillsides of river valleys (Figure 10.7.4).

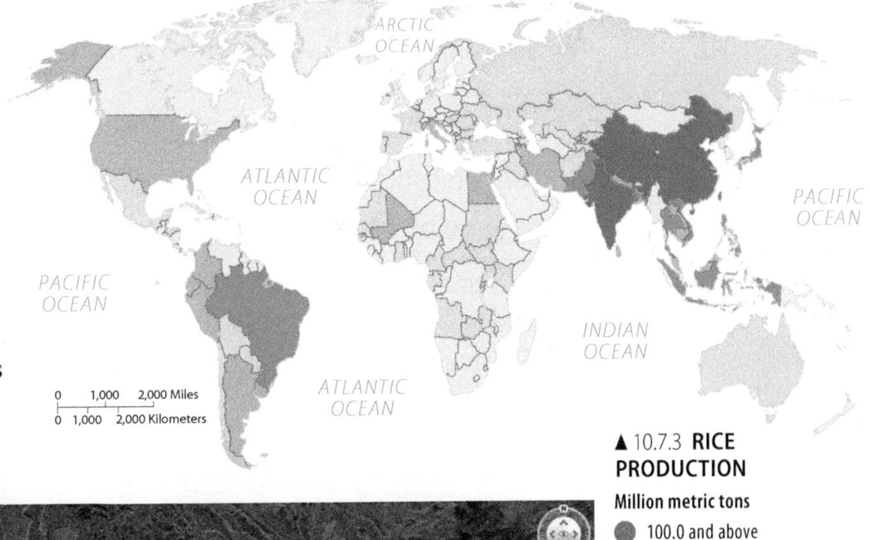

▲ 10.7.3 **RICE PRODUCTION**

Million metric tons
- 100.0 and above
- 10.0–99.9
- 1.0–9.9
- below 1.0
- no data

▲ 10.7.4 **INTENSIVE SUBSISTENCE FARMING**

Use Google Earth to explore rice farming in Southeast Asia.

Fly to: *Banaue, Philippines.*

Use the mouse to zoom in near the Banaue label until you see a series of brown swirling stripes.

Drag to: *Street view* on one of the brown swirling stripes.

Exit *Ground level view*.

1. Is the topography of this region flat or hilly?

2. What are the brown swirling stripes?

PLANTATION AGRICULTURE

A **plantation** is a form of commercial agriculture in developing regions that specializes in one or two crops. They are found primarily in the tropics and subtropics, especially in Latin America, sub-Saharan Africa, and Asia (Figure 10.7.5).

Although situated in developing countries, plantations are often owned or operated by Europeans or North Americans and grow crops for sale primarily in developed countries. Among the most important crops grown on plantations are cotton, sugarcane, coffee, rubber, and tobacco.

Until the Civil War, plantations were important in the U.S. South, where the principal crop was cotton, followed by tobacco and sugarcane. Slaves brought from Africa performed most of the labor until the abolition of slavery and the defeat of the South in the Civil War. Thereafter, plantations declined in the United States; they were subdivided and either sold to individual farmers or worked by tenant farmers.

▼ 10.7.5 **SUGARCANE PLANTATION**
Thailand.

10.8 Commercial Agriculture Regions

- Six main types of commercial agriculture are found in developed countries.
- The type of agriculture is influenced by physical geography.

Commercial agriculture in developed countries can be divided into six main types. Each type is predominant in distinctive regions within developed countries, depending largely on climate.

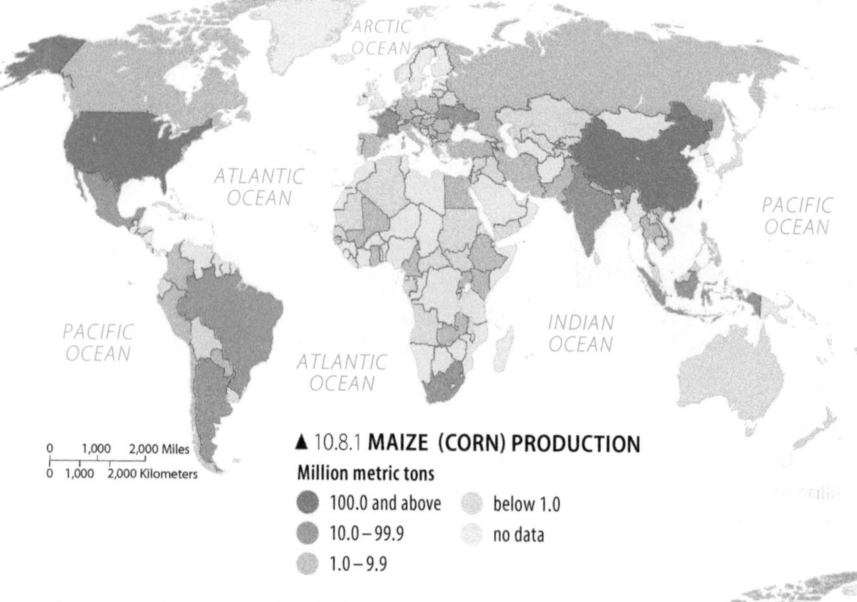

▲ 10.8.1 **MAIZE (CORN) PRODUCTION**
Million metric tons
- 100.0 and above
- 10.0–99.9
- 1.0–9.9
- below 1.0
- no data

MIXED CROP AND LIVESTOCK

The most distinctive characteristic of mixed crop and livestock farming is its integration of crops and livestock. Maize (corn) is the most commonly grown crop, (Figure 10.8.1) followed by soybeans. Most of the crops are fed to animals rather than consumed directly by humans. A typical mixed commercial farm devotes nearly all land area to growing crops but derives more than three-fourths of its income from the sale of animal products, such as beef, milk, and eggs.

Mixed crop and livestock farming typically involves **crop rotation.** The farm is divided into a number of fields, and each field is planted on a planned cycle, often of several years duration.

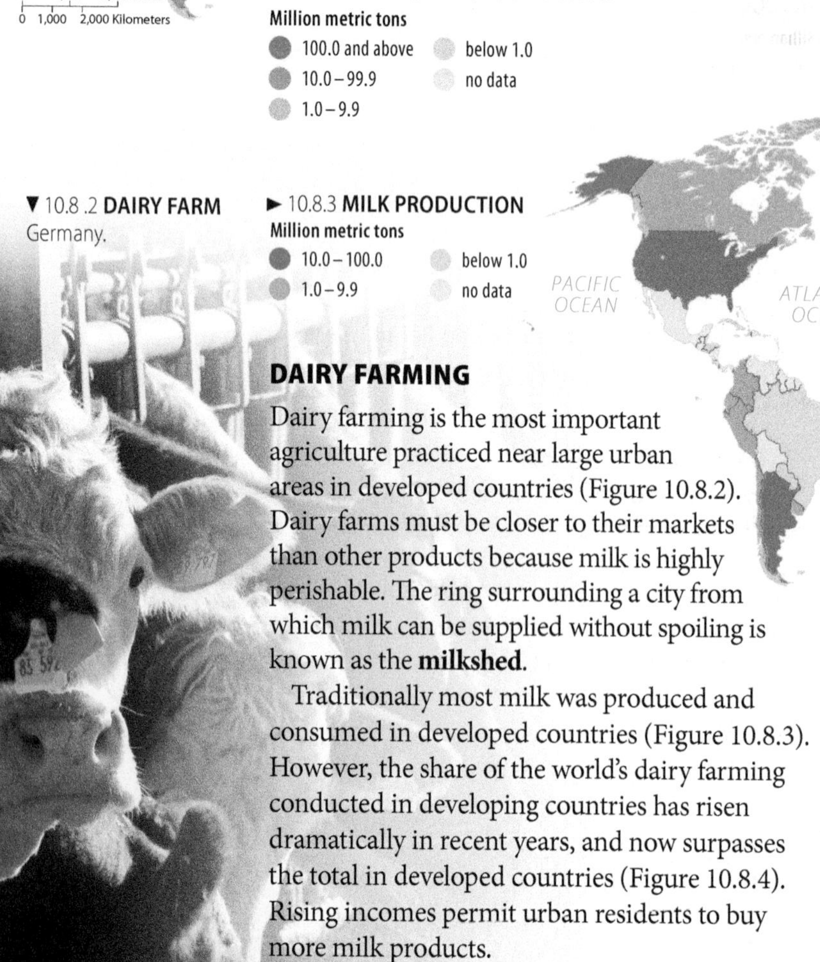

▼ 10.8.2 **DAIRY FARM** Germany.

► 10.8.3 **MILK PRODUCTION**
Million metric tons
- 10.0–100.0
- 1.0–9.9
- below 1.0
- no data

DAIRY FARMING

Dairy farming is the most important agriculture practiced near large urban areas in developed countries (Figure 10.8.2). Dairy farms must be closer to their markets than other products because milk is highly perishable. The ring surrounding a city from which milk can be supplied without spoiling is known as the **milkshed.**

Traditionally most milk was produced and consumed in developed countries (Figure 10.8.3). However, the share of the world's dairy farming conducted in developing countries has risen dramatically in recent years, and now surpasses the total in developed countries (Figure 10.8.4). Rising incomes permit urban residents to buy more milk products.

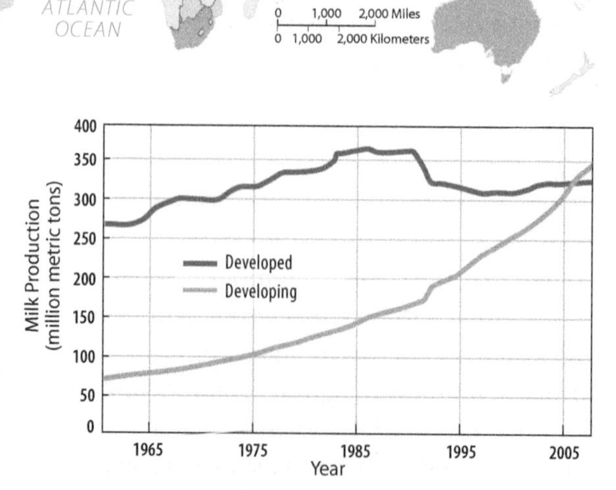

▲ 10.8.4 **MILK PRODUCTION**

GRAIN FARMING

Commercial grain farms are generally located in regions that are too dry for mixed crop and livestock farming, such as the Great Plains of North America (Figure 10.8.5). Unlike mixed crop and livestock farming, crops on a grain farm are grown primarily for consumption by humans rather than by livestock.

The most important crop grown is wheat, used to make flour. It can be stored relatively easily without spoiling and can be transported a long distance. Because wheat has a relatively high value per unit weight, it can be shipped profitably from remote farms to markets.

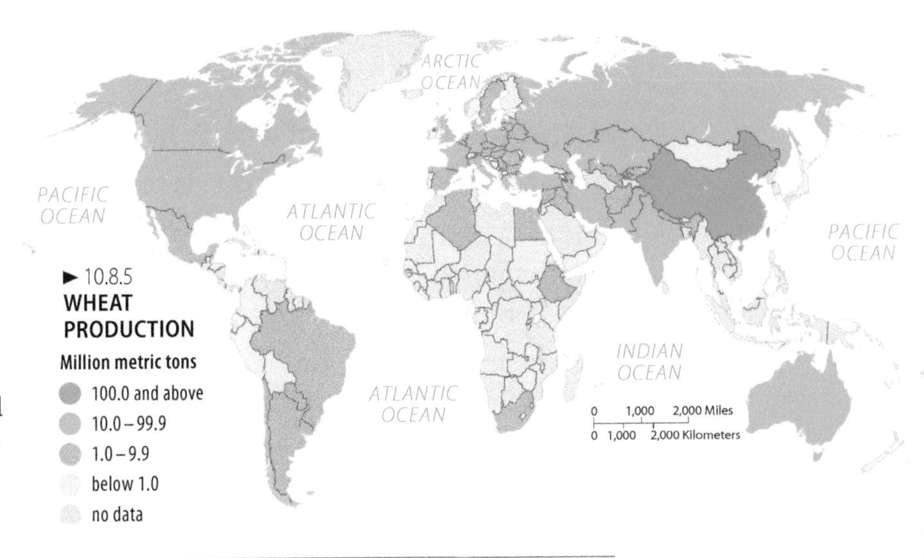

► 10.8.5 **WHEAT PRODUCTION**

LIVESTOCK RANCHING

Ranching is the commercial grazing of livestock over an extensive area. It is practiced primarily on semiarid or arid land where the vegetation is too sparse and the soil too poor to support crops. China is the leading producer of pig meat, the United States of chicken and beef (Figure 10.8.6).

Ranching has been glamorized in novels and films, although the cattle drives and "Wild West" features of this type of farming actually lasted only a few years in the mid-nineteenth century. Contemporary ranching has become part of the meat-processing industry, rather than carried out on isolated farms.

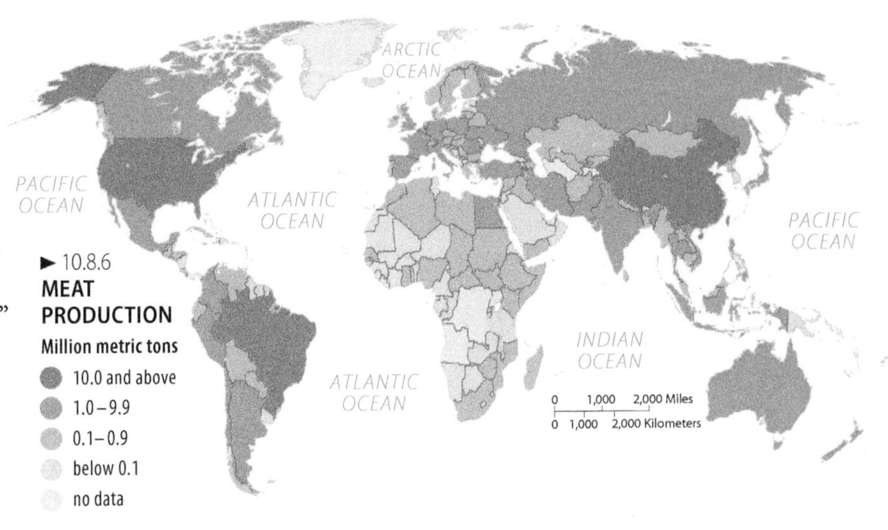

► 10.8.6 **MEAT PRODUCTION**

COMMERCIAL GARDENING AND FRUIT FARMING

Commercial gardening and fruit farming are the predominant types of agriculture in the U.S. Southeast (Figure 10.8.7). The region has a long growing season and humid climate and is accessible to the large markets in the big cities along the East Coast. It is frequently called **truck farming,** because "truck" was a Middle English word meaning bartering or the exchange of commodities.

Truck farms grow many of the fruits and vegetables that consumers demand in developed countries, such as apples, cherries, lettuce, and tomatoes. A form of truck farming called specialty farming has spread to New England. Farmers are profitably growing crops that have limited but increasing demand among affluent consumers, such as asparagus, mushrooms, peppers, and strawberries.

MEDITERRANEAN AGRICULTURE

Mediterranean agriculture exists primarily on lands that border the Mediterranean Sea and other places that share a similar physical geography, such as California, central Chile, the southwestern part of South Africa, and southwestern Australia. Winters are moist and mild, summers hot and dry. The land is very hilly, and mountains frequently plunge directly to the sea, leaving very little flat land. The two most important crops are olives (primarily for cooking oil) and grapes (primarily for wine).

▼ 10.8.7 **COMMERCIAL GARDENING**
Peanut farm, Georgia, U.S.A..

10.9 Fishing

- ► **Fish are either caught wild or farmed.**
- ► **Increasing fish consumption is resulting in overfishing.**

The agriculture discussed thus far in this chapter is land-based. Humans also consume food acquired from Earth's waters, including fish, crustaceans (such as shrimp and crabs), molluscs (such as clams and oysters), and aquatic plants (such as watercress).

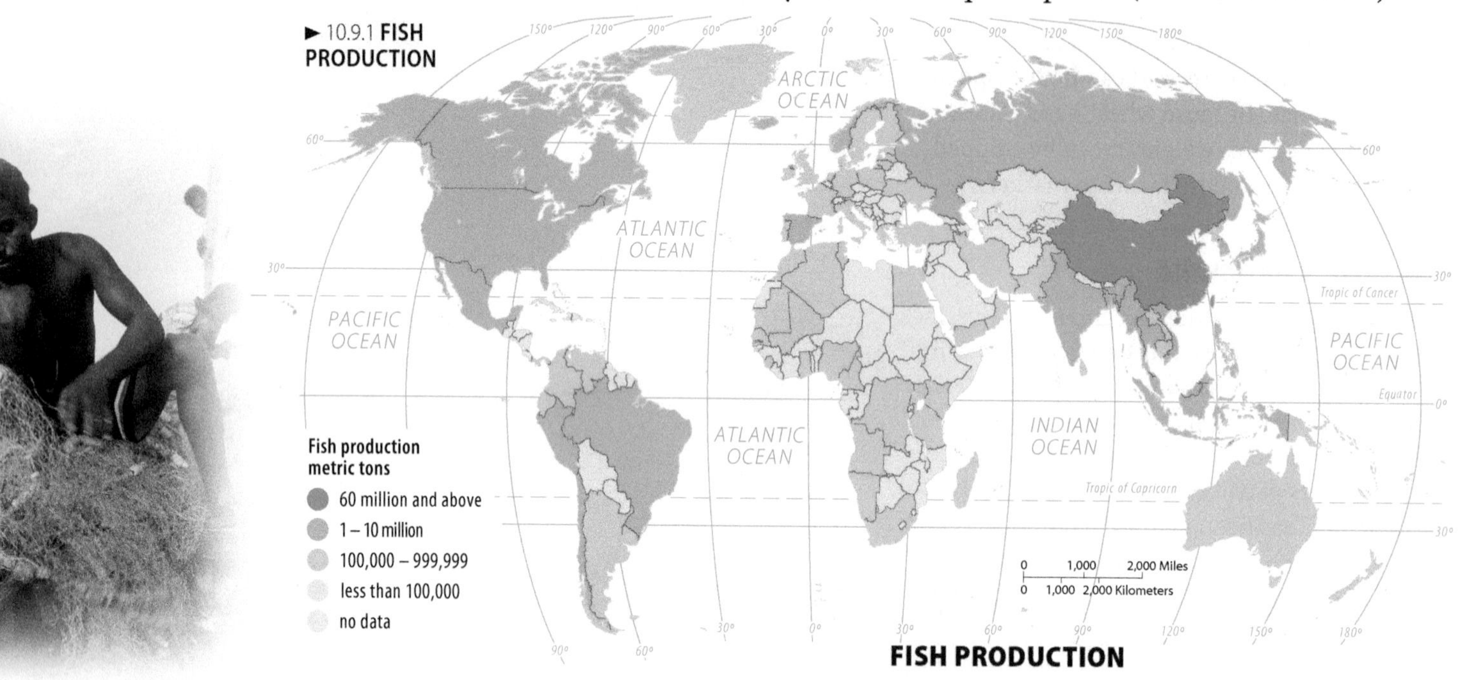

► 10.9.1 **FISH PRODUCTION**

▲ 10.9.2 **FISHING**
Mauritania.

▲ 10.9.3 **MAJOR FISHING REGIONS**

FISH PRODUCTION

Water-based food is acquired in two ways:

- Fishing, which is the capture of wild fish and other seafood living in the waters.
- **Aquaculture**, or **aquafarming**, which is the cultivation of seafood under controlled conditions.

About two-thirds of the fish caught from the ocean is consumed directly by humans, whereas the remainder is converted to fish meal and fed to poultry and hogs.

China is responsible for 40 percent of the world's yield of fish (Figure 10.9.1). The other leading countries are naturally those with extensive ocean boundaries, including Peru, Indonesia, India, Chile, Japan, and the United States.

The world's oceans are divided into 18 major fishing regions, including seven each in the Atlantic and Pacific oceans and four in the Indian Ocean (Figure 10.9.2). The three areas with the largest yield are all in the Pacific (Figure 10.9.3). Fishing is also conducted in inland waterways, such as lakes and rivers.

FISH CONSUMPTION

At first glance, increased use of food from the sea is attractive. Oceans are vast, covering nearly three-fourths of Earth's surface and lying near most population concentrations. Historically the sea has provided only a small percentage of the world food supply. Increased fish consumption was viewed as a way to meet the needs of a rapidly growing global population.

In fact, during the past half-century per capita consumption of fish has doubled worldwide, and tripled in developing countries (Figure 10.9.4). Still, fish accounts for only 6 percent of all protein consumed by humans, though a rapidly increasing source in developing countries if not in developed ones (Figure 10.9.5).

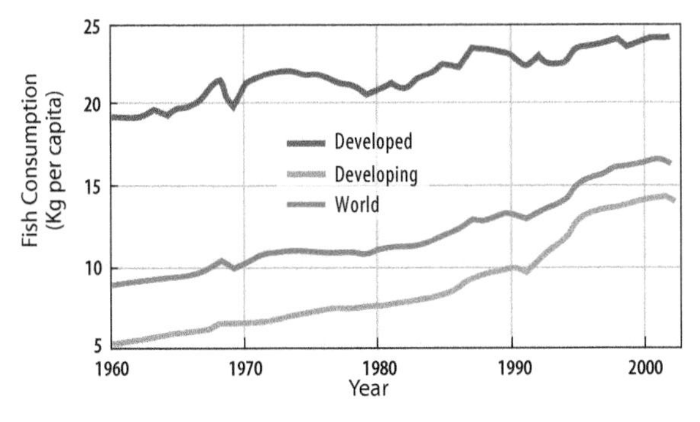

▲ 10.9.4 **FISH CONSUMPTION PER CAPITA**

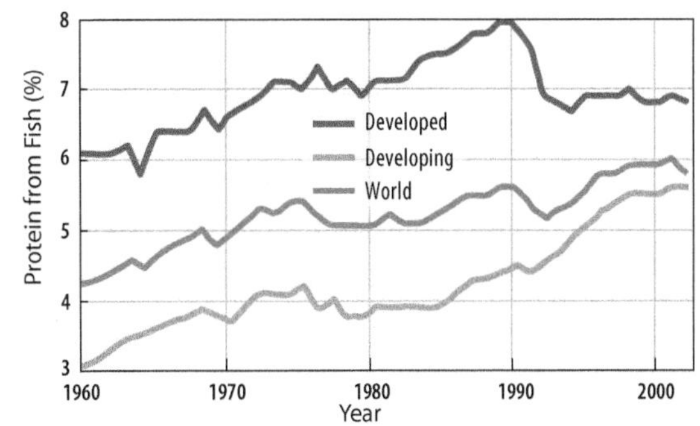

▲ 10.9.5 **PERCENT PROTEIN FROM FISH**

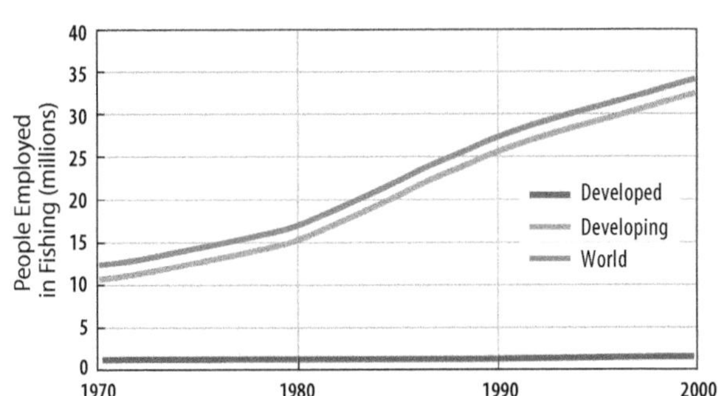

▲ 10.9.6 **EMPLOYMENT IN FISHING AND AQUACULTURE**

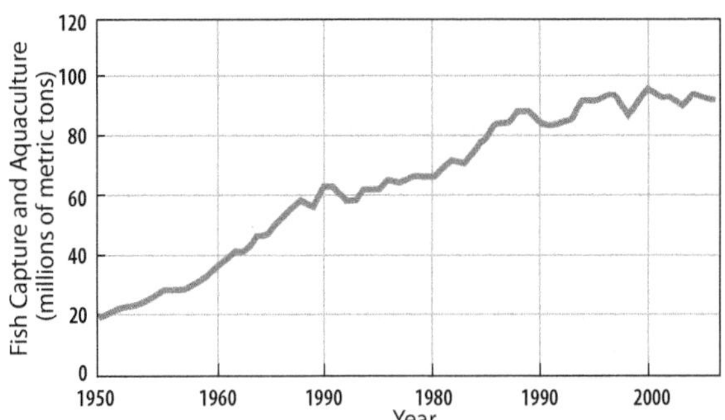

▲ 10.9.7 **WORLD FISH CAPTURE AND AQUACULTURE**

OVERFISHING

Worldwide, 35 million people are employed in fishing and aquaculture, nearly all in developing countries (Figure 10.9.6). Production of fish is increasing worldwide (Figure 10.9.7). The growth results entirely from expansion of aquaculture (Figure 10.9.8). The capture of wild fish in the oceans and lakes has stagnated since the 1990s despite population growth and increased demand to consume fish.

The population of some fish species in the oceans and lakes has declined because of **overfishing**, which is capturing fish faster than they can reproduce. Overfishing has been particularly acute in the North Atlantic and Pacific oceans. Overfishing has reduced the population of tuna and swordfish by 90 percent in the past half-century, for example. The United Nations estimates that one-quarter of fish stocks have been overfished and one-half fully exploited, leaving only one-fourth underfished.

▼ 10.9.8 **AQUACULTURE FISH FARMING**
Japan.

10.10 Subsistence Agriculture and Population Growth

- Four strategies can increase food supply in developing countries.
- Increasing productivity and finding new sources are most promising.

Two issues discussed in earlier chapters influence the challenges faced by subsistence farmers. First, because of rapid population growth in developing countries (discussed in Chapter 5), subsistence farmers must feed an increasing number of people. Second, because of adopting the international trade approach to development (discussed in Chapter 9), subsistence farmers must grow food for export instead of for direct consumption. Four strategies have been identified to increase food supply.

EXPAND AGRICULTURAL LAND

Historically, world food production increased primarily by expanding the amount of land devoted to agriculture. When the world's population increased more rapidly during the Industrial Revolution beginning in the eighteenth century, pioneers could migrate to sparsely inhabited territory and cultivate the land.

New land might appear to be available, because only 11 percent of the world's land area is currently used for agriculture. But excessive or inadequate water makes expansion difficult. The expansion of agricultural land has been much slower than the increase of the human population for several decades (Figure 10.10.1).

▼10.10.1 **AGRICULTURAL LAND AND POPULATION GROWTH**

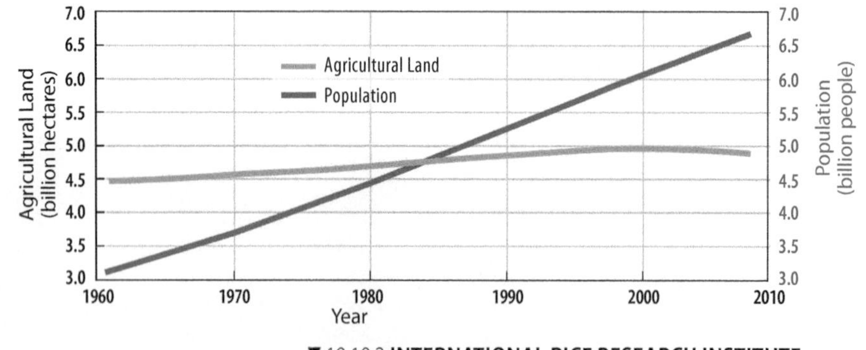

▼ 10.10.2 **INTERNATIONAL RICE RESEARCH INSTITUTE, HOME OF THE "GREEN REVOLUTION"**

INCREASE AGRICULTURAL PRODUCTIVITY

New agricultural practices have permitted farmers worldwide to achieve much greater yields from the same amount of land. The invention and rapid diffusion of more productive agricultural techniques during the 1960s and 1970s is called the **green revolution**.

Scientists began experiments during the 1950s to develop a higher-yield form of wheat. A decade later, the International Rice Research Institute created a "miracle" rice seed (Figure 10.10.2). The Rockefeller and Ford foundations sponsored many of the studies, and the program's director, Dr. Norman Borlaug, won the Nobel Peace Prize in 1970. More recently, scientists have developed new high-yield maize (corn). Scientists have continued to create higher-yield hybrids that are adapted to environmental conditions in specific regions.

The **green revolution** was largely responsible for preventing a food crisis in developing countries during the 1970s and 1980s. The new miracle seeds were diffused rapidly around the world. India's wheat production, for example, more than doubled in 5 years. After importing 10 million tons of wheat annually in the mid-1960s, India by 1971 had a surplus of several million tons.

Will these scientific breakthroughs continue in the twenty-first century? To take full advantage of the new "miracle seeds," farmers must use more fertilizer and machinery, both of which depend on increasingly expensive fossil fuels. To maintain the green revolution, governments in developing countries must allocate scarce funds to subsidize the cost of seeds, fertilizers, and machinery.

IMPROVED FOOD SOURCES

Improved food sources could come from:

- Higher protein cereal grains. People in developing countries depend on grains that lack certain proteins. Hybrids with higher protein content could achieve better nutrition without changing food-consumption habits.
- Palatability of rarely consumed foods. Some foods are rarely consumed because of taboos, religious values, and social customs. In developed countries, consumers avoid consuming recognizable soybean products like tofu and sprouts, but could be induced to eat soybeans shaped like burgers and franks (Figure 10.10.3).

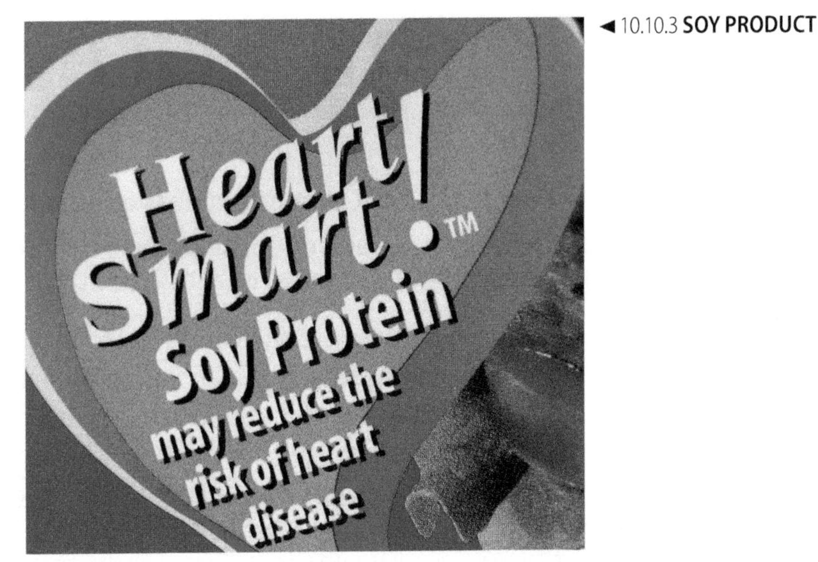

◄ 10.10.3 **SOY PRODUCT**

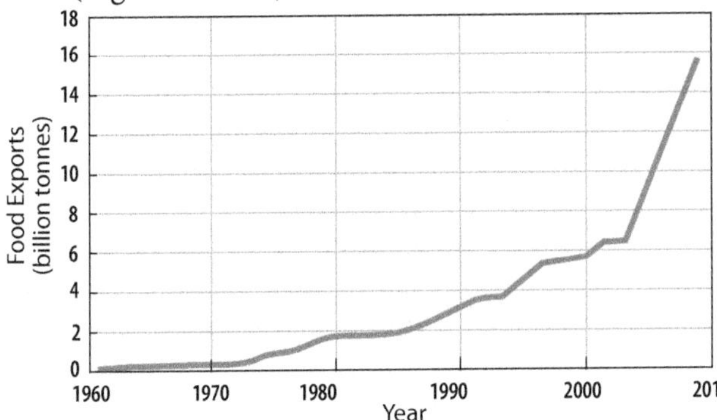

▲ 10.10.4 **WORLD FOOD EXPORTS**

EXPAND EXPORTS

Trade in food has increased rapidly, especially since 2000 (Figure 10.10.4). The three top export grains are wheat, maize (corn), and rice. Argentina, Brazil, the Netherlands and the United States are the four leading net exporters of agricultural products (Figures 10.10.5 and Figure 10.10.6). Japan, China, Russia, and the United Kingdom are the leading net importers.

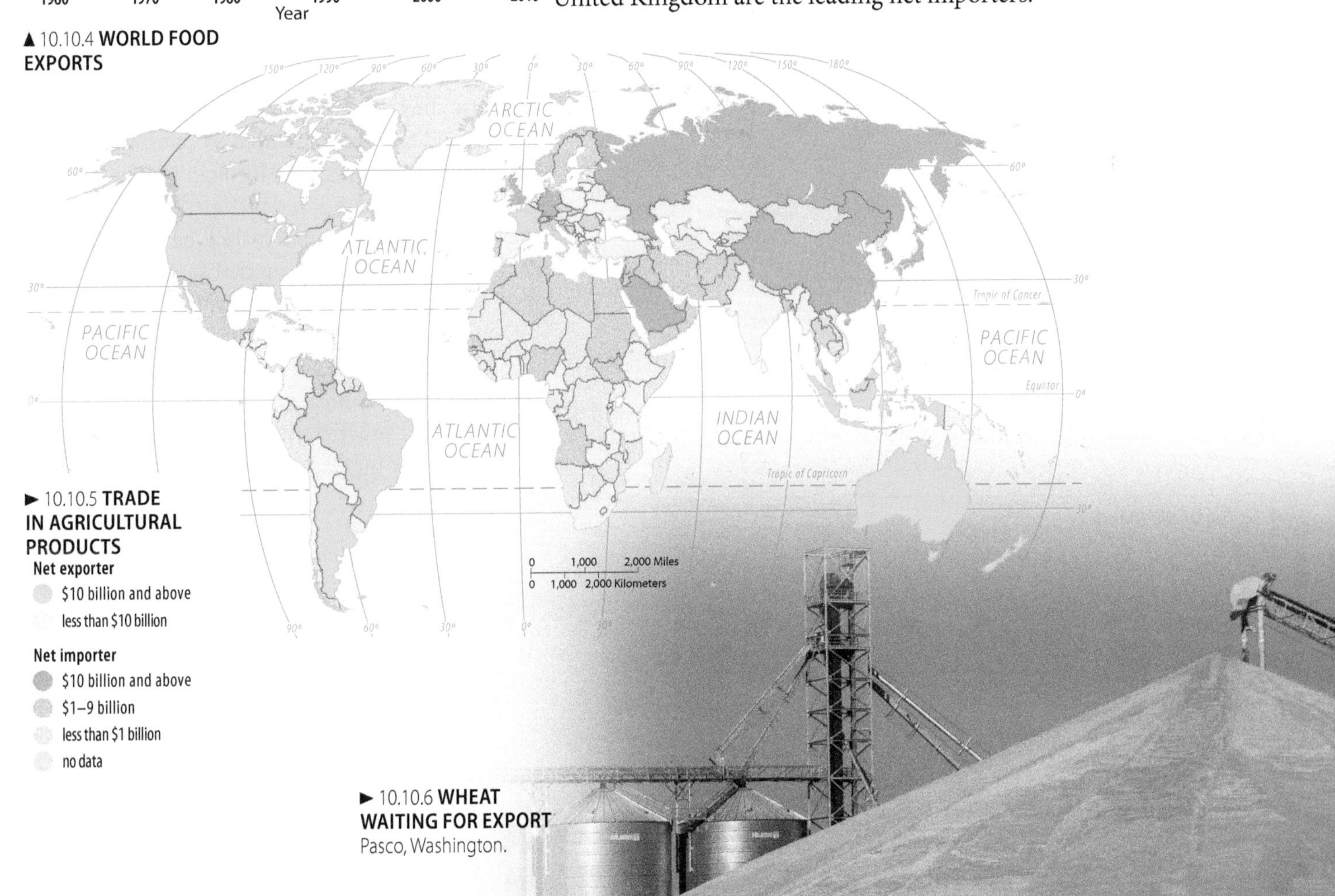

► 10.10.5 **TRADE IN AGRICULTURAL PRODUCTS**

Net exporter
- $10 billion and above
- less than $10 billion

Net importer
- $10 billion and above
- $1–9 billion
- less than $1 billion
- no data

► 10.10.6 **WHEAT WAITING FOR EXPORT** Pasco, Washington.

10.11 Commercial Agriculture and Market Forces

- **Farming is part of agribusiness in developed countries.**
- **Because of overproduction, farmers in developed countries may receive government subsidies to reduce output.**

The system of commercial farming found in developed countries is called **agribusiness**, because the family farm is not an isolated activity but is integrated into a large food-production industry. Agribusiness encompasses such diverse enterprises as tractor manufacturing, fertilizer production, and seed distribution. This type of farming responds to market forces rather than to feeding the farmer. Geographers use the von Thünen model to help explain the importance of proximity to market in the choice of crops on commercial farms (Figure 10.11.1).

Farmers are less than 2 percent of the U.S. labor force, but around 20 percent of U.S. labor works in food production and services related to agribusiness—food processing, packaging, storing, distributing, and retailing. Although most farms are owned by individual families, many other aspects of agribusiness are controlled by large corporations.

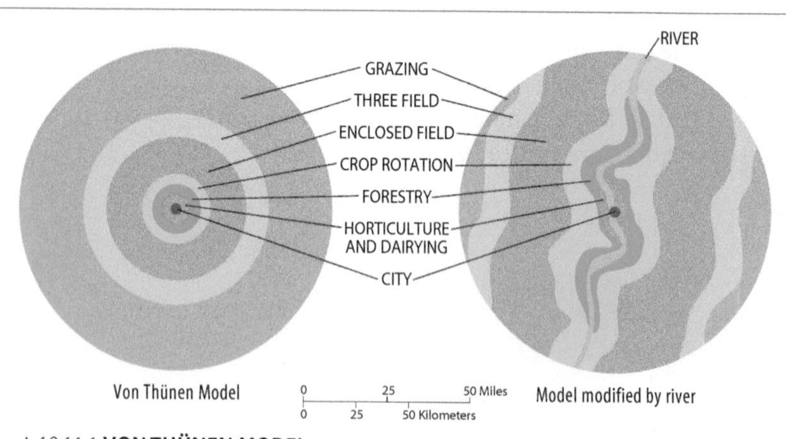

▲ 10.11.1 **VON THÜNEN MODEL**
Johann Heinrich von Thünen, a farmer in northern Germany, proposed a model to explain the importance of proximity to market in the choice of crops on commercial farms. The von Thünen model was first proposed in 1826 in a book titled *The Isolated State*. According to the model, which was later modified by geographers, a commercial farmer initially considers which crops to cultivate and which animals to raise based on market location. Von Thünen based his general model of the spatial arrangement of different crops on his experiences as owner of a large estate in northern Germany. He found that specific crops were grown in different rings around the cities in the area.

▼ 10.11.2 **DAIRY COWS**
California.

PRODUCTIVITY CHALLENGES

The experience of dairy farming in the United States demonstrates the growth in productivity (Figure 10.11.2). The number of dairy cows has declined since 1960 but production has increased, because yield per cow has tripled (Figure 10.11.3).

Commercial farmers suffer from low incomes because they are capable of producing much more food than is demanded by consumers in developed countries. Although the food supply has increased in developed countries, demand has remained constant, because of low population growth and market saturation.

A surplus of food can be produced because of widespread adoption of efficient agricultural practices (Figure 10.11.4). New seeds, fertilizers, pesticides, mechanical equipment, and management practices have enabled farmers to obtain greatly increased yields per area of land.

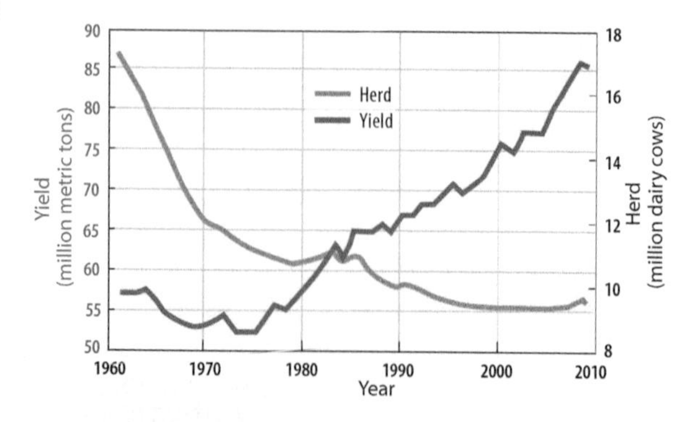

▲ 10.11.3 **U.S. DAIRY PRODUCTIVITY**

GOVERNMENT SUBSIDIES

The U.S. government has three policies that are supposed to address the problem of excess productive capacity:

- Farmers are encouraged to avoid producing crops that are in excess supply. Because soil erosion is a constant threat, the government encourages planting fallow crops, such as clover, to restore nutrients to the soil and to help hold the soil in place. These crops can be used for hay, forage for pigs, or to produce seeds for sale.
- The government pays farmers when certain commodity prices are low. The government sets a target price for the commodity and pays farmers the difference between the price they receive in the market and a target price set by the government as a fair level for the commodity. The target prices are calculated to give farmers the same price for the commodity today as in the past, when compared to other consumer goods and services.
- The government buys surplus production and sells or donates it to foreign governments. In addition, low-income Americans receive food stamps in part to stimulate their purchase of additional food.

Farming in Europe is subsidized even more than in the United States. Government policies in developed countries point out a fundamental irony in worldwide agricultural patterns. In developed countries, farmers are encouraged to grow less food, whereas some developing countries struggle to increase food production to match the rate of growth in the population.

◄ 10.11.4 **AGRIBUSINESS IN DAIRY FARMING**
(top) Large-scale milking; (below) transportation from farm to processing; (below left) processing; (below right) bottling; and (bottom) retailing.

10.12 Sustainable Agriculture

- Sustainable agriculture and organic farming rely on sensitive land management.
- Sustainable agriculture also limits use of chemicals and integrates crops and livestock.

Some commercial farmers are converting their operations to **sustainable agriculture**, an agricultural practice that preserves and enhances environmental quality. An increasingly popular form of sustainable agriculture is organic farming.

▼ 10.12.1 **SHARE OF THE WORLD'S ORGANIC FARMING**

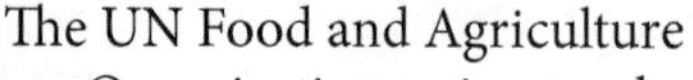

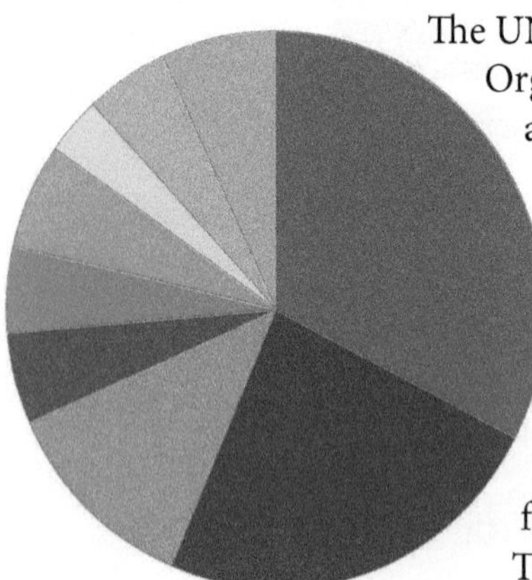

The UN Food and Agriculture Organization estimates the share of agricultural land farmed through organic practices at 0.29 percent worldwide (3/10 of 1 percent). Australia is the world leader, with one-third of the world's total organic farmland (Figure 10.12.1). Europe has the highest percentage of farmland devoted to organic farming (Figure 10.12.2).

Three principal practices distinguish sustainable agriculture (and at its best, organic farming) from conventional agriculture.

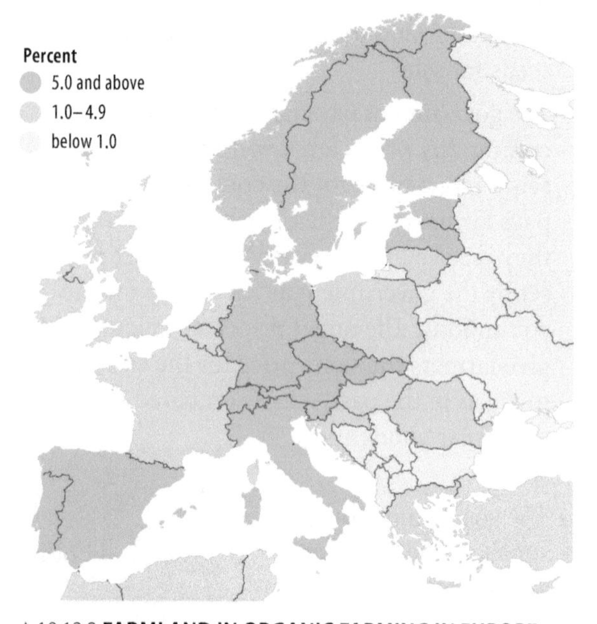

▲ 10.12.2 **FARMLAND IN ORGANIC FARMING IN EUROPE**

SENSITIVE LAND MANAGEMENT

Sustainable agriculture protects soil in part through **ridge tillage**, which is a system of planting crops on ridge tops. Crops are planted on 10- to 20-centimeter (4- to 8-inch) ridges that are formed during cultivation or after harvest. The crop is planted on the same ridges, in the same rows, year after year. Ridge tillage is attractive for two main reasons—lower production costs and greater soil conservation.

Production costs are lower with ridge tillage in part because it requires less investment in tractors and other machinery than conventional planting. An area that would be prepared for planting under conventional farming with three to five tractors can be prepared for ridge tillage with only one or two tractors. The primary tillage tool is a row-crop cultivator that can form ridges. There is no need for a plow, or field cultivator, or a 300-horsepower four-wheel-drive tractor.

With ridge tillage, the space between rows needs to match the distance between wheels of the machinery. If 75 centimeters (30 inches) are left between rows, tractor tires will typically be on 150-centimeter (60-inch) centers and combine wheels on 300-centimeter (120-inch) centers. Wheel spacers are available from most manufacturers to fit the required spacing (Figure 10.12.3).

Ridge tillage features a minimum of soil disturbance from harvest to the next planting. A compaction-free zone is created under each ridge and in some row middles. Keeping the trafficked area separate from the crop-growing area improves soil properties. Over several years the soil will tend to have increased organic matter, greater water holding capacity, and more earthworms. The channels left by earthworms and decaying roots enhance drainage.

Ridge tillage compares favorably with conventional farming for yields while lowering the cost of production. Although more labor intensive than other systems, it is profitable on a per-acre basis. In Iowa, for example, ridge tillage has gained favor for production of organic and herbicide-free soybeans, which sell for more than regular soybeans.

LIMITED USE OF CHEMICALS

In conventional agriculture, seeds are often genetically modified to survive when herbicides and insecticides are sprayed on the fields to kill weeds and insects. These are known as "Roundup-Ready" seeds, because the herbicide's creator Monsanto Corp. sells it under the brand name "Roundup." Aside from adverse impacts on soil and water quality, widespread use of "Roundup-Ready" seeds is causing some weeds to become resistant to the herbicide.

Sustainable agriculture, on the other hand, involves application of limited if any herbicides to control weeds. In principle, farmers can control weeds without chemicals, although it requires additional time and expense that few farmers can afford. Researchers have found that combining mechanical weed control with some chemicals yields higher returns per acre than relying solely on one of the two methods.

Ridge tilling also promotes decreased use of chemicals, which can be applied only to the ridges and not the entire field. Combining herbicide banding—which applies chemicals in narrow bands over crop rows—with cultivating may be the best option for many farmers.

INTEGRATED CROP AND LIVESTOCK

Sustainable agriculture attempts to integrate the growing of crops and the raising of livestock as much as possible at the level of the individual farm. Animals consume crops grown on the farm and are not confined to small pens.

In conventional farming, integration between crops and livestock generally takes place through intermediaries rather than inside an individual farm. That is, many farmers in the mixed crop and livestock region actually choose to only grow crops or only raise animals. They sell their crops off the farm or purchase feed for their animals from outside suppliers.

Sustainable agriculture is sensitive to the complexities of interdependencies between crops and livestock:

- Herd size and distribution. The correct number and distribution of livestock for the area is determined based on the landscape and forage sources. Prolonged concentration of livestock in a specific location can result in permanent loss of vegetative cover, so the farmer needs to move the animals.
- Animal confinement. The moral and ethical debate regarding the welfare of confined livestock is particularly intense. From a practical perspective, manure from non-confined animals can contribute to soil fertility.
- Management in extreme weather. Herd size may need to be reduced during periods of short- and long-term droughts.
- Flexible feeding and marketing. Feed costs are the largest single variable cost in livestock operation. Feed costs can be kept to a minimum by monitoring animal condition and performance and understanding seasonal variations in feed and forage quality on the farm.

▼ 10.12.3 **ORGANIC FARMING IN EUROPE**

A country's agriculture remains one of the best measures of the level of development and standard of living. Despite major changes, agriculture in developing countries still employs a large percentage of the population, and producing food for local survival is still paramount.

Key Questions

What do people eat?

- Agriculture originated in multiple hearths and diffused to numerous places simultaneously.
- What people eat is influenced by a combination of level of development, cultural preferences, and environmental constraints.
- One in eight humans is undernourished.

How is agriculture distributed?

- Several agricultural regions can be identified based on farming practices.
- Subsistence agriculture, typical of developing regions, involves growing food for one's own consumption.
- Commercial agriculture, typical of developed countries, involves growing food to sell off the farm.
- Commercial agriculture involves larger farms, fewer farmers, and more mechanization than does subsistence agriculture.

What challenges does agriculture face?

- Subsistence agriculture faces distinctive economic challenges resulting from rapid population growth and pressure to adopt international trade strategies to promote development.
- Commercial agriculture faces distinct challenges resulting from access to markets and overproduction.
- Sustainable farming plays an increasing role in the preservation and enhancement of environmental quality.

Thinking Geographically

Assume that the United States constitutes one agricultural market centered around the largest city New York.

1. To what extent can the major agricultural regions of the United States be viewed as irregularly shaped rings around the market center, as von Thünen applied to southern Germany?

Review in Chapter 5 the concept of overpopulation: the number of people in an area exceeds the capacity of the environment to support life at a decent standard of living (Figure 10.CR.1).

2. What agricultural regions face rapid population growth but have relatively limited capacities to support intensive food production?

Compare world distributions of maize, wheat, and rice production.

3. To what extent do differences in the distribution of these crops derive from environmental conditions and to what extent from food preferences and other cultural values?

On the Internet

The United Nations Food and Agricultural Organization maintains a comprehensive database at **www.fao.org** or scan the QR on the first page of this chapter.

The principal source of data about U.S. agriculture is the U.S. Department of Agriculture's National Agricultural Statistical Service, at **www.nass.usda.gov.**

Global organic farming statistics are at the Research Institute of Organic Agriculture, at **www.organic-world.net.**

Information about sustainable agriculture is found through Sustainable Agriculture Research and Education at **www.sare.org.**

◄ 10.CR.1 **OVERPOPULATION IN MALI**
A desert region can be sparsly inhabited yet overpopulated if it has rapid population growth and limited resources, as in the case of Mali.

InteractiveMapping

AGRICULTURE AND CLIMATE REGIONS

Agriculture prcatices vary within East Asia.

Launch MapMaster East Asia in MasteringGEOGRAPHY™

Select *Climate* from the *Physical Environment* menu.

Next select *Agricultural Regions* from the *Economic* menu.

Which practice matches most closely with which climate region?

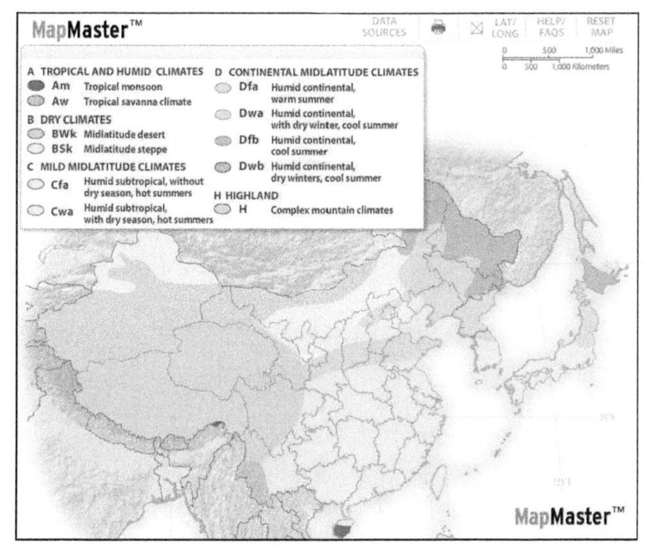

Explore

BENTON COUNTY, INDIANA

Use Google Earth to explore the agricultural landscape of Benton County, Indiana.

Fly to: *Freeland Park, Indiana*

Click to show historical imagery

Move the cursor to: *4/3/2005*, then to *8/12/2007*, then to *10/06/2009*.

Note the change in the predominant color of the fields at these three dates.

Why does the predominant color of the fields change from April to August and then to October?

Key Terms

Agribusiness
Commercial agriculture characterized by the integration of different steps in the food-processing industry, usually through ownership by large corporations.

Agriculture
The deliberate effort to modify a portion of Earth's surface through the cultivation of crops and the raising of livestock for sustenance or economic gain.

Aquaculture (or aquafarming)
The cultivation of seafood under controlled conditions.

Cereal grain
A grass yielding grain for food.

Commercial agriculture
Agriculture undertaken primarily to generate products for sale off the farm.

Crop
Grain or fruit gathered from a field as a harvest during a particular season.

Crop rotation
The practice of rotating use of different fields from crop to crop each year, to avoid exhausting the soil.

Dietary energy consumption
The amount of food that an individual consumes.

Food security
Physical, social, and economic access at all times to safe and nutritious food sufficient to meet dietary needs and food preferences for an active and healthy life.

Grain
Seed of a cereal grass.

Green revolution
Rapid diffusion of new agricultural technology, especially new high-yield seeds and fertilizers.

Intensive subsistence agriculture
A form of subsistence agriculture in which farmers must expend a relatively large amount of effort to produce the maximum feasible yield from a parcel of land.

Milkshed
The ring surrounding a city from which milk can be supplied without spoiling.

Overfishing
Capturing fish faster than they can reproduce.

Pastoral nomadism
A form of subsistence agriculture based on herding domesticated animals.

Plantation
A large farm in tropical and subtropical climates that specializes in the production of one or two crops for sale, usually to a more developed country.

Ranching
A form of commercial agriculture in which livestock graze over an extensive area.

Ridge tillage
System of planting crops on ridge tops in order to reduce farm production costs and promote greater soil conservation.

Shifting cultivation
A form of subsistence agriculture in which people shift activity from one field to another; each field is used for crops for a relatively few years and left fallow for a relatively long period.

Slash-and-burn agriculture
Another name for shifting cultivation, so named because fields are cleared by slashing the vegetation and burning the debris.

Subsistence agriculture
Agriculture designed primarily to provide food for direct consumption by the farmer and the farmer's family.

Sustainable agriculture
Farming methods that preserve long-term productivity of land and minimize pollution, typically by rotating soil-restoring crops with cash crops and reducing inputs of fertilizer and pesticides.

Swidden
A patch of land cleared for planting through slashing and burning.

Taboo
A restriction on behavior imposed by social custom.

Terroir
French term for the contribution of a location's distinctive physical features to the way food tastes, similar to the English expressions "grounded" or "sense of place."

Truck farming
Commercial gardening and fruit farming, so named because truck was a Middle English word meaning bartering or the exchange of commodities.

Undernourishment
Dietary energy consumption that is continuously below the minimum requirement for maintaining a healthy life and carrying out light physical activity.

Wet rice
Rice planted on dryland in a nursery and then moved to a deliberately flooded field to promote growth.

▶ LOOKING AHEAD

Agriculture is practiced throughout the inhabited world, because the need for food is universal. Industry—the manufacturing of goods in factories—is much more highly clustered in a handful of regions.

11 Industry

Manufacturing jobs are viewed as a special asset by communities around the world. They are seen as the "engine" of economic growth and prosperity. Different communities possess distinctive assets in attracting particular types of industries as well as challenges in retaining them.

A generation ago, industry was highly clustered in a handful of communities within developed countries, but industry has diffused to more communities, including some in developing countries. Meanwhile, a loss of manufacturing jobs has caused economic problems for communities in developed countries, like the United States, traditionally dependent on them.

Where is industry clustered?

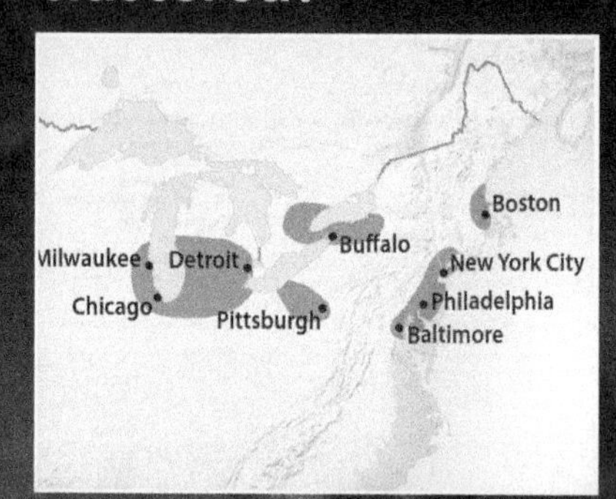

STEEL WORKS, LIAONING, CHINA

What situation factors influence industrial location?

- 11.3 **Situation Factors in Locating Industry**
- 11.4 **Changing Steel Production**
- 11.5 **Changing Auto Production**
- 11.6 **Ship by Boat, Rail, Truck, or Air?**

SCAN TO ACCESS U.S. LABOR STATISTICS

What site factors influence industrial location?

- 11.7 **Site Factors in Industry**
- 11.8 **Textile and Apparel Production**
- 11.9 **Emerging Industrial Regions**

11.1 The Industrial Revolution

- The Industrial Revolution transformed how goods are produced for society.
- The United Kingdom was home to key events in the Industrial Revolution.

The modern concept of industry—meaning the manufacturing of goods in a factory—originated in northern England and southern Scotland during the second half of the eighteenth century. From there, industry diffused in the nineteenth century to Europe and to North America and in the twentieth century to other regions.

ORIGINS OF THE INDUSTRIAL REVOLUTION

The **Industrial Revolution** was a series of improvements in industrial technology that transformed the process of manufacturing goods. Prior to the Industrial Revolution, industry was geographically dispersed across the landscape. People made household tools and agricultural equipment in their own homes or obtained them in the local village. Home-based manufacturing was known as the **cottage industry** system (Figure 11.1.1).

The term *Industrial Revolution* is somewhat misleading:

- The transformation was far more than industrial, and it did not happen overnight.
- The Industrial Revolution resulted in new social, economic, and political inventions, not just industrial ones.
- The changes involved a gradual diffusion of new ideas and techniques over decades, rather than an instantaneous revolution.

▲▼ 11.1.1 **TRANSFORMATION OF AN INDUSTRY**
(top) In the early nineteenth century, the textile industry was a cottage industry based on people spinning and weaving by hand in their homes. (bottom) By the middle of the century, the industry had become based in factories and mills. In this interior view of a cotton mill in 1835 girls and women tend carding, drawing, and roving machinery.

Nonetheless, the term is commonly used to define the process that began in the United Kingdom in the late 1700s.

The invention most important to the development of factories was the steam engine, patented in 1769 by James Watt, a maker of mathematical instruments in Glasgow, Scotland (Figure 11.1.2). Watt built the first useful steam engine, which could pump water far more efficiently than the watermills then in common use, let alone human or animal power. The large supply of steam power available from James Watt's steam engines induced firms to concentrate all steps in a manufacturing process in one building attached to a single power source.

▲ 11.1.2 **JAMES WATT'S STEAM ENGINE**
Steam injected in a cylinder (left side of engine) pushes a piston attached to a crankshaft that drives machinery (right side of engine).

TRANSFORMATION OF KEY INDUSTRIES

Industries impacted by the Industrial Revolution included:

- **Coal:** The source of energy to operate the ovens and the steam engines. Wood, the main energy source prior to the Industrial Revolution, was becoming scarce in England because it was in heavy demand for construction of ships, buildings, and furniture, as well as for heat. Manufacturers turned to coal, which was then plentiful in England.
- **Iron:** The first industry to benefit from Watt's steam engine. The usefulness of iron had been known for centuries, but it was difficult to produce because ovens had to be constantly heated, something the steam engine could do (Figure 11.1.3).
- **Transportation:** Critical for diffusing the Industrial Revolution. First canals and then railroads enabled factories to attract large numbers of workers, bring in bulky raw materials such as iron ore and coal, and ship finished goods to consumers (Figure 11.1.4).
- **Textiles:** Transformed from a dispersed cottage industry to a concentrated factory system during the late eighteenth century, as illustrated in Figure 11.1.1. In 1768, Richard Arkwright, a barber and wigmaker in Preston, England, invented machines to untangle cotton prior to spinning. Too large to fit inside a cottage, spinning frames were placed inside factories near sources of rapidly flowing water, which supplied the power. Because the buildings resembled large watermills, they were known as mills.
- **Chemicals:** An industry created to bleach and dye cloth. In 1746, John Roebuck and Samuel Garbett established a factory to bleach cotton with sulfuric acid obtained from burning coal. When combined with various metals, sulfuric acid produced another acid called vitriol, useful for dying clothing.
- **Food processing:** Essential to feed the factory workers no longer living on farms. In 1810, French confectioner Nicholas Appert started canning food in glass bottles sterilized in boiling water.

▲ 11.1.3 **IRON ORE SMELTING**
Coalbrookdale by Night, an 1801 painting by Philip James de Loutherbourg, depicts the Coalbrookdale Company's iron ore smelter in Ironbridge, England. The painting is in London's Science Museum.

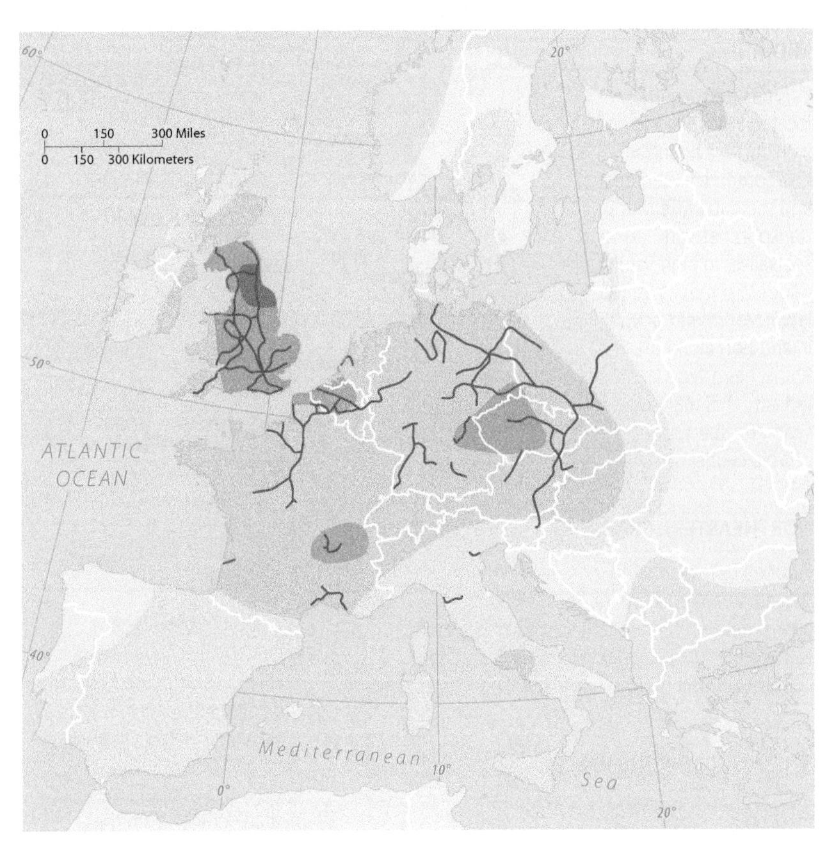

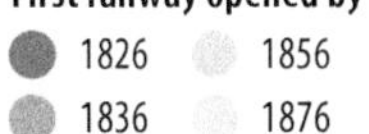

▲ 11.1.4 **DIFFUSION OF RAILROADS**
Europe's political problems retarded the diffusion of the railroad. Cooperation among small neighboring states was essential to build an efficient rail network and to raise money for constructing and operating the system. Because such cooperation could not be attained, railroads in some parts of Europe were delayed 50 years after their debut in the United Kingdom.

11.2 Distribution of Industry

► **Three-fourths of the world's manufacturing is clustered in three regions.**
► **The major industrial regions are divided into subareas.**

Industry is concentrated in three of the nine world regions discussed in Chapter 9: Europe (Figures 11.2.1 and 11.2.2), East Asia (Figure 11.2.3), and North America (Figure 11.2.4). Each of the three regions accounts for roughly one-fourth of the world's total industrial output. Outside these three regions the leading industrial producers are Brazil and India.

▼ 11.2.1 **EUROPE'S INDUSTRIAL AREAS**
Europe was the first region to industrialize during the nineteenth century. Numerous industrial centers developed in Europe as countries competed with each other for supremacy.

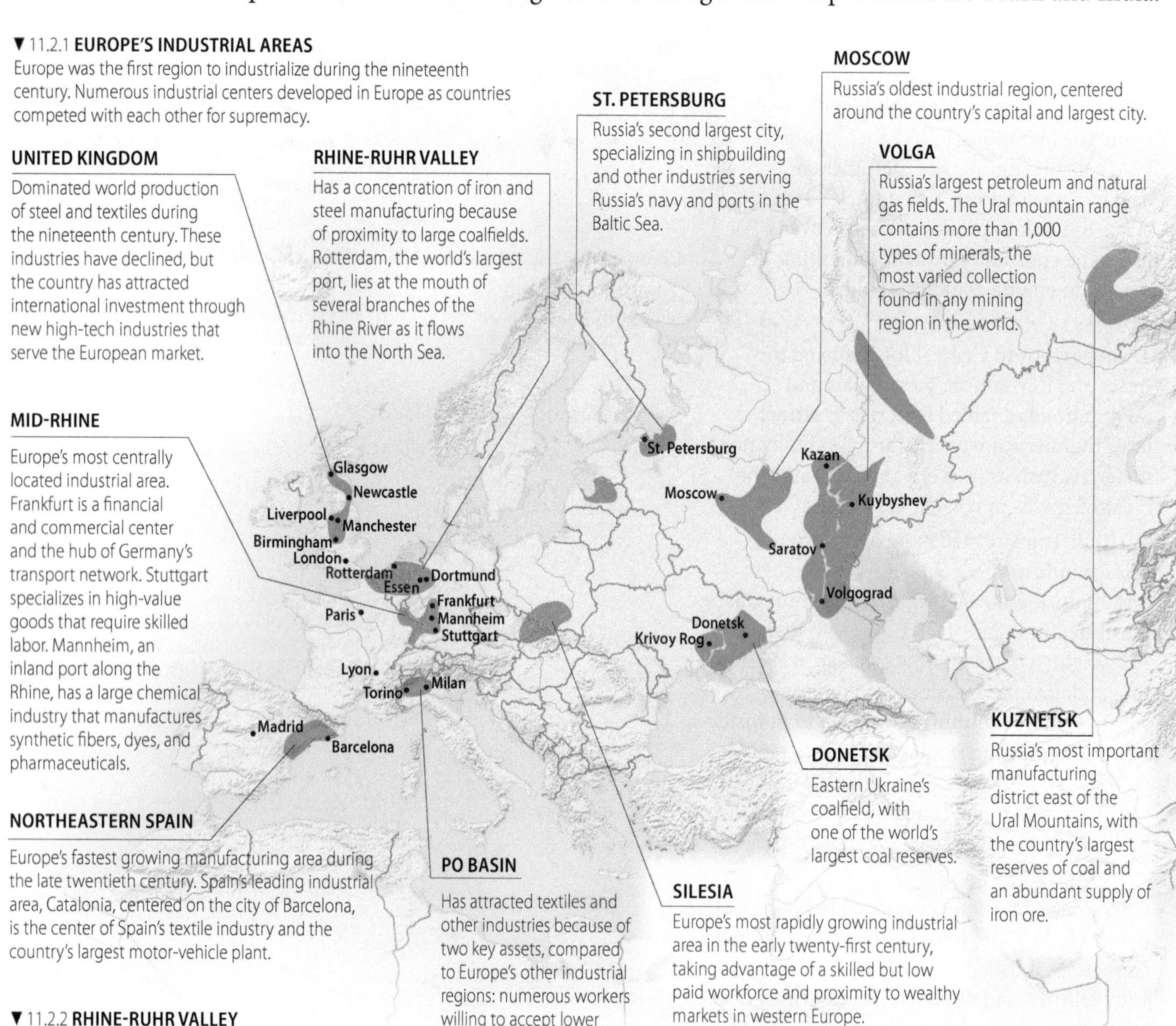

▼ 11.2.2 **RHINE-RUHR VALLEY**

▼ 11.2.3 **EAST ASIA'S INDUSTRIAL AREAS**

East Asia became an important industrial region in the second half of the twentieth century, beginning with Japan. Into the twenty-first century, China has emerged as the world's leading manufacturing country by most measures.

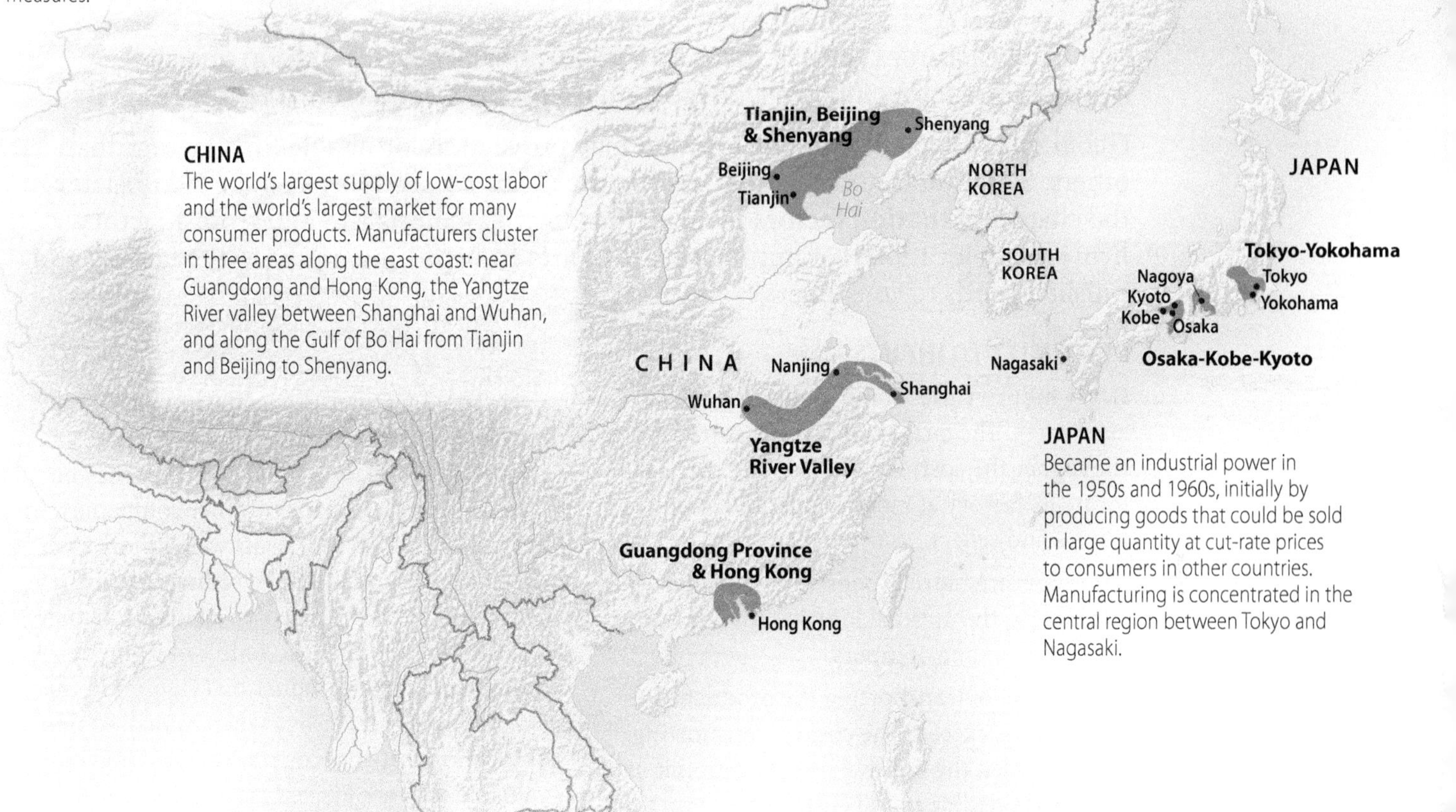

▼ 11.2.4 **NORTH AMERICA'S INDUSTRIAL AREAS**

Industry arrived a bit later in North America than in Europe, but it grew much faster in the nineteenth century. North America's manufacturing was traditionally highly concentrated in northeastern United States and southeastern Canada. In recent years, manufacturing has relocated to the South, lured by lower wages and legislation that has made it difficult for unions to organize factory workers.

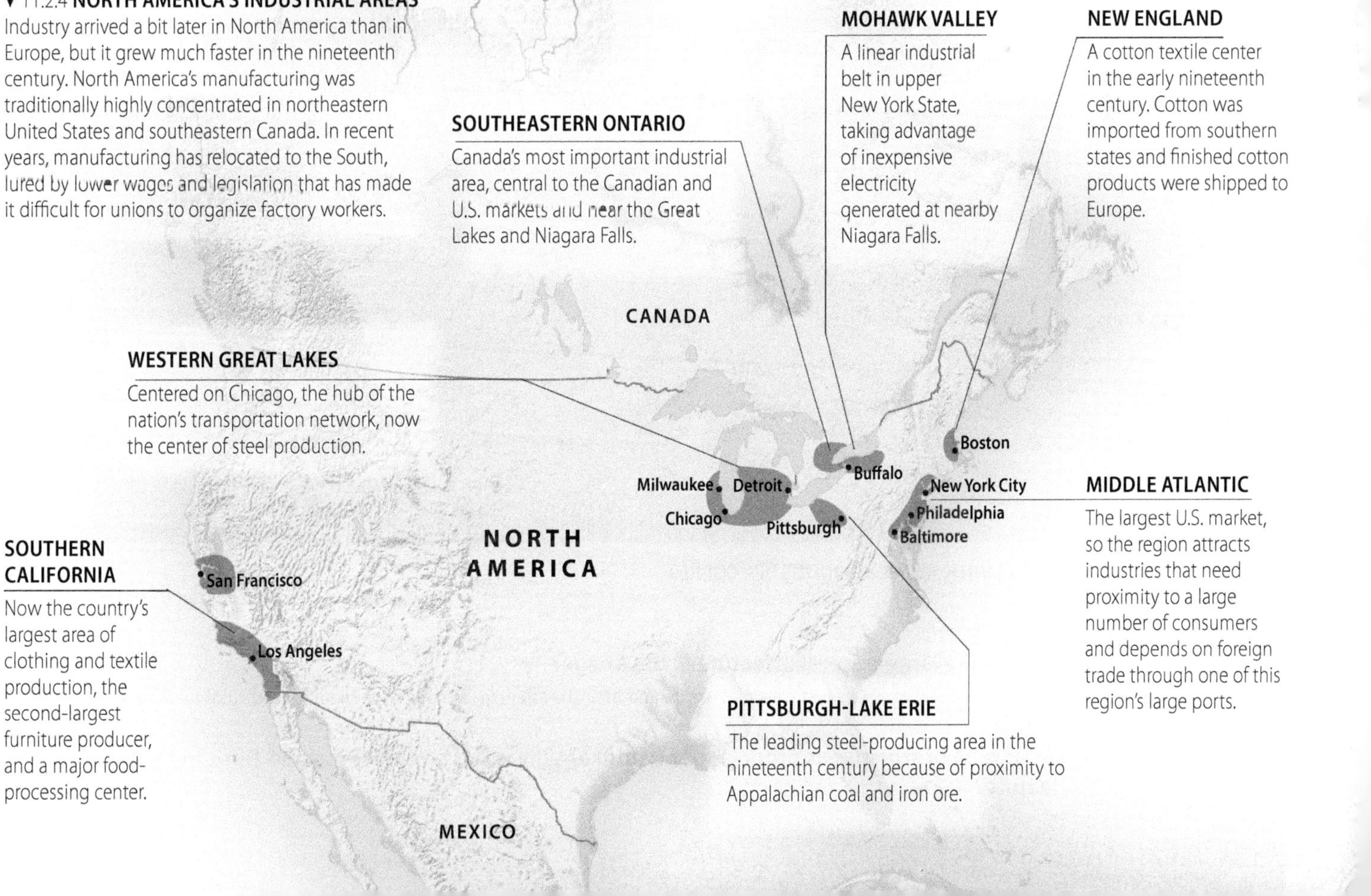

11.3 Situation Factors in Locating Industry

- **A manufacturer typically faces two geographical costs: situation and site.**
- **Situation factors involve transporting materials to and from a factory.**

Geographers explain why one location may prove more profitable for a factory than others. Situation factors are discussed in the next four sections, and site factors later in the chapter. **Situation factors** involve transporting materials to and from a factory. A firm seeks a location that minimizes the cost of transporting inputs to the factory and finished goods to consumers.

PROXIMITY TO INPUTS

Every industry uses some inputs and sells to customers. The farther something is transported, the higher the cost, so a manufacturer tries to locate its factory as close as possible to both buyers and sellers.

- If inputs are more expensive to transport than products, the optimal location for a factory is near the source of inputs.
- If the cost of transporting the product to customers exceeds the cost of transporting inputs, then the optimal plant location is as close as possible to the customer.

Every manufacturer uses some inputs. These may be resources from the physical environment (minerals, wood, or animals), or they may be parts or materials made by other companies. An industry in which the inputs weigh more than the final products is a **bulk-reducing industry**. To minimize transport costs, a bulk-reducing industry locates near the source of its inputs. An example is copper production (Figure 11.3.1). Copper ore is very heavy when mined, so mills that concentrate the copper by removing less valuable rock are located close to the mine.

▲ 11.3.1 **BULK-REDUCING INDUSTRY: COPPER**

Use Google Earth to explore Mount Isa, Australia's leading copper production center, which includes mining and smelting.

Fly to: *Parkside, Mount Isa, Australia.*

What is the large crater-like feature in the image?

Locate the plume of smoke just west of Parkside and drag the street view icon to the main road west of the Parkside label.

What type of structure is producing the smoke? Why is this structure located close to the other feature?

► 11.3.2 **BULK-GAINING INDUSTRY: BEER BOTTLING**
The two best-selling beer companies in the United States together operate 21 breweries.
Launch MapMaster North America in MasteringGEOGRAPHY
Select: *Political then Cities*
Select: *Economic* then *Breweries*
Select: *Population* then *Population Density*
Draw a line between Winnipeg and San Antonio.
How many of the 21 breweries are east of the line? Is population density higher or lower east of the line?

PROXIMITY TO MARKETS

For many firms, the optimal location is close to markets, where the product is sold. The cost of transporting goods to consumers is a critical location factor for three types of industries:

- **Bulk-gaining industries** make something that gains volume or weight during production. A prominent example is beverage bottling. Empty cans and bottles are brought to the bottler, filled with the soft drink or beer, and shipped to consumers. The principal input placed in the beverage container is water, which is relatively bulky, heavy, and expensive to transport (Figure 11.3.2).
- **Single-market manufacturers** make products sold primarily in one location, so they also cluster near their markets. For example, the manufacturers of parts for motor vehicles are specialized manufacturers often with only one or two customers—the major carmakers such as General Motors and Toyota (Figure 11.3.3).
- **Perishable products** must be located near their markets so their products can reach consumers as rapidly as possible (Figure 11.3.4).

▲ 11.3.3 **SINGLE-MARKET MANUFACTURER: CAR PARTS**
General Motors worker at a plant in Parma, Ohio, stamps car parts destined for GM assembly plants.

◄ 11.3.4 **PERISHABLE PRODUCT: MILK PRODUCTION**
Food producers such as bakers and milk bottlers must locate near their customers to assure rapid delivery, because people do not want stale bread or sour milk.

11.4 Changing Steel Production

- Steel production has traditionally been a prominent example of a bulk-reducing industry.
- Restructuring has made steel production more sensitive to market locations.

The two principal inputs in steel production are iron ore and coal. Steelmaking is a bulk-reducing industry that traditionally located to minimize the cost of transporting these two inputs.

U.S. STEEL MILLS

The distribution of steelmaking in the United States demonstrates that when the source of inputs or the relative importance of inputs changes, the optimal location for the industry changes:

- **Mid-nineteenth century:** The U.S. steel industry concentrated around Pittsburgh in southwestern Pennsylvania, where iron ore and coal were both mined. The area no longer has steel mills, but it remains the center for research and administration.
- **Late nineteenth century:** Steel mills were built around Lake Erie, in the Ohio cities of Cleveland, Youngstown, and Toledo, and near Detroit (Figure 11.4.1). The locational shift was largely influenced by the discovery of rich iron ore in the Mesabi Range, a series of low mountains in northern Minnesota. This area soon became the source for virtually all iron ore used in the U.S. steel industry. The ore was transported by way of Lake Superior, Lake Huron, and Lake Erie. Coal was shipped from Appalachia by train.
- **Early twentieth century:** Most new steel mills were located near the southern end of Lake Michigan. The main raw materials continued to be iron ore and coal, but changes in steelmaking required more iron ore in proportion to coal. Thus, new steel mills were built closer to the Mesabi Range to minimize transportation cost. Coal was available from nearby southern Illinois, as well as from Appalachia.

- Integrated steel mills
- Historical location of steel industry
- Major iron ore deposit
- Major bituminous coal deposit

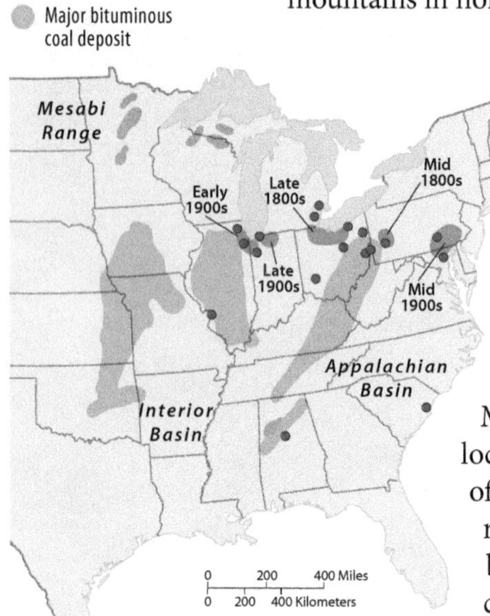

▲ 11.4.1 **INTEGRATED STEEL MILLS IN THE UNITED STATES**
Integrated steel mills are densely clustered near the southern Great Lakes, especially Lake Erie and Lake Michigan. Historically, the most critical factor in siting a steel mill was to minimize transportation cost for raw materials, especially heavy, bulky iron ore and coal. Most surviving mills are in the Midwest to maximize access to consumers.

- **Mid-twentieth century:** Most large U.S. steel mills built during the first half of the twentieth century were located in communities near the East and West coasts, including Baltimore, Los Angeles, and Trenton, New Jersey. These coastal locations partly reflected further changes in transportation cost. Iron ore increasingly came from other countries, especially Canada and Venezuela, and locations near the Atlantic and Pacific oceans were more accessible to those foreign sources. Further, scrap iron and steel—widely available in the large metropolitan areas of the East and West coasts—had become an important input in the steel-production process.
- **Late twentieth century:** Many steel mills in the United States closed (Figure 11.4.2). Most of the survivors were around southern Lake Michigan and along the East Coast.

▼ 11.4.2 **CLOSED STEEL MILL, BETHLEHEM, PENNSYLVANIA**
A large percentage of integrated steel mills in North America and Europe have closed in recent decades. Steel production has moved to less developed countries and to mini-mills.

RESTRUCTURING THE STEEL INDUSTRY

The shift of world manufacturing to new industrial regions can be seen clearly in steel production (Figure 11.4.3). World steel production doubled between 1980 and 2010, from around 700 million to around 1,400 million metric tons. China was responsible for 600 million of the 700 million increase, and other developing countries (primarily India and South Korea) for another 100 million metric tons (Figure 11.4.4). Production in developed countries remained unchanged at approximately 500 million metric tons (Figure 11.4.5).

China's steel industry has grown in part because of access to the primary inputs of iron ore and coal. However, the primary factor in recent years has been increased demand by growing industries in China that use a lot of steel, such as motor vehicles

▲ 11.4.3 **STEEL MILL IN CHINA**
Shanxi Halxin steel mill.

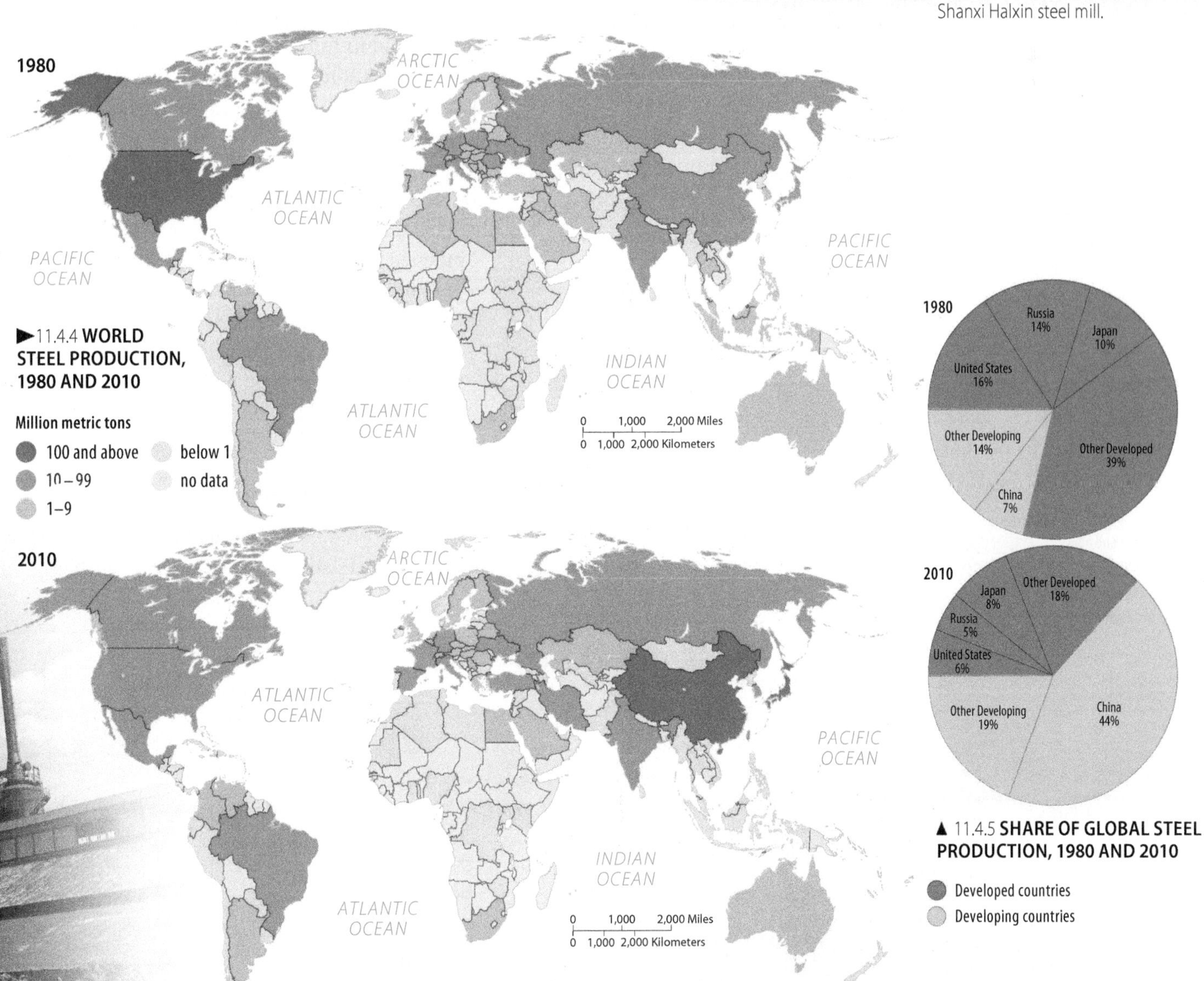

►11.4.4 **WORLD STEEL PRODUCTION, 1980 AND 2010**

▲ 11.4.5 **SHARE OF GLOBAL STEEL PRODUCTION, 1980 AND 2010**

11.5 Changing Auto Production

- ► **Motor vehicles are bulk-gaining products that are made near their markets.**
- ► **In the United States, most carmaking operations have clustered in auto alley.**

The motor vehicle is a prominent example of a fabricated metal product, described earlier as one of the main types of bulk-gaining industries. As a bulk-gaining industry, most motor vehicle production is concentrated near the markets for the vehicles.

GLOBAL DISTRIBUTION OF PRODUCTION

Carmakers put together vehicles at final assembly plants, using thousands of parts supplied by independent companies. Sixty percent of the world's final assembly plants are controlled by ten carmakers:

- 2 U.S.-based: Ford and GM.
- 4 Europe-based: Germany's Volkswagen, Italy's Fiat (which controls Chrysler), and France's Renault (which controls Nissan) and Peugeot.
- 4 Asia-based: Japan's Toyota, Honda, and Suzuki, and South Korea's Hyundai.

These companies operate assembly and parts plants in many countries. Nationality matters in terms of location of corporate headquarters, top managers, research facilities, and shareholders.

The world's three major industrial regions house 80 percent of the world's final assembly plants, including 40 percent in East Asia, 25 percent in Europe, and 15 percent in North America (Figure 11.5.1). Three-fourths of vehicles sold in North America are assembled in North America (Figure 11.5.2). Similarly most vehicles sold in Europe are assembled in Europe, most vehicles sold in Japan are assembled in Japan, and most vehicles sold in China are assembled in China.

Carmakers' assembly plants account for only around 30 percent of the value of the vehicles that bear their names. As a result of outsourcing, independent parts makers supply the other 70 percent of the value. Many of these parts are also made near their markets—the final assembly plants—especially the steel parts, which comprise more than half of the weight of vehicles (Figure 11.5.3).

On the other hand, many parts do not need to be manufactured close to the customer. For them, changing site factors are more important. Some locate in countries that have relatively low labor costs, such as Mexico, China, and Czech Republic.

▼ 11.5.1 **MOTOR VEHICLE PRODUCTION**

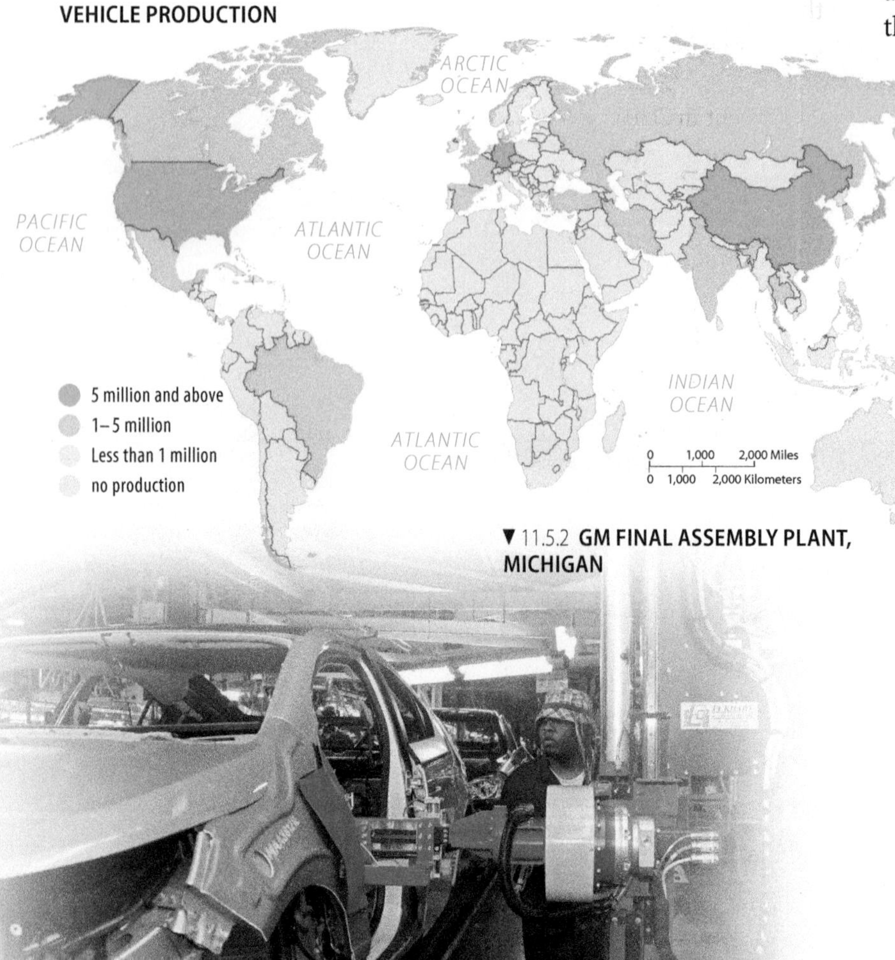

▼ 11.5.2 **GM FINAL ASSEMBLY PLANT, MICHIGAN**

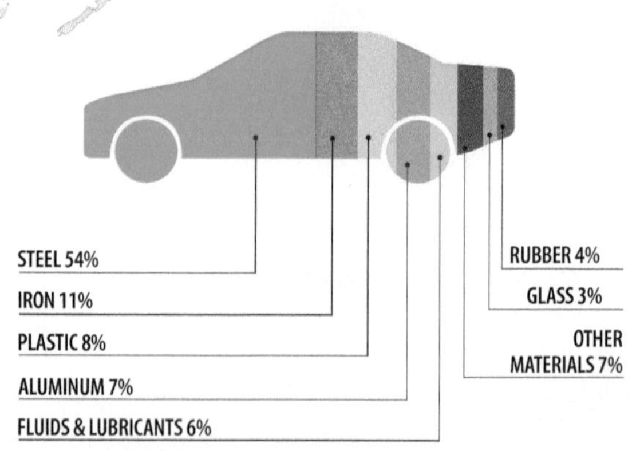

▲ 11.5.3 **COMPOSITION OF A CAR**

U.S. VEHICLE PRODUCTION

In the United States, vehicles are fabricated at about 50 final assembly plants, from parts made at several thousand other plants. Most of the assembly and parts plants are located in the interior of the country, between Michigan and Alabama, centered in a corridor known as "auto alley," formed by north-south interstate highways 65 and 75 (Figure 11.5.4).

For a bulk-gaining operation, such as a final assembly plant, the critical location factor is minimizing transportation to the market, in this case the 15 million North Americans who buy new vehicles each year. If a company has a product that is made at only one plant, and the critical location factor is to minimizing the cost of distribution throughout North America, then the optimal factory location is in the U.S. interior rather than on the East or West Coast.

Most parts makers also locate in auto alley to be near the final assembly plants. Seats, for example, are invariably manufactured within an hour of the final assembly plant. A seat is an especially large and bulky object, and carmakers do not want to waste valuable space in their assembly plants by piling up an inventory of them. Most engines, transmissions, and metal body parts are also produced within a couple hours of an assembly plant.

Within auto alley, U.S.-owned carmakers and suppliers have clustered in Michigan and nearby northern states, whereas foreign-owned carmakers and parts suppliers have clustered in the southern portion of auto alley (Figure 11.5.4). Chrysler, Ford, and General Motors are known as the Detroit 3 because their headquarters and research facilities are clustered in the Detroit area.

The share of U.S. sales accounted for by the Detroit 3 has declined from 75 percent in 1995 to 45 percent in 2010. Some "foreign" cars have turned out to have higher U.S. content than some cars sold by the Detroit 3 (Figure 11.5.4). Meanwhile, the declining fortunes of the Detroit 3 have resulted in closure of many of the plants in the northern part of auto alley.

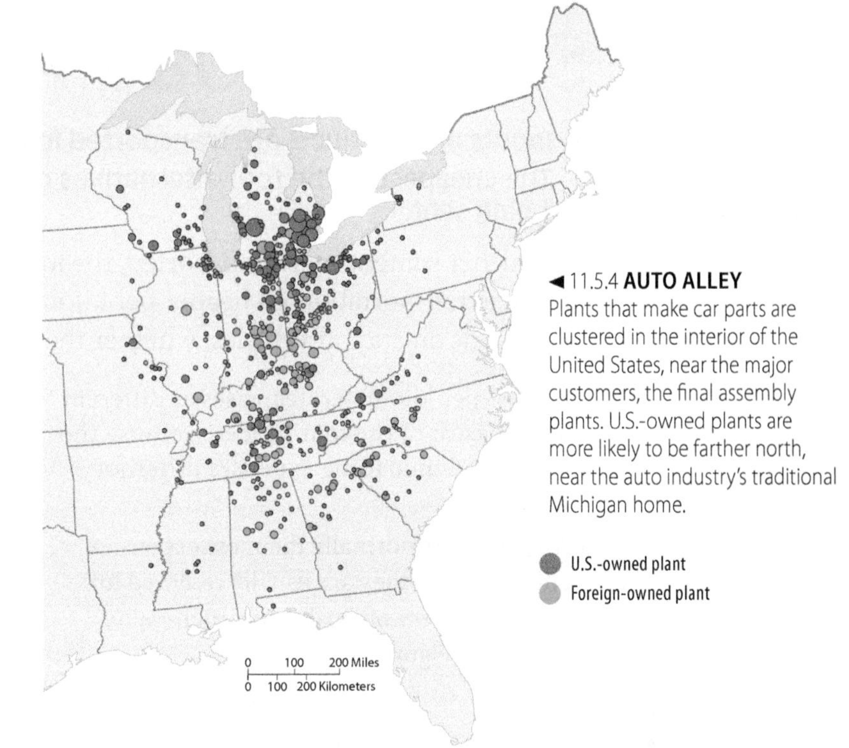

◄ 11.5.4 **AUTO ALLEY**
Plants that make car parts are clustered in the interior of the United States, near the major customers, the final assembly plants. U.S.-owned plants are more likely to be farther north, near the auto industry's traditional Michigan home.

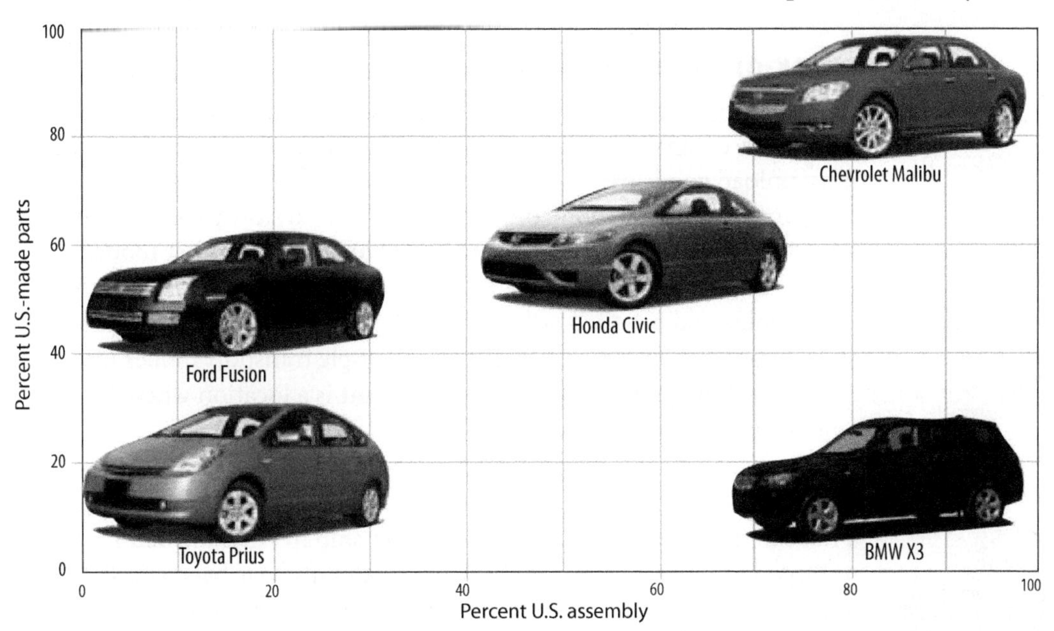

◄ 11.5.5 **"AMERICAN" AND "FOREIGN" CARS**
The x axis shows the percentage of these vehicles sold in the United States that were assembled in the United States in 2011. The y axis shows the percentage of U.S.-made parts in these vehicles. GM's Chevrolet Malibu was assembled entirely in the United States with all but a handful of U.S.-made parts. Toyota's Prius was imported from Japan with Japanese-made parts. Ford's Fusion was assembled in Mexico with about one-half U.S. parts. BMW's X3 was assembled in the United States with parts mostly imported from Germany. Some Honda Civics were assembled in the United States with mostly U.S.-made parts, and some Civics were imported from Japan with mostly Japanese-made parts.

11.6 Ship by Boat, Rail, Truck, or Air?

- **Inputs and products are transported in one of four ways: ship, rail, truck, or air.**
- **The cheapest of the four alternatives changes with the distance that goods are being sent.**

The farther something is transported, the lower is the cost per kilometer (or mile). Longer-distance transportation is cheaper per kilometer in part because firms must pay workers to load goods on and off vehicles, whether the material travels 10 kilometers or 10,000.

▲ 11.6.1 **SHIP BY AIR**
Air transport is used to ship packages long distances in a hurry.

The cost per kilometer decreases at different rates for each of the four modes, because the loading and unloading expenses differ for each mode.

- Airplanes are normally the most expensive alternative, so they are usually reserved for speedy delivery of small-bulk, high-value packages (Figure 11.6.1).
- Ships are attractive for very long distances because the cost per kilometer is very low (Figure 11.6.2).
- Trains are often used to ship to destinations that take longer than one day to reach. Trains take longer than trucks to load, but once under way are not required to make daily rest stops like truck drivers.
- Trucks are most often used for short-distance delivery because they can be loaded and unloaded quickly and cheaply.

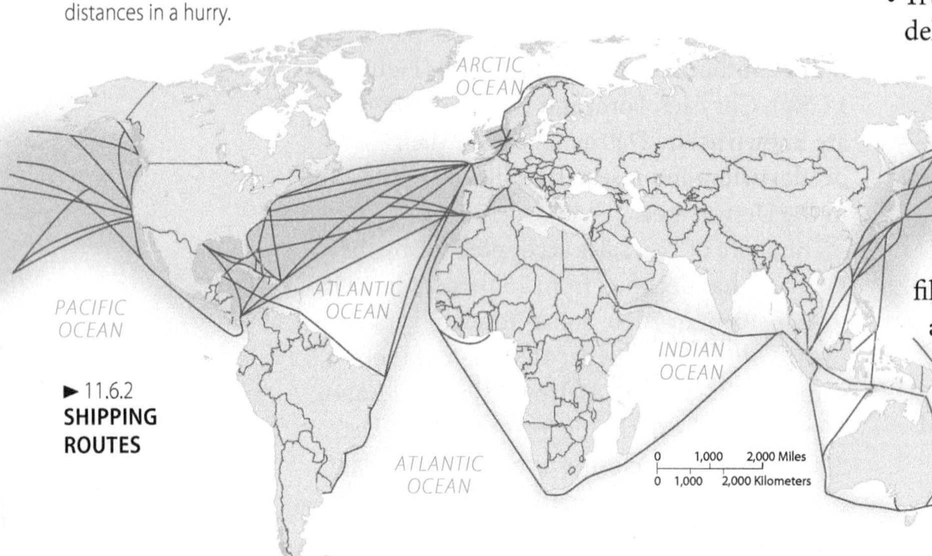

► 11.6.2 **SHIPPING ROUTES**

Air-cargo companies such as FedEx and UPS promise overnight delivery for most packages. They pick up packages in the afternoon and transport them by truck to the nearest airport. Late at night, planes filled with packages are flown to a central hub airport in the interior of the country, such as Memphis, Tennessee, and Louisville, Kentucky. The packages are then transferred to other planes, flown to airports nearest their destination, transferred to trucks, and delivered the next morning.

BREAK-OF-BULK POINTS

Regardless of transportation mode, cost rises each time that inputs or products are transferred from one mode to another. For example, workers must unload goods from a truck and then reload them onto a plane. The company may need to build or rent a warehouse to store goods temporarily after unloading from one mode and before loading to another mode.

Some companies may calculate that the cost of one mode is lower for some inputs and products, whereas another mode may be cheaper for other goods. Many companies that use multiple transport modes locate at a break-of-bulk point. A **break-of-bulk point** is a location where transfer among transportation modes is possible.

Containerization has facilitated transfer of packages between modes at break-of-bulk points (Figure 11.6.3). Containers may be packed into a rail car, transferred quickly to a container ship to cross the ocean, and unloaded into trucks at the other end. Large ships have been specially built to accommodate large numbers of rectangular, box-like containers.

▼ 11.6.3. **BREAK-OF-BULK POINT, PORT OF LOS ANGELES**
Many goods that are shipped long distances are packed in uniformly sized containers, which can be quickly transferred between ships and trucks or trains.

JUST-IN-TIME DELIVERY

Proximity to market has long been important for many types of manufacturers, as discussed earlier in this chapter. The factor has become even more important in recent years because of the rise of just-in-time delivery.

As the name implies, **just-in-time** is shipment of parts and materials to arrive at a factory moments before they are needed. Just-in-time delivery is especially important for delivery of inputs, such as parts and raw materials, to manufacturers of fabricated products, such as cars and computers.

Under just-in-time, parts and materials arrive at a factory frequently, in many cases daily if not hourly. Suppliers of the parts and materials are told a few days in advance how much will be needed over the next week or two, and first thing each morning exactly what will be needed at precisely what time that day.

Just-in-time delivery reduces the money that a manufacturer must tie up in wasteful inventory (Figure 11.6.4). The percentage of the U.S. economy tied up in inventory has been cut in half during the past quarter-century. Manufacturers also save money through just-in-time delivery by reducing the size of the factory, because space does not have to be wasted on piling up a mountain of inventory.

To meet a tight timetable, a supplier of parts and materials must locate factories near its customers. If only an hour or two notice is given, a supplier has no choice but to locate a factory within 50 miles or so of the customer.

▲ 11.6.4 **ELIMINATING INVENTORY**
Leading computer manufacturers have cut costs in part through eliminating the need to store inventory in warehouses. These computers are being built in China only after the buyer has placed the order.

Just-in-time delivery sometimes merely shifts the burden of maintaining inventory to suppliers. Walmart, for example, holds low inventories but tells its suppliers to hold high inventories "just in case" a sudden surge in demand requires restocking on short notice.

JUST-IN-TIME DISRUPTIONS

Just-in-time delivery means that producers have less inventory to cushion against disruptions in the arrival of needed parts. Three kinds of disruptions can result from reliance on just-in-time delivery:

- **Labor unrest.** A strike at one supplier plant can shut down the entire production within a couple of days. Also disrupting deliveries could be a strike in the logistics industry, such as truckers or dockworkers.
- **Traffic.** Deliveries may be delayed when traffic is slowed by accident, construction, or unusually heavy volume. Trucks and trains are both subject to these types of delays, especially crossing international borders (Figure 11.6.5).
- **Natural hazards.** Poor weather conditions can afflict deliveries anywhere in the world. Blizzards and floods can close highways and rail lines. The 2011 earthquake and tsunami in Japan put many factories and transportation lines out of service for months. Carmakers around the world had to curtail production because key parts had been made at the damaged factories.

► 11.6.5 **DELIVERY DISRUPTIONS**
These vehicles on a highway in Ontario are backed up trying to cross the border into Michigan.

11.7 Site Factors in Industry

► **Site factors result from the unique characteristics of a location.**
► **The three main site factors are labor, land, and capital.**

Site factors are industrial location factors related to the costs of factors of production inside the plant, notably labor, land, and capital.

LABOR

A **labor-intensive** industry is one in which wages and other compensation paid to employees constitute a high percentage of expenses. Labor costs an average of 11 percent of overall manufacturing costs in the United States, so a labor-intensive industry would have a much higher percentage than that (Figure 11.7.1).

The average annual wage paid to male workers exceeds $30,000 or $15 per hour in most developed countries, compared to less than $5,000 or $2.50 per hour in most developing countries (Figure 11.7.2). Health care, retirement pensions, and other benefits add substantially to the wage compensation in developed countries, but not in developing countries.

For some manufacturers—but not all—the difference between paying workers $2.50 and $15 per hour is critical. For example, most of the cost of an iPhone is in the parts (made mostly in Japan, Germany, and South Korea) and the gross profit to Apple (based in the United States). One step in the production process is labor intensive—snapping all the parts together at an assembly plant—and this step is done in China with relative low-wage workers (Figure 11.7.3).

▲ 11.7.1 **LABOR** Chinese workers in a packaging factory. Around the world, approximately 150 million people are employed in manufacturing, according to the UN, International Labor Organization (ILO). China has around 20 percent of the world's manufacturing workers and the United States around 10 percent.

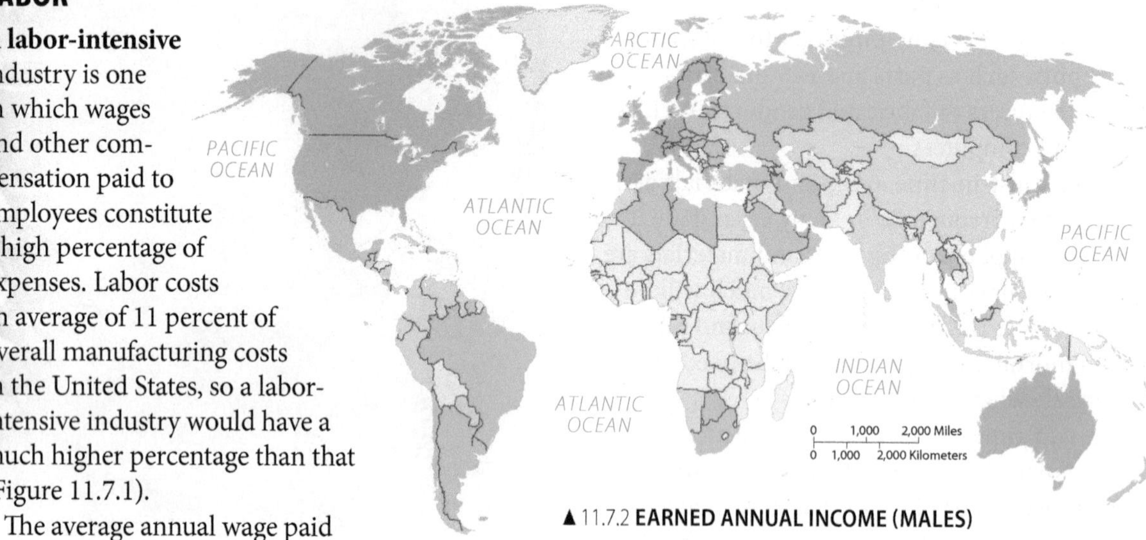

▲ 11.7.2 **EARNED ANNUAL INCOME (MALES)**

- $30,000 and above
- $10,000 – $29,999
- $5,000 – $9,999
- below $5,000
- no data

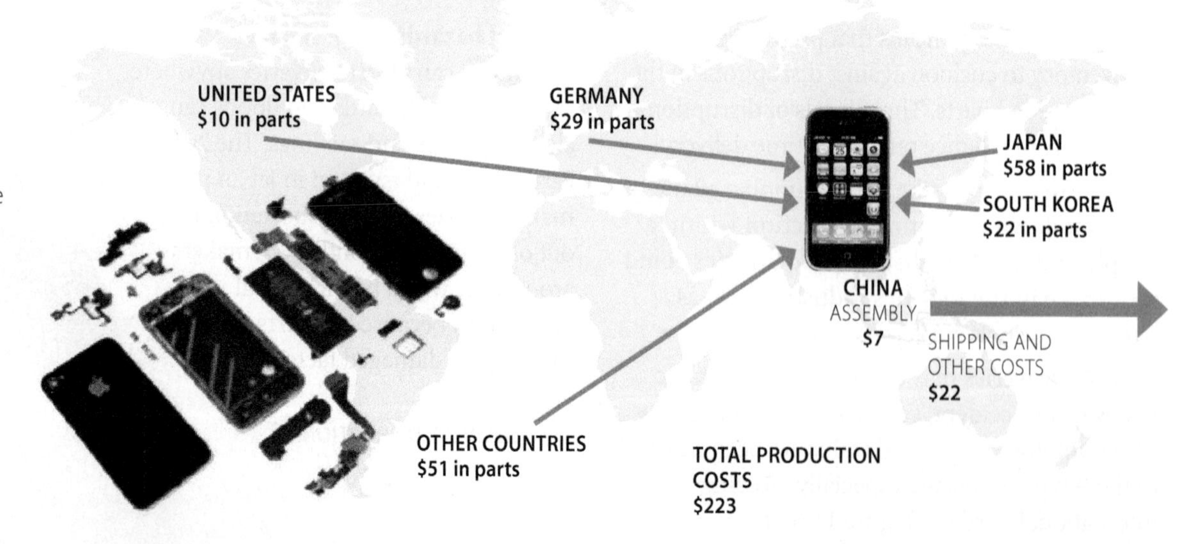

► 11.7.3 **COST STRUCTURE OF AN iPHONE** The cost of manufacturing an iPhone is substantially less than the price that consumers pay.

LAND

In the early years of the Industrial Revolution, multistory factories were constructed in the heart of the city. Now, they are more likely to be built in suburban or rural areas, in part to provide enough space for one-story buildings.

Raw materials are typically delivered at one end and moved through the factory on conveyors or forklift trucks. Products are assembled in logical order and shipped out at the other end.

Locations on the urban periphery are also attractive for factories to facilitate delivery of inputs and shipment of products. In the past, when most material moved in and out of a factory by rail, a central location was attractive because rail lines converged there.

With trucks now responsible for transporting most inputs and products, proximity to major highways is more important for a factory. Especially attractive is the proximity to the junction of a long-distance route and the beltway or ring road that encircles most cities. Factories cluster in industrial parks located near suburban highway junctions (Figure 11.7.4).

Also, land is much cheaper in suburban or rural locations than near the center of a city. A hectare (or an acre) of land in the United States may cost only a few thousand dollars in a rural area, tens of thousands in a suburban location, and hundreds of thousands near the center.

▲ 11.7.4 **LAND**
Factory outside Vic, Spain.

CAPITAL

Manufacturers typically borrow capital—the funds to establish new factories or expand existing ones. One important factor in the clustering in California's Silicon Valley of high-tech industries has been availability of capital. One-fourth of all capital in the United States is spent on new industries in the Silicon Valley (Figure 11.7.5).

Banks in Silicon Valley have long been willing to provide money for new software and communications firms even though lenders elsewhere have hesitated. High-tech industries have been risky propositions—roughly two-thirds of them fail—but Silicon Valley financial institutions have continued to lend money to engineers with good ideas so that they can buy the software, communications, and networks they need to get started.

The ability to borrow money has become a critical factor in the distribution of industry in developing countries. Financial institutions in many developing countries are short of funds, so new industries must seek loans from banks in developed countries. But enterprises may not get loans if they are located in a country that is perceived to have an unstable political system, a high debt level, or ill-advised economic policies.

▼ 11.7.5 **CAPITAL**
San Jose, California, in Silicon Valley.

11.8 Textile and Apparel Production

- **Textile and apparel production is a prominent example of a labor-intensive industry.**
- **Textile and apparel production generally requires less skilled, low-wage workers.**

Production of apparel and textiles, which are woven fabrics, is a prominent example of an industry that generally requires less-skilled, low-cost workers. The textile and apparel industry accounts for 6 percent of the dollar value of world manufacturing but a much higher 14 percent of world manufacturing employment, an indicator that it is a labor-intensive industry. The percentage of the world's women employed in this type of manufacturing is even higher.

▲ 11.8.1 COTTON YARN PRODUCTION, CHINA

Textile and apparel production involves three principal steps. All are labor-intensive compared to other industries, but the importance of labor varies somewhat among them. As a result, their global distributions are not identical, because the three steps are not equally labor-intensive.

SPINNING OF FIBERS TO MAKE YARN

Fibers can be spun from natural or synthetic elements (Figure 11.8.1). Cotton is the principal natural fiber—three-fourths of the total—followed by wool. Historically, natural fibers were the sole source, but today synthetics account for three-fourths and natural fibers only one-fourth of world thread production. Because it is a labor-intensive industry, spinning is done primarily in low-wage countries, primarily China (Figure 11.8.2).

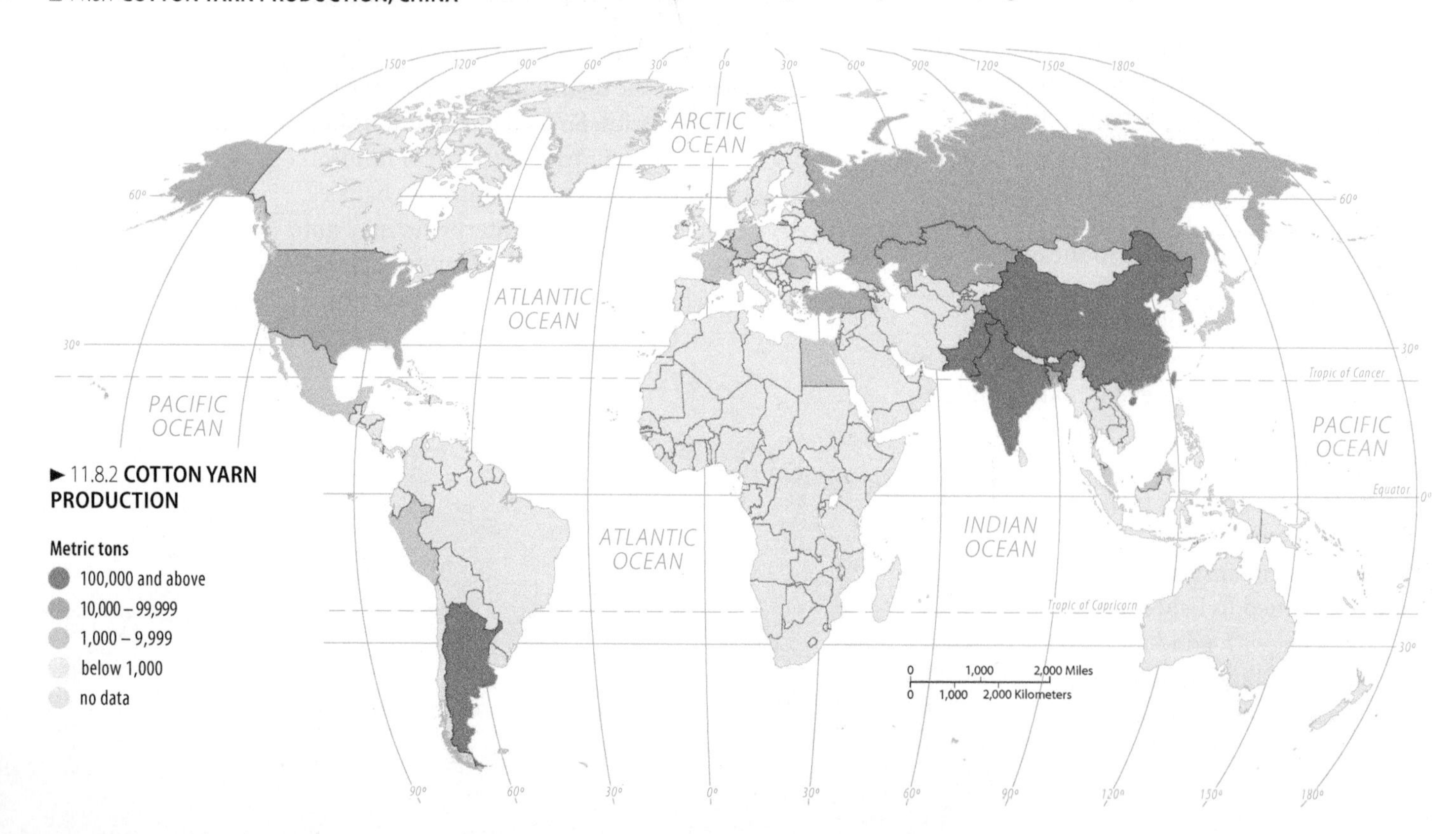

► 11.8.2 COTTON YARN PRODUCTION

WEAVING OR KNITTING YARN INTO FABRIC

Labor constitutes an even higher percentage of total production cost for weaving than for the spinning and assembly steps. China alone accounts for nearly 60 percent of the world's woven cotton fabric production, and India another 30 percent (Figure 11.8.3) Fabric has been woven or laced together by hand for thousands of years on a loom, which is a frame on which two sets of threads are placed at right angles to each other. Even on today's mechanized looms, a loom has one set of threads, called a warp, which is strung lengthwise. A second set of threads, called a weft, is carried in a shuttle that moves over and under the warp (Figure 11.8.4).

▲ 11.8.3 **WOVEN COTTON FABRIC PRODUCTION**

Square meters
- 20 billion and above
- 1 billion – 3 billion
- 0.1 billion – 1.0 billion
- below 0.1 billion
- no data

◄ 11.8.4 **WOVEN COTTON FABRIC PRODUCTION, CHINA**

CUTTING AND SEWING OF FABRIC FOR ASSEMBLING INTO CLOTHING AND OTHER PRODUCTS

Textiles are assembled into four main types of products: garments, carpets, home products such as bed linens and curtains, and industrial uses such as headliners inside motor vehicles. Developed countries play a larger role in assembly than in spinning and weaving because most of the consumers of assembled products are located there. For example, two-thirds of the women's blouses sold worldwide in a year are sewn in developed countries (Figure 11.8.5). However, the percentage of clothing produced in developing countries has been increasing.

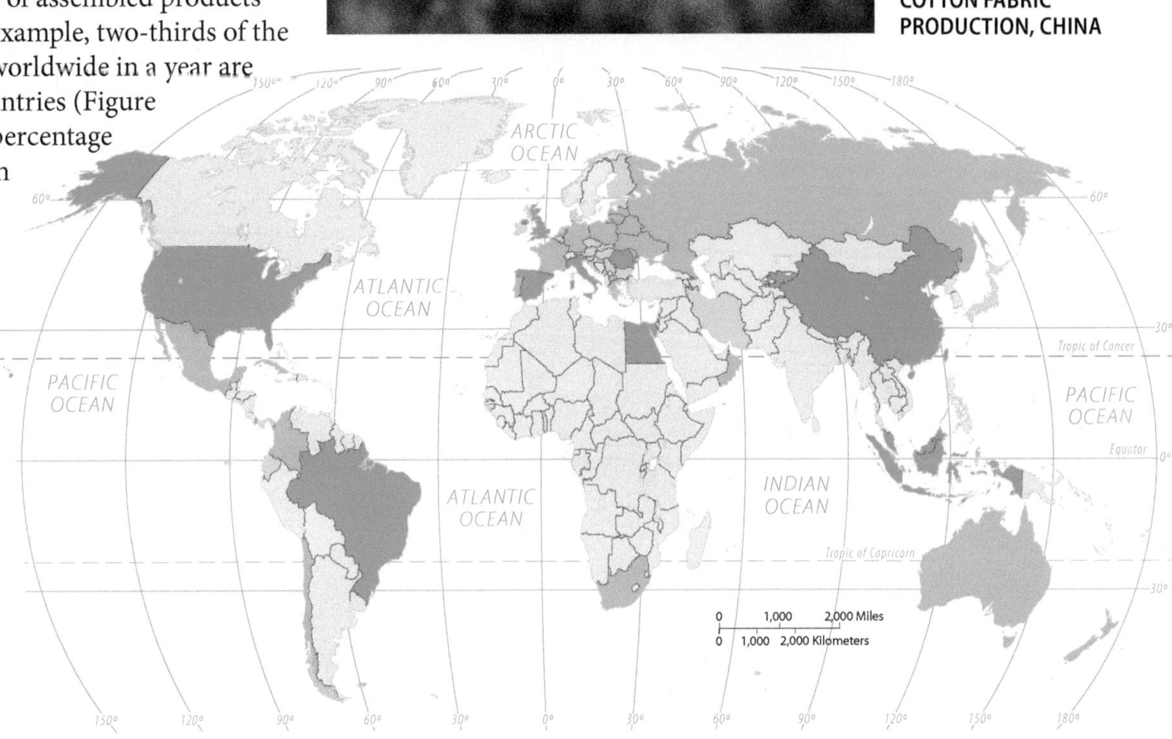

► 11.8.5 **WOMEN'S BLOUSE PRODUCTION**

- 10 million and above
- 1 million – 9 million
- 100,000 – 999,999
- below 100,000
- no data

11.9 Emerging Industrial Regions

- **Manufacturing is growing in locations not traditionally considered as industrial centers.**
- **The four BRIC countries are expected to be increasingly important industrial centers.**

Industry is on the move around the world. Site factors, especially labor costs, have stimulated industrial growth in new regions, both internationally and within developed regions. Situation factors, especially proximity to growing markets, have also played a role in the emergence of new industrial regions.

INTERREGIONAL SHIFTS IN THE UNITED STATES

Manufacturing jobs have been shifting within the United States from the North and East to the South and West (Figure 11.9.1). Between 1950 and 2009, the North and East lost 6 million manufacturing jobs and the South and West gained 2 million.

The principal site factor for many manufacturers was labor-related: enactment of **right-to-work** laws by a number of states, especially in the South. A right-to-work law requires a factory to maintain a so-called "open shop" and prohibits a "closed shop."

- In a "closed shop," a company and a union agree that everyone must join the union to work in the factory.
- In an "open shop," a union and a company may not negotiate a contract that requires workers to join a union as a condition of employment.

By enacting right-to-work laws, southern states made it much more difficult for unions to organize factory workers, collect dues, and bargain with employers from a position of strength. As a result, the percentage of workers who are members of a union is much lower in the South than elsewhere in the United States. Car plants, steel, textiles, tobacco products, and furniture industries have dispersed through smaller communities in the South, many in search of a labor force willing to work for less pay than in the North and to forgo joining a union (Figure 11.9.2).

▼ 11.9.1 **CHANGING U.S. MANUFACTURING, 1950 AND 2010**

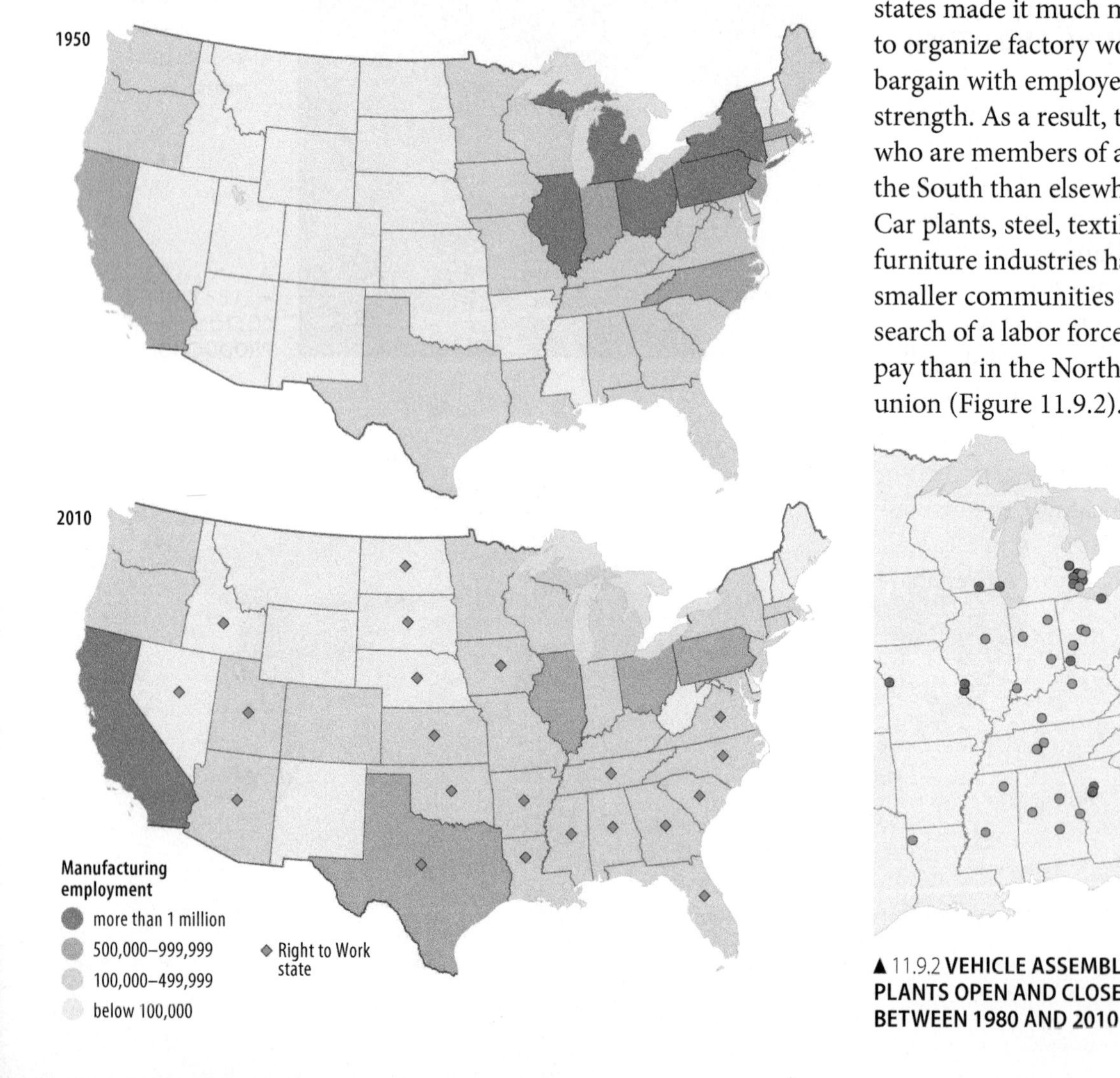

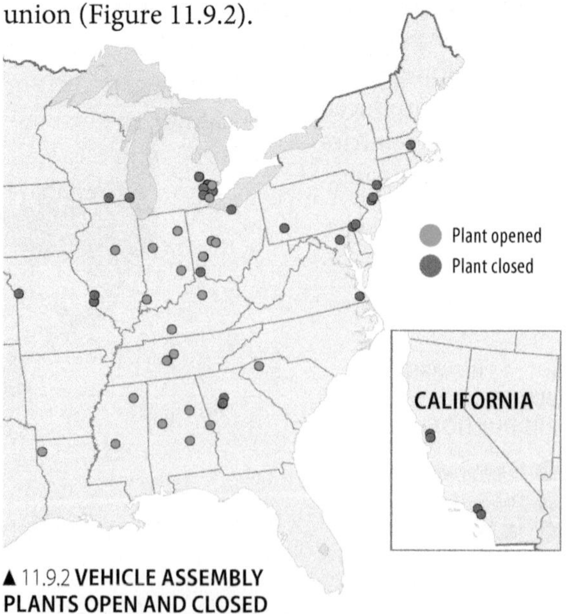

▲ 11.9.2 **VEHICLE ASSEMBLY PLANTS OPEN AND CLOSED BETWEEN 1980 AND 2010**

INDUSTRY IN MEXICO

Manufacturing has been increasing in Mexico. The North Atlantic Free Trade Agreement (NAFTA), effective 1994, eliminated most barriers to moving goods between Mexico and the United States.

Because it is the nearest low-wage country to the United States, Mexico attracts labor-intensive industries that also need proximity to the U.S. market. Although the average wage is higher in Mexico than in most developing countries (refer to Figure 11.7.2), the cost of shipping from Mexico to the United States is lower than from other developing countries.

Mexico City, the country's largest market, is the center for industrial production for domestic consumption (Figure 11.9.3). Other factories have located in Mexico's far north to be as close as possible to the United States.

▲ 11.9.3 **CAR PARTS PLANTS IN MEXICO**

EMERGING INDUSTRIAL POWERS: THE "BRIC" COUNTRIES

Much of the world's future growth in manufacturing is expected to locate outside the principal industrial regions described earlier in section 11.2. The financial analysis firm Goldman Sachs has coined the acronym BRIC to indicate the countries it expects to dominate global manufacturing during the twenty-first century. BRIC is an acronym for four countries—Brazil, Russia, India, and China (Figure 11.9.4). They are also known as the newly emerging economies.

The four BRIC countries together currently control one-fourth of the world's land and two-fifths of the world's population, but the four combined account for only one-sixth of world GDP. All four countries have made changes to their economies in recent years, embracing international trade with varying degrees of enthusiasm. By mid-twenty-first century, the four BRIC countries, plus the United States and Mexico, are expected to have the world's six largest economies.

The four BRIC countries have different advantages for industrial location. Russia and Brazil (Figure 11.9.5), currently classified by the United Nations as having high levels of development, are especially rich in inputs critical for industry. China and India, classified as having medium levels of development, have the two largest labor forces and potential markets for consumer goods.

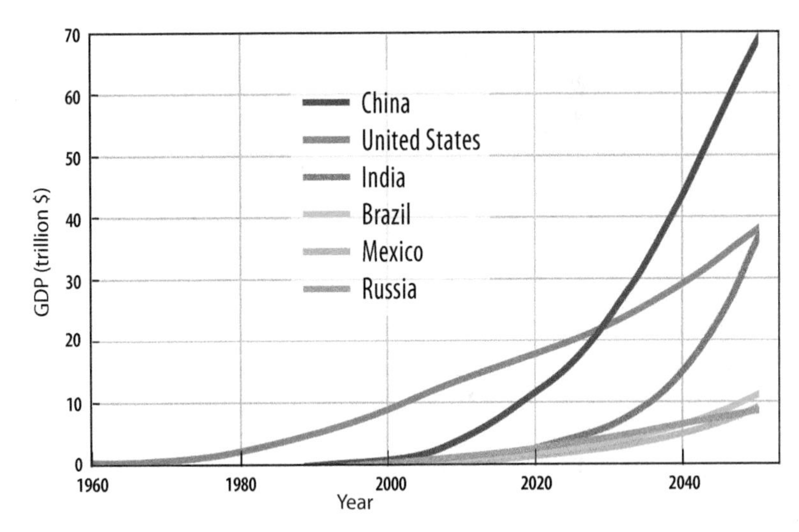

▲ 11.9.4 **GDP HISTORY AND FORECAST FOR BRIC COUNTRIES, UNITED STATES, AND MEXICO**

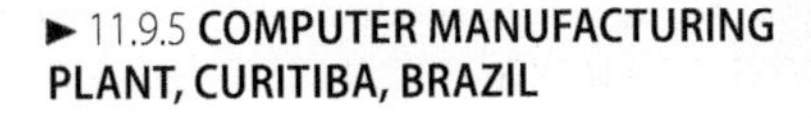

► 11.9.5 **COMPUTER MANUFACTURING PLANT, CURITIBA, BRAZIL**

Three recent changes in the structure of manufacturing have geographic consequences:

- Factories have become more productive through introduction of new machinery and processes. A factory may continue to operate at the same location but require fewer workers to produce the same output.
- Companies are locating production in communities where workers are willing to adopt more flexible work rules. Firms are especially attracted to smaller towns where low levels of union membership reduce vulnerability to work stoppages, even if wages are kept low and layoffs become necessary.
- By spreading production among many countries, or among many communities within one country, large corporations have increased their bargaining power with local governments and labor forces. Production can be allocated to locations where the local government is especially helpful and generous in subsidizing the costs of expansion, and the local residents are especially eager to work in the plant.

Key Questions

Where is industry clustered?

- The Industrial Revolution originated in the United Kingdom and diffused to Europe and North America in the twentieth century.
- World industry is highly clustered in three regions—Europe, North America, and East Asia.

What situation factors influence industrial location?

- A company tries to identify the optimal location for a factory through analyzing situation and site factors.
- Situation factors involve the cost of transporting both inputs into the factory and products from the factory to consumers.
- Steel and motor vehicle industries have traditionally located factories primarily because of situation factors.

What site factors influence industrial location?

- Three site factors—land, labor, and capital—control the cost of doing business at a location.
- Production of textiles and apparel has traditionally been located primarily because of site factors.
- New industrial regions are emerging because of their increased importance for site and situation factors.

Thinking Geographically

The North American Free Trade Agreement (NAFTA) among Canada, Mexico, and the United States was implemented in 1994.

1. What have been the benefits and the drawbacks to Canada, Mexico, and the United States as a result of NAFTA?

To induce Hyundai to open a plant in 2010 in West Point, Georgia, to assemble its Kia models, the state spent $36 million to buy the site and donate it to Hyundai, $61 million to build infrastructure such as roads and rail lines, $73 million to train the workers, and $90 million in tax benefits (Figure 11.CR.1).

2. Why would the state of Georgia spend $260 million to get the Kia factory? Did Georgia overpay?

Manufacturing is more dispersed than in the past, both within and among countries.

3. What are the principal manufacturers in your community or area? How have they been affected by increasing global competition?

► 11.CR.1 **KIA ASSEMBLY PLANT, WEST POINT, GEORGIA**

Interactive Mapping

SITUATION FACTORS AND RUSSIAN INDUSTRY

Russia's principal industrial location asset is proximity to inputs.

Launch MapMaster Russian Domain in MasteringGEOGRAPHY™

Select *Economic* then *Industrial Regions.*

Next select *Economic* then *Major Natural resources* then *Coal and Iron only* (deslect others).

1. Coal and iron are the two principal inputs into steel production. Which of these inputs is close to Russia's principal industrial areas, and which must be transported relatively far?

Deselect coal and iron and instead select other natural resources to see which are near the industrial areas and which have to be transported.

2. What overall pattern of industrial location do you see?

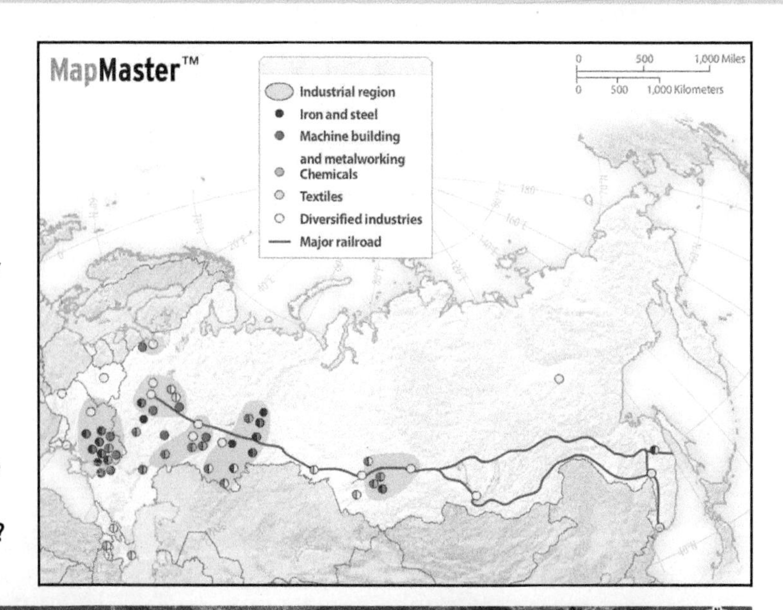

Explore

SAN ANTONIO, TEXAS

Use Google Earth to explore the changing industrial landscape in the United States.

Fly to: *1 Lone Star Pass, San Antonio, Texas, USA.*

Zoom out until the entire factory complex is visible.

Click: *Show Historical Imagery.*

Move the time line to: *9/27/2002.*

1. What has changed from 9/27/2002 until today?

2. Move the time line forward from 9/27/2002; at what date can a change in the land first be seen?

Zoom out again until the city of San Antonio is visible.

3. What are some advantages of 1 Lone Star Pass as an industrial location?

Key Terms

Break-of-bulk point
A location where transfer is possible from one mode of transportation to another.

Bulk-gaining industry
An industry in which the final product weighs more or comprises a greater volume than the inputs.

Bulk-reducing industry
An industry in which the final product weighs less or comprises a lower volume than the inputs.

Cottage industry
Manufacturing based in homes rather than in a factory, commonly found prior to the Industrial Revolution.

Industrial Revolution
A series of improvements in industrial technology that transformed the process of manufacturing goods.

Just-in-time delivery
Shipment of parts and materials to arrive at a factory moments before they are needed.

Labor-intensive industry
An industry for which labor costs comprise a high percentage of total expenses.

Right-to-work state
A U.S. state that has passed a law preventing a union and company from negotiating a contract that requires workers to join a union as a condition of employment.

Site factors
Location factors related to the costs of factors of production inside the plant, such as land, labor, and capital.

Situation factors
Location factors related to the transportation of materials into and from a factory.

On the Internet

Statistics on employment in manufacturing, as well as other sectors of the U.S. economy, are at the U.S. Department of Labor's Bureau of Labor Statistics website, at **www.bls.gov**, or scan the QR on the first page of this chapter.

▶ LOOKING AHEAD

Most of the growth in jobs in the United States and in the world are in the service (or tertiary) sector, and most service jobs are located in urban settlements.

12 Services and Settlements

Flying across the United States on a clear night, you look down on the lights of settlements, large and small. You see small clusters of lights from villages and towns, and large, brightly lit metropolitan areas. Geographers apply economic geography concepts to explain regularities in the pattern of settlements.

The regular pattern of settlements in the United States and other developed countries reflects where services are provided. Three-fourths of the workers in developed countries are employed in the service sector of the economy. These services are provided in settlements.

The regular distribution of settlements observed over developed countries is not seen in developing countries. Geographers explain that the pattern in developing countries results from having much lower percentages of workers in services.

Where are consumer services distributed?

INTERNET CAFE, THAILAND

Where are business services distributed?

12.5 **Hierarchy of Business Services**

12.6 **Business Services in Developing Countries**

12.7 **Economic Base**

Where are settlements distributed?

12.8 **Rural Settlements**

12.9 **Settlements in History**

12.10 **Urbanization**

SCAN TO GET THE POPULATION FOR THE WORLD'S 500 LARGEST CITIES

12.1 Types of Services

- ▶ **Three types of services are consumer, business, and public.**
- ▶ **Employment has grown more rapidly in some services than in others.**

A **service** is any activity that fulfills a human want or need and returns money to those who provide it. Services generate more than two-thirds of GDP in most developed countries, compared to less than one-half in most developing countries (Figure 12.1.1).

▶ 12.1.1 **PERCENTAGE OF GDP FROM SERVICES**

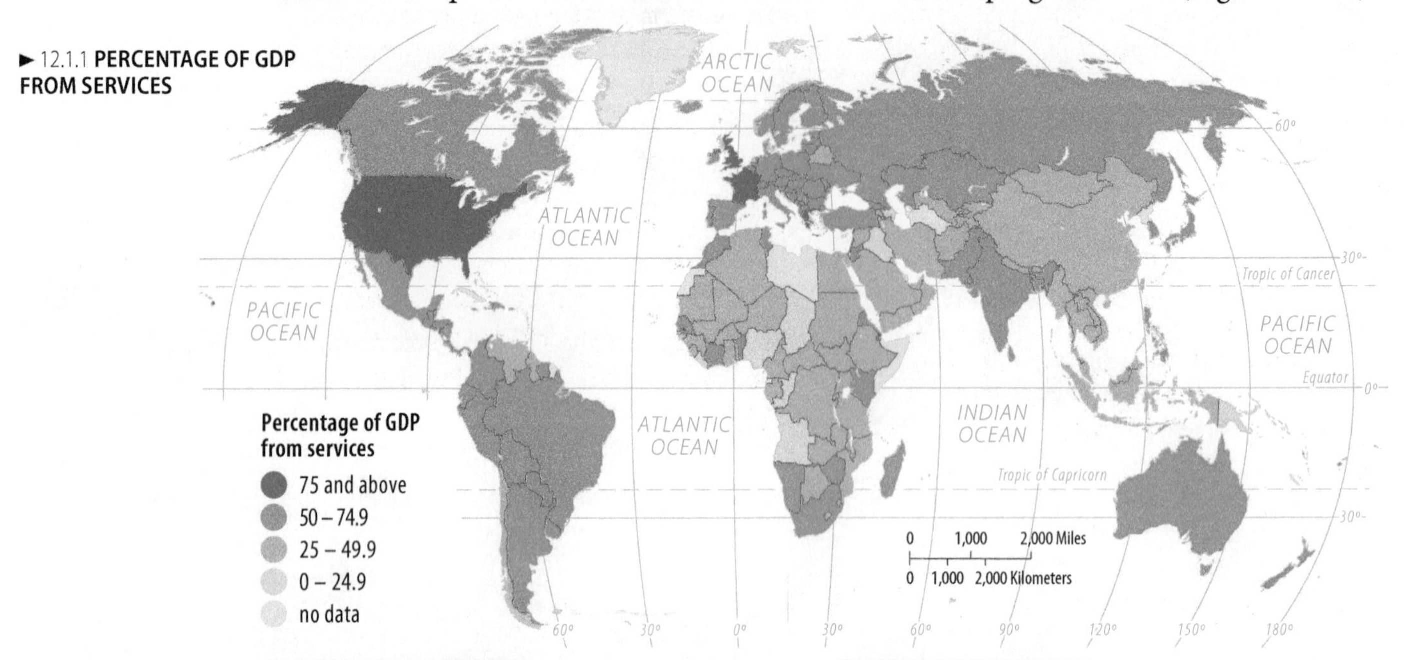

CONSUMER SERVICES

The service sector of the economy is subdivided into three types—consumer services, business services, and public services. Each of these sectors is divided into several major subsectors.

Consumer services provide services to individual consumers who desire them and can afford to pay for them. Nearly one-half of all jobs in the United States are in consumer services. Four main types of consumer services are retail, education, health, and leisure (Figure 12.1.2).

▼ 12.1.2 **U.S. CONSUMER SERVICES**

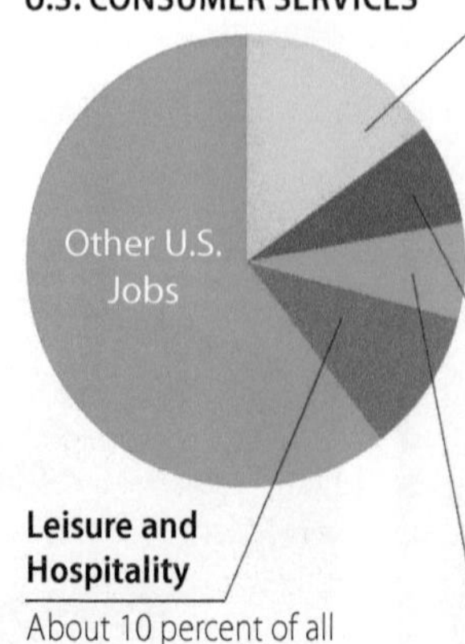

Retail and Wholesale
About 15 percent of all U.S. jobs. Department stores, grocers, and motor vehicle sales and service account for nearly one-half of these jobs. Another one-fourth are wholesalers who provide merchandise to retailers.

Education
About 7 percent of all U.S. jobs. Two-thirds of educators work in public schools, the other one-third in private schools. In Figure 12.1.5, educators at public schools are counted in public-sector employment.

Leisure and Hospitality
About 10 percent of all U.S. jobs, primarily in restaurants and bars.

Health Care
About 7 percent of all U.S. jobs, primarily hospitals, doctors' offices, and nursing homes.

BUSINESS SERVICES

Business services facilitate other businesses. One-fourth of all jobs in the United States are in business services. Professional services, financial services, and transportation services are the three main types of business services (Figure 12.1.3).

▼ 12.1.3 **U.S. BUSINESS SERVICES**

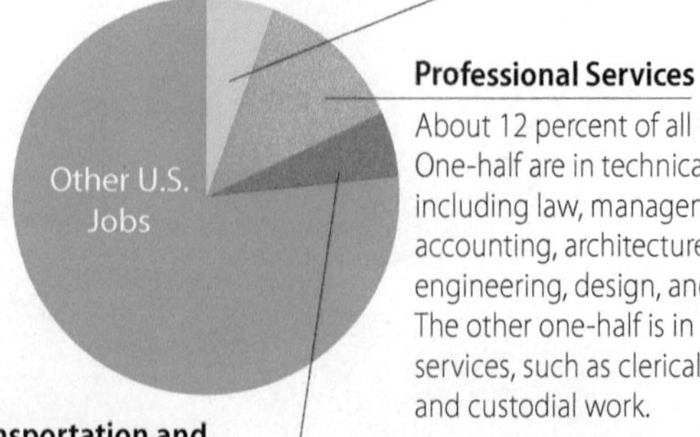

Financial Services
About 6 percent of all U.S. jobs. One-half of these jobs are in banks and other financial institutions, one-third in insurance companies, and the remainder in real estate.

Professional Services
About 12 percent of all U.S. jobs. One-half are in technical services, including law, management, accounting, architecture, engineering, design, and consulting. The other one-half is in support services, such as clerical, secretarial, and custodial work.

Transportation and Information Services
About 5 percent of U.S. jobs. One-half are in transportation, primarily trucking. The other one-half are in information services such as publishing and broadcasting as well as utilities such as water and electricity.

PUBLIC SERVICES

Public services provide security and protection for citizens and businesses. About 17 percent of all U.S. jobs are in the public sector, 9 percent if public school employees were excluded from the total and counted instead under education (consumer) services (Figure 12.1.4). Excluding educators, one-sixth of public-sector employees work for the federal government, one-fourth for one of the 50 state governments, and three-fifths for one of the tens of thousands of local governments. When educators are counted, the percentages for state and local governments would be higher.

The distinction among services is not absolute. For example, individual consumers use business services, such as consulting lawyers and keeping money in banks, and businesses use consumer services, such as purchasing stationery and staying in hotels. A public service worker at a national park may provide the same service as a consumer service worker at Disneyland. Geographers find the classification useful, because the various types of services have different distributions, and different factors influence locational decisions.

▼ 12.1.4 **U.S. PUBLIC SERVICES**

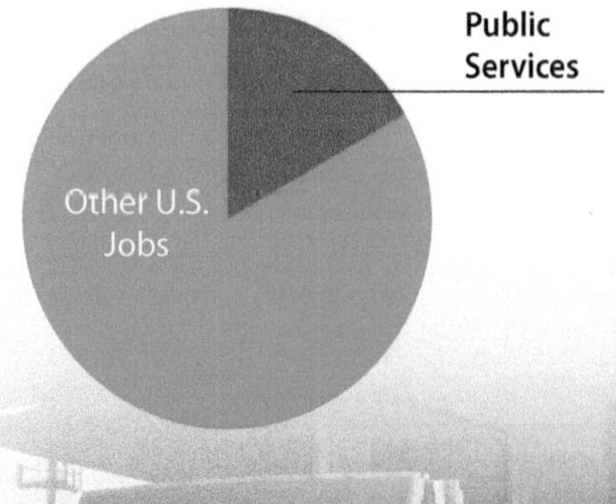

EMPLOYMENT CHANGE IN THE UNITED STATES

The growth in employment in the United States has been in services, whereas employment in primary- and secondary-sector activities has declined (Figure 12.1.5).

- **Business services**: Jobs expanded most rapidly in professional services (such as engineering, management, and law), data processing, advertising, and temporary employment agencies.
- **Consumer services** (Figure 12.1.6): The most rapid increase has been in the provision of health care, including hospital staff, clinics, nursing homes, and home health-care programs. Other large increases have been recorded in education, recreation, and entertainment.
- **Public services**: The share of employment in public services has declined during the past two decades.

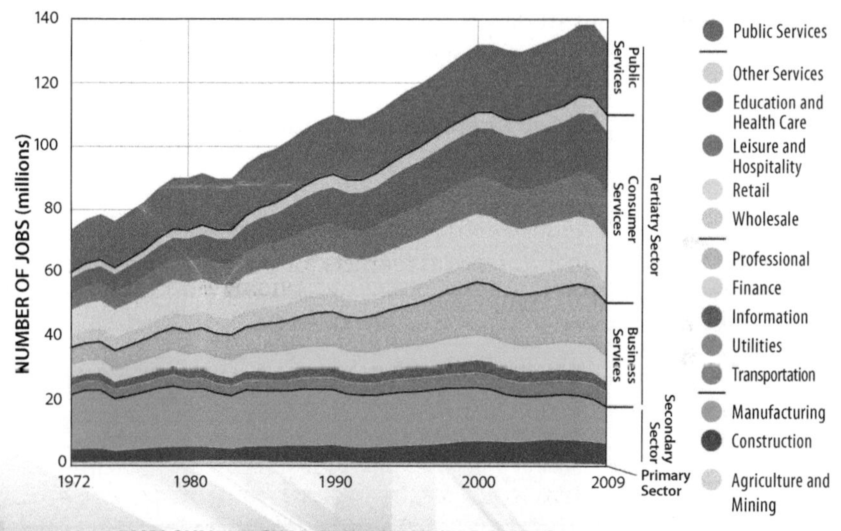

▲ 12.1.5 **EMPLOYMENT CHANGE IN THE UNITED STATES**

◄ 12.1.6 **EMPLOYMENT IN CONSUMER SERVICES**
Unemployed residents of the Detroit area lined up to apply for 200 jobs at a new Meijer store.

12.2 Central Place Theory

- Central place theory explains the location of consumer services.
- A central place has a market area, range, and threshold.

Consumer services and business services do not have the same distributions. The various types of consumer services—wholesale, retail, hospitality, education, and health care—generally follow a regular pattern based on size of settlements, with larger settlements offering more consumer services than smaller ones (Figure 12.2.1).

MARKET AREA OF A SERVICE

Selecting the right location for a new shop is probably the single most important factor in the profitability of a consumer service. **Central place theory** explains how services are distributed and why a regular pattern of settlements exists—at least in developed countries such as the United States. Central place theory was first proposed in the 1930s by German geographer Walter Christaller, based on his studies of southern Germany.

A **central place** is a market center for the exchange of goods and services by people attracted from the surrounding area. The central place is so called because it is centrally located to maximize accessibility from the surrounding region. Central places compete against each other to serve as markets for goods and services. This competition creates a regular pattern of settlements, according to central place theory.

The area surrounding a service from which customers are attracted is the **market area** or **hinterland**. A market area is a good example of a nodal region—a region with a core where the characteristic is most intense. To establish the market area, a circle is drawn around the node of service on a map. The territory inside the circle is its market area.

Because most people prefer to get services from the nearest location, consumers near the center of the circle obtain services from local establishments. The closer to the periphery of the circle, the greater is the percentage of consumers who will choose to obtain services from other nodes. People on the circumference of the market-area circle are equally likely to use the service, or go elsewhere (Figure 12.2.2).

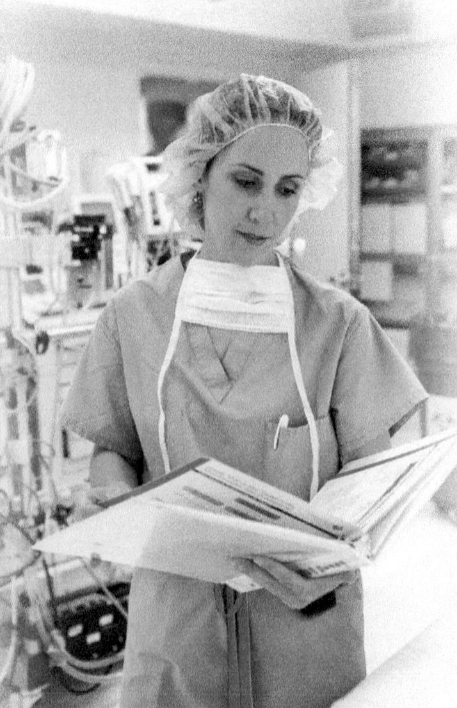

► 12.2.1 **TYPES OF CONSUMER SERVICES** Clockwise from top left **Wholesale**, such as a distribution center. **Retail**, such as a department store. **Health Care**, such as a hospital. **Education**, such as a school. **Hospitality**, such as a restaurant.

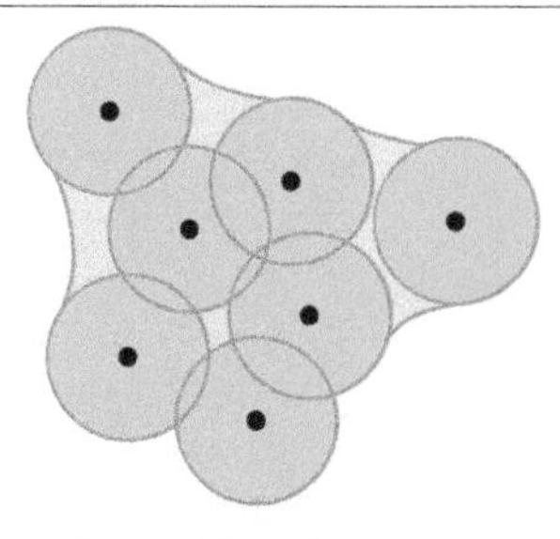

The problem with circles.
Circles are equidistant from center to edge, but they overlap or leave gaps. An arrangement of circles that leaves gaps indicates that people living in the gaps are outside the market area of any service, which is obviously not true. Overlapping circles are also unsatisfactory, for one service or another will be closer, and people will tend to patronize it.

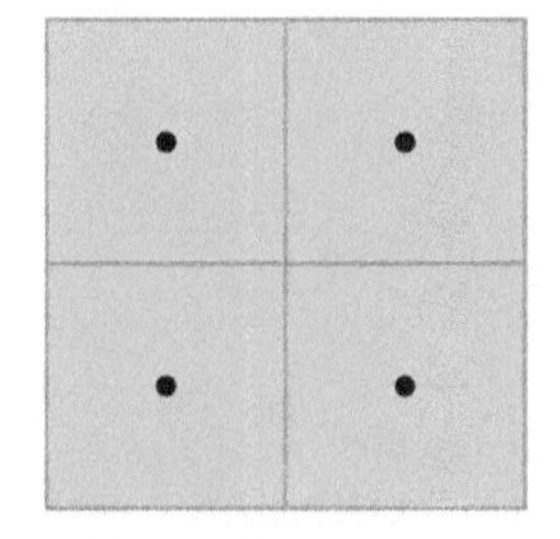

The problem with squares.
Squares nest together without gaps, but their sides are not equidistant from the center. If the market area is a circle, the radius—the distance from the center to the edge—can be measured, because every point around a circle is the same distance from the center. But in a square the distance from the center varies among points along a square.

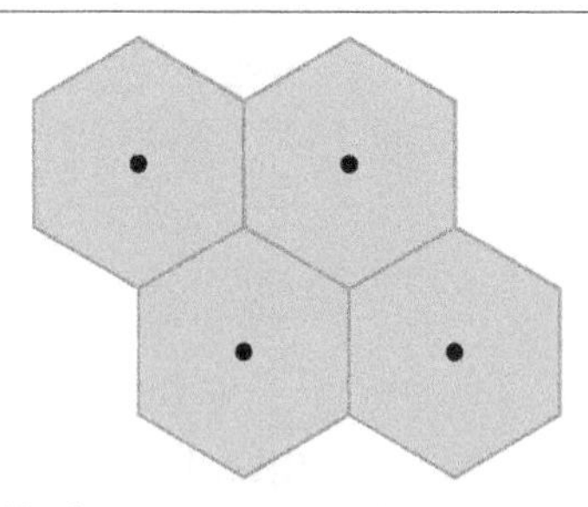

The hexagon compromise.
Geographers use hexagons to depict the market area of a good or service because hexagons offer a compromise between the geometric properties of circles and squares.

▲ 12.2.2 **WHY CENTRAL PLACE THEORY USES HEXAGONS TO DELINEATE MARKET AREAS**
Geographers use hexagons to represent market areas because of their geometric properties compared with those of circles and squares.

RANGE OF A SERVICE

The market area of every service varies. To determine the extent of a market area, geographers need two pieces of information about a service—its range and its threshold (Figure 12.2.3).

How far are you willing to drive for a pizza? To see a doctor for a serious problem? To watch a ballgame? The **range** is the maximum distance people are willing to travel to use a service. The range is the radius of the circle drawn to delineate a service's market area.

People are willing to go only a short distance for everyday consumer services, like groceries and pharmacies. But they will travel a long distance for other services, such as a major league baseball game or a concert. Thus a convenience store has a small range, whereas a stadium has a large range.

If firms at other locations compete by providing the service, the range must be modified. As a rule, people tend to go to the nearest available service: someone in the mood for a McDonald's hamburger is likely to go to the nearest McDonald's. Therefore, the range of a service must be determined from the radius of a circle that is irregularly shaped rather than perfectly round. The irregularly shaped circle takes in the territory for which the proposed site is closer than the competitors' sites.

The range must be modified further because most people think of distance in terms of time, rather than in terms of a linear measure like kilometers or miles. If you ask people how far they are willing to travel to a restaurant or a baseball game, they are more likely to answer in minutes or hours than in distance.

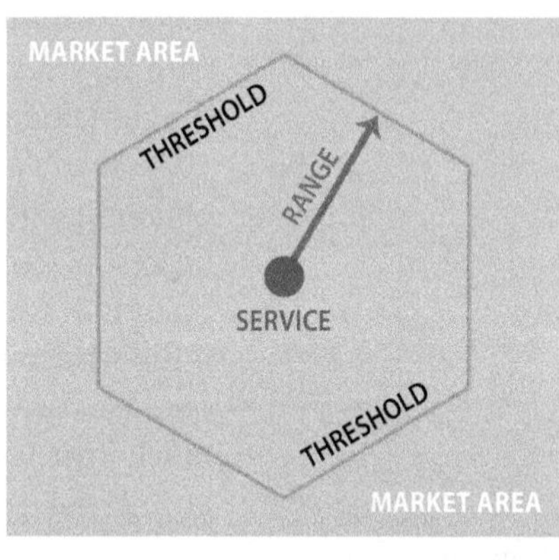

▲ 12.2.3 **MARKET AREA OF A SERVICE**
The range is the radius, and the threshold is a sufficient number of people inside the area to support the service.

THRESHOLD OF A SERVICE

The second piece of geographic information needed to compute a market area is the **threshold**, which is the minimum number of people needed to support the service. Every enterprise has a minimum number of customers required to generate enough sales to make a profit. Once the range has been determined, a service provider must determine whether a location is suitable by counting the potential customers inside the irregularly shaped circle.

How potential consumers inside the range are counted depends on the product. Convenience stores and fast-food restaurants appeal to nearly everyone, whereas other goods and services appeal primarily to certain consumer groups. Movie theaters attract younger people; chiropractors attract older folks. Poorer people are drawn to thrift stores; wealthier ones might frequent upscale department stores. Amusement parks attract families with children, but nightclubs appeal to singles. If a good or service appeals to certain customers, then only the type of good or service that appeals to them should be counted inside the range.

12.3 Hierarchy of Consumer Services

- **Small settlements provide services with small thresholds, ranges, and market areas.**
- **In developed countries, the size of settlements follows the rank-size rule.**

Small settlements are limited to consumer services that have small thresholds, short ranges, and small market areas, because too few people live in small settlements to support many services. A large department store or specialty store cannot survive in a small settlement, because the minimum number of people needed exceeds the population within range of the settlement.

Larger settlements provide consumer services having larger thresholds, ranges, and market areas. In addition, neighborhoods within large settlements also provide services having small thresholds and ranges. Services patronized by a small number of locals can coexist in a neighborhood ("mom-and-pop stores") along with services that attract many from throughout the settlement.

We spend as little time and effort as possible in obtaining consumer services and thus go to the nearest place that fulfills our needs. There is no point in traveling to a distant department store if the same merchandise is available at a nearby one. We travel greater distances only if the price is much lower or if the item is unavailable locally.

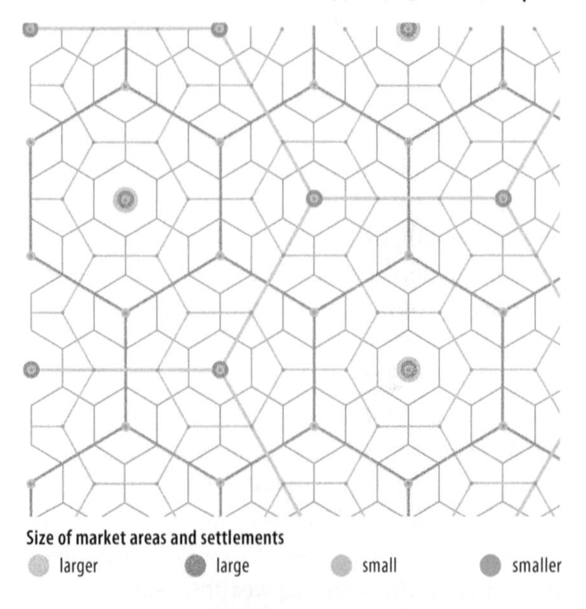

▲ 12.3.1 **NESTING OF SETTLEMENTS AND SERVICES**
According to central place theory, market areas are arranged in a regular pattern. Larger market areas, based in large settlements, are fewer in number and farther apart from each other than smaller market areas and settlements. Larger settlements also provide services with smaller market areas; consequently larger settlements have both larger and smaller market areas drawn around them.

NESTING OF SETTLEMENTS AND SERVICES

According to central place theory, market areas across a developed country would be a series of hexagons of various sizes, unless interrupted by physical features such as mountains and bodies of water. Developed countries have numerous small settlements with small thresholds and ranges, and far fewer large settlements with large thresholds and ranges. In his original study, Walter Christaller showed that the distances between settlements in southern Germany followed a regular pattern.

The nesting pattern can be illustrated with overlapping hexagons of different sizes. Four different levels of market area—for hamlet, village, town, and city—are shown in Figure 12.3.1. Hamlets with very small market areas are represented by the smallest contiguous hexagons. Larger hexagons represent the market areas of larger settlements and are overlaid on the smaller hexagons, because consumers from smaller settlements obtain some services in larger settlements.

Across much of the interior of the United States, a regular pattern of settlements can be observed, even if not precisely the same as the generalized model shown in Figure 12.3.1. In north central North Dakota, for example, Minot—the largest city in the area, with 41,000 inhabitants—is surrounded by seven small towns with between 1,000 and 5,000 inhabitants, fifteen villages with between 100 and 999 inhabitants, and nineteen hamlets with less than 100 inhabitants (Figure 12.3.2). The small towns have average ranges of 30 kilometers (20 miles) and market areas of around 2,800 square kilometers (1,200 square miles). The hamlets have ranges of around 15 kilometers (10 miles) and market areas of around 800 square kilometers (300 square miles).

RANK-SIZE DISTRIBUTION OF SETTLEMENTS

In many developed countries, geographers observe that ranking settlements from largest to smallest (population) produces a regular pattern or hierarchy. This is the **rank-size rule**, in which the country's nth-largest settlement is 1/n the population of the largest settlement. In other words, the second-largest city is one-half the size of the largest, the fourth-largest city is one-fourth the size of the largest, and so on. When plotted on logarithmic paper, the rank-size distribution forms a fairly straight line. The distribution of settlements closely follows the rank-size rule in the United States and a handful of other countries.

If the settlement hierarchy does not graph as a straight line, then the society does not have a rank-size distribution of settlements. Several developed countries in Europe follow the rank-size distribution among smaller settlements but not among the largest ones. Instead, the largest settlement in these countries follows the **primate city rule**. According to the primate city rule, the largest settlement has more than twice as many people as the second-ranking settlement. In this distribution, the country's largest city is called the **primate city** (Figure 12.3.3).

The existence of a rank-size distribution of settlements is not merely a mathematical curiosity. It has a real impact on the quality of life for a country's inhabitants. A regular hierarchy—as in the United States—indicates that the society is sufficiently wealthy to justify the provision of goods and services to consumers throughout the country. Conversely, the primate city distribution in a developing country indicates that there is not enough wealth in the society to pay for a full variety of services (Figure 12.3.4).

▲ 12.3.2 **SETTLEMENTS IN NORTH DAKOTA**
Central place theory helps to explain the distribution of settlements of varying sizes in North Dakota. Larger settlements are fewer and farther apart, whereas smaller settlements are more frequent and closer together.

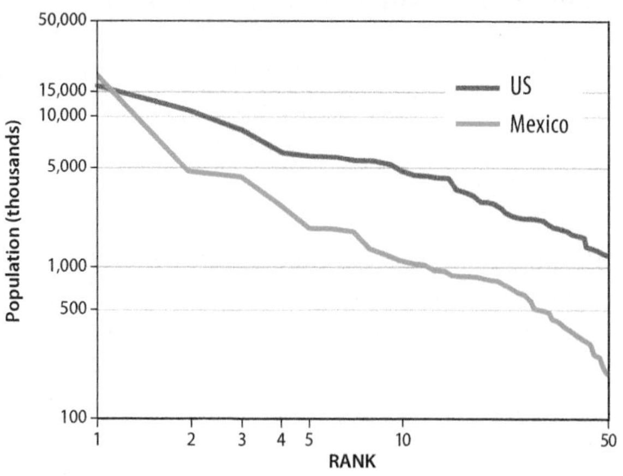

▲ 12.3.3 **RANK-SIZE AND PRIMATE CITY DISTRIBUTIONS OF SETTLEMENTS**
Mexico follows the primate city distribution. Its largest city, Mexico City, is five times larger than its second largest city, Guadalajara. The United States follows more closely the rank-size distribution, because the largest city, New York, is not that much larger than the second largest city, Los Angeles.

► 12.3.4 **MORELIA (ABOVE) AND BALTIMORE (BELOW)**
Morelia and Baltimore are the twentieth largest urban settlements in Mexico and the United States, respectively. Baltimore has a population of 2.7 million compared to only 800,000 for Morelia.

12.4 Market Area Analysis

- Retailers determine profitability of a site by calculating the range and threshold.
- Site selection is facilitated through the use of GIS.

Geographers apply central place theory to create market area studies that assist service providers with opening and expanding their facilities. And in a severe economic downturn, market area analysis helps determine where to close facilities.

Service providers often say that their three most important location factors are "location, location, and location." What they actually mean is that proximity to customers is the only critical geographical factor in locating a service. This contrasts with manufacturers, who must balance a variety of site and situation factors, as discussed in Chapter 11.

The best location for a factory is typically described as a region of the world, or perhaps a large area within a region. For example, auto alley—the optimal location for most U.S. motor vehicle factories—is an area of roughly 100,000 square kilometers. For service providers, the optimal location is much more precise: one corner of an intersection can be profitable and another corner of the same intersection unprofitable.

PROFITABILITY OF A LOCATION

Would a new department store be profitable in your community (Figure 12.4.1)? The two components of central place theory described in section 12.2—range and threshold—together determine the answer. Here's how:

1. **Compute the range.** You might survey local residents and determine that people are generally willing to travel up to 15 minutes to reach a department store.
2. **Compute the threshold.** A department store typically needs roughly 250,000 people living within a 15-minute radius.
3. **Draw the market area.** Draw a circle with a 15-minute travel radius around the proposed location. Count the number of people within the circle. If more than 250,000 people are within the radius, then the threshold may be high enough to justify locating the new convenience store in your community. However, your store may need a larger threshold and range to attract some of the available customers if competitors are located nearby.

The threshold must also be adjusted to the fact that the further customers are from the service the less likely they are to patronize it. Geographers have adapted the gravity model from physics. The **gravity model** predicts that the optimal location of a service is directly related to the number of people in the area and inversely related to the distance people must travel to access it. The best location will be the one that minimizes the distances that all potential customers must travel to reach the service.

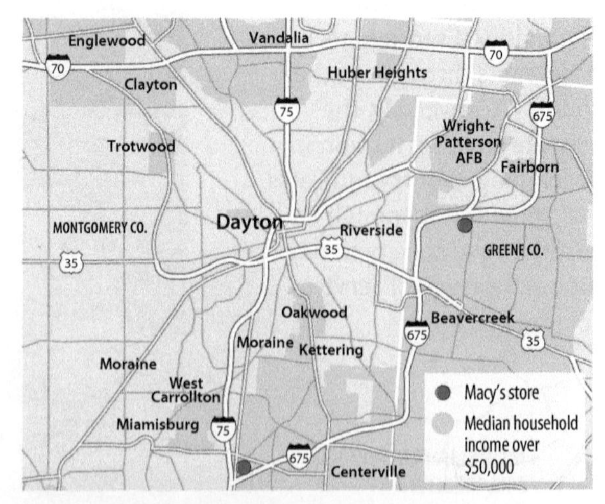

▲ 12.4.1 MARKET AREA, RANGE, AND THRESHOLD FOR MACY'S DEPARTMENT STORES IN THE DAYTON, OHIO, METROPOLITAN AREA

LOCATING A NEW RETAIL STORE

Major U.S. department store chains, mall developers, and other large retailers employ geographers to determine the best locations to build new stores. A large retailer has many locations to choose from when deciding to build new stores. A suitable site is one with the potential for generating enough sales to justify using the company's scarce capital to build it. Here are the steps for a large supermarket:

1. **Define market area.** The first step in forecasting sales for a proposed new retail outlet is to define the market or trade area where the store would derive most of its sales. Analysis relies heavily on the company's records of their customers' credit card transactions at existing stores. What are the zip codes of customers who paid by credit card? The market area of a department store is typically defined as the zip codes where two-thirds to three-fourths of the customers live. Walmart locates most of its stores on the edge of the city, because that is where most of its customers live (Figure 12.4.2).
2. **Estimate range.** Based on the zip codes of credit card customers, geographers estimate that the range for a large supermarket is about a 10-minute driving time.
3. **Estimate threshold.** The threshold for a large supermarket is about 25,000 people living within the 15-minute range with appropriate income levels. Walmart typically is attracted to areas of modest means, whereas supermarkets like Kroger, Publix, and Safeway prefer to be near higher income people. In the Dayton, Ohio, area, for example, Kroger has most of its stores in the relatively affluent south and east (Figure 12.4.3).
4. **Market share.** The proposed new supermarket will have to share customers with competitors. Geographers typically predict market share through the so-called analog method. One or more existing stores are identified in locations that the geographer judges to be comparable to the location of the proposed store. The market share of the comparable stores is applied to the proposed new store.

Information about the viability of a proposed new store is depicted through GIS. One layer of the GIS depicts the trade area of the proposed store. Other layers display characteristics of the people living in the area, such as distribution of households, average income, and competitors' stores.

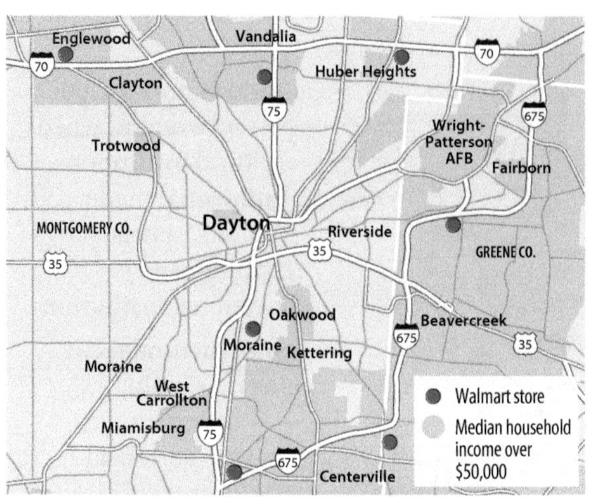

▲ 12.4.2 **MARKET AREA, RANGE, AND THRESHOLD FOR WALMART STORES IN THE DAYTON, OHIO, METROPOLITAN AREA**

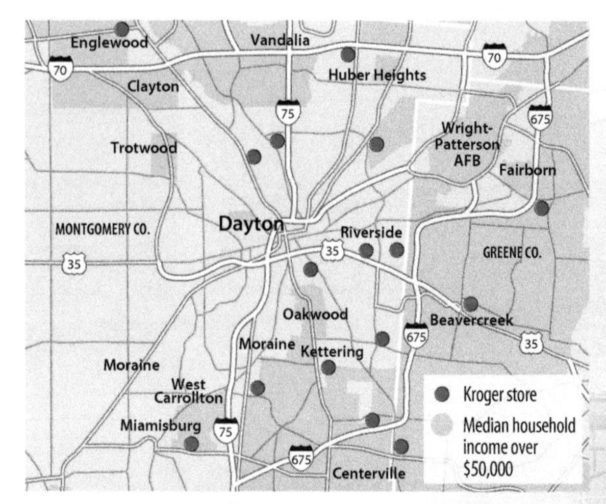

▲ 12.4.3 **MARKET AREA, RANGE, AND THRESHOLD FOR KROGER SUPERMARKETS IN THE DAYTON, OHIO, METROPOLITAN AREA**

12.5 Hierarchy of Business Services

- A hierarchy of world cities can be identified based on business services.
- A hierarchy of cities also exists inside the United States.

Every settlement provides consumer services to people in a surrounding area, but not every settlement of a given size has the same number and types of business services. Business services disproportionately cluster in a handful of settlements.

BUSINESS SERVICES IN WORLD CITIES

Geographers distinguish settlements according to their importance in the provision of business services. At the top of the hierarchy are settlements known as world cities or global cities that play an especially important role in global business services. World cities are most closely integrated into the global economic system because they are at the center of the flow of information and capital.

- Headquarters of large corporations are clustered in world cities, and shares of these corporations are bought and sold on the stock exchanges located in world cities. Obtaining information in a timely manner is essential in order to buy and sell shares at attractive prices.
- Lawyers, accountants, and other professionals cluster in world cities to provide advice to major corporations and financial institutions. Advertising agencies, marketing firms, and other services concerned with style and fashion locate in world cities to help corporations anticipate changes in taste and to help shape those changes.
- As centers for finance, world cities attract the headquarters of the major banks, insurance companies, and specialized financial institutions where corporations obtain and store funds for expansion of production.
- World cities also contain a disproportionately high share of the world's arts, culture, consumer spending on luxury goods, and political power.

Global cities are divided into three levels, called alpha, beta, and gamma. These three levels in turn are further subdivided (Figure 12.5.1). A combination of economic, political, cultural, and infrastructure factors are used to identify world cities and to distinguish among the various ranks.

- Economic factors include the number of headquarters for multinational corporations, financial institutions, and law firms that influence the world economy.
- Political factors include hosting headquarters for international organizations and capitals of countries that play a leading role in international events.
- Cultural factors include presence of renowned cultural institutions, influential media outlets, sports facilities, and educational institutions.
- Infrastructural factors include a major international airport, health-care facilities, and advanced communications systems.

▼ 12.5.1 HIERARCHY OF WORLD CITIES

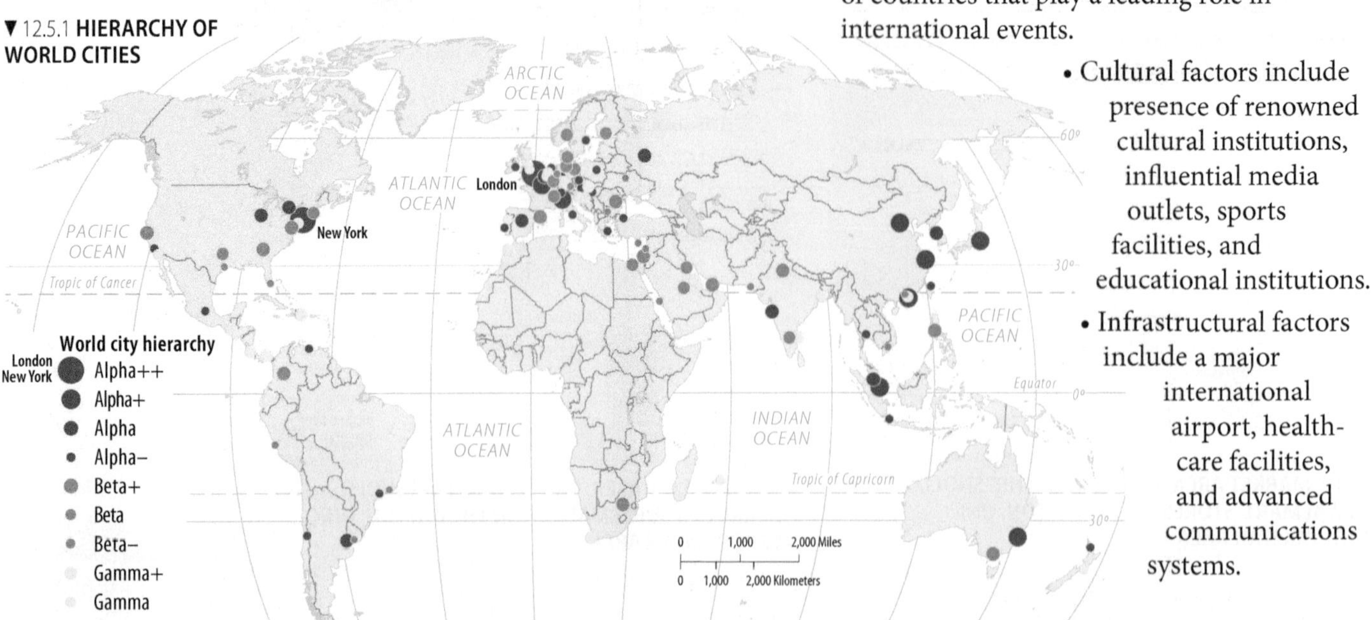

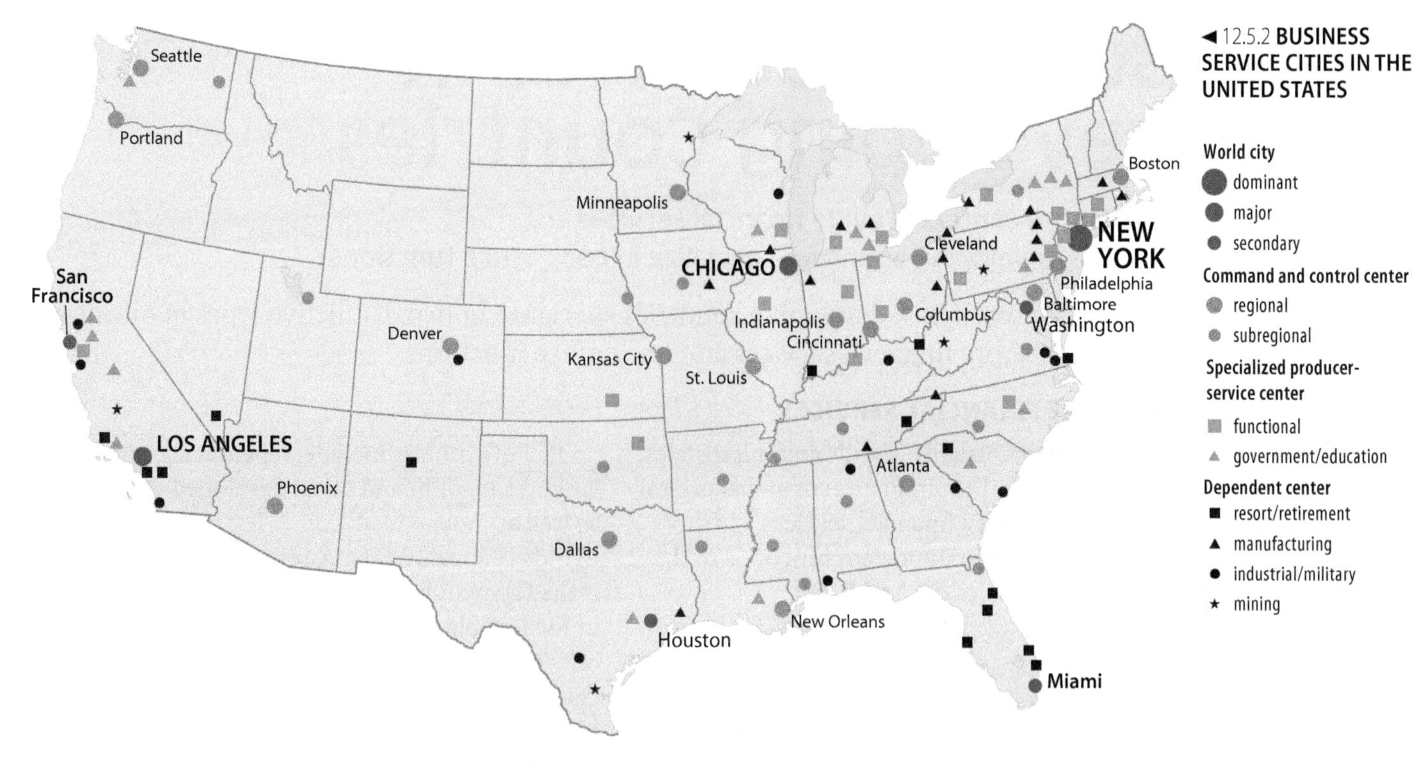

◄ 12.5.2 **BUSINESS SERVICE CITIES IN THE UNITED STATES**

THREE LOWER TIERS OF BUSINESS SERVICES

Below the first tier of world cities are three other tiers of settlements according to type and extent of business services. Examples of each can be seen in the United States (Figure 12.5.2). The world cities in the United States are shown in Figure 12.5.3.

Second tier: command and control centers. These contain the headquarters of many large corporations, well-developed banking facilities, and concentrations of other business services, including insurance, accounting, advertising, law, and public relations. Important educational, medical, and public institutions can be found in these command and control centers. Examples include Baltimore, Cleveland, Phoenix, and St. Louis.

Third tier: specialized producer-service centers. These offer narrower and more highly specialized services. One group of these cities specializes in the management and R&D (research and development) activities related to specific industries, such as Detroit (motor vehicles), Pittsburgh (steel), and Rochester (office equipment). A second group of these cities specializes as centers of government and education, notably state capitals that also have a major university, such as Albany, Lansing, and Madison.

Fourth tier: dependent centers. These provide relatively unskilled jobs. Four subgroups include:

- Resort, retirement, and residential centers, such as Albuquerque, Fort Lauderdale, Las Vegas, and Orlando, clustered in the South and West.
- Manufacturing centers, such as Buffalo, Chattanooga, Erie, and Rockford, clustered mostly in the old northeastern manufacturing belt.
- Military centers, such as Huntsville, Newport News, and San Diego, clustered mostly in the South and West.
- Mining and industrial centers, such as Charleston, West Virginia, and Duluth, Minnesota, located in mining areas.

► 12.5.3 **THE U.S. WORLD CITIES: LOS ANGELES (top), CHICAGO (middle), NEW YORK (bottom)**

12.6 Business Services in Developing Countries

- **Offshore centers provide financial services.**
- **Some developing countries specialize in back office functions.**

In the global economy, developing countries specialize in two distinctive types of business services: offshore financial services and back office functions.

OFFSHORE FINANCIAL SERVICES

Small countries, usually islands and microstates, exploit niches in the circulation of global capital by offering offshore financial services. Offshore centers provide two important functions in the global circulation of capital:

- **Taxes.** Taxes on income, profits, and capital gains are typically low or nonexistent in these locations. Companies incorporated in an offshore center also have tax-free status regardless of the nationality of the owners.
- **Privacy.** Bank secrecy laws help individuals and businesses evade disclosure in their home countries. People and corporations can protect their assets from lawsuits by depositing their money in offshore banks. Creditors cannot reach such assets in bankruptcy hearings. The privacy laws and low tax rates in offshore centers can also provide havens to tax dodges and other illegal schemes.

Offshore centers include dependencies of the United Kingdom and other developed countries, as well as independent countries. Many are islands (Figure 12.6.1). A prominent example is the Cayman Islands, a British Crown Colony in the Caribbean near Cuba. The Caymans have only 40,000 inhabitants, but there are 70,000 companies there, including several hundred banks and the world's four largest legal and accounting firms (Figure 12.6.2).

In the Caymans, it is a crime to discuss confidential business—defined as matters learned on the job—in public. Assets placed in an offshore center by an individual or corporation in a trust are not covered by lawsuits originating in other countries. To get at those assets, additional lawsuits would have to be filed in the offshore centers, where privacy laws would shield the individual or corporation from undesired disclosures.

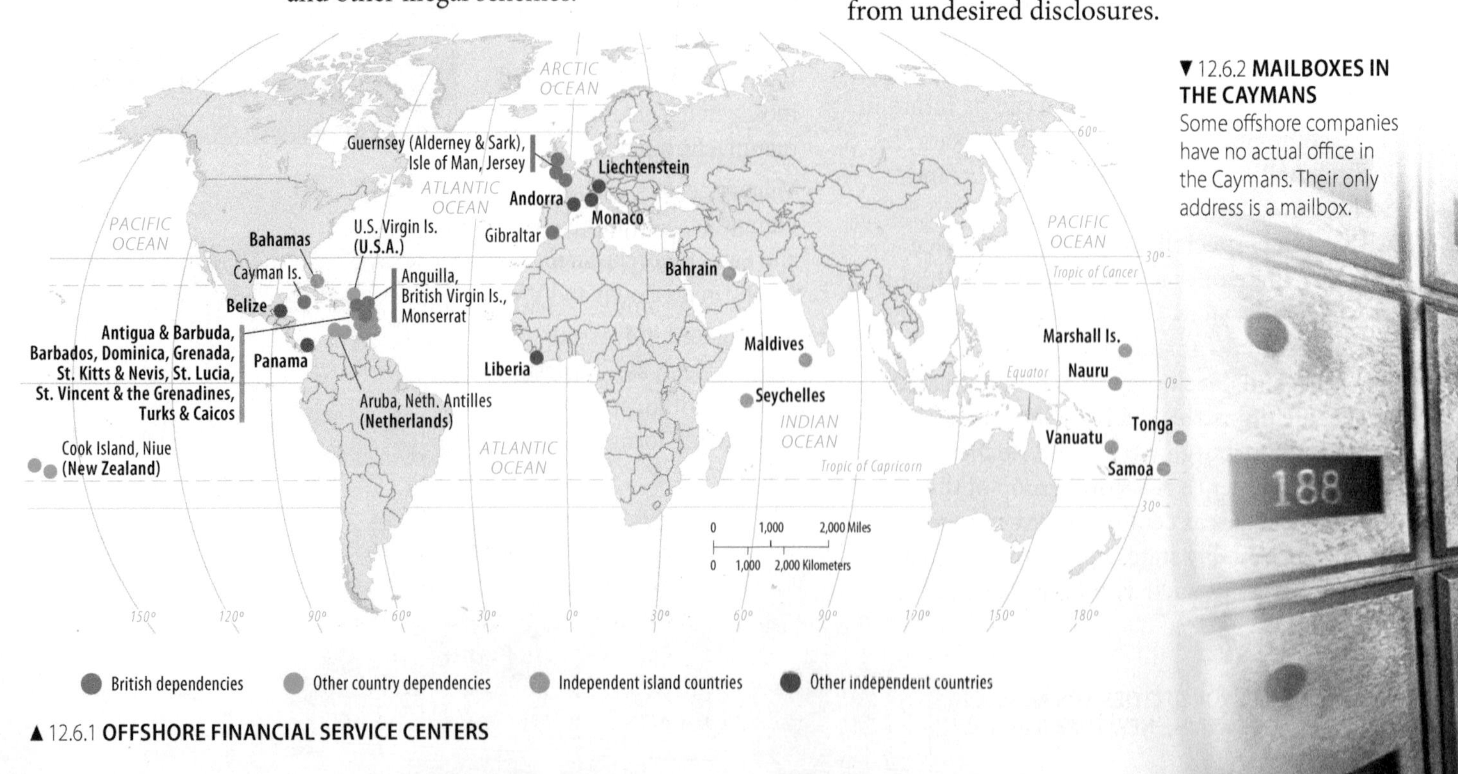

▲ 12.6.1 **OFFSHORE FINANCIAL SERVICE CENTERS**

▼ 12.6.2 **MAILBOXES IN THE CAYMANS**
Some offshore companies have no actual office in the Caymans. Their only address is a mailbox.

BACK OFFICES

Developing countries are increasingly centers for back-office functions, also known as business-process outsourcing. Typical back-office functions include processing insurance claims, payroll management, transcription work, and other routine clerical activities. Back-office work also includes centers for responding to billing inquiries related to credit cards, shipments, and claims, or technical inquiries related to installation, operation, and repair. Need to have your computer fixed? Correct a mistake on your credit card bill? Change your plane reservation? The human you have reached is probably at a call center in a developing country (Figure 12.6.3).

Traditionally, companies housed their back-office staff in the same office building as their management staff, or at least in nearby buildings. A large percentage of the employees in a downtown bank building, for example, would be responsible for sorting paper checks and deposit slips. Proximity was considered important to assure close supervision of routine office workers and rapid turnaround of information. For many business services, improved telecommunications have eliminated the need for spatial proximity.

Selective developing countries have attracted back offices for two reasons related to labor:

- **Low wages.** Most back-office workers earn a few thousand dollars per year—higher than wages paid in most other sectors of the economy, but only one-tenth the wages paid for workers performing similar jobs in developed countries.

▲ 12.6.3 **CALL CENTER, BANGALORE, INDIA**

- **Ability to speak English.** A handful of developing countries possess a large labor force fluent in English (Figure 12.6.4). In Asia, countries such as India, Malaysia, and the Philippines have substantial numbers of workers with English-language skills, a legacy of British and American colonial rule. The ability to communicate in English over the telephone is a strategic advantage in competing for back offices with neighboring countries, such as Indonesia and Thailand, where English is less commonly used. Familiarity with English is an advantage not only for literally answering the telephone but also for gaining a better understanding of the preferences of American consumers through exposure to English-language music, movies, and television.

▲ 12..6.4 **ADVERTISEMENT, BANGALORE, INDIA**

Call-center employees must be able to understand what a customer located in North America is trying to say and must be able to respond clearly in language understood by a "typical" North American. Call centers in Asia pretend that they are located in North America and are employing Americans. But in one respect, they can't escape the "tyranny" of geography. Refer to Figure 1.4.4, the map of world time zones. In the middle of the day, when most Americans are placing calls, it is the middle of the night in Asia. So call center employees in Asia typically work all night.

12.7 Economic Base

- Settlements can be classified by their economic base.
- Talent is not distributed uniformly among cities.

A settlement's distinctive economic structure derives from its **basic industries**, which export primarily to businesses and individuals outside the settlement. **Nonbasic industries** are enterprises whose customers live in the same community—essentially, consumer services. A community's unique collection of basic industries defines its **economic base.**

A settlement's economic base is important, because exporting by the basic industries brings money into the local economy, thus stimulating the provision of more nonbasic consumer services for the settlement. New basic industries attract new workers to a settlement, and they bring their families with them. The settlement then attracts additional consumer services to meet the needs of the new workers and their families. Thus a new basic industry stimulates establishment of new supermarkets, laundromats, restaurants, and other consumer services. But a new nonbasic service, such as a supermarket, will not induce construction of new basic industries.

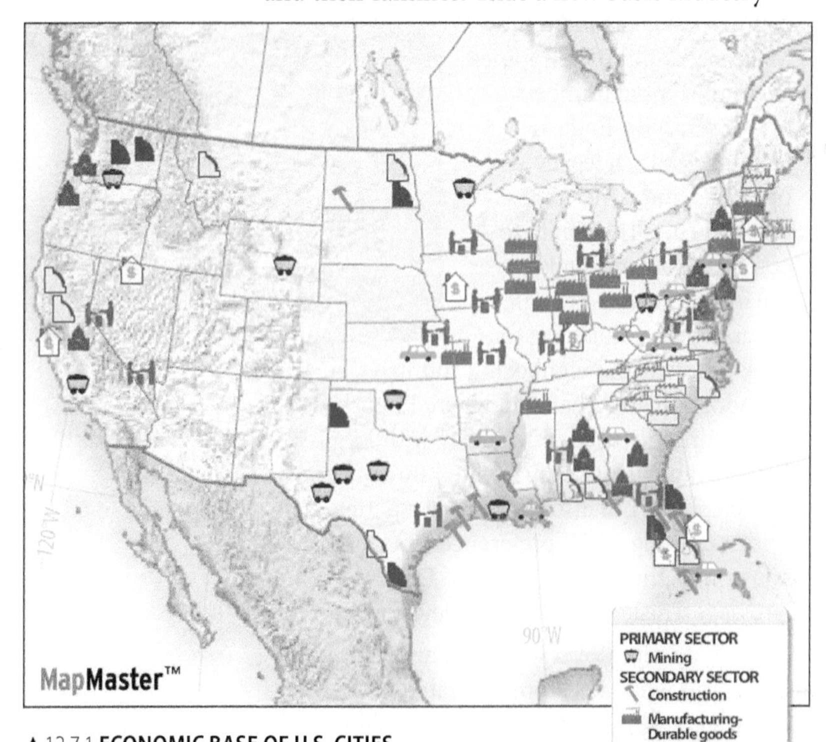

▲ 12.7.1 **ECONOMIC BASE OF U.S. CITIES**
Cities have distinctive economic bases.

Launch MapMaster North America in MasteringGEOGRAPHY

Select: *Economic* then *Economic base of U.S. cities*

Select: *Political* then *Cities*

What is the name of a city whose economic base is mining? Construction? Manufacturing durable goods? Manufacturing nondurable goods? Retail? Wholesale? Personal services? Finance? Transportation? Public services?

SPECIALIZING IN SPECIFIC SERVICES

Settlements in the United States can be classified by their type of basic activity (Figure 12.7.1). Each type of basic activity has a different spatial distribution. The concept of basic industries originally referred to manufacturing. In a postindustrial society, such as the United States, increasingly the basic economic activities are in services.

Examples of settlements specializing in business services include:

- General business: New York, Los Angeles, Chicago, and San Francisco.
- Computing and data processing services: Boston and San Jose.
- High-tech industries support services: Austin, Orlando, and Raleigh-Durham.
- Military activity support services: Albuquerque, Colorado Springs, Huntsville, Knoxville, and Norfolk.
- Management consulting services: Washington, D.C.

Examples of settlements specializing in consumer services:

- Entertainment and recreation: Atlantic City, Las Vegas, and Reno.
- Medical services: Rochester, Minnesota.

Examples of settlements specializing in public services:

- State capitals: Sacramento and Tallahassee.
- Large universities: Tuscaloosa.
- Military: Arlington.

DISTRIBUTION OF TALENT

Individuals possessing special talents are not distributed uniformly among cities. Some cities have a higher percentage of talented individuals than others (Figure 12.7.2). Talent was measured by Richard Florida as a combination of the percentage of people in the city with college degrees, the percentage employed as scientists or engineers, and the percentage employed as professionals or technicians.

Florida found a significant positive relationship between the distribution of talent and the distribution of cultural diversity in the largest U.S. cities (Figure 12.7.3). In other words, cities with high cultural diversity tended to have relatively high percentages of talented individuals (Figure 12.7.4). Attracting talented individuals is important for a city, because these individuals are responsible for promoting economic innovation. They are likely to start new businesses and infuse the local economy with fresh ideas.

▲ 12.7.3 **THE ATTRACTION OF URBAN NIGHTLIFE**
A lively nightlife scene attracts young talented people to some cities, such as Miami, Florida.

◄ 12.7.2 **GEOGRAPHY OF TALENT**

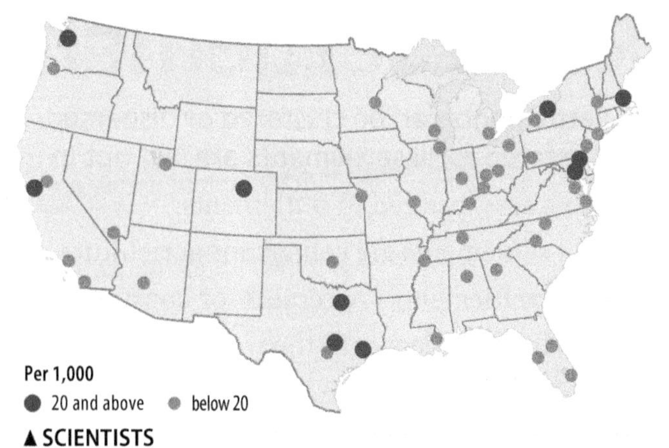

▲ **SCIENTISTS**

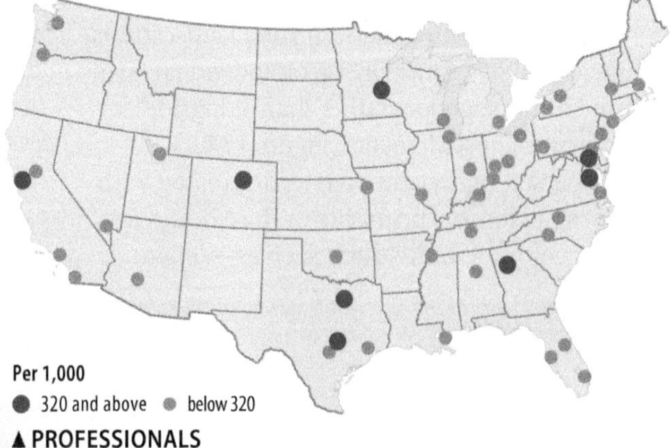

▲ **PROFESSIONALS**

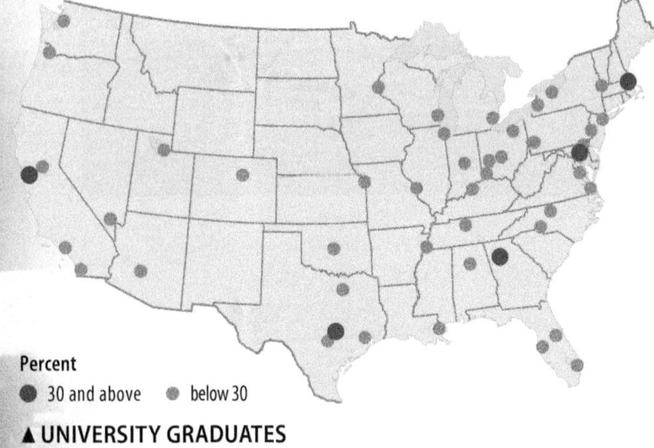

▲ **UNIVERSITY GRADUATES**

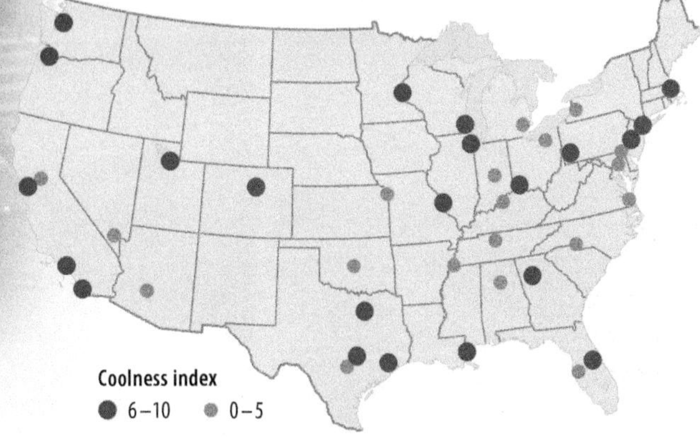

◄ 12.7.4 **GEOGRAPHY OF CULTURAL DIVERSITY**
The map is based on a "coolness" index, developed by POV Magazine, combined the percentage of population in their 20s, the number of bars and other nightlife places per capita, and the number of art galleries per capita.

12.8 Rural Settlements

- ▶ **Settlements can be clustered or dispersed.**
- ▶ **Clustered rural settlements are laid out in many types of patterns.**

Services are clustered in settlements.

- Rural settlements are centers for agriculture and provide a small number of services.
- Urban settlements are centers for consumer and business services.

One-half of the people in the world live in a rural settlement and the other half in an urban settlement.

DISPERSED RURAL SETTLEMENTS

A **dispersed rural settlement,** typical of the contemporary North American rural landscape, is characterized by farmers living on individual farms isolated from neighbors rather than alongside other farmers in settlements (Figure 12.8.1).

A dispersed settlement pattern originated with American colonists, primarily in the Middle Atlantic colonies. Individuals such as William Penn (Pennsylvania), Lord Baltimore (Maryland), and Sir George Carteret (the Carolinas) received large land grants by the King of England and in turn sold tracts to individual colonists. Pioneers, primarily from the Middle Atlantic colonies, crossed the Appalachian Mountains and established dispersed farms on the frontier. Land was plentiful and cheap, and people bought as much as they could manage.

◄ 12.8.1 **DISPERSED RURAL SETTLEMENT**

Use Google Earth to explore the dispersed settlement pattern in northern North Dakota.

Fly to: *Russell, North Dakota*

Zoom out until your screen begins to resemble a checkerboard.

1. **What is one personal quality a resident of this landscape would need to possess? Explain your answer.**
2. **What physical features interrupt the checkerboard pattern in the east?**

(below) Farm in northern, North Dakota.

CLUSTERED RURAL SETTLEMENTS

A **clustered rural settlement** is an agricultural-based community in which a number of families live in close proximity to each other, with fields surrounding the collection of houses and farm buildings. A clustered rural settlement typically includes homes, barns, tool sheds, and other farm structures, plus consumer services, such as religious structures, schools, and shops. A handful of public and business services may also be present in the clustered rural settlement, often centered on an open area called a common.

Much of rural England was laid out in clustered settlements (Figure 12.8.2). When early English settlers reached New England they originally built clustered settlements. They typically traveled to the New World in a group and wanted to live close together to reinforce common cultural and religious values. The contemporary New England landscape contains remnants of the old clustered rural settlement pattern.

▲ 12.8.2 **CLUSTERED RURAL SETTLEMENT**
Use Google Earth to explore an English village.
Fly to: *Finchingfield, England*
Zoom in until Finchingfield occupies most of the Google Earth screen.
1. Where is the center of Finchingfield?
2. How were you able to identify it?

CLUSTERED LINEAR RURAL SETTLEMENTS

Clustered rural settlements are sometimes arranged in a geometric pattern. Linear rural settlements feature buildings clustered along a road or body of water to facilitate transportation and communications. The fields extend behind the buildings in long narrow strips (Figure 12.8.3). Long-lot farms can be seen today along the St. Lawrence River in Québec.

▲ 12.8.3 **CLUSTERED LINEAR RURAL SETTLEMENT**
Use Google Earth to explore settlement patterns along the St. Lawrence River.
Fly to: *Les Bricailles, Québec*
Drag the Street View icon west from the Les Bricailles placemark to the highway where it passes open fields and explore the landscape along the highway.
1. How would you describe the landscape of Les Bricailles?
Exit street view and zoom out until you can see both banks of the St. Lawrence River.
2. Why might it be an advantage to farmers to own narrow strips of land extending far inland from the river?

CLUSTERED CIRCULAR RURAL SETTLEMENTS

The clustered circular rural settlement consists of a central open space surrounded by structures. Von Thünen observed this circular rural pattern in Germany in his landmark agricultural studies in the early nineteenth century (refer to section 10.11). Germany's Gewandorf settlements consisted of a core of houses, barns, and churches encircled by different types of agricultural activities.

In sub-Saharan Africa, the Maasi people, who are pastoral nomads, build circular settlements known as kraal; women have the principal responsibility for constructing them. The kraal villages have enclosures for livestock in the center, surrounded by a ring of houses. Compare *kraal* to the English word *corral* (Figure 12.8.4).

▼ 12.8.4 **CLUSTERED CIRCULAR RURAL SETTLEMENT**
Maasi kraal settlement, Kenya.

12.9 Settlements in History

- **Settlements originated in multiple hearths and diffused in multiple directions.**
- **Through history, the world's largest settlement has usually been in Southwest Asia, Egypt, or China.**

Permanent settlements existed prior to the beginning of recorded history around 5,000 years ago. The earliest settlements may have been established as service centers:

- **Consumer services**. The first permanent settlements may have been places for nomads to bury and honor their dead. They were also places to house women and children while males hunted for food. Women made tools, clothing, and containers.
- **Business services**. Early settlements were places where groups could store surplus food and trade with other groups.
- **Public services**. Early settlements housed political leaders, as well as military forces to guard the residents of the settlement.

Settlements may have originated in Mesopotamia, part of the Fertile Crescent of Southwest Asia, and diffused at an early date west to Egypt and east to China and South Asia's Indus Valley. Or they may have originated independently in each of the four hearths. In any case, from these four hearths, the concept of settlements diffused to the rest of the world.

Until around 300 B.C., the world's largest settlements were in Mesopotamia and Egypt (Figure 12.9.1). Ancient Memphis, Egypt, may have been the first settlement to exceed 30,000 inhabitants around 5,000 years ago (Figure 12.9.2). Beginning around 2,400 years ago, as settlements diffused from Southwest Asia, settlements in India, China, and Europe frequently emerged as the world's largest. Constantinople (now Istanbul, Turkey) was the world's largest settlement for the longest stretch of time during the Middle Ages (Figures 12.9.3 and 12.9.4).

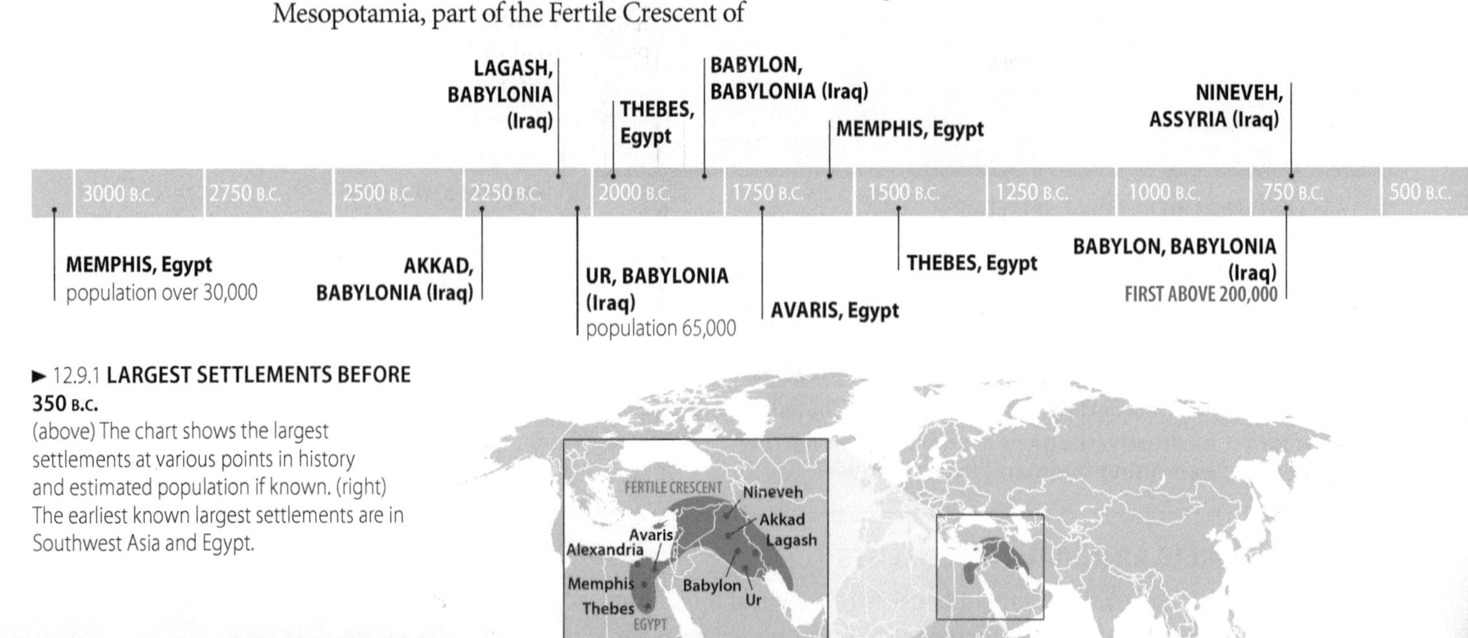

► 12.9.1 **LARGEST SETTLEMENTS BEFORE 350 B.C.**
(above) The chart shows the largest settlements at various points in history and estimated population if known. (right) The earliest known largest settlements are in Southwest Asia and Egypt.

◄ 12.9.2 **MEMPHIS, EGYPT**
The Alabaster Sphinx was constructed around 3,500 years ago near Memphis, Egypt, which at the time was probably the world's largest urban settlement.

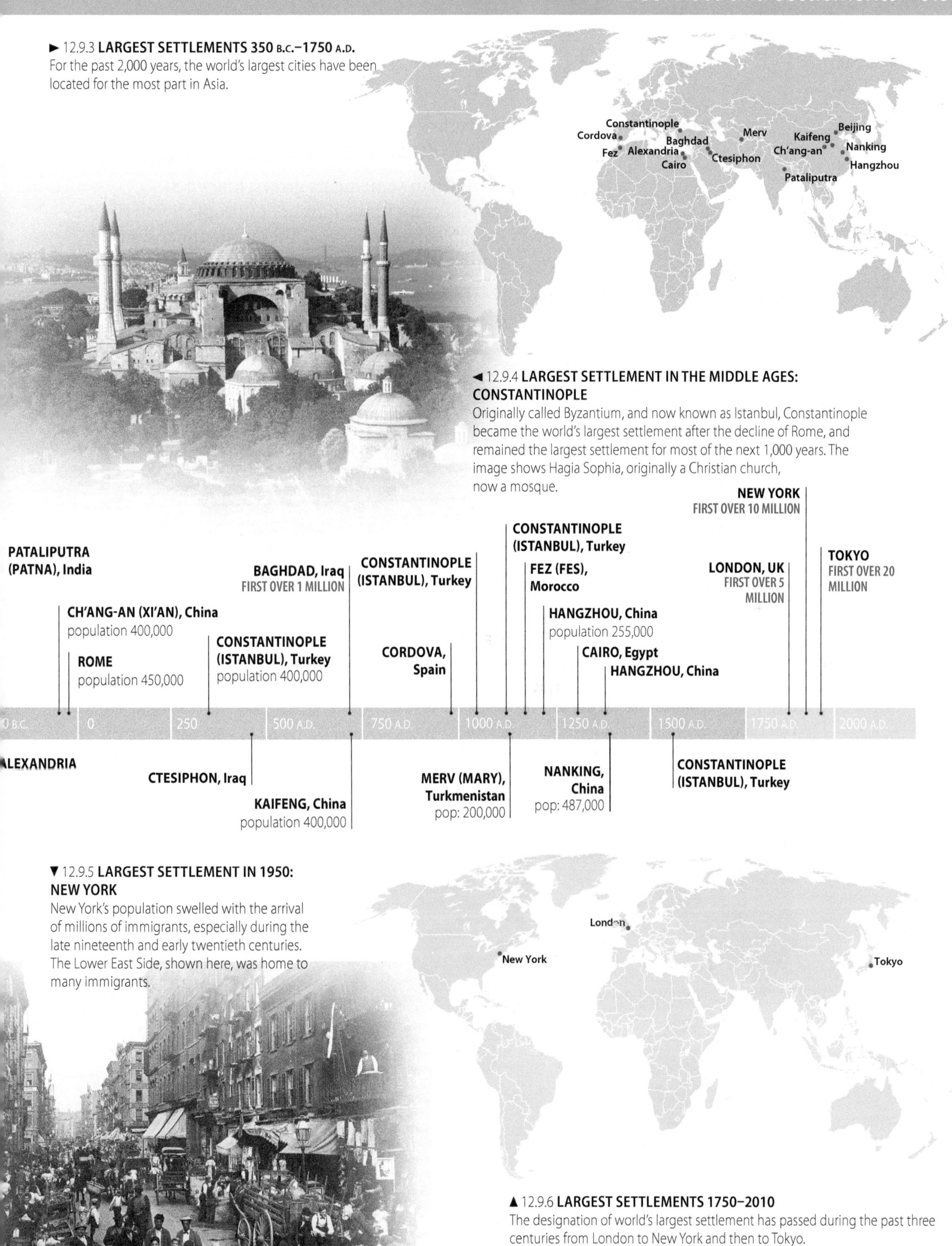

► 12.9.3 **LARGEST SETTLEMENTS 350 B.C.–1750 A.D.**
For the past 2,000 years, the world's largest cities have been located for the most part in Asia.

◄ 12.9.4 **LARGEST SETTLEMENT IN THE MIDDLE AGES: CONSTANTINOPLE**
Originally called Byzantium, and now known as Istanbul, Constantinople became the world's largest settlement after the decline of Rome, and remained the largest settlement for most of the next 1,000 years. The image shows Hagia Sophia, originally a Christian church, now a mosque.

▼ 12.9.5 **LARGEST SETTLEMENT IN 1950: NEW YORK**
New York's population swelled with the arrival of millions of immigrants, especially during the late nineteenth and early twentieth centuries. The Lower East Side, shown here, was home to many immigrants.

▲ 12.9.6 **LARGEST SETTLEMENTS 1750–2010**
The designation of world's largest settlement has passed during the past three centuries from London to New York and then to Tokyo.

12.10 Urbanization

- **Developed countries have a higher percentage of people living in urban areas, a consequence of economic restructuring.**
- **Most of the world's largest cities are in developing countries.**

The process by which the population of urban settlements grows, known as urbanization, has two dimensions—an increase in the number of people living in cities and an increase in the percentage of people living in cities. These two dimensions of urbanization occur for different reasons and have different global distributions.

PERCENTAGE OF PEOPLE IN CITIES

The world's population of urban settlements exceeded that of rural settlements for the first time in human history in 2008. The percentage of people living in cities increased from 3 percent in 1800 to 6 percent in 1850, 14 percent in 1900, and 30 percent in 1950.

A large percentage of people living in urban areas is a measure of a country's level of development. Three-fourths of people live in urban settlements in developed countries, compared to about two-fifths in developing countries (Figure 12.10.1). The major exception to the global pattern is Latin America, where the urban percentage is comparable to the level of developed countries (Figure 12.10.2).

The higher percentage of urban residents in developed countries is a consequence of changes in economic structure during the past two centuries—first the Industrial Revolution in the nineteenth century and then the growth of services in the twentieth century. During the past 200 years rural residents in developed countries have migrated from the countryside to work in the factories and services that are concentrated in cities.

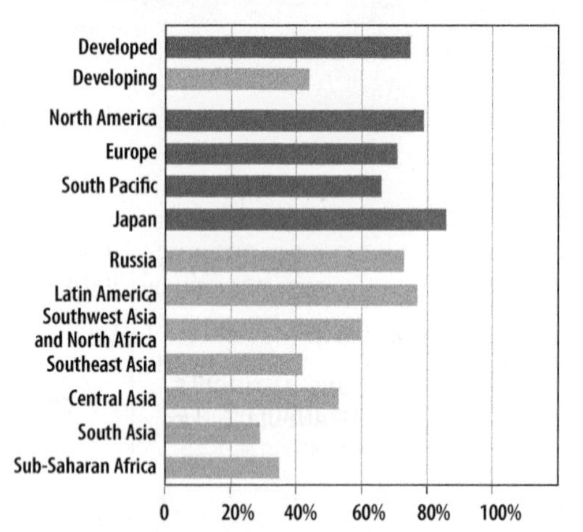

▲ 12.10.2 **PERCENT LIVING IN URBAN SETTLEMENTS BY REGION**

► 12.10.1 **PERCENT LIVING IN URBAN SETTLEMENTS**

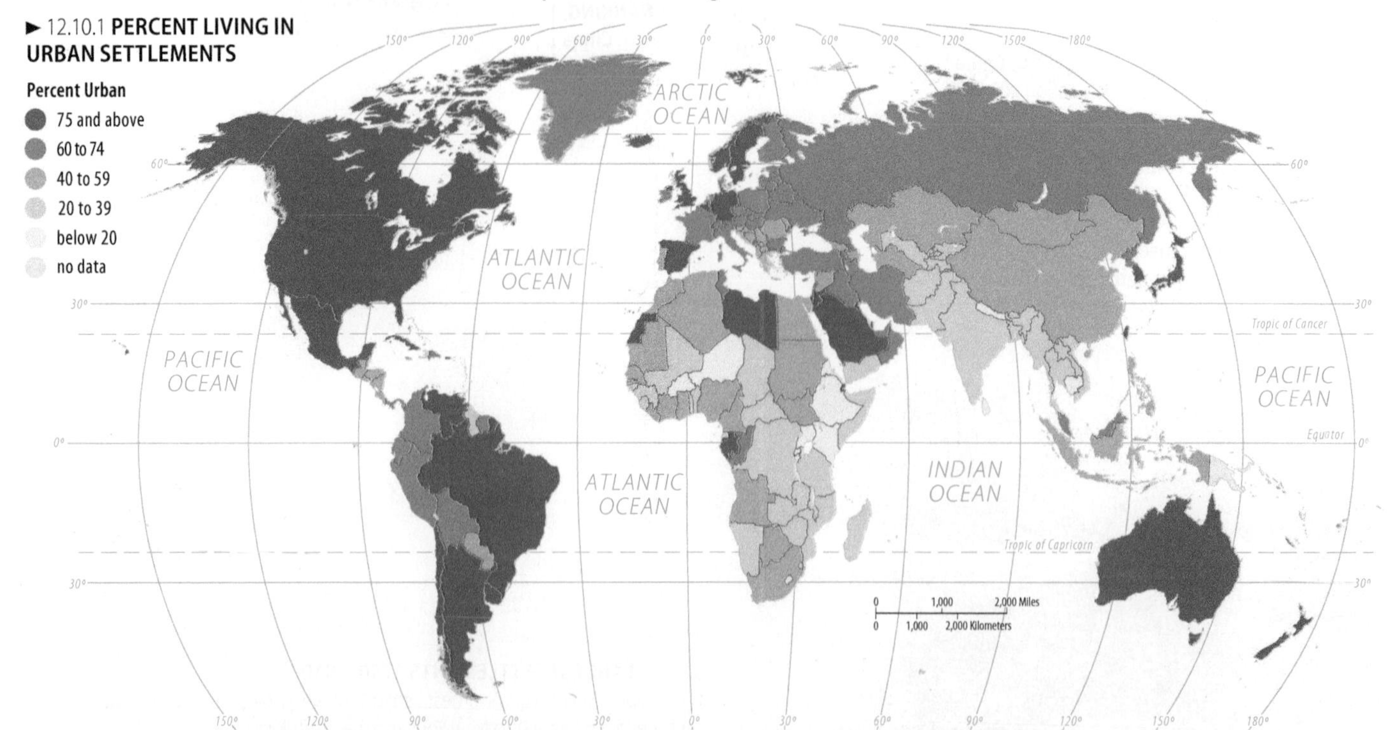

▲ 12.10.3 **URBAN SETTLEMENTS WITH AT LEAST 3 MILLION INHABITANTS**

NUMBER OF PEOPLE IN CITIES

Developed countries have a higher percentage of urban residents, but developing countries have more of the very large urban settlements (Figure 12.10.3). Seven of the ten (and sixteen of the twenty) most populous cities are in developing countries.

Identifying the world's largest cities is difficult, because each country defines cities in a unique manner. *Demographia* uses maps and satellite imagery to delineate urban areas consistently regardless of country. According to *Demographia*, 171 urban areas have at least 2 million inhabitants, 105 at least 3 million, 55 at least 5 million, 22 at least 10 million, and 4 (Tokyo, Jakarta, New York, and Seoul) at least 20 million.

That developing countries dominate the list of largest urban settlements is remarkable because urbanization was once associated with economic development. In 1900, after diffusion of the Industrial Revolution from Great Britain to Europe and North America, all ten of the world's largest cities were in developed countries.

Compare the world's most populous cities to the most important business service centers (refer to Figure 12.5.1). Several of the world's most populous cities in developing countries—including Jakarta, Manila, and Mexico City—do not rank among the world's most important business service centers (Figure 12.10.4). On the other hand, cities in developed countries such as London, Paris, Chicago, and Toronto rank among the world's twenty most important settlements for business services but are not among the twenty most populous.

▼ 12.10.4 **MEXICO CITY**

Geographers do not merely observe the distribution of services; they play a major role in creating it. Shopping center developers, large department store and supermarket chains, and other retailers employ geographers to identify new sites for stores and assess the performance of existing stores. Geographers conduct statistical analyses based on the gravity model to delineate underserved market areas where new stores could be profitable, as well as to identify overserved market areas where poorly performing stores are candidates for closure.

Developers of new retail services obtain loans from banks and financial institutions to construct new stores and malls. Lending institutions want assurance that the proposed retail development has a market area with potential to generate sufficient profits to repay the loan. They employ geographers to make objective market-area analyses independent of the excessively optimistic forecasts submitted by the retailer.

Many service providers make location decisions on the basis of instinct, intuition, and tradition. In an increasingly competitive market, retailers and other services that place themselves in the optimal location secure a critical advantage.

Key Questions

Where are consumer services distributed?

- Three types of services are consumer, business, and public.
- In developed countries, the distribution of consumer services follows a regular pattern, explained through central place theory.
- Services have market areas, ranges, and thresholds that can be measured.
- Geographers apply central place theory to identify profitable locations for services.

Where are business services distributed?

- Business services are disproportionately clustered in world cities.
- Distinctive business services in developing countries include offshore financial services and back offices.
- Talented people are attracted to world cities by cultural diversity.

Where are settlements distributed?

- Outside North America, most rural settlements are clustered.
- The first settlements predate recorded history.
- Developed countries have higher percentages of urban dwellers, whereas developing countries have most of the world's largest cities.

Thinking Geographically

Consult Wikipedia's Global Cities article. Several indexes are included, with somewhat varying criteria. Some cities (such as London and Singapore) are included in the top ten global cities on all lists, whereas others (such as Sydney and Shanghai) are not (Figure 12.CR.1).

1. Identify a city in a developed country in addition to Sydney and a city in a developing country in addition to Shanghai that do not appear among the top ten global cities on all lists. What factors have caused these two cities to be included or excluded?

Your community's economy is expanding or contracting as a result of the performance of its basic industries. Two factors can explain the performance of your community's basic industries. One is that the sector is expanding or contracting nationally. The other is that the sector is performing much better or worse in your community than in the country as a whole.

2. Which of the two factors better explains the performance of your community's basic industries?

In a developed region like North America, even cities not classified as world cities are connected to the global economy.

3. What evidence can you find in your community of economic ties to world cities located elsewhere in North America? In other regions?

▼ 12.CR.1 **SYDNEY, AUSTRALIA**

Interactive Mapping

CITIES AND PHYSICAL FEATURES IN EAST ASIA

China's largest cities are not distributed uniformly across the country.

Launch MapMaster East Asia in

MasteringGEOGRAPHY

Select *Physical features*

Select *Cities*

1. **In which physical regions are China's cities clustered?**
2. **Where are there fewer cities?**
3. **How can you explain this pattern?**

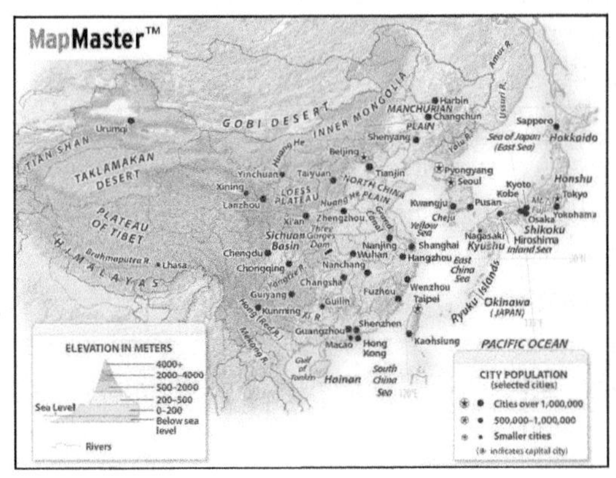

Explore

UR, IRAQ

Use Google Earth to explore one of the world's oldest cities.

Fly to: *Ur, Nassriya, Iraq*

Click 3D Buildings layer in the Primary Database.

Zoom in until the runways of an airstrip appear to the northeast. Then center on the runways.

Notice the dark oval shape about 1.25 miles to the northeast. Recenter and zoom in on the oval until a 3D building appears.

1. **Measure the dimensions of the 3D building. Is it larger or smaller than your house?**
2. **What can you infer about the history of this site from the size and shape of the building and the dark, oval area around it?**

On the Internet

The population of cities organized by country can be accessed at **http://www.citypopulation.de/World.html** or scan the QR on the first page of the chapter.

An attempt to apply the same definition to measuring the population of cities everywhere has been published by *Demographia*, accessed at ***www.demographia.com***. Wendell Cox, a private consultant, is the sole owner of Demographia.

Key Terms

Basic industries
Industries that sell their products or services primarily to consumers outside the settlement.

Business services
Services that primarily meet the needs of other businesses, including professional, financial, and transportation services.

Central place
A market center for the exchange of services by people attracted from the surrounding area.

Central place theory
A theory that explains the distribution of services, based on the fact that settlements serve as centers of market areas for services; larger settlements are fewer and farther apart than smaller settlements and provide services for a larger number of people who are willing to travel farther.

Clustered rural settlement
An agricultural based community in which a number of families live in close proximity to each other, with fields surrounding the collection of houses and farm buildings.

Consumer services
Businesses that provide services primarily to individual consumers, including retail services and education, health, and leisure services.

Dispersed rural settlement
A rural settlement pattern in which farmers live on individual farms isolated from neighbors.

Economic base
A community's collection of basic industries.

Gravity model
A model that holds that the potential use of a service at a particular location is directly related to the number of people in a location and inversely related to the distance people must travel to reach the service.

Market area (or hinterland)
The area surrounding a central place, from which people are attracted to use the place's goods and services.

Nonbasic industries
Industries that sell their products primarily to consumers in the community.

Primate city
The largest settlement in a country, if it has more than twice as many people as the second-ranking settlement.

Primate city rule
A pattern of settlements in a country, such that the largest settlement has more than twice as many people as the second-ranking settlement.

Public services
Services offered by the government to provide security and protection for citizens and businesses.

Range (of a service)
The maximum distance people are willing to travel to use a service.

Rank-size rule
A pattern of settlements in a country, such that the nth largest settlement is 1/n the population of the largest settlement.

Service
Any activity that fulfills a human want or need and returns money to those who provide it.

Threshold
The minimum number of people needed to support the service.

▸ LOOKING AHEAD

This chapter has looked at the distribution of cities across the world. The next chapter focuses on the distribution of people and activities within cities.

13 Urban Patterns

When you stand at the corner of Fifth Avenue and 34th Street in New York City, staring up at the Empire State Building, you know that you are in a city. When you are standing in an Iowa cornfield, you have no doubt that you are in the country. Geographers help explain what makes city and countryside different places.

A large city is stimulating and agitating, entertaining and frightening, welcoming and cold. A city has something for everyone, but a lot of those things are for people different from you. Urban geography helps to sort out the complexities of familiar and unfamiliar patterns in urban areas. Models help to explain where different people and activities are distributed within urban areas, and why those differences occur.

Where are people distributed within urban areas?

URBAN SPRAWL INTO THE DESERT, PHOENIX, ARIZONA.

How are urban areas expanding?

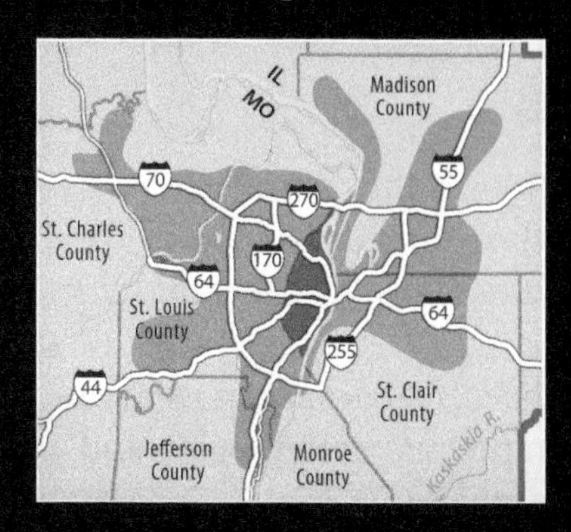

What challenges do cities face?

SCAN FOR CENSUS MAPS OF EVERY U.S. CITY

13.1 The Central Business District

- Downtown is known as the central business district (CBD).
- The CBD contains consumer, business, and public services.

The best-known and most visually distinctive area of most cities is the central area, commonly called downtown and known to geographers by the more precise term **central business district (CBD)**. The CBD is usually one of the oldest districts in a city, often the original site of the settlement (Figure 13.1.1).

Consumer, business, and public services are attracted to the CBD because of its accessibility. The center is the easiest part of the city to reach from the rest of the region and is the focal point of the region's transportation network (Figure 13.1.2).

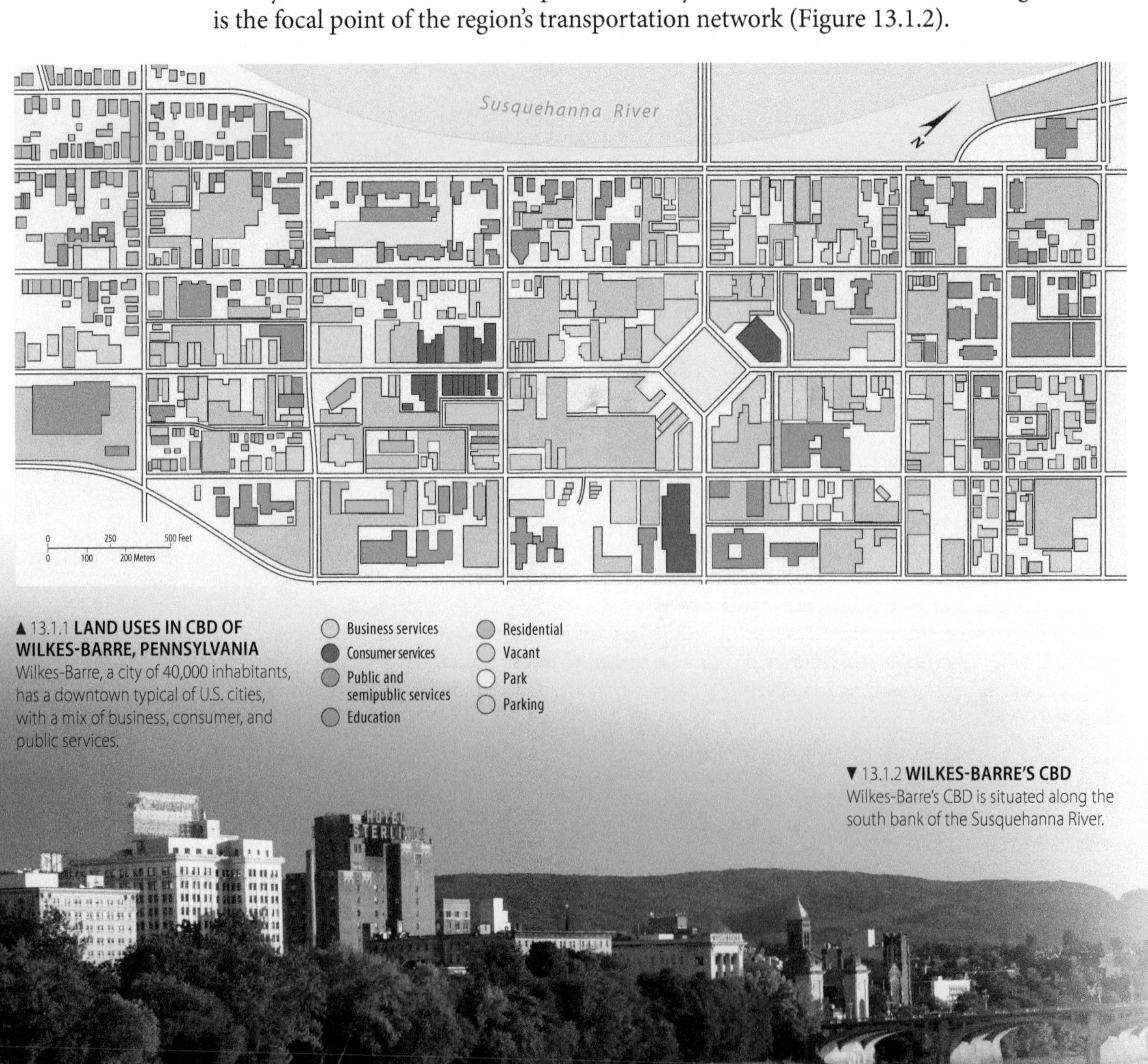

▲ 13.1.1 **LAND USES IN CBD OF WILKES-BARRE, PENNSYLVANIA**
Wilkes-Barre, a city of 40,000 inhabitants, has a downtown typical of U.S. cities, with a mix of business, consumer, and public services.

▼ 13.1.2 **WILKES-BARRE'S CBD**
Wilkes-Barre's CBD is situated along the south bank of the Susquehanna River.

BUSINESS SERVICES

Even with modern telecommunications, many professionals still exchange information primarily through face-to-face contact. Business services, such as advertising, banking, finance, journalism, and law, are centrally located to facilitate rapid communication of fast-breaking news (Figure 13.1.3). Face-to-face contact also helps to establish a relationship of trust based on shared professional values.

People in such businesses particularly depend on proximity to professional colleagues. Lawyers, for example, locate near government offices and courts. Services such as temporary secretarial agencies and instant printers locate downtown to be near lawyers, forming a chain of interdependency that continues to draw offices to the center city.

Extreme competition for limited building sites results in very high land values in the CBD. Because of its high value, land is used more intensively in the center than elsewhere in the city.

Compared to other parts of the city, the central area uses more space below and above ground level. Beneath most central cities runs a vast underground network of garages, loading docks, utilities, walkways, and transit lines. Demand for space in the central city has also made high-rise structures economically feasible.

▲ 13.1.3 **BUSINESS SERVICES IN WILKES-BARRE'S CBD**
Downtown office buildings house several banks.

▲ 13.1.4 **CONSUMER SERVICES IN WILKES-BARRE'S CBD**
F.M. Kirby Center for the Performing Arts.

CONSUMER SERVICES

Consumer services in the CBD serve the many people who work in the center and shop during lunch or working hours. These businesses sell office supplies, computers, and clothing, or offer shoe repair, rapid photocopying, dry cleaning, and so on.

Large department stores once clustered in the CBD, often across the street from one another, but most have relocated to suburban malls. In several CBDs, new shopping areas attract suburban shoppers as well as out-of-town tourists with unique recreation and entertainment experiences (Figure 13.1.4).

PUBLIC SERVICES

Public services typically located downtown include City Hall, courts, and libraries (Figure 13.1.5). These facilities cluster in the CBD to facilitate access for people living in all parts of town.

Sports facilities and convention centers have been constructed or expanded downtown in many cities. These structures attract a large number of people, including many suburbanites and out-of-towners. Cities place these facilities in the CBD because they hope to stimulate more business for downtown restaurants, bars, and hotels.

▼ 13.1.5 **PUBLIC SERVICES IN WILKES-BARRE'S CBD**
Much of downtown Wilkes-Barre is devoted to public services, such as Luzerne County Courthouse.

13.2 Models of Urban Structure

- Three models of urban structure describe where groups typically cluster within urban areas.
- The three models demonstrate that cities grow in rings, wedges, and nodes.

Sociologists, economists, and geographers have developed three models to help explain where different types of people tend to live in an urban area—the concentric zone, sector, and multiple nuclei models.

The three models describing the internal structure of cities were developed in Chicago. Since Chicago developed on a flat prairie, few physical features (except for Lake Michigan to the east) interrupted the growth of the city and its region. Chicago includes a central business district (CBD) known as the Loop, because elevated railway lines loop around it. Surrounding the Loop are residential areas to the south, west, and north. The three models were later applied to cities elsewhere in the United States and in other countries.

CONCENTRIC ZONE MODEL

According to the **concentric zone** model, created in 1923 by sociologist E. W. Burgess, a city grows outward from a central area in a series of five concentric rings, like the growth rings of a tree (Figure 13.2.1).

- The innermost zone is the CBD, where nonresidential activities are concentrated.
- A second ring, the zone in transition, contains industry and poorer-quality housing. Immigrants to the city first live in this zone in high-rise apartment buildings (Figure 13.2.2).
- The third ring, the zone of working-class homes, contains modest older houses occupied by stable, working-class families(Figure 13.2.3).
- The fourth zone has newer and more spacious houses for middle-class families.
- A commuters' zone beyond the continuous built-up area of the city is inhabited by people who work in the center but choose to live in "bedroom communities" for commuters.

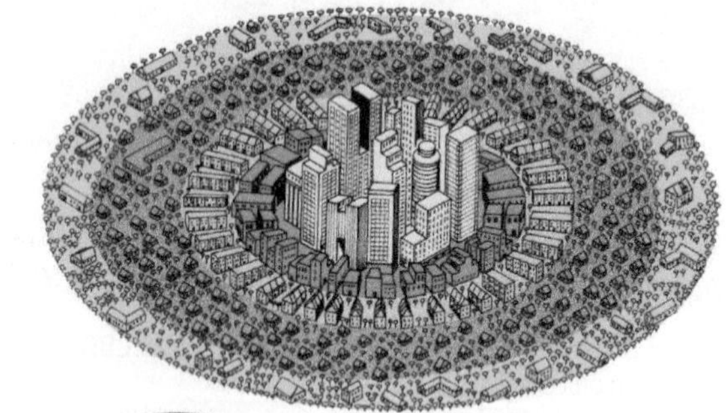

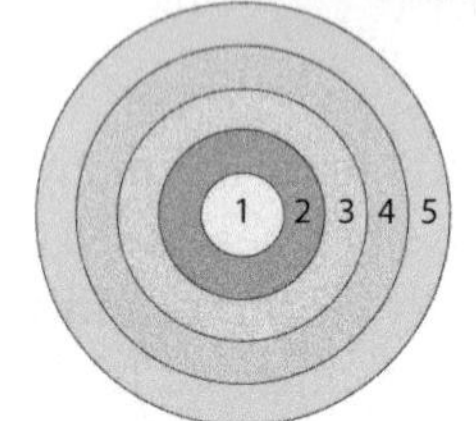

1 Central business district
2 Zone of transition
3 Zone of independent workers' homes
4 Zone of better residences
5 Commuter's zone

▲ 13.2.1 CONCENTRIC ZONE MODEL

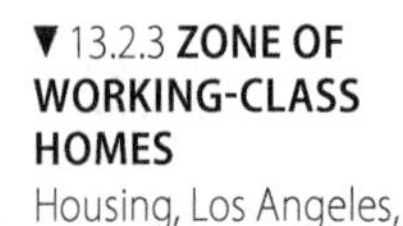

▼ 13.2.3 **ZONE OF WORKING-CLASS HOMES** Housing, Los Angeles, California.

▼ 13.2.2 **ZONE OF TRANSITION** Apartments for rent in an old building, Bronx, New York.

SECTOR MODEL

According to the **sector** model, developed in 1939 by land economist Homer Hoyt, the city develops in a series of sectors (Figure 13.2.4). As a city grows, activities expand outward in a wedge, or sector, from the center. Hoyt mapped the highest-rent areas for a number of U.S. cities at different times and showed that the highest social-class district usually remained in the same sector, although it moved farther out along that sector over time.

Once a district with high-class housing is established, the most expensive new housing is built on the outer edge of that district, farther out from the center. The best housing is therefore found in a corridor extending from downtown to the outer edge of the city. Industrial and retailing activities develop in other sectors, usually along good transportation lines (Figure 13.2.5).

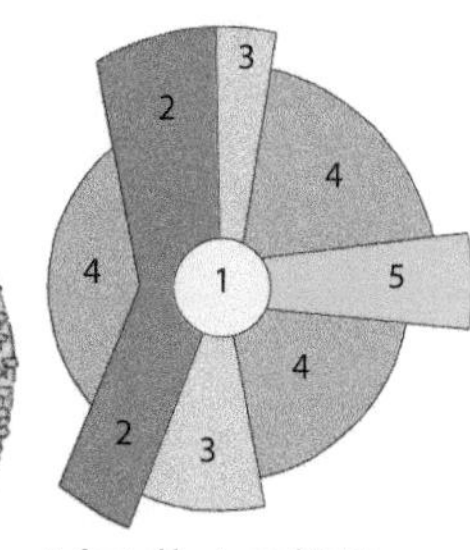

▲ 13.2.4 **SECTOR MODEL**

▲ 13.2.5 **TRANSPORTATION SECTOR** Los Angeles.

MULTIPLE NUCLEI MODEL

Geographers C. D. Harris and E. L. Ullman developed the **multiple nuclei** model in 1945 (Figure 13.2.6). According to the multiple nuclei model, a city is a complex structure that includes more than one center around which activities revolve. Examples of these nodes include a port, neighborhood business center, university, airport, and park.

The multiple nuclei theory states that some activities are attracted to particular nodes, whereas others try to avoid them. For example, a university node may attract well-educated residents, pizzerias, and bookstores (Figures 13.2.7 and 13.2.8). An airport may attract hotels and warehouses. On the other hand, incompatible land-use activities will avoid clustering in the same locations. Heavy industry and high-class housing, for example, rarely exist in the same neighborhood.

▲ 13.2.7 **UNIVERSITY NODE** Harvard University, Cambridge, Massachusetts.

▼ 13.2.6 **MULTIPLE NUCLEI MODEL**

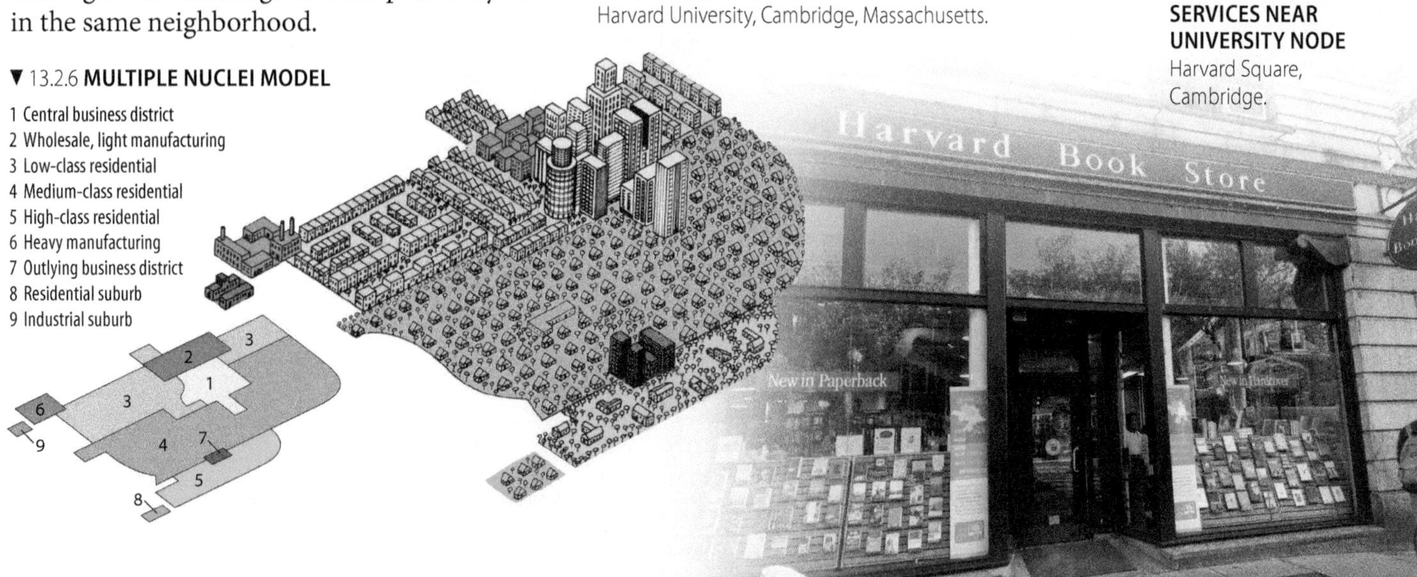

▼ 13.2.8 **CONSUMER SERVICES NEAR UNIVERSITY NODE** Harvard Square, Cambridge.

13.3 Social Area Analysis

- Census data can be used to map the distribution of social characteristics.
- The three models together explain where people live within U.S. cities.

The three models help us understand where people with different social characteristics tend to live within an urban area. They can also help to explain why certain types of people tend to live in particular places.

THE CENSUS

Effective use of the models depends on the availability of data at the scale of individual neighborhoods. In the United States and many other countries, that information comes from a national census. Urban areas in the United States are divided into **census tracts** that contain approximately 5,000 residents and correspond, where possible, to neighborhood boundaries.

Every decade the U.S. Bureau of the Census publishes data summarizing the characteristics of the residents living in each tract. Examples of information the bureau publishes include the number of nonwhites, the median income of all families, and the percentage of adults who finished high school.

The spatial distribution of any of these social characteristics can be plotted on a map of the community's census tracts. Computers have become invaluable in this task, because they permit rapid creation of maps and storage of voluminous data about each census tract. Social scientists can compare the distributions of characteristics and create an overall picture of where various types of people tend to live. This kind of study is known as **social area analysis**.

COMBINING THE THREE MODELS

The three models taken individually do not explain why different types of people live in distinctive parts of the city. But if the models are combined rather than considered independently, they help geographers explain where different types of people live in a city, such as Dallas, Texas.

- The sector theory suggests that a family with a higher income will not live in the same sector of the city as a family with a lower income (Figures 13.3.1 and 13.3.2).
- One family owns its home, whereas the other rents. The concentric zone model suggests that the owner-occupant is much more likely to live in an outer ring and the renter in an inner ring (Figure 13.3.3).
- The multiple nuclei theory suggests that people with the same ethnic or racial background are likely to live near each other (Figure 13.3.4).

LIMITATIONS OF THE MODELS

Critics point out that the models are too simple and fail to consider the variety of reasons that lead people to select particular residential locations. Because the three models are all based on conditions that existed in U.S. cities between the two world wars, critics also question their relevance to contemporary urban patterns in the United States or in other countries.

People tend to reside in certain locations depending on their particular personal characteristics. This does not mean that everyone with the same characteristics must live in the same neighborhood, but the models say that most people prefer to live near others who have similar characteristics.

◄ 13.3.1 **DALLAS: WESTERN SECTOR**

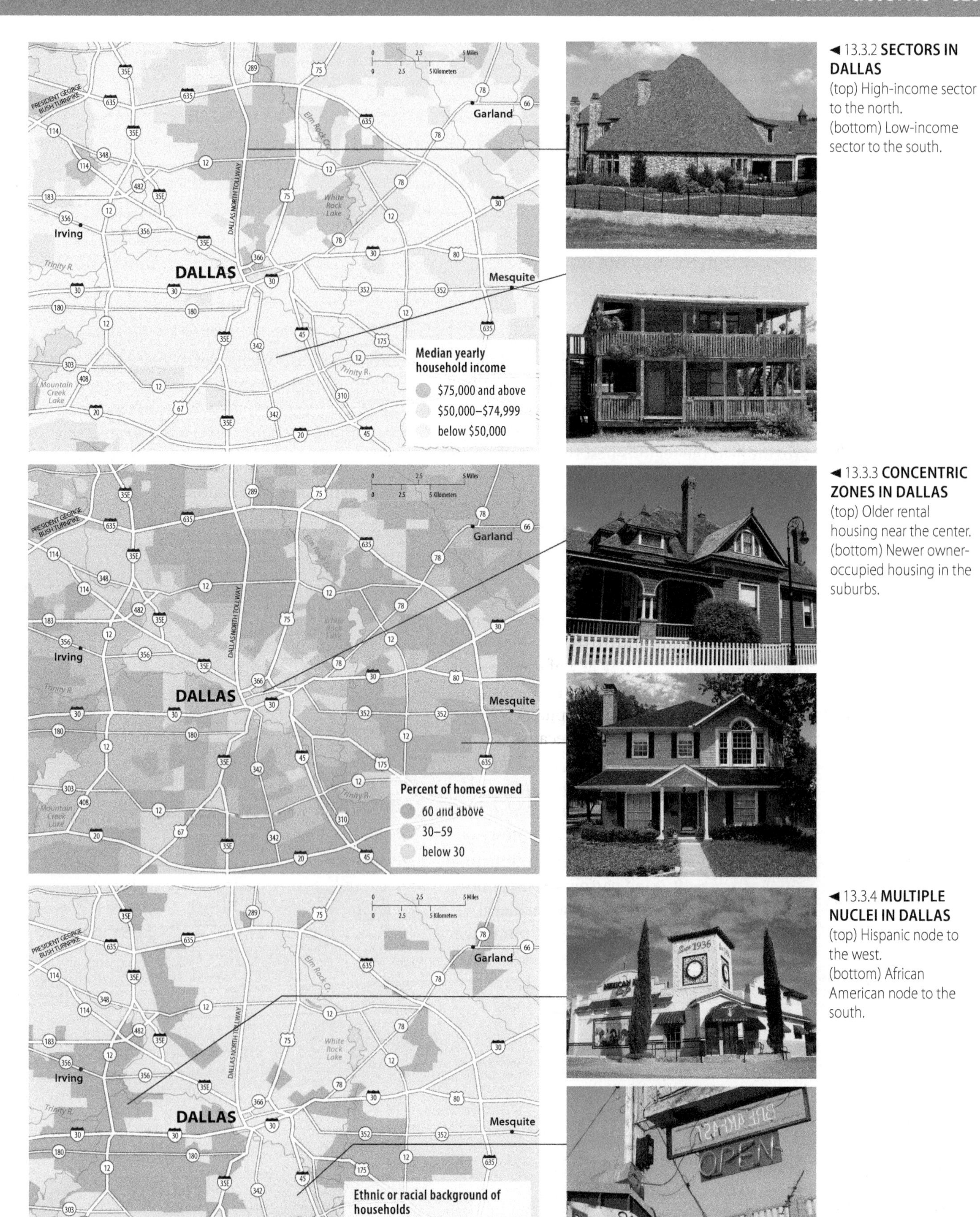

◄ 13.3.2 **SECTORS IN DALLAS**
(top) High-income sector to the north.
(bottom) Low-income sector to the south.

◄ 13.3.3 **CONCENTRIC ZONES IN DALLAS**
(top) Older rental housing near the center.
(bottom) Newer owner-occupied housing in the suburbs.

◄ 13.3.4 **MULTIPLE NUCLEI IN DALLAS**
(top) Hispanic node to the west.
(bottom) African American node to the south.

13.4 Urban Patterns in Europe

- European CBDs contain more residents and consumer services than do U.S. CBDs.
- Poor people are more likely to live in outer rings in European cities.

The three models may describe the spatial distribution of social classes in the United States, but American urban areas differ from those elsewhere in the world. These differences do not invalidate the models, but they do point out that social groups in other countries may not have the same reasons for selecting particular neighborhoods within their cities.

EUROPE'S CBDS

More people live in the CBDs of European cities than those of the United States. Wealthy people are especially attracted to residences in European CBDs. A central location provides proximity to the region's best shops, restaurants, cafés, and cultural facilities. Wealthy people are also attracted by the opportunity to occupy elegant residences in carefully restored, beautiful old buildings.

To serve these residents, European CBDs contain consumer services, such as markets, bakeries, and butchers (Figure 13.4.1). Some European CBDs ban motor vehicles from busy streets, thus emulating one of the most attractive attributes of large shopping malls—pedestrian-only walkways (Figure 13.4.2).

On the other hand, European CBDs are less dominated by business services than U.S. CBDs. European CBDs display a legacy of low-rise structures and narrow streets, built as long ago as medieval times. The most prominent structures may be churches and former royal palaces (Figure 13.4.3). Some European cities try to preserve their historic CBDs by limiting high-rise buildings. After several high-rise offices were permitted to be built in Paris, for example, the public outcry was so great that officials banned further ones (Figure 13.4.4).

▼ 13.4.1 CONSUMER SERVICES FOR RESIDENTS IN PARIS CBD

▼ 13.4.2 PEDESTRIAN-ONLY ZONE, PLACE GEORGES POMPIDOU, PARIS

▼ 13.4.3 LOUVRE MUSEUM, PARIS, A FORMER ROYAL PALACE

▲ 13.4.4 MONTPARNASSE TOWER, PARIS
After construction of this tower, Europe's tallest, public opposition limited construction of further high-rises in Paris.

SECTOR MODEL IN EUROPEAN CITIES

As in the United States, wealthier people in European cities cluster along a sector extending out from the CBD. In Paris, for example, high-income residents moved from the royal palace at the Louvre west towards another royal palace at Versailles (Figure 13.4.5).

Monthly household income (euros)
above 1,800
1,301 – 1,800
1,000 – 1,300
less than 1,000

Cergy
Nanterre
Bobigny
Paris
Versailles
Créteil
Evry
Melun

◄▲ 13.4.5 **SECTORS IN PARIS**
The sector to the west (left) has higher average income households than the sector to the east (above).

CONCENTRIC ZONE MODEL IN EUROPEAN CITIES

In the United States, outer rings are more likely to contain owner-occupied detached houses for families, whereas the inner rings are more likely to contain rented apartments for individuals. A similar pattern exists in European cities (Figure 13.4.6). European cities have relatively few free-standing owner-occupied houses, but to the extent that they exist they are in outer rings.

In contrast with the United States, though, outer rings of European cities also house most of the urban area's poor people. Vast suburbs containing dozens of high-rise apartment buildings house these people who were displaced from the inner city.

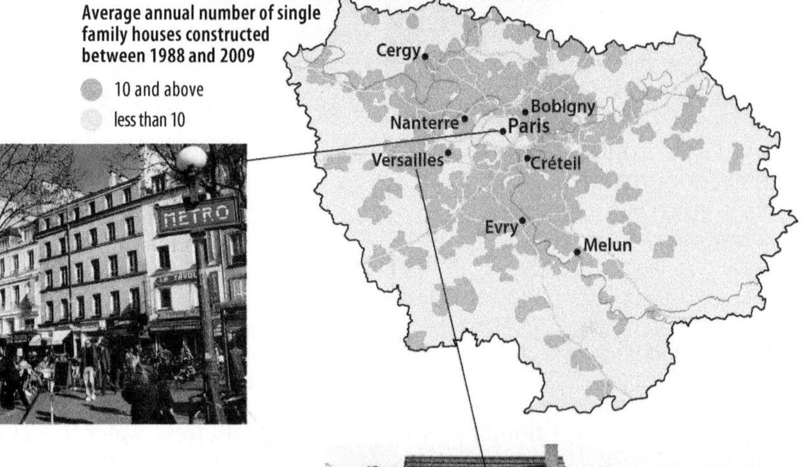

▲► 13.4.6 **CONCENTRIC ZONES IN PARIS**
(above) Older housing near the center. (right) Newer housing in the suburbs.

In the past, low-income people lived in the center of European cities. Before the invention of electricity in the nineteenth century, social segregation was vertical: wealthier people lived on the first or second floors, whereas poorer people occupied the dark, dank basements, or they climbed many flights of stairs to reach the attics. Today, low-income people are less likely to live in European inner-city neighborhoods. Poor-quality housing has been renovated for wealthy people or demolished and replaced by offices or luxury apartment buildings.

European suburban residents face the prospect of long commutes by public transportation to reach jobs and other downtown amenities. Shops, schools, and other services are worse than in inner neighborhoods, and the suburbs have high rates of crime, violence, and drug dealing. Because the housing is mostly in high-rise buildings, people lack large private yards.

MULTIPLE NUCLEI MODEL IN EUROPEAN CITIES

European cities also show evidence of the multiple nuclei model. Many residents of the suburbs are persons of color or recent immigrants from Africa or Asia who face discrimination and prejudice by white Europeans (Figure 13.4.7).

▼ 13.4.7 **MULTIPLE NUCLEI IN PARIS**
Much of the housing for low-income minorities and immigrants is in high-rise buildings in the suburbs (top of the page). Immigrants protest in Paris against legislation aimed at limiting immigration (below).

13.5 Urban Patterns in Latin America

- Cities in developing countries follow patterns similar to those of European cities.
- Cities in developing countries have been influenced by colonial rule.

In developing countries, as in Europe, the poor tend to be accommodated in the outer rings, whereas the wealthy live near the center of cities as well as in a sector extending from the center. The similarity between cities in Europe and developing countries is not a coincidence: European colonial planning left a heavy mark on the design of cities in developing countries.

Many cities in developing countries have passed through three stages of design:

- Pre-European colonization.
- European colonial period.
- Postcolonial independence.

▲ 13.5.1 **PRECOLONIAL CITY: TENOCHTITLÁN** (left) The Aztecs built the city on an island in Lake Texcoco. (right) Templo Mayor dominated the center of the city. The twin shrines on the top of the temple were dedicated to the Aztec God of rain and agriculture (in blue) and to the Aztec God of war (in red).

PRECOLONIAL CITIES

Before the Europeans established colonies in Africa, Asia, and Latin America, most people lived in rural settlements. Cities were often laid out with a central area including marketplace, religious structures, government buildings, and homes of wealthy families. Families with less wealth and recent migrants to the city lived on the edge.

In Mexico, the Aztecs founded Mexico City—which they called Tenochtitlán—on a hill known as Chapultepec ("the hill of the grasshopper"). When forced by other people to leave the hill, they migrated a few kilometers south, near the present-day site of the University of Mexico, and then in 1325 to a marshy 10-square-kilometer (4-square-mile) island in Lake Texcoco (Figure 13.5.1). The node of religious life was the Templo Mayor (Great Temple).

Three causeways with drawbridges linked Tenochtitlán to the mainland and also helped to control flooding. An aqueduct brought fresh water from Chapultepec. Most food, merchandise, and building materials crossed from the mainland to the island by canoe, barge, or other type of boat, and the island was laced with canals to facilitate pickup and delivery of people and goods. Over the next two centuries the Aztecs conquered the neighboring peoples and extended their control through the central area of present-day Mexico. As their wealth and power grew, Tenochtitlán grew to a population of a half-million.

▼ 13.5.2 **COLONIAL CITY: MEXICO CITY ZÓCALO**
After the Spanish conquered Tenochtitlán in 1521, they destroyed the city and dispersed or killed most of the inhabitants. The city, renamed Mexico City, was rebuilt around a main square, called the Zócalo, in the center of the island, on the site of the Aztecs' sacred precinct. The Spanish reconstructed the streets in a grid pattern extending from the Zócalo. A Roman Catholic cathedral was built near the site of the demolished Great Temple, and the National Palace was erected on the site of the Aztec emperor Moctezuma's destroyed palace.

COLONIAL CITIES

When Europeans gained control of Africa, Asia, and Latin America, they expanded existing cities to provide colonial services, such as administration, military command, and international trade, as well as housing for Europeans who settled in the colony. Existing native towns were either left to one side or demolished because they were totally at variance with European ideas.

Colonial cities followed standardized plans. All Spanish cities in Latin America, for example, were built according to the Laws of the Indies, drafted in 1573. The laws explicitly outlined how colonial cities were to be constructed—a gridiron street plan centered on a church and central plaza, walls around individual houses, and neighborhoods built around central, smaller plazas with parish churches or monasteries (Figure 13.5.2). Compared to precolonial cities, these European districts typically contained wider streets and public squares, larger houses surrounded by gardens, and much lower densities.

CITIES SINCE INDEPENDENCE

Following independence, Latin American cities have grown in accordance with the sector and concentric zone models (Figure 13.5.3).

- **Sectors.** An elite sector forms along a narrow "spine" that contains offices, shops, and amenities attractive to wealthy people, such as restaurants, theaters, and parks. In Mexico City, the "spine" is a 14-lane, tree-lined boulevard called the Paseo de la Reforma, designed by Emperor Maximilian in the 1860s. The wealthy built imposing palacios (palaces) along it (Figure 13.5.4).
- **Concentric zones.** Cities in developing countries have expanded rapidly as millions of people immigrate in search of work. In Mexico City, most of Lake Texcoco was drained in 1903 to permit expansion of the city, including the airport (Figure 13.5.5). A large percentage of poor immigrants to urban areas in developing countries live in squatter settlements on the periphery, especially on hillsides (Figure 13.5.6). Squatter settlements lack such services as paved roads and sewers, because neither the city nor the residents can afford them. Electricity service may be stolen by running a wire from the nearest power line. The United Nations estimated that 1 billion people worldwide lived in squatter settlements in 2005.

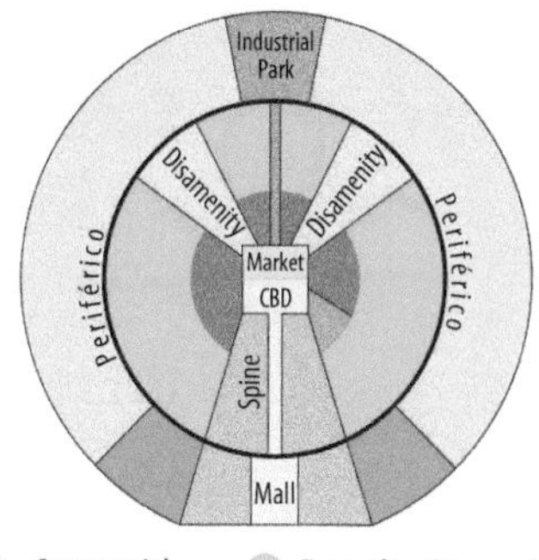

▲ 13.5.3 **MODEL OF A LATIN AMERICAN CITY**
Geographers Ernest Griffin and Larry Ford show that in Latin American cities wealthy people push out from the center in a well-defined elite residential sector. The poor live in the outer or (Periférico) ring.

▼ 13.5.4 **SECTOR MODEL IN MEXICO CITY**
Wealthy people live in a sector along the Paseo de la Reforma, a wide boulevard to the west of downtown.

▲ 13.5.5 **CONCENTRIC ZONE MODEL IN MEXICO CITY**
Mexico City has grown rapidly. Low income areas are on the periphery, especially near the airport (top of image).

◄ 13.5.6 **MEXICO CITY SQUATTER SETTLEMENT**
Squatter settlements have developed on the hillsides surrounding Mexico City.

13.6 Defining Urban Settlements

- ▶ **Urban settlements can be defined legally, or as urbanized or metropolitan areas.**
- ▶ **Urban growth has caused adjacent metropolitan areas to overlap.**

Urban settlements can be defined in three ways.

LEGAL DEFINITION OF CITY

The term **city** defines an urban settlement that has been legally incorporated into an independent, self-governing unit. In the United States, a city surrounded by suburbs is sometimes called a **central city.**

A city has locally elected officials, the ability to raise taxes, and the responsibility for providing essential services. The boundaries of the city define the geographic area within which the local government has legal authority (Figure 13.6.1).

URBANIZED AREA

With the rapid growth of urban settlements, many urban residents live in suburbs, beyond the boundaries of the central city. In the United States, the central city and the surrounding built-up suburbs are called an **urbanized area**. Approximately 70 percent of Americans live in urbanized areas, including about 30 percent in central cities and 40 percent in surrounding jurisdictions.

Working with urbanized areas is difficult because few statistics are available about them. Most data in the United States and other countries are collected for cities, counties, and other local government units, but urbanized areas do not correspond to government boundaries.

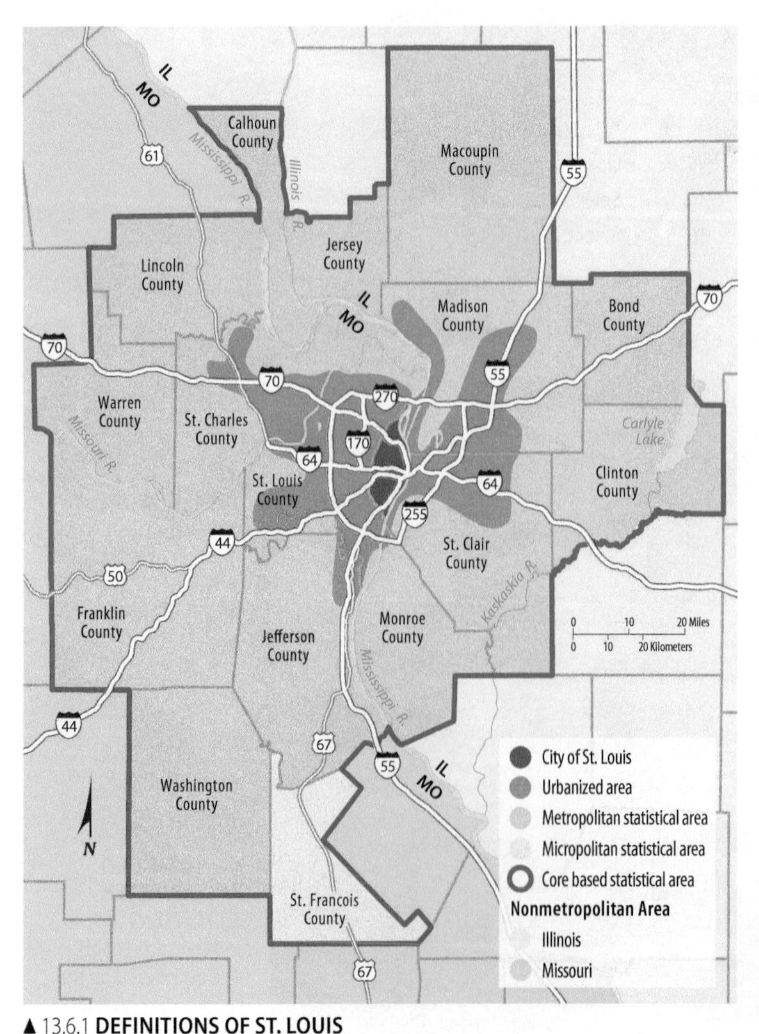

▲ 13.6.1 **DEFINITIONS OF ST. LOUIS**

METROPOLITAN STATISTICAL AREA

The area of influence of a city extends beyond legal boundaries and adjacent built-up jurisdictions. For example, commuters may travel a long distance to work and shop in the city or built-up suburbs. People in a wide area watch the city's television stations, read the city's newspapers, and support the city's sports teams.

The U.S. Bureau of the Census has created a method of measuring the functional area of a city, known as the **metropolitan statistical area (MSA)**. An MSA includes the following:

1. An urbanized area of at least 50,000 inhabitants.
2. The county within which the city is located.
3. Adjacent counties with a high population density and a large percentage of residents working in the central city's county.

The census designated 366 MSAs as of 2009, encompassing 84 percent of the U.S. population.

The census has also designated smaller urban areas as **micropolitan statistical areas (μSAs)**. The Greek letter μ or "mu" stands for *micro-*. These include an urbanized area of between 10,000 and 50,000 inhabitants, the county in which it is found, and adjacent counties tied to the city. The United States had 576 micropolitan statistical areas in 2009, for the most part found around southern and western communities previously considered rural in character. About 10 percent of Americans live in a micropolitan statistical area.

▲ 13.6.2 **MEGALOPOLIS**

Metropolitan statistical area
Micropolitan statistical area

▼ 13.6.3 **OVERLAPPING METROPOLITAN AREAS IN EUROPE**

Open MapMaster Europe in MasteringGEOGRAPHY

Select: *Population* then *Population Density*. Set Layer Opacity at 50%.

Select: *Physical Environment* then *Physical Features*

Select: *Political* then *Cities*

Europe's most extensive population concentration with overlapping metropolitan areas follows what major river? What cities are located in this area?

OVERLAPPING METROPOLITAN AREAS

The 366 MSAs and 576 μSAs together are known as **core based statistical areas (CBSAs)**. Recognizing that many MSAs and μSAs have close ties, the census has combined some of them into 125 **combined statistical areas (CSAs).** A CSA is defined as two or more contiguous CBSAs tied together by commuting patterns. The 125 CSAs plus the remaining 186 MSAs and 407 μSAs together are known as **primary census statistical areas (PCSAs)**.

In the northeastern United States, metropolitan areas are so close together that they now form one continuous urban complex, extending from north of Boston to south of Washington, D.C. (Figure 13.6.2). Geographer Jean Gottmann named this region Megalopolis, a Greek word meaning "great city." Overlapping metropolitan areas exist in other developed regions, including Europe and Japan (Figure 13.6.3).

13.7 Fragmented Government

- ► Cities traditionally grew through annexation.
- ► Today most urban areas have a large number of local governments.

As they became more populous in the nineteenth century, U.S. cities expanded by adding peripheral land. Now cities are surrounded by a collection of suburban jurisdictions whose residents prefer to remain legally independent of the large city. The fragmentation of local government in the United States makes it difficult to address such issues as traffic congestion, affordable housing, and good schools that transcend local government boundaries.

ANNEXATION

The process of legally adding land area to a city is **annexation**. Rules concerning annexation vary among states. Normally, land can be annexed into a city only if a majority of residents in the affected area vote in favor of doing so.

Peripheral residents generally desired annexation in the nineteenth century, because the city offered better services, such as water supply, sewage disposal, trash pickup, paved streets, public transportation, and police and fire protection. Thus, as U.S. cities grew rapidly in the nineteenth century, the legal boundaries frequently changed to accommodate newly developed areas. For example, the city of Chicago expanded from 26 square kilometers (10 square miles) in 1837 to 492 square kilometers (190 square miles) in 1900 (Figure 13.7.1).

Today, however, cities are less likely to annex peripheral land because the residents prefer to organize their own services rather than pay city taxes for them. As a result, today's cities are surrounded by a collection of suburban jurisdictions, whose residents prefer to remain legally independent of the large city.

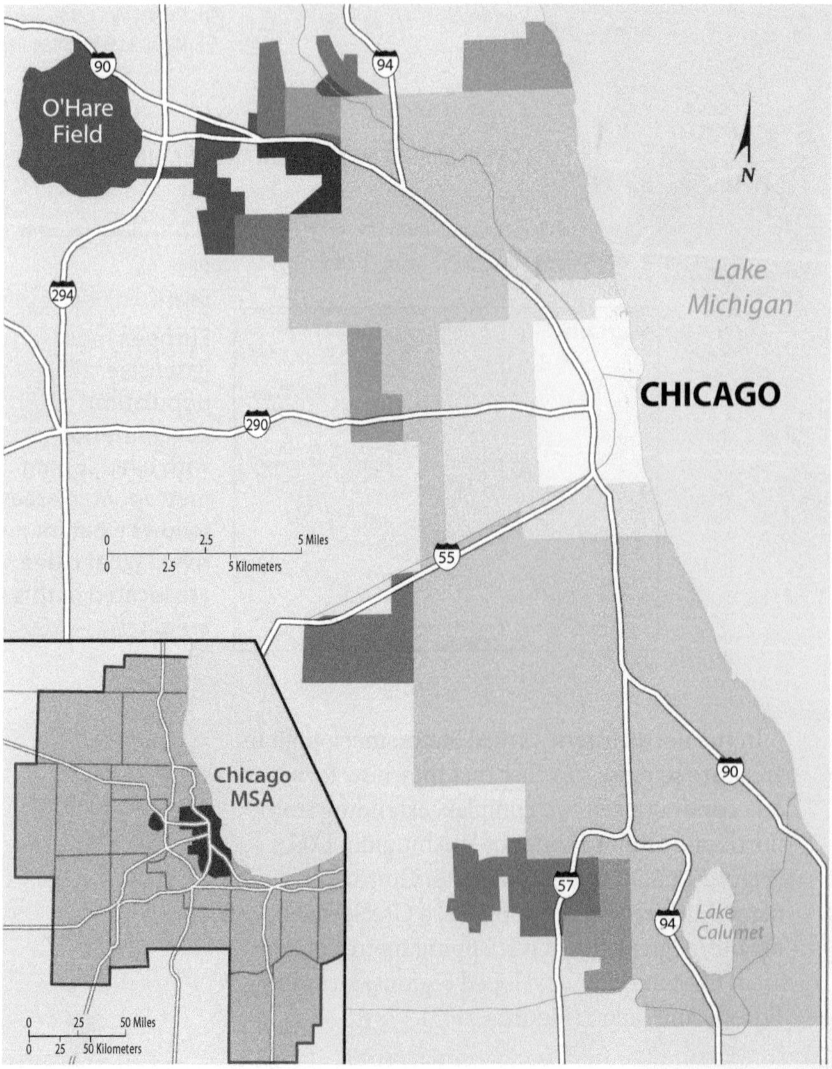

◄ 13.7.1 ANNEXATION IN CHICAGO

City limits in 1837

Annexed by
- 1870
- 1890
- 1900
- 1930
- 1960
- 1990

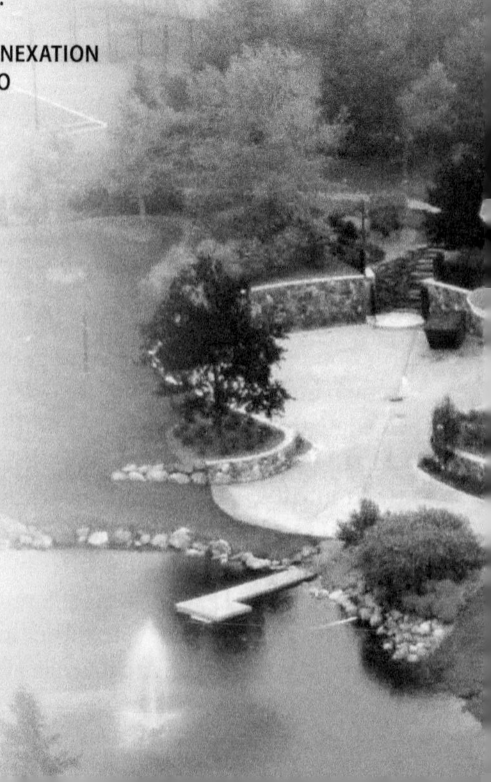

METROPOLITAN GOVERNMENT

The number of local governments exceeds 20,000 throughout the United States, including several hundred in the Detroit area alone (Figure 13.7.2). The fragmentation of local governments in the Detroit area has surrounded an impoverished city of Detroit (Figure 13.7.3) with wealthier suburbs (Figure 13.7.4).

Originally, some of these peripheral jurisdictions were small, isolated towns that had a tradition of independent local government before being swallowed up by urban growth. Others are newly created communities whose residents wish to live close to the large city but not legally be part of it.

The large number of local government units has led to calls for a metropolitan government that could coordinate—if not replace—the numerous local governments in an urban area. Strong metropolitan-wide governments have been established in a few places in North America. Two kinds exist:

- **Federations.** Examples include Toronto and other large Canadian cities. Toronto's metropolitan government was created in 1953 through federation of 13 municipalities. A two-tier system existed until 1998, when the municipalities were amalgamated into a single government.
- **Consolidations of City and County Governments.** Examples include Indianapolis and Miami. The boundaries of Indianapolis were changed to match those of Marion County. Government functions that were once handled separately now are combined into a joint operation in the same office building. In Florida, the city of Miami and surrounding Dade County have combined some services, but the city boundaries have not been changed to match those of the county.

▲ 13.7.3 INNER-CITY DETROIT

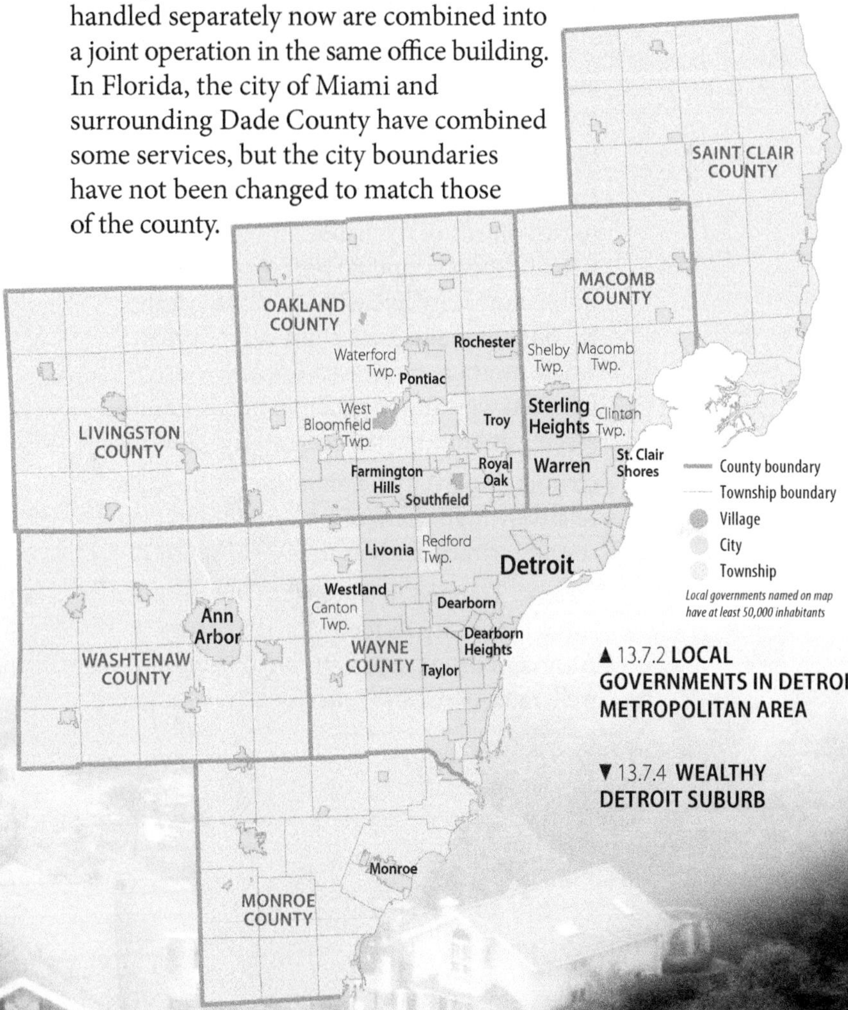

▲ 13.7.2 LOCAL GOVERNMENTS IN DETROIT METROPOLITAN AREA

▼ 13.7.4 WEALTHY DETROIT SUBURB

13.8 Decline and Renewal

- Low-income residents concentrate in the inner-city neighborhoods in U.S. cities.
- Some U.S. inner-city neighborhoods have been gentrified.

Inner cities in the United States contain concentrations of low income people who face a variety of economic, social, and physical challenges very different from those faced by suburban residents.

INNER-CITY CHALLENGES

Inner city residents are frequently referred to as a permanent underclass because they are trapped in an unending cycle of hardships:

- **Inadequate job skills.** Inner city residents are increasingly unable to compete for jobs. They lack technical skills needed for most jobs because fewer than half complete high school.
- **Culture of poverty.** Unwed mothers give birth to two-thirds of the babies in U.S. inner-city neighborhoods, and 80 percent of children in the inner city live with only one parent. Because of inadequate child-care services, single mothers may be forced to choose between working to generate income and staying at home to take care of the children.
- **Crime.** Inner-city neighborhoods have a relatively high share of a metropolitan area's serious crimes, such as murder (Figure 13.8.1).
- **Drugs.** Trapped in a hopeless environment, some inner-city residents turn to drugs. Although drug use is a problem in suburbs as well, rates of use have increased most rapidly in inner cities. Some drug users obtain money through criminal activities.
- **Homelessness.** Several million people are homeless in the United States. Most people are homeless because they cannot afford housing and have no regular income. Homelessness may have been sparked by family problems or job loss (Figure 13.8.2).
- **Lack of services.** The concentration of low-income residents in inner-city neighborhoods of central cities has produced financial problems. These people require public services, but they can pay very little of the taxes to support the services. Central cities face a growing gap between the cost of needed services in inner-city neighborhoods and the availability of funds to pay for them.
- **Deteriorated housing.** Inner-city housing is subdivided by absentee landlords into apartments for low-income families, a process known as filtering. Landlords stop maintaining houses when the rent they collect becomes less than the maintenance cost. In such a case, the building soon deteriorates and grows unfit for occupancy.

▲ 13.8.2 **HOMELESSNESS**
Homeless people camp under I-75 bridge across the Ohio River.

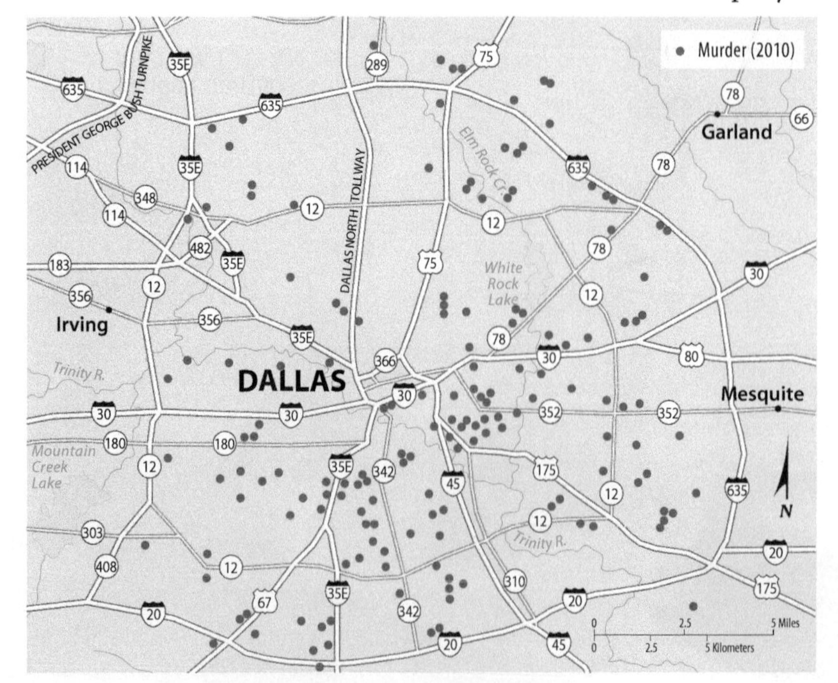

▼ 13.8.1 **MURDERS IN DALLAS**
Compare with Figures 13.3.2, 13.3.3, and 13.3.4. Most of the murders in Dallas occurred in low-income minority areas, and most victims, as well as those arrested for murder in Dallas, were minorities.

▲ 13.8.3
GENTRIFICATION
Spitalfields neighborhood in London before gentrification (left) and after gentrification (right).

GENTRIFICATION

Gentrification is the process by which middle-class people move into deteriorated inner-city neighborhoods and renovate the housing. Most cities have at least one gentrified inner-city neighborhood. In a few cases, inner-city neighborhoods never deteriorated, because the community's social elite maintained them as enclaves of expensive property. In most cases, inner-city neighborhoods have only recently been renovated by the city and by private investors (Figure 13.8.3).

Middle-class families are attracted to deteriorated inner-city housing for a number of reasons. First, houses may be larger, more substantially constructed, yet cheaper in the inner city than in the suburbs. Inner-city houses may also possess attractive architectural details such as ornate fireplaces, cornices, high ceilings, and wood trim.

Gentrified inner-city neighborhoods also attract middle-class individuals who work downtown. Inner-city living eliminates the strain of commuting on crowded freeways or public transit. Others seek proximity to theaters, bars, restaurants, and other cultural and recreational facilities located downtown. Renovated inner-city housing appeals to single people and couples without children, who are not concerned with the quality of inner-city schools (Figure 13.8.4).

Because renovating an old inner-city house can be nearly as expensive as buying a new one in the suburbs, cities encourage the process by providing low-cost loans and tax breaks. Public expenditures for renovation have been criticized as subsidies for the middle class at the expense of people with lower incomes, who are forced to move out of the gentrified neighborhoods because the rents in the area are suddenly too high for them.

▲ 13.8.4 **INNER CITY REDEVELOPMENT**

Fly to: *3600 S State St, Chicago, IL*

Above is an image from 9/26/2000.

Drag to enter *Street view* at the corner of 45th and State St.

How does the current use of land compare to what was there in 2000? Which time period appears to provide more attractive and better living conditions?

13.9 Suburban Sprawl

- Suburbs sprawl outside American cities.
- Retailing as well as housing has grown in suburbs.

In 1950, only 20 percent of Americans lived in suburbs. After more than a half-century of rapid suburban growth, 50 percent of Americans now live in suburbs. U.S. suburbs are characterized by **sprawl**, which is the progressive spread of development over the landscape.

THE PERIPHERAL MODEL

North American urban areas follow what Chauncey Harris (creator of the multiple nuclei model) called the **peripheral model**. According to the peripheral model, an urban area consists of an inner city surrounded by large suburban residential and business areas tied together by a beltway or ring road (Figure 13.9.1).

Peripheral areas lack the severe physical, social, and economic problems of inner-city neighborhoods. But the peripheral model points to problems of sprawl and segregation that characterize many suburbs.

ATTRACTIONS OF SUBURBS

Public opinion polls in the United States show people's strong desire for suburban living. In most polls, more than 90 percent of respondents prefer the suburbs to the inner city. Suburban living offers many attractions:

- A detached single-family dwelling rather than a row house or apartment.
- A yard surrounding the house for children to play.
- Space to park several cars at no cost.
- A greater opportunity for home ownership.
- Protection from inner-city crime and congestion.
- Proximity to good schools.

▼ 13.9.1 **PERIPHERAL MODEL OF URBAN AREAS**

1 Central city
2 Suburban residential area
3 Shopping mall
4 Industrial district
5 Office park
6 Service center
7 Airport complex
8 Combined employment & shopping center

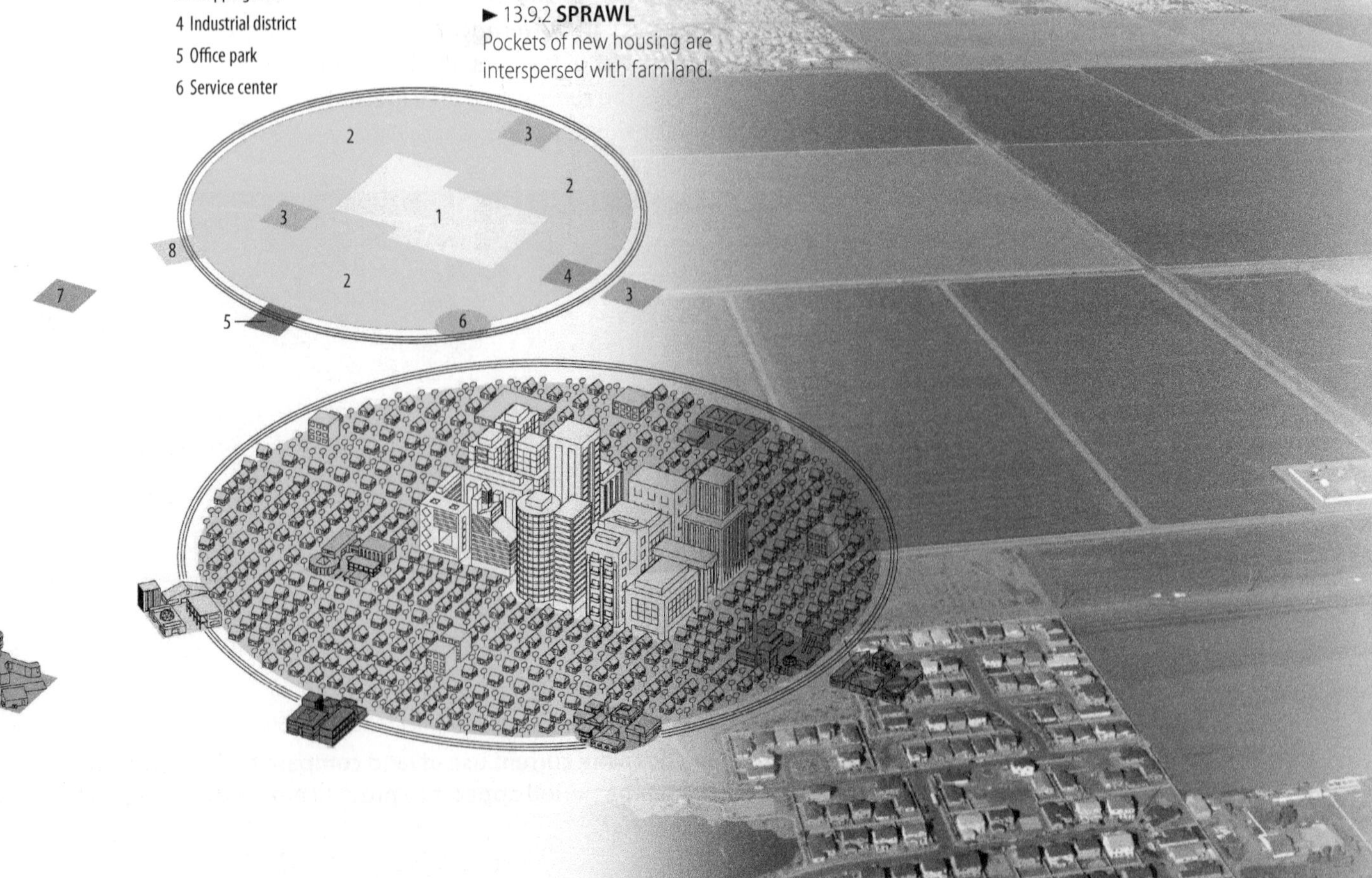

► 13.9.2 **SPRAWL**
Pockets of new housing are interspersed with farmland.

THE COSTS OF SPRAWL

When private developers select new housing sites, they seek cheap land that can easily be prepared for construction—land often not contiguous to the existing built-up area. Land is not transformed immediately from farms to housing developments. Instead, developers buy farms for future construction of houses by individual builders. Developers frequently reject land adjacent to built-up areas in favor of sites outside the urbanized area, depending on the price and physical attributes of the alternatives. The periphery of U.S. cities therefore looks like Swiss cheese, with pockets of development and gaps of open space (Figure 13.9.2).

Urban sprawl has some undesirable traits:

- Roads and utilities must be extended to connect isolated new developments to nearby built-up areas.
- Motorists must drive longer distances and consume more fuel.
- Agricultural land is lost to new developments, and other sites lie fallow while speculators await the most profitable time to build homes on them.
- Local governments typically spend more on services for these new developments than they collect in additional taxes.

SEGREGATION

The modern residential suburb is segregated in two ways.

- Housing in a given suburb is usually built for people of a single social class, with others excluded by virtue of the cost, size, or location of the housing. Segregation by race and ethnicity also persists in many suburbs.
- Residents are separated from commercial and manufacturing activities that are confined to compact, distinct areas.

SUBURBAN RETAILING

Suburban residential growth has fostered change in the distribution of consumer services. Historically, urban residents bought food and other daily necessities at small neighborhood shops in the midst of housing areas and shopped in the CBD for other products. CBD sales have stagnated because suburban residents won't make the long journey there.

Instead, retailing has been increasingly concentrated in planned suburban shopping malls, auto-friendly strip malls, and big-box stores, surrounded by generous parking lots (Figure 13.9.3). These nodes of consumer services are called **edge cities.** Edge cities originated as suburban residences for people who worked in the central city, and then shopping malls were built to be near the residents. Edge cities now also serve as nodes of business services (Figure 13.9.4).

A shopping center is built by a developer, who buys the land, builds the structures, and leases space to individual merchants. The key to a successful large shopping center is the inclusion of one or more anchors. Most consumers go to a center to shop at an anchor and, while there, patronize the smaller shops. The anchors may be a supermarket and discount store in a smaller center or several department stores in a larger center.

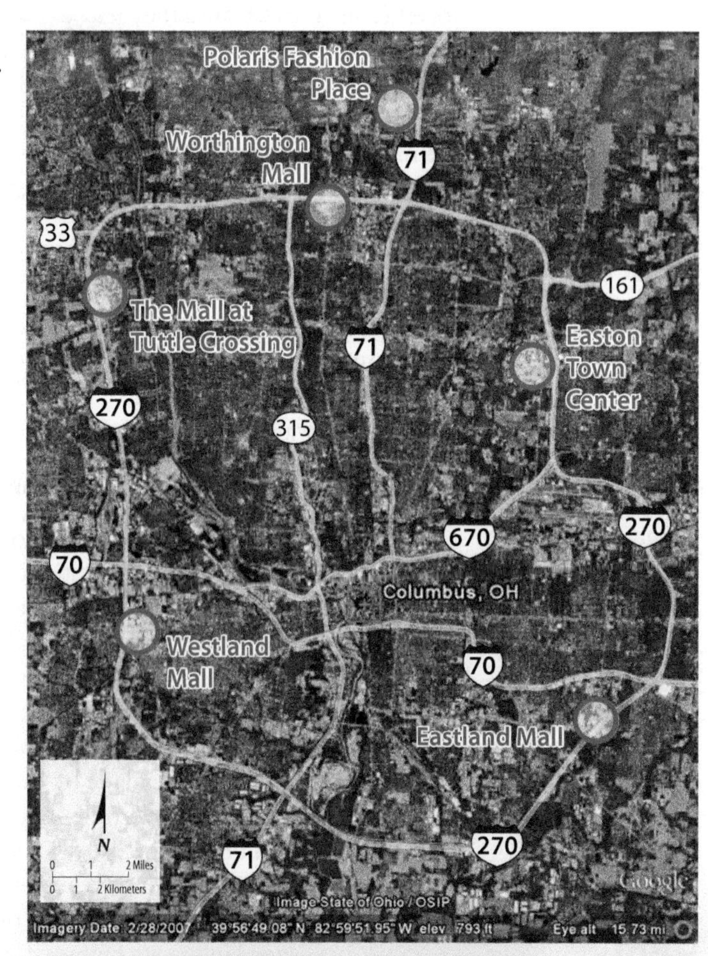

▲ 13.9.3 **SHOPPING MALLS NEAR COLUMBUS, OHIO**

▼ 13.9.4 **EDGE CITY**
Easton Town Center, outside Columbus, Ohio.

13.10 Urban Transportation

- Most trips in the U.S. are by private motor vehicle.
- Public transportation has made a modest comeback in some cities.

People do not travel aimlessly; their trips have a precise point of origin, destination, and purpose. More than half of all trips are work related. Shopping or other personal business and social journeys each account for approximately one-fourth of all trips. Sprawl makes people more dependent on motor vehicles for access to work, shopping, and social activities.

DEVELOPMENT OF URBAN TRANSPORTATION

Historically, people lived close together in cities because they had to be within walking distance of shops and places of employment. The invention of the railroad in the nineteenth century enabled people to live in suburbs and work in the central city. Cities then built street railways (called trolleys, streetcars, or trams) and underground railways (subways) to accommodate commuters. Rail and trolley lines restricted suburban development to narrow strips within walking distance of the stations.

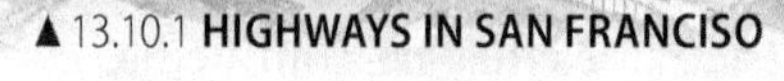

▲ 13.10.1 HIGHWAYS IN SAN FRANCISCO

MOTOR VEHICLES

Until the twentieth century, the growth of suburbs was constrained by poor transportation. Motor vehicles have permitted large-scale development of suburbs at greater distances from the center, in the gaps between the rail lines. More than 95 percent of all trips within U.S. cities are made by car.

The U.S. government has encouraged the use of cars and trucks by paying 90 percent of the cost of limited-access high-speed interstate highways, which crisscross 74,000 kilometers (46,000 miles) across the country (Figure 13.10.1). The use of motor vehicles is also supported by policies that keep the price of fuel below the level found in most other countries.

The motor vehicle is an important user of land in the city. An average city allocates about one-fourth of its land to roads and parking lots (refer to Figure 13.1.1). Valuable land is devoted to parking cars and trucks, although expensive underground and multistory parking structures can reduce the amount of ground-level space needed. Freeways cut a wide path through the heart of cities, and elaborate interchanges consume even more space.

Motor vehicles have costs beyond their purchase and operation: delays imposed on others, increased need for highway maintenance, construction of new highways, and pollution. The average American loses 36 hours per year sitting in traffic jams and wastes 55 gallons of gasoline.

Technological improvements may help traffic flow. Computers mounted on the dashboards alert drivers to traffic jams and suggest alternate routes. On freeways, vehicle speed and separation from other vehicles can be controlled automatically rather than by the driver.

Motorists can be charged for using congested roads or pay high tolls to drive on uncongested roads (Figure 13.10.2). The inevitable diffusion of such technology in the twenty-first century will reflect the continuing preference of most people to use private motor vehicles.

▼ 13.10.2 CONGESTION CHARGING IN LONDON

PUBLIC TRANSIT

In larger cities, public transportation is better suited than motor vehicles to moving large numbers of people, because each traveler takes up far less space. Public transportation is cheaper, less polluting, and more energy efficient than the automobile. It also is particularly suited to rapidly bringing a large number of people into a small area. Despite the obvious advantages of public transportation for commuting, only 5 percent of trips in U.S. cities are by public transit. Outside of big cities, public transportation is extremely rare or nonexistent.

Public transportation has been expanded in some U.S. cities to help reduce air pollution and conserve energy. New subway lines and existing systems expanded in a number of cities (Figure 13.10.3). The federal government has permitted Boston, New York, and other cities to use funds originally allocated for interstate highways to modernize rapid transit service instead. The trolley—now known by the more elegant term of light-rail transit—is making a modest comeback in North America (Figure 13.10.4). California, the state that most symbolizes the automobile-oriented American culture, is the leader in construction of new light-rail transit lines, as well as retention of historic ones (Figure 13.10.5).

Despite modest recent successes, most public transportation systems are caught in a vicious circle, because fares do not cover operating costs. As patronage declines and expenses rise, the fares are increased, which drives away passengers and leads to service reduction and still higher fares.

▲ 13.10.3 **PUBLIC TRANSIT OPTIONS IN SAN FRANCISCO: BART SUBWAY**

◄ 13.10.4 **PUBLIC TRANSIT OPTIONS IN SAN FRANCISCO: MUNI LIGHT RAIL**

▼ 13.10.5 **PUBLIC TRANSIT OPTIONS IN SAN FRANCISCO: CABLE CARS**

What is the future for cities? As shown in this chapter, contradictory trends are at work simultaneously. Why does one inner-city neighborhood become a slum and another an upper-class district? Why does one city attract new shoppers and visitors while another languishes?

The suburban lifestyle as exemplified by the detached single-family house with surrounding yard attracts most people. Yet inner-city residents may rarely venture out to suburbs. Lacking a motor vehicle, they have no access to most suburban locations. Lacking money, they do not shop in suburban malls or attend sporting events at suburban arenas. The spatial segregation of inner-city residents and suburbanites lies at the heart of the stark contrasts so immediately observed in any urban area.

Several U.S. states have taken strong steps in the past few years to curb sprawl, reduce traffic congestion, and reverse inner-city decline. The goal is to produce a pattern of compact and contiguous development, while protecting rural land for agriculture, recreation, and wildlife.

Key Questions

Where are people distributed within urban areas?

- The Central Business District (CBD) contains a large share of a city's business and public services.
- The concentric zone, sector, and multiple nuclei models describe where different types of people live within urban areas.
- The three models together foster understanding that people live in different rings, sectors, and nodes depending on their stage in life, social status, and ethnicity.

How are urban areas expanding?

- Urban areas have expanded beyond the legal boundaries of cities to encompass urbanized areas and metropolitan areas that are functionally tied to the cities.
- With suburban growth, most metropolitan areas have been fragmented into a large number of local governments.

What challenges do cities face?

- Most Americans now live in suburbs that surround cities.
- Low-income inner-city residents face a variety of economic, social, and physical challenges.
- Tying together sprawling American urban areas is dependency on motor vehicles.

Thinking Geographically

Draw a sketch of your community or neighborhood. In accordance with Kevin Lynch's *The Image of the City*, place five types of information on the map—districts (homogeneous areas), edges (boundaries that separate districts), paths (lines of communication), nodes (central points of interaction), and landmarks (prominent objects on the landscape).

1. How clear an image does your community have for you?

Jane Jacobs wrote in *Death and Life of Great American Cities* that an attractive urban environment is one that is animated with an intermingling of a variety of people and activities, such as found in many New York City neighborhoods (Figure 13.CR.1).

2. What are the attractions and drawbacks to living in such environments?

Officials of rapidly growing cities in developing countries discourage the building of houses that do not meet international standards for sanitation and construction. Also discouraged are private individuals offering transportation in vehicles that lack decent tires, brakes, and other safety features. Yet the residents prefer substandard housing to no housing, and they prefer unsafe transportation to no transportation.

3. What would be the advantages and problems for a city if health and safety standards for housing, transportation, and other services were relaxed?

► 13.CR.1 **NEW YORK'S GREENWICH VILLAGE**

Interactive Mapping

OVERLAPPING METROPOLITAN AREAS IN NORTH AMERICA

Overlapping metropolitan areas are emerging in North America in addition to Megalopolis.

Open MapMaster North America in MasteringGEOGRAPHY

Select: *Political* then *Cities*

Select: *Population* then *Population Density.*

Where in North America other than Megalopolis do there appear to be overlapping metropolitan areas?

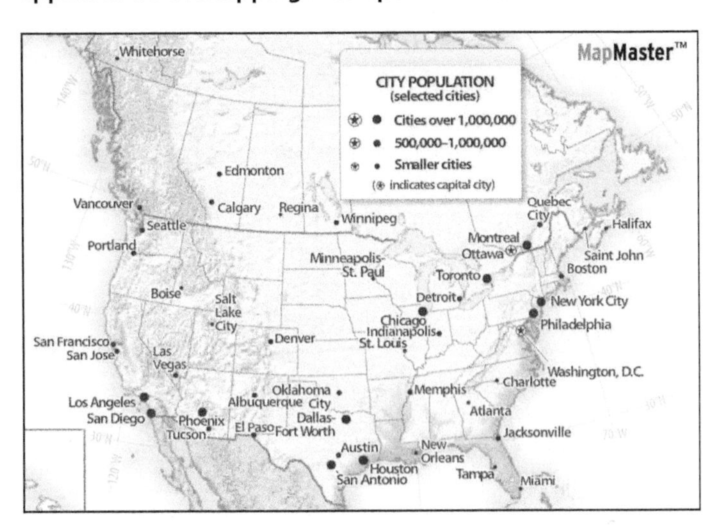

Explore

CHICAGO, ILLINOIS

Use Google Earth to explore Chicago's changing lakefront.

Fly to: *Soldier Field, Chicago.*

Drag to enter *Street view* on top of Soldier Field.

Exit *Ground level view.*

Rotate compass so that North is at the top.

Zoom out and move the image until the lakefront and green peninsula (Northerly Island Park) are visible to the east and buildings to the west.

Click Historical Imagery and slide date back to 4/23/2000

1. **What changes have occurred to the east of Soldier Field, along the lakefront?**
2. **What changes have occurred to the west of Soldier Field?**
3. **How do you think the development of this node of the city will influence people's activities?**

Key Terms

Annexation
Legally adding land area to a city in the United States.

Census tract
An area delineated by the U.S. Bureau of the Census for which statistics are published; in urbanized areas, census tracts are often delineated to correspond roughly to neighborhoods.

Central business district (CBD)
The area of a city where consumer, business, and public services are clustered.

City
An urban settlement that has been legally incorporated into an independent, self-governing unit.

Combined statistical area (CSA)
In the United States, two or more contiguous core based statistical areas tied together by commuting patterns.

Concentric zone model
A model of the internal structure of cities in which social groups are spatially arranged in a series of rings.

Core based statistical area (CBSA)
In the United States, a term referring to either a metropolitan statistical area or a micropolitan statistical area.

Edge city
A large node of office and retail activities on the edge of an urban area.

Gentrification
A process of converting an urban neighborhood from a predominantly low-income renter-occupied area to a predominantly middle-class owner-occupied area.

Metropolitan statistical area (MSA)
In the United States, a central city of at least 50,000 population, the county within which the city is located, and adjacent counties meeting one of several tests indicating a functional connection to the central city.

Micropolitan statistical area (μSA)
In the United States, an urban area of between 10,000 and 50,000 inhabitants, the county in which it is found, and adjacent counties tied to the city.

Multiple nuclei model
A model of the internal structure of cities in which social groups are arranged around a collection of nodes of activities.

Peripheral model
A model of North American urban areas consisting of an inner city surrounded by large suburban residential and business areas tied together by a beltway or ring road.

Primary census statistical area (PCSA)
In the United States, all of the combined statistical areas plus all of the remaining metropolitan statistical areas and micropolitan statistical areas.

Sector model
A model of the internal structure of cities in which social groups are arranged around a series of sectors, or wedges, radiating out from the central business district (CBD).

Social area analysis
Statistical analysis used to identify where people of similar living standards, ethnic background, and lifestyle live within an urban area.

Sprawl
Development of new housing sites at relatively low density and at locations that are not contiguous to the existing built-up area.

Squatter settlement
An area within a city in a developing country in which people illegally establish residences on land they do not own or rent and erect homemade structures.

Urbanized area
In the United States, a central city plus its contiguous built-up suburbs.

On the Internet

Social Explorer provides access to census data at all scales, including urban, at **www.socialexplorer.com**, or scan the QR at the beginning of the chapter. An interactive map enables users to choose the area of interest from among hundreds of census variables.

▶ LOOKING AHEAD

Our journey ends with an examination of the use, misuse, and reuse of resources.

14 Resources

It is almost impossible to imagine a day, or even a few hours, without water or electricity. Natural resources have always been central to the human condition, but as the world population grows beyond 7 billion and economic growth proceeds, strains on natural resources are becoming more acute.

Many of the major areas of tension in the world today are related to natural resource issues. Oil is a focus of international concerns, and variations in oil prices have immediate and large effects on our economies. Fresh water is becoming increasingly scarce. Local and regional-scale water problems are closely linked to global demand for agricultural commodities, and this demand along with water scarcity contributes to rising food prices.

At the same time, we know that history is filled with examples of technological developments that have responded to natural resource needs. Rather than allowing resource stresses to worsen, we can imagine a wide variety of innovations and changing human habits that could make present resource crises irrelevant, and shift attention to commodities that we barely notice today.

NUCLEAR POWER PLANT, BADEN-WÜRTTEMBERG, GERMANY

What is a natural resource and what determines the value of resources?

How do we use energy and mineral resources?

SCAN FOR DATA ON ENERGY FROM THE US ENERGY INFORMATION ADMINISTRATION

How can we manage our resources for future generations?

14.1 Resource Concepts

- Natural resources are created by natural processes, but they are defined by culture, technology, and economic conditions.
- Over time, changing cultural, technological, and economic conditions can cause resources to become more valuable or even to cease to have value.

A **natural resource** is anything created through natural processes that people use and value. Examples include plants, animals, coal, water, air, land, metals, sunlight, and wilderness.

If the resource is naturally produced at rates similar to our use we call it a **renewable resource.** Natural resources are especially important to geography because they are the specific elements of the atmosphere, biosphere, hydrosphere, and lithosphere with which people interact. Natural resources can be distinguished from human-made resources, which are human creations such as money, factories, computers, information, and labor. A substance is merely part of nature until a society has a use for it. Consequently, a natural resource is defined by culture, technology, and economic system:

- Cultural values influence demand for commodities, and also affect society's willingness to influence supply and demand through policy.
- Technology has a tremendous influence on our ability to use certain resources, and on the relative costs and benefits of using those resources.
- A society's economic system affects whether a resource is affordable and accessible. In a market economy, supply and demand are the principal factors determining affordability.

The same elements of society apply to the study of any example of a natural resource, whether it be oil, diamonds, forests, or clean air. In every case, a combination of the three factors is necessary for a substance to be valued as a natural resource.

CULTURAL VALUES AND NATURAL RESOURCES

To survive, humans need shelter, food, and clothing, and make use of a variety of resources to meet these needs. We can build homes of grass, wood, mud, stone, or brick. We can eat the flesh of fish, cattle, pigs, or mice—or we can consume grains, fruit, and vegetables. We can make clothing from animal skins, cotton, silk, or polyester. Cultural values help us identify things as resources to sustain life (Figure 14.1.1).

▲ 14.1.1 **VALUING RESOURCES**
Which is better—an urban park (left: Central Park, New York) or a wilderness (right: Olympic National Park, Washington State)? Obviously we would like both. If resources are scarce and choices have to be made, which should we support? Typically urban parks serve many people, while few visit wilderness areas. Wilderness harbors natural communities that are not found elsewhere. Does that give them value comparable to urban parks?

A swamp is a good example of how shifting cultural values can turn an unused feature into a resource (Figure 14.1.2). A century ago, swamps were seen in the United States as noxious, humid, buggy places where diseases thrived instead of places that provide usable commodities. Swamps were valued only as places to dump waste or to convert into agricultural land. Eliminating swamps was good, because it removed the breeding ground for mosquitoes while simultaneously creating productive and valuable land. During the twentieth century, cultural values changed in the United States. We became more aware of the value of natural ecosystems, and the role of swamps in controlling floods, providing habitat for wildlife, and reducing water pollution. Changing public attitude toward swamps is reflected in our vocabulary: instead of calling them swamps, now we use a more positive term, wetlands. Today wetlands are a valued land resource, protected by law. We restore damaged wetlands, create new ones, and restrict activities that might harm them.

▲ 14.1.2 **PROTECTING WETLANDS: NEW JERSEY**
In the United States wetlands, formerly called swamps, used to be seen in negative terms—places with no positive values and even some negative ones. Wetlands like these near New York City are now protected for their biodiversity and positive effects on water quality.

TECHNOLOGY AND NATURAL RESOURCES

The utility of a natural substance depends on a society's technological ability to obtain it and to adapt it to that society's purposes. Metals, for example, are elements that can be formed into materials that have high strength in relation to their weight, can generally withstand high temperatures, and are good conductors of heat and electricity. However, a metal ore is not a resource if the society lacks knowledge to recover its metal content and to shape the metal into a useful object, such as a tool, structural beam, coin, or automobile fender.

Earth has many substances that we do not use today, because we lack the means to extract them or the knowledge of how to use them. Things that might become resources in the near future are potential resources. Radioactive uranium had little value until we developed the technology to use it in weapons and generating electricity.

ECONOMICS AND NATURAL RESOURCES

Natural resources acquire a monetary value through exchange in a marketplace. The price of a substance in the marketplace, as well as the quantity that is bought and sold, is determined by supply and demand. Common sense tells us some principles of supply and demand. A commodity that requires less labor, machinery, and raw material to produce will sell for less than a commodity that is harder to produce. The greater the supply, the lower the price; the greater the demand, the higher the price. Consumers will pay more for a commodity if they strongly desire it than if they have only a moderate desire. If a product's price is low, consumers will demand more than if the cost is high.

Externalities are exchanges of commodities that take place outside the marketplace and thus have no price attached at the time of exchange. For example, a coal-fired power plant generates air pollution in the process of producing electricity. The people who buy the electricity pay for the electricity directly. The price they pay reflects their desire to have the electricity and their willingness to pay for it. But people downwind from the power plant receive the pollution, whether they like it or not. The power plant does not directly pay for the privilege of discharging pollution to the atmosphere, although it does bear some of the costs of pollution control. This unpriced pollution thus constitutes a hidden cost. If the power plant were charged for its pollution and the people who receive it were compensated, perhaps the power plant would emit less pollution and thereby have less impact on the people exposed to it. Many important natural resource problems result from the inability of markets to account for pollution.

14.2 Balancing Competing Interests

- Most resources can be used for several purposes, and most resource needs can be met in several different ways.
- If a resource is used for one purpose, its utility may be limited or enhanced for other purposes.

Many natural resources are valued for specific properties—coal for the heat it releases when burned, wood for its strength and beauty as a building material, fish as a source of protein, clean water for its healthiness. In most cases, several substances may serve the same purpose, so if one is scarce or expensive, another can be substituted. Copper is an excellent conductor of electricity, but it may be expensive relative to wire made of other metals that can be substituted. For information transmission, such as in computer networks, using light in a fiber-optic cable is more efficient than using electrons in a copper wire.

The substitutability of one substance for another is important in stabilizing resource prices and limiting problems caused by resource scarcity. If one commodity becomes scarce and expensive, cheaper alternatives usually are found. Such substitution is central to our ability to use resources over extended periods without exhausting them and without decline in our standard of living.

However, many resources have no substitutes. There is only one Old Faithful Geyser and only one species of sperm whale, so if we destroy Old Faithful or force extinction of the sperm whale, we have no substitutes waiting to be tapped (Figure 14.2.1). Other geysers and whales exist, but they are not the same as those we now know. The uniqueness of these resources is the essence of their value.

▼ 14.2.1 **OLD FAITHFUL** This geyser in Yellowstone National Park serves no practical purpose, but is highly valued simply for its uniqueness.

BALANCING COMPETING INTERESTS

What happens with people who don't agree on a resource's best use? Political and economic relations are central to any situation in which competing interests battle for control of a scarce resource. For example, in 2011 the U.S. Environmental Protection Agency issued new rules for restricting pollution from power plants in the eastern United States. The debate over the rules was mainly between two interest groups:

- Environmentalists, who argued that the rules would reduce the health and ecological effects of the pollution,
- The electric utility industry, which argued that increased costs to utilities would hurt the economy.

Environmentalists and electric utilities each have their own power base and allies in federal and state governments, and the debate between these traditional adversaries lasted several years. In the end, the economic argument for the rules was a strong one. Reduced health care costs and reduced sick days were valued at $120 to $280 billion per year, while the pollution-control devices to be installed in power plants were estimated to cost $0.8 to $2.4 billion.

Government regulators often make use of economic principles in deciding how to manage resources. In market economies, the relative values that society places on different uses of resources are reflected in the prices of resource commodities. The demand for electricity affects the price we are willing to pay for coal that is used to generate it. The cost of

▲ 14.2.2 **REFORESTATION IN PACIFIC FORESTS OF ECUADOR** Funded as a carbon offset, this forest is intended to permanently sequester carbon, offsetting emissions from fossil fuels.

health care affects the amount we are willing to spend on pollution controls that will protect our health.

One market-based approach to pollution control is a system called cap-and-trade, in which a maximum is set on the amount of pollution that will be allowed, and different polluters can buy and sell permission to emit the polluting substance in question. A company that is able to operate profitably at low emission rates will sell pollution permits to companies that must emit more to function. In this way a cleaner company benefits from its efficient practices, while a dirtier company is required to pay more. A cap-and-trade system for controlling CO_2 emissions has been considered in the United States for years, and the Chicago Climate Exchange has operated since 2003 as a voluntary market for such a system. Concerned individuals or companies that wish to reduce greenhouse gas emissions purchase credits in proportion to the amount of CO_2 they emit, and the money may be used to pay for carbon sequestration projects such as reforestation or for investments in new technologies that reduce CO_2 emissions (Figure 14.2.2).

Making resource-use decisions on the basis of market values alone can be difficult when the commodities involved cannot easily be assigned monetary values. What is the value of clean air, or of biodiversity, for example? Sometimes we can determine one value by considering the value of something else that is closely linked to the first, hard-to-price value. For example, the value of the solitude or beauty one may enjoy in a certain forest may be difficult to determine. But the amount of money tourists pay to visit that forest may be a good indicator of how much these aspects of the forest are worth (Figure 14.2.3).

Whether we consider forests, water and air resources, energy, minerals, farmland, or any other natural resource, it is clear that most resources are coming under increasing pressure worldwide, and the environmental impacts of resource use are increasing. As long as population grows and/or per capita resource use increases (and in general these trends have been ongoing for centuries), competition for increasingly scarce natural resources will increase.

◀▼ 14.2.3 **FOREST USES** Forests have many uses, ranging from fiber (lumber, paper) to wildlife (below: Pahang State, Malaysia), recreation (left: Jasper National Park, Alberta, Canada), and water quality.

14.3 Energy Use

- As technology has changed, so have our needs for energy and the forms in which we use that energy.
- Our demand for electricity is growing, but we still rely on fossil fuels to generate most of that electricity.

CHANGES IN ENERGY USE

From the time that humans first began to use fire until the mid-1800s, wood was the most important source of energy. In North America, the native population was generally low enough that forest depletion was not a problem. When Europeans settled in the mainland of North America, beginning in the seventeenth century, they harvested the forests for fuel and lumber and cleared the land for agriculture. By the end of the nineteenth century, most of the forests near populated areas of the eastern United States had been cut down, and fuel wood became very expensive. It was also inadequate for providing the large amounts of energy demanded by a growing industrial economy. Wood still provides a large portion of energy in developing countries, though supplies are dwindling in areas of dense, rural populations.

In the early twenty-first century, three fossil fuels provide more than 85 percent of the world's energy and more than 90 percent in developed countries (Figure 14.3.1).

- Coal, a **fossil fuel** created through photosynthesis by plants that grew millions of years ago, served as a substitute for fuel wood during the nineteenth century.
- Oil, also derived ultimately from photosynthesis, was a minor resource until the diffusion of motor vehicles early in the 1900s. The power and reduced weight of the internal combustion engine made oil a much more attractive fuel than coal for transportation. As the infrastructure of refineries, pipelines, and distribution facilities (including gas stations) grew, oil quickly replaced coal for most purposes except electricity generation and some industrial uses. It is currently the world's most important energy resource.
- Natural gas was once burned off during oil drilling as a waste product because it was too difficult to handle, and markets for it were not established. In recent years, however, it has become an important energy source.

Just as we abandoned wood for coal and coal for oil and natural gas, we are now in the midst of a transition from fossil fuels to electricity (Figure 14.3.2). Basic needs such as heating and cooking are increasingly supplied by electricity, and air conditioning is regarded as a necessity rather than a luxury in much of the world. The rapid expansion of use of electronic devices also has dramatically increased our electricity use. We are beginning to power our automobiles with electricity. However, this transition is occurring at the consumption end of the system more than the production end—we are still heavily reliant on fossil fuels for generating electricity.

The tendency of society to rapidly adopt a new energy source when the technology and economy favor it, and simultaneously abandon an old source, leads to rapid transformations in our energy economy. Such transitions are encouraged by scarcity of a resource as it is exploited and depleted. Today we are wondering when we will make the transition away from oil.

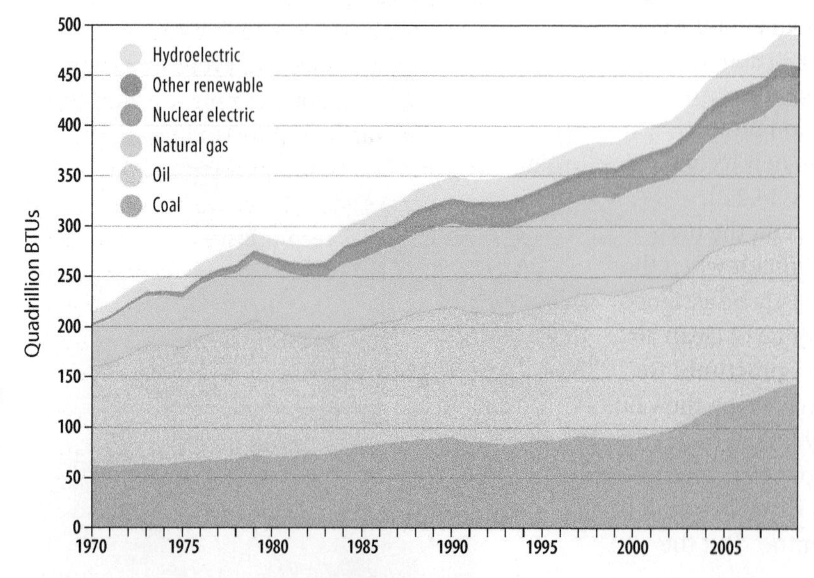

► 14.3.1 **TRENDS IN WORLD ENERGY USE, 1970–2007**
World energy use has grown dramatically in recent decades. Events such as the 1970s oil crises, the collapse of the Soviet Union, and the recession that began in 2008 have a clear impact on energy use, as seen in periods of reduced growth or even decline in energy use.

▲► 14.3.2 **SUNSET BOULEVARD 1960 (ABOVE), 2010 (RIGHT)**
Electricity use has grown very rapidly in the last few decades.

DEMAND VS. RESERVES

Deposits of a resource that have been identified and are commercially extractable at present prices are called **reserves**. If we divide Earth's current proven oil reserves (about 1,300 billion barrels) by current annual consumption (about 26 billion barrels), we see that we have about 50 years of oil supply remaining, at present rates of consumption. These rates of consumption will change, and new reserves will be discovered, but we see that current proven oil reserves will probably last a few decades. Thus Earth's oil reserves are likely to be significantly depleted during the twenty-first century, possibly in your lifetime.

Every discovery of new deposits extends the life of the resource. But extracting oil is becoming harder—and therefore more expensive. At some point, the combination of a high cost of finding and extracting oil, and growing attractiveness of other energy technologies such as electricity, are likely to drive a transition away from oil and toward a new energy source. When that happens, we will shift our energy infrastructure toward the new energy source(s) and the demand for oil will drop, reducing prices below what it costs to extract the oil. This will cause oil production to reach a peak and begin a decline. These ideas have become known as "peak oil."

If we accept that a peak must happen sometime, the key question is: When? Some argue that the peak will occur very soon—in the next few years. These individuals, sometimes referred to as "peakists," point to declining production in major oil-producing regions such as the United States and the North Sea, and suggest that previous estimates of oil reserves may have been too high. Others argue that undiscovered or undeveloped oil fields are many and that higher prices will spur development of these resources and provide oil for several decades into the future.

Interestingly, each of these predictions can be self-fulfilling. If we believe that the peak will occur very soon, then we will avoid investing in things that use oil. For example, we wouldn't build a new gas station but instead we might build charging facilities in parking lots for plug-in electric vehicles. Once we begin to invest in alternative energy technologies, economies of scale will begin to make these technologies more affordable, and the transition away from oil will be under way. On the other hand, if we believe that oil will be with us for 20 or 30 years at least, then we will go ahead and build that gas station and the gasoline-powered cars that will use it, and people will pay the higher prices necessary to keep driving their cars.

14.4 Fossil Energy

- **Fossil fuels are unevenly distributed, with some countries producing much more than they consume and others heavily dependent on imports.**
- **Existing known fossil fuel deposits are sufficient to last at least decades into the future.**

The world relies heavily on fossil fuels for nearly all its energy needs. This reliance, however, cannot last indefinitely. Fossil fuels are consumed at rates far exceeding the rates at which they are created in nature. Once they are consumed they are not replaced, which means they are **nonrenewable resources.**

UNEVEN DISTRIBUTION AND CONSUMPTION

Wealthy and industrialized countries typically consume large amounts of energy, while poorer countries usually consume much less. The United States imports roughly half of its needs, western European countries more than half, and Japan more than 90 percent (Figure 14.4.1). The major producer nations vary depending on the fuel (Figure 14.4.2).

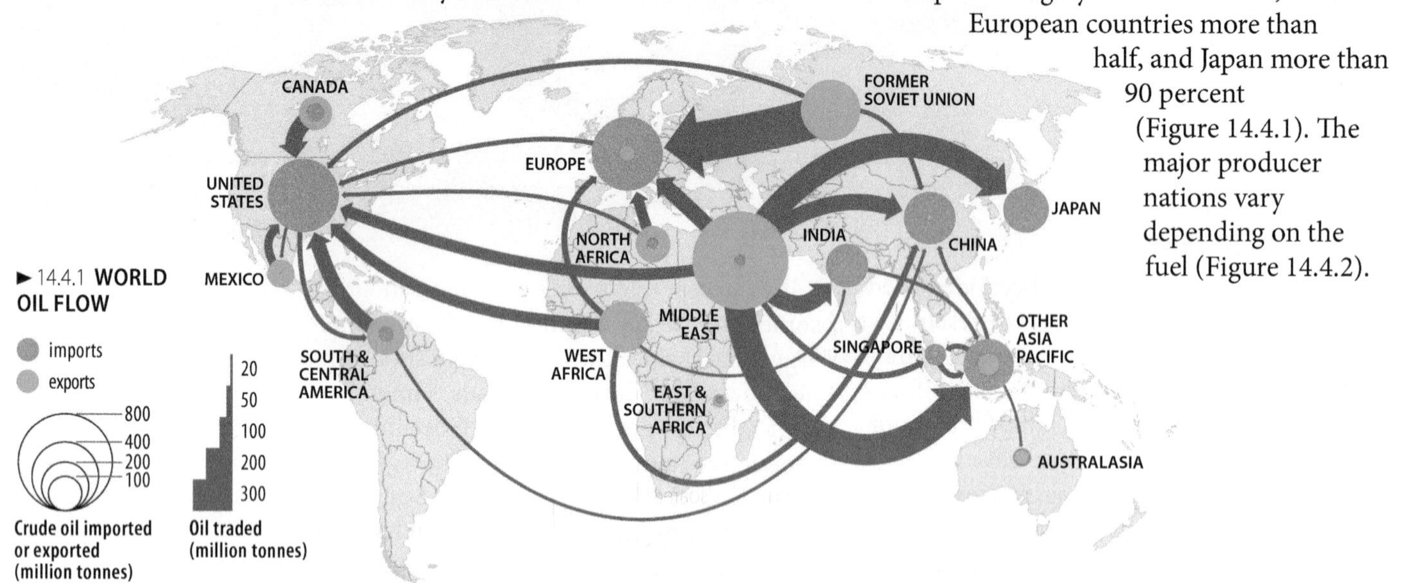

► 14.4.1 **WORLD OIL FLOW**

Given our dependence on nonrenewable fossil fuels, it is important to know for how long they will be available. Estimates of the amount of fossil fuels remaining carry a large amount of uncertainty because new deposits remain to be discovered and some deposits may be in forms or places that make them too expensive to recover. However, we do have some idea what is available. The ratio of reserves to annual production (R/P) is an indication of the number of years the known reserves will last. World R/P ratios for oil and gas are currently 51 and 57 years, respectively, while the ratio for coal is 125 years. Thus we appear to have enough fossil fuels to last decades at least, should we choose to use them.

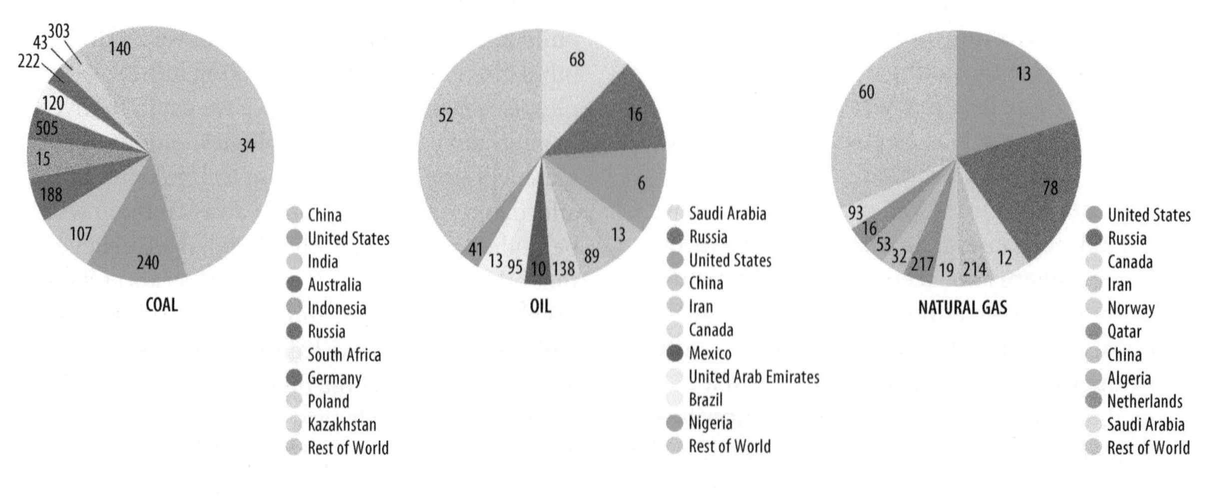

► 14.4.2 **PRODUCTION AND RESERVES OF FOSSIL FUELS**
For each resource, only the top 10 countries are shown. The countries listed account for 88 percent of all coal reserves, 75 percent of all oil reserves, and 72 percent of all gas reserves. Numbers are ratios of reserves to annual production, in years. Data are for 2009, except coal data, which are for 2008.

OIL

The United States was once a net oil exporter but since the mid-twentieth century has imported more than half its needs. A significant disruption of supply can cause a rise in prices. This price volatility increases international tension and interest in other energy sources (Figure 14.4.3).

▼ 14.4.3 **OIL PRICE HISTORY**

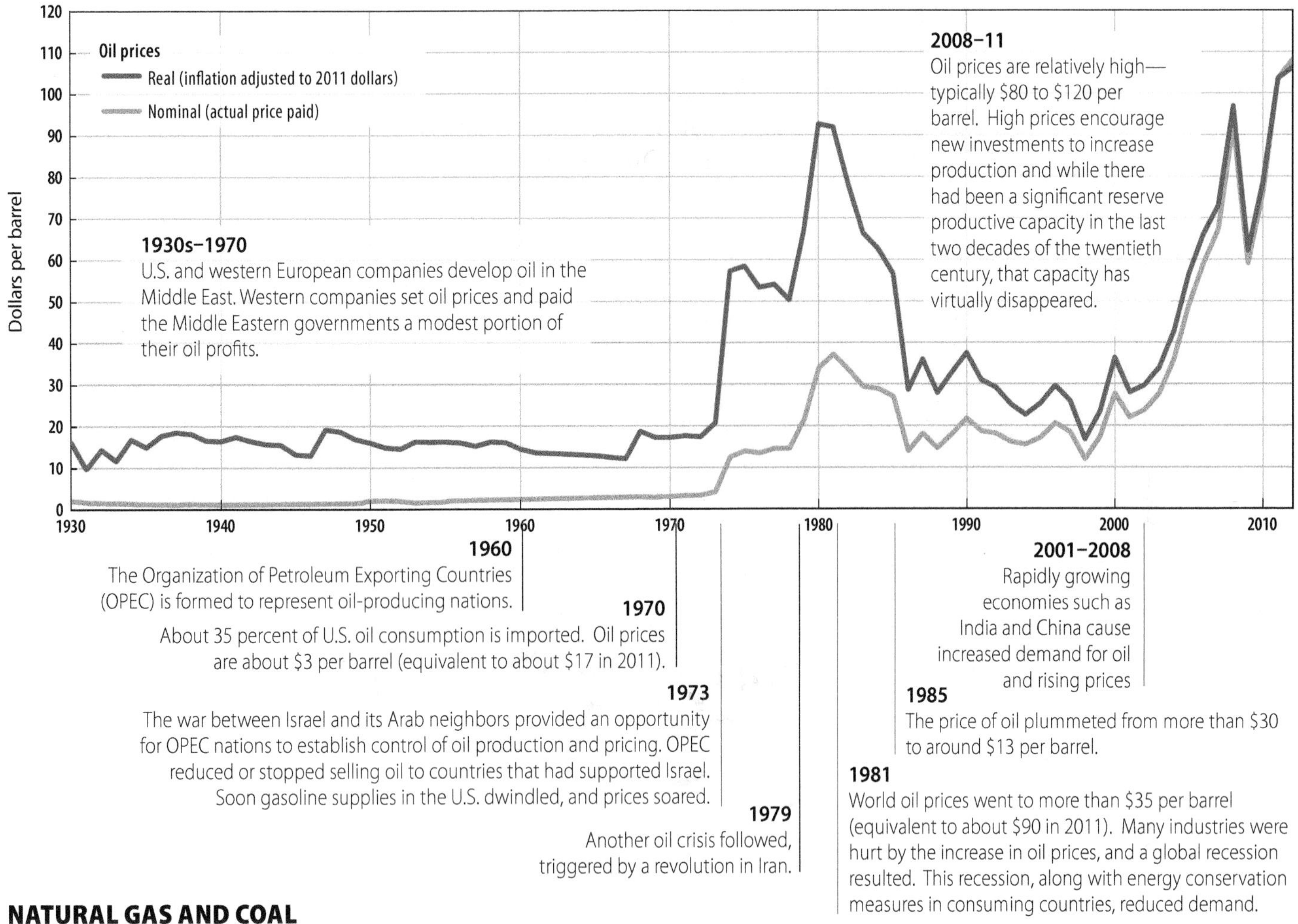

NATURAL GAS AND COAL

Natural gas is important in many areas as the clean-burning fuel of choice for the residential, commercial, and industrial sectors. Gas has absorbed a large part of the recent increase in U.S electricity production.

Gas resources are related to oil resources in that regions that have one tend to have the other, although the quantities vary considerably. Gas is harder to ship across the sea than is oil, so proximity to markets is especially important. Gas production has provided fuel for much of the growth in electricity production that has occurred in the last 30 years. In Europe, Russia's role in supplying the energy needs of western Europe has grown dramatically. In recent years, gas production has expanded in the U.S. using a technique called hydrofracking that allows extraction of gas from rocks that previously were not considered usable resources. The method is controversial because it may cause contamination of groundwater. On the other hand, gas is cleaner to burn than coal or oil, and emits less carbon dioxide per unit of heat than other fossil fuels.

World coal reserves are more abundant than oil or natural gas. Coal can play an especially important role in providing coal-rich countries such as China and the United States with energy because it is used mainly to produce electricity. Several problems hinder expanded use of coal, including air pollution and land and water impacts. Coal is a major contributor to CO_2 emissions and to **acid deposition**, which occurs when sulfur and nitrogen oxides combine with water to form acidic precipitation. Mining also damages the land from which coal is extracted. Technologies may be developed that will allow us to use coal more cleanly, but they will be expensive and may not be fully effective.

14.5 Alternative Energy

- Nuclear power is widely used, does not contribute to global warming, and has much potential for growth, but safety concerns are a major barrier.
- Renewable energy resources make a relatively minor contribution to our energy supply, while energy conservation has the greatest potential to balance needs with supply.

The rising costs of fossil fuels combined with concerns about global warming are stimulating interest in alternative sources of energy—especially those that produce electricity.

NUCLEAR POWER

Nuclear power supplies about one-third of all electricity in Europe, and about 20 percent of U.S. electricity. Japan, South Korea, and Taiwan also rely on nuclear-generated electricity (Figure 14.5.1).

Despite the fact that nuclear power is relatively clean, it presents serious problems of potential accidents and the generation of radioactive waste. Three major incidents have had dramatic influences on the nuclear power industry.

- In March 1979, a power plant at Three Mile Island experienced a loss of coolant that contributed to a partial meltdown of the reactor core (Figure 14.5.2). The incident was relatively minor in terms of radioactive releases and no one was injured, but it caused a dramatic increase in public awareness of the hazards involved.
- The greatest nuclear catastrophe to date occurred in 1986, in a nuclear power plant at Chernobyl, Ukraine, near the border with Belarus. (Both countries at the time were part of the Soviet Union.) The accident has caused over 3,500 deaths, and potentially many more from subsequent cancers. The impact of this accident extended through Europe, causing bans on the sale of milk and fresh vegetables because of possible contamination by radioactive fallout. These incidents virtually ended the construction of new nuclear power plants in the United States and many other countries, and the growth of nuclear power slowed dramatically for the next two decades.
- The catastrophe at the Fukushima Daiichi plant in Japan in March 2011 has once again raised fears surrounding nuclear power. It came at a time when, in many areas including the United States, interest in building new nuclear power plants was starting to grow. It remains to be seen whether public faith in nuclear power will recover sufficiently to justify building new plants.

▼ 14.5.1 **WORLD PRODUCTION OF NUCLEAR POWER**
The worldwide total is about 14 percent of total electric power generation.

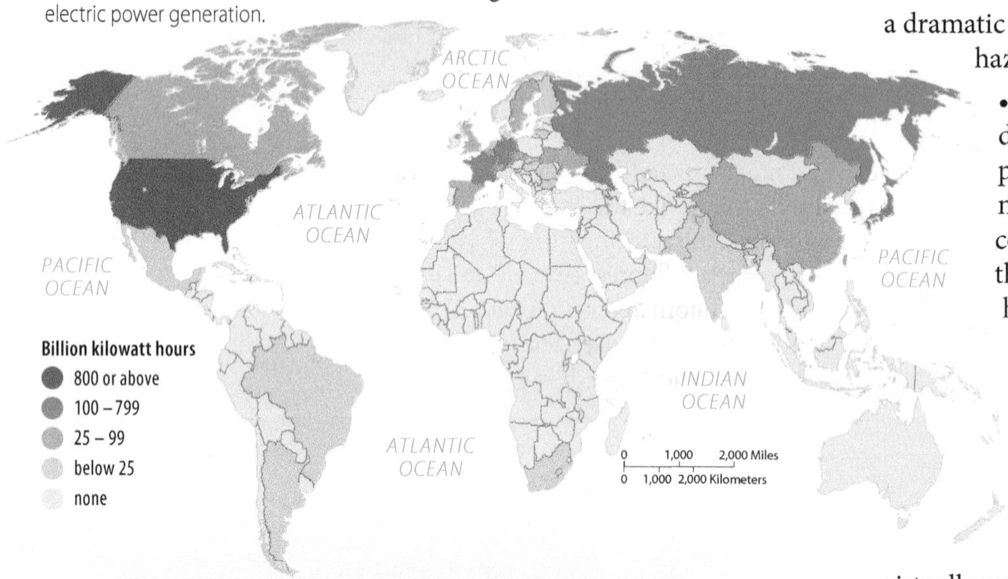

◄ 14.5.2 **THREE MILE ISLAND NUCLEAR GENERATING STATION, NEAR HARRISBURG, PENNSYLVANIA**
An accident at this plant in 1979 caused no significant release of radiation, but increased concerns about safety.

RENEWABLE ENERGY SOURCES

Renewable energy includes a diverse range of technologies, from traditional wood-burning, to hydroelectricity that has been used for more than a century, to technologies such as wind-generated electricity, ethanol and biodiesel, photovoltaics, tidal energy, and solar thermal power production that are emerging today or are still in development (Figure 14.5.3). Wood fuel supplies a substantial amount of energy for cooking and heating, especially in poor countries. Hydroelectricity is the most important renewable source for commercial power, producing about 6 percent of global commercial energy. All the other renewable technologies together are responsible for less than 2 percent of global energy production.

Many forms of plant matter can be converted to **biofuels**, liquid or gaseous fuels that can be burned, for example, in modern automobiles. Sugarcane, corn, and soybeans are readily converted to ethanol and diesel fuel. However, it takes a significant amount of land to produce the feedstock for biofuels, and the increase in production has taken much land that could have been used for food. Also, the amount of fuel energy that is produced from manufacturing ethanol from corn or biodiesel from soybeans is small after one accounts for the amount of fuel consumed for tractor fuel, fertilizer, and other energy inputs to biofuel production.

Canada, China, Brazil, and the United States are the largest producers of hydroelectric power in the world. Most of the best sites for hydroelectric generation are already in use in the United States and Europe, but in many areas, there is considerable undeveloped potential. Opposition to construction of big dams and reservoirs is strong among environmentalists who fear the damage they cause, such as loss of farmland, animal habitat, or natural beauty.

At present, solar energy is used in two principal ways: thermal energy and photovoltaic electricity production. Solar thermal energy is heat collected directly from sunshine in collectors on the roofs of buildings. The heat absorbed by these collectors is then carried in water or other liquids to the places where it is needed. **Photovoltaic electric** production is a direct conversion of solar energy to electricity in photovoltaic cells. Photovoltaic power is competitive in price with conventional electricity in much of the world, and offers considerable potential as economies of scale bring prices down. However, it suffers from the problem that solar energy is not reliable: it varies with season and with weather. Geothermal energy use in the United States is mainly in the form of electricity production at a handful of sites, led by California's Geysers plants, although building-scale heating and cooling systems are becoming increasingly common.

Wind generation of electricity is one of the fastest-growing renewable energy technologies today (Figure 14.5.4), although it still amounts to less than 2 percent of total U.S. electricity production. Environmentalists are divided on wind energy: some favor wind generation because it is renewable and pollution-free, while others oppose it on the grounds of visual impacts and its hazards to birds.

Conservation—reducing waste and doing more work with the same amount of energy—is equivalent to a new energy source.

Conservation technologies are highly diverse and most are readily available. They range from simple things like turning off lights or insulating buildings, to replacing incandescent lightbulbs with compact fluorescents, to "smart" buildings that have automated systems for monitoring and controlling energy use. Some of these changes are actively encouraged by government policy, but the one force that will most certainly spur conservation is an increase in energy prices. Such an increase can be managed by tax policy, or it can happen as a consequence of trends in the marketplace driven by supply and demand.

▼ 14.5.3 **U.S. RENEWABLE ENERGY PRODUCTION, 2010**

Biomass
Hydropower
Wind
Geothermal
Solar

▼ 14.5.4 **WIND ENERGY PRODUCTION**
Wind farms like this one in California have been built in many areas in recent years.

14.6 Mineral Resources

- Demand for minerals is closely tied to technology and price, and is highly variable over time.
- For most minerals, only a few countries produce a large precentage of total world supply.

We value most minerals for their properties of strength, malleability (ability to be shaped), weight, and chemical reactivity rather than for their aesthetic characteristics. Few car owners care if the engine is made from aluminum or iron; what matters is that it is powerful, durable, and efficient. Gold is a rare exception of a mineral valued mostly for its beauty (more than three-quarters of gold use is in jewelry), although even gold is increasingly demanded for industrial uses, especially electronics.

Because we value a mineral primarily for its mechanical or chemical properties, our use of mineral resources is continually changing in response to technological and economic change. Most minerals have substitutes available. In addition, as new technological processes and products are invented, demand can suddenly increase for materials that had little use in the past.

The substitutability of one mineral for another has an important consequence: even though world supply of a mineral resource may be limited, we will never run out of it. The reason is that if the supply of a resource dwindles relative to demand, its price will rise. As price increases:

1. Demand for the mineral will decrease, slowing its rate of depletion.
2. Mining companies will have added incentive to locate and extract new deposits of the mineral, especially deposits that might have been neglected when prices were lower.
3. Recycling of the mineral will become more feasible.
4. Research to find substitute materials will intensify, and as use of the substitute increases, demand for the scarce mineral will cease before the supply is exhausted.

In the short run mineral prices can be highly variable, but in the long run these factors help to stabilize mineral prices and prevent crises associated with depletion of nonrenewable resources.

DISTRIBUTION OF MINERAL RESOURCES

Mineral deposits are not uniformly distributed around the world. Most of the world's supply of particular minerals is concentrated in a handful of countries. For example, although aluminum is one of the most abundant elements on Earth, in 2010, just six countries—Australia, Brazil, China, Guinea, India, and Jamaica—produced about 94 percent of the world's total aluminum ore (bauxite) (Figure 14.6.1).

The concentration of mineral resources and production in a few countries favors the establishment of cartels. A **cartel** is a group of producers that agree to control a particular market by limiting production and/or setting prices. Occasionally, a cartel is able to control world markets briefly but usually not for an extended period of time.

▼ 14.6.1 **PRODUCTION OF BAUXITE, NICKEL, ZINC, AND COPPER IN 2010**

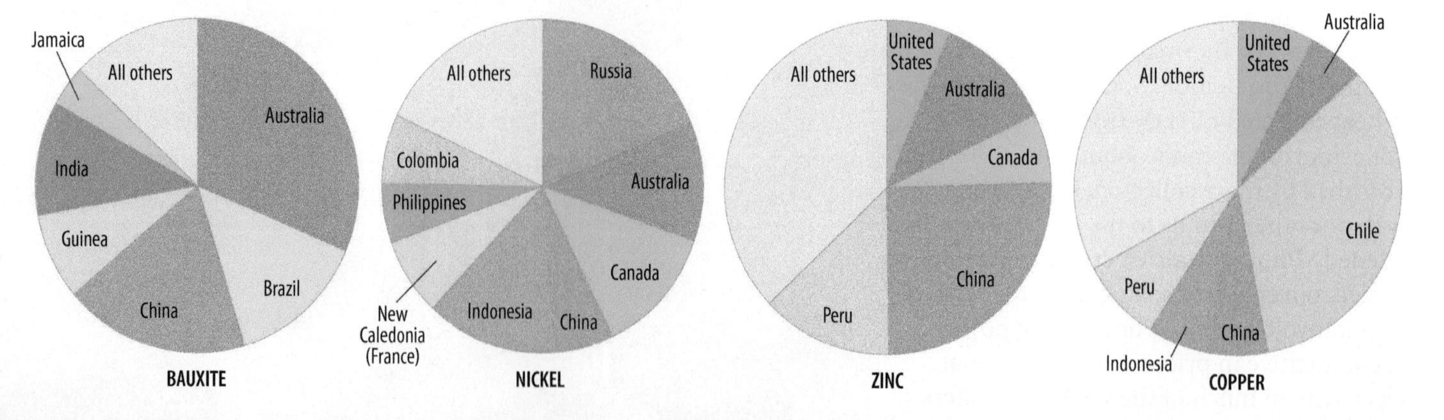

Fluctuations in the price of a mineral as a result of actions by a cartel, a political dispute, or a limited supply rarely continue long. If high prices persist, technological innovations usually enable the substitution of cheaper minerals for more expensive ones. The likelihood that substitute products will be developed means that countries (and companies) run a great risk if they are overly dependent on the production of only one specific mineral.

LITHIUM

Lithium is a metal with a wide range of uses, but has received considerable attention in recent years because of its growing importance in batteries used in a wide variety of devices such as cell phones, laptop computers, and hybrid and electric-powered vehicles. Lithium-based batteries are attractive because of their light weight, and they also are capable of providing relatively high voltages. Commercially valuable lithium deposits are found in several countries (Figure 14.6.2). Chile and Australia each produce about one-third of global output, and China and Argentina are also major producers. Bolivia has vast resources that it is only beginning to exploit, and several other countries including the United States have the potential to increase production. At present rates of production the world's reserves would last about 500 years. However, if the market for electric and hybrid vehicles expands dramatically and if those vehicles use lithium-based batteries, prices will rise and production will increase. Given the speed of technology change and the potential for new deposits to be developed it is unlikely that world supplies of lithium will be severely limited, but it will be a mineral of considerable importance in coming years.

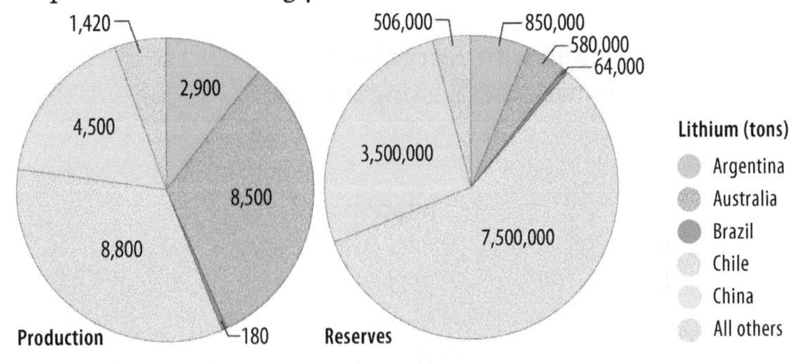

▲ 14.6.2 **LITHIUM PRODUCTION AND RESERVES**
(left) Production: Argentina, Australia, Chile, and China together are responsible for over 90% of world production. (right) Reserves: Bolivia is not included in the reserve data, but it is believed to have reserves similar to or greater than those of Chile. Total world reserves are sufficient to last 500 years at current consumption rates, but consumption is expected to increase considerably.

RARE EARTHS

The term "rare earths" refers to a group of elements called lanthanides and actinides that are not often found in large commercially valuable quantities. They have a wide range of industrial uses, mainly as chemical catalysts but also including metallurgical applications, and they are also used in some specialized consumer products such as motors in electric vehicles and in electronics. China has about half of the world's reserves, and is responsible for over 90 percent of world production (Figure 14.6.3). China is thus in a position to control the world market. Export restrictions have made some rare earth minerals more expensive outside of China, giving an advantage to Chinese manufacturers. Some countries have built factories in China to make sure they have access to Chinese rare earths. However, because other potential sources of rare earths exist, if Chinese export restrictions and resulting high prices continue it is likely that alternatives will be developed, reducing Chinese control of the market.

▼ 14.6.3 **RARE EARTH MINING IN CHINA**
China is the world's largest producer of rare earth minerals, which are used for a wide variety of purposes including chemical processes and electronics applications.

14.7 Water Resources

- **Renewable water resources are derived from precipitation and must be recharged from the surface.**
- **Most water used in agriculture is lost to evaporation, while most water used in homes, commerce, and industry can be cleaned and used again.**

Our renewable water resources are derived from precipitation. Most precipitation is used to supply plants with their needs, through transpiration, and some is lost to evaporation. The remainder ultimately becomes runoff, with a large portion of that water passing through groundwater storages en route to streams.

Because precipitation and evapotranspiration vary widely from place to place, so do water resources (Figure 14.7.1). Large areas of the planet, including some areas of the humid tropics, have large amounts of water available and relatively small populations. For example, about 18 percent of all the freshwater runoff in the world flows to the sea via the Amazon River, but the Amazon basin's population is less than 0.1 percent of world population Other areas have modest precipitation but high evapotranspiration. China and India together are about 37 percent of world population, but have less than 9 percent of the world's renewable fresh water (Figure 14.7.2).

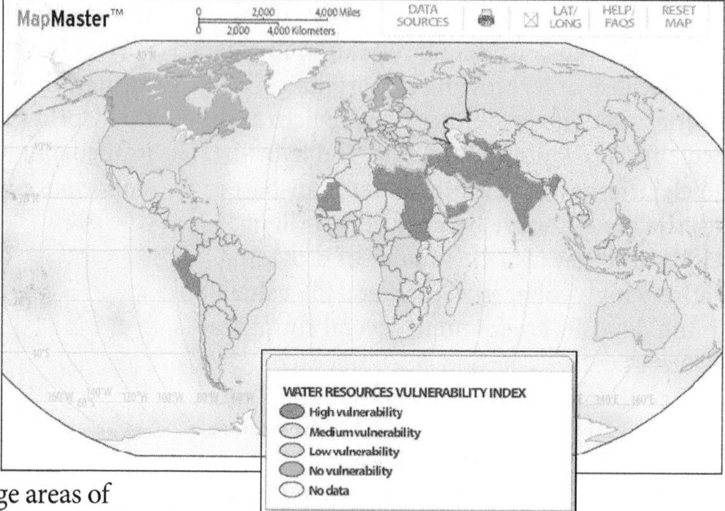

▲ 14.7.1 **WATER RESOURCES VULNERABILITY**

Open Mapmaster World in MasteringGEOGRAPHY

Select *Water Resources Vulnerability* from the *Physical Environment* menu.

Then select *Population Density* from the *Population* menu.

1. **What countries around the world have severe water stresses?**
2. **Which region has the strongest correlation between high population density and severe water stress?**

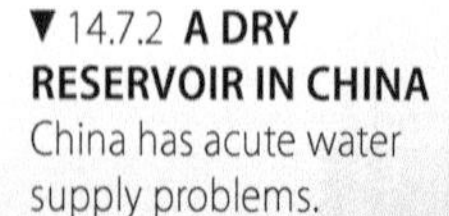

▼ 14.7.2 **A DRY RESERVOIR IN CHINA**
China has acute water supply problems.

In the United States, about 483 billion cubic meters of freshwater, or about 16 percent of the renewable supply, are withdrawn from the ground or from rivers and lakes per year. About 14 percent of this water is used in homes and businesses, 46 percent in industry (most of which is for cooling thermal electric power plants and is almost immediately returned to the river or lake from which it was taken), and 40 percent in agriculture (Figure 14.7.3). About two-thirds of the water we use is returned to rivers and lakes in liquid form, while the remaining one-third is evaporated, mostly from irrigated fields. This evaporative use, also called **consumptive use**, is especially important because the water cannot be treated and reused. Agriculture is the most heavily consumptive water use because the main purpose of the water is to supply plants that transpire it.

Even in relatively well-watered parts of the world, use of water is reaching levels that strain available resources. Many areas of humid climate have seasons in which precipitation and soil moisture do not meet the evapotranspiration needs of plants, let alone what humans remove from rivers or groundwater. In some regions, water use has reached the point that major rivers sometimes no longer reach the sea. This is the case for the Colorado River in the southwestern United States, and the Huang He (Yellow River) in China. In some densely populated coastal regions, groundwater use is causing salty seawater to contaminate fresh groundwater and in some cases has made the groundwater unusable. And because we use some water to carry away our wastes, not all water in streams and lakes is available for irrigation, washing, or drinking. Around the world, millions of reservoirs have been built to capture water in seasons with ample runoff and store it for use in other times of the year. However, these reservoirs have negative as well as positive effects, and opposition to future reservoir construction is growing.

▲ 14.7.3 **CENTER-PIVOT IRRIGATION IN WESTERN KANSAS**
Irrigation is critical to world agricultural output. Use Google Earth to explore irrigated fields in Kansas.

Fly to: *Colby, Kansas*, USA

Zoom to an eye altitude of about 20 km.

The circles that you see filling the landscape are center-pivot irrigation systems. Without them, this region would be mostly short-grass prairie; with them it produces crops usually grown in more humid areas. The circles align with the square-mile-based survey system used in most of the United States.

Using the ruler tool, what is the diameter of most of the circles?

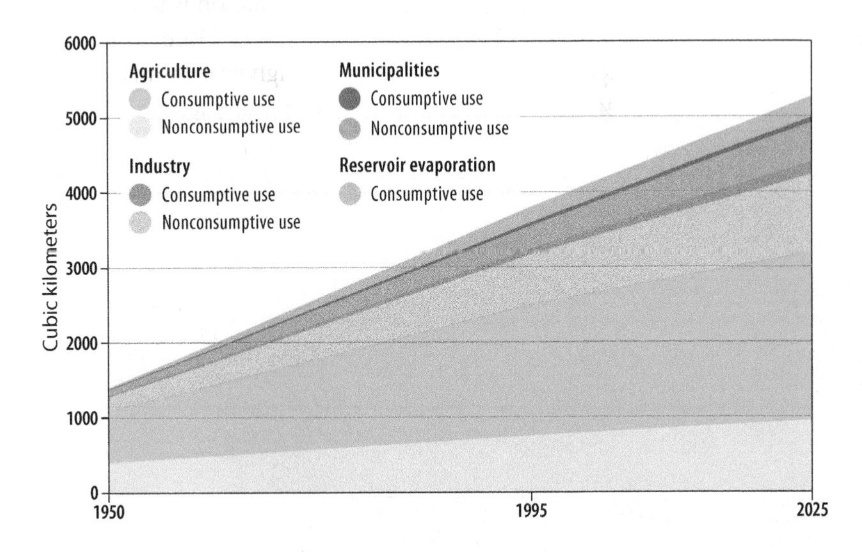

▲ 14.7.4 **INCREASING DEMAND FOR WATER**
Increased demand for agricultural commodities is stressing water supplies.

Projections of future water needs for a growing human population show that we will face serious problems. More people on the planet means more food, and that will mean greater demands for irrigation water (Figure 14.7.4). In addition, diets are changing, with meat consumption increasing rapidly in some parts of the world, including China. Most meat production depends on feeding animals crops such as soybeans and corn, and a significant portion of the food value of those crops is lost to respiration in the process. Therefore total crop production needs will grow faster than the human population. Furthermore, much of the water used in crop production is evapotranspired and therefore not available for treatment and reuse. Thus total water needs are expected to increase dramatically in the coming decades.

14.8 Water Pollution

- **Water pollution comes from diverse sources, including agriculture, urban waste, and industry.**
- **Dramatic improvements in water quality have been made in some areas, and with investment much more can be done.**

Water is a renewable resource, but unlike sunshine it is one that can be significantly harmed by misuse. Water **pollution** (elevated levels of impurities in the environment caused by human activity) results when more waste is added than the receiving water can accommodate. Water can dissolve a wide range of substances, and it can transport bacteria, plants, fish, sediment, toxic chemicals, and trash of all kinds. Water pollution results when substances enter the water faster than they can be carried off, diluted, or decomposed.

▼ 14.8.1 **URBAN RUNOFF, LONDON** Water that runs off parking lots, roads, and building roofs is a source of nonpoint pollution.

Water pollutants have diverse sources. Some come from a **point source**—they enter a stream at a specific location, such as a wastewater discharge pipe. Others come from a **nonpoint source**—they come from a large diffuse area, as happens when organic matter or fertilizer washes from a field during a storm (Figure 14.8.1). Nonpoint sources usually pollute in greater quantities and are much harder to control. Agricultural lands are the dominant source of nonpoint pollution in most of the world, although atmospheric sources such as acid deposition derived from air pollution can also be important.

In streams, natural biological and physical processes break down pollutants, disperse them, or cause them to accumulate in sediment. As a result, the concentration of a pollutant usually declines downstream from the place where waste is discharged. Prior to the twentieth century, this was the only way our pollutants were removed from streams. As urban populations grew and we became more aware of the health and ecological consequences of pollution, treatment systems were built that receive wastewater from sewer systems and clean it before returning it to the environment.

Groundwater pollution is also a serious problem. Pesticides applied to lawns and golf courses find their way into streams and groundwater. Landfills and underground tanks can leak pollutants into groundwater. Sometimes toxic substances are deliberately pumped into the ground as a way of disposing of them. Groundwater flow velocities are low and storage times are long. Groundwater is also not exposed to the air, nor are biological processes that might break down pollutants as active in groundwater as they are in streams. As a result, if groundwater becomes contaminated it can take a long time to be cleaned.

WASTEWATER

The most acute water pollution problems in the world come from **wastewater**—water that has been used for some purpose and is returned to the environment as a liquid. Wastewater problems are greatest when the volume is large in proportion to the amount of stream flow that is available to dilute it. Wastewater comprises about 15 percent of the flow of U.S. rivers, and so good wastewater treatment is critical. Rich, industrial countries like the United States generate more wastewater than do poorer nations, but they also have greater capacity to treat this wastewater. In rich countries, strict legislation requires treatment facilities to be upgraded, making their rivers cleaner than a few decades ago.

In developing countries, untreated sewage often goes directly into rivers that also supply drinking water. A combination of poor general sanitation, nutrition, and medical care can make drinking this water deadly. Waterborne diseases such as cholera, typhoid, and dysentery are major causes of death in developing countries. Because of improper sanitation, millions of people in Asia, Africa, and South America die each year from waterborne diseases. As people in these rapidly growing regions crowd into urban areas, drinking water becomes less safe, and waterborne pathogens flourish (Figure 14.8.2).

In a typical sewage-treatment plant, large solid particles are screened from the water or allowed to settle and the remaining water is oxygenated to allow bacteria to break down organic matter. The water is then discharged to the environment. In advanced systems, some of the products of that breakdown, such as nitrogen and phosphorus, are removed. Most of the world's population living in developed countries is served by some kind of sewage collection and treatment system. In developing countries, however, sewage systems are less common, particularly in rural areas.

Water quantity and water quality are closely linked. Flowing water is needed to dilute and carry away pollution, while pollution limits the usefulness of existing water resources. As we anticipate increasing water scarcity in the world, improving water treatment is essential. In some areas that have critical water shortages but sufficient capital to make the investment, advanced treatment plants now allow direct reuse of the wastewater from treatment plants (Figure 14.8.3). Depending on the degree of treatment, this water can be used for purposes such as irrigation, or potentially for human consumption.

▲ 14.8.2 **POLLUTED RIVER ON THE OUTSKIRTS OF XIANYANG, CHINA** Many poorer countries lack resources for basic water treatment and so suffer severe water quality problems.

▼ 14.8.3 **ADVANCED WATER TREATMENT PLANT IN ORANGE COUNTY, CALIFORNIA** The output of this plant is of sufficiently high quality to be used as supplemental drinking water.

14.9 Air Pollution

- ▶ **Air pollution is highest in concentrated areas, especially large and dense cities.**
- ▶ **Technical solutions to air pollution problems are readily available, but cost and enforcement of regulations can be problematic, especially in poorer countries.**

The atmosphere is constantly stirred by temperature and pressure differences that mix air vertically and horizontally. As it moves from one place to another, air carries with it various wastes. The more waste we discharge to the atmosphere, or the less the air circulates, the greater the concentration of pollution.

Average air at the surface contains about 78 percent nitrogen, 21 percent oxygen, and less than 1 percent argon. The remaining 0.04 percent of air's composition includes several trace gases. Air pollution is a human-caused concentration of trace substances at a greater level than occurs in average air. In addition to the carbon dioxide (CO_2) emitted by combustion of fossil fuels, the most common air pollutants include carbon monoxide (CO), nitrogen oxides (NO_X), **particulates** (very small particles of dust, ash, and other materials), sulfur oxides (SO_X), and hydrocarbons (Figure 14.9.1).

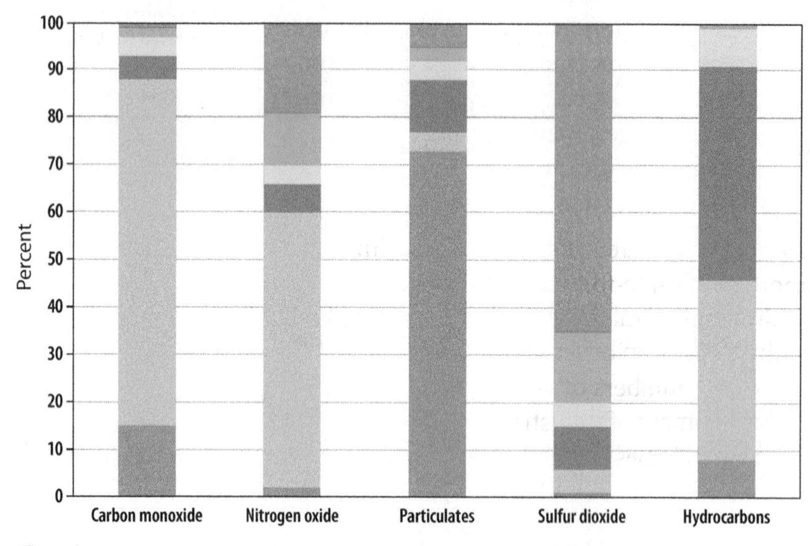

Electric utilities
Other industrial fuel combustion
Other fuel combustion
Industrial processes
Vehicles
Miscellaneous rural and natural resources

▲ 14.9.1 **MAJOR HUMAN-CAUSED AIR POLLUTION SOURCES IN THE UNITED STATES, 2008**
Nationwide, particulates come mostly from soil erosion, but in urban areas they are derived from vehicles and industrial sources. Sulfur oxides are mostly derived from industrial sources. Transport, mostly cars and trucks, is the most important source of carbon monoxide as well as a major source of nitrogen oxides and hydrocarbons.

URBAN AIR POLLUTION

Although air pollution can be found everywhere on Earth, the worst problems are in densely populated areas because large quantities of pollutants are discharged into the air (Figure 14.9.2). The problem is aggravated in cities when the wind cannot adequately disperse these pollutants. Some of the main components of urban air pollution are:

- Carbon monoxide (CO) caused by incomplete combustion of fossil fuels. Breathing CO reduces the oxygen level in blood, impairing vision and alertness and threatening persons who have chronic respiratory problems.
- Hydrocarbons, also resulting from improper fuel combustion, as well as from evaporation of solvents as in paint. Hydrocarbons and NO_X in the presence of sunlight form photochemical smog, which causes respiratory problems and stinging in the eyes.

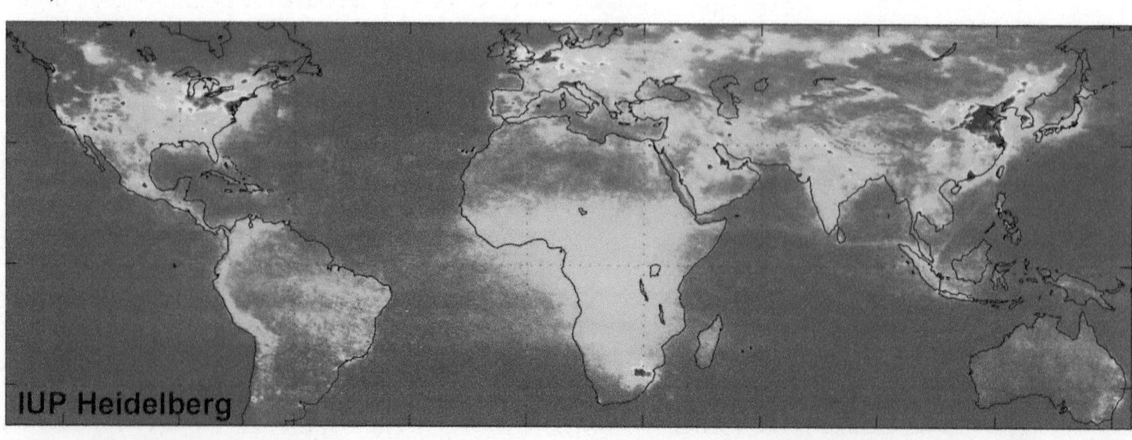

▶ 14.9.2 **AIR POLLUTION SEEN FROM A SATELLITE**
This map produced from data from a European satellite shows the average nitrogen dioxide concentration in the troposphere between January 2003 and June 2004.

10^{15} molecules/cm^2
-1 0 1 2 3 4 5 6

◀ 14.9.3 **TRAFFIC CONGESTION IN TORONTO, CANADA**
Numbers of vehicles in many cities have grown faster than the capacity of the road system, leading to congestion and increased air pollution. The purity of air is paramount to life on Earth. Some air pollutants come from natural processes unrelated to human actions, such as dust, forest fire smoke, and volcanic discharges, but humans add to this by discharging into the atmosphere smoke and gas from burning fossil fuels, incinerators, evaporating solvents, and industrial processes.

- Particulates, including dust and smoke particles. You can see particulates as a dark plume of smoke emitted from a smoke stack or a diesel truck—not a white plume, which is condensed water vapor. Many particles are too small to see, however.

Weather plays a critical role in air pollution. First, the faster the wind, the faster pollution is carried away and replaced with clean air. Second, vertical circulation is sometimes inhibited by warm air lying above cool air in what is known as a **temperature inversion**. This reduces pollutant dispersal. Finally, sunlight promotes formation of photochemical smog. The worst urban air pollution occurs where a combination of slight winds, temperature inversions, and clear skies allow pollutants to accumulate. Cities that experience these conditions more frequently have worse pollution problems than cities that do not experience them.

Mexico City is notorious for severe air pollution, especially in winter, when high pressure often dominates, and the surrounding mountains discourage dispersal of pollutants by wind. In the eastern United States, pollution problems are worst in summer and autumn, because weather conditions that limit pollutant dispersal are more common then. In Los Angeles and San Francisco, the pollution season is also summer and autumn because inversions and bright sunshine are more persistent then.

Progress in controlling urban air pollution is mixed. In relatively wealthy developed countries with strict regulations, air quality has improved. Controls on use of coal and improvements in automobile engines, and manufacturing processes have all contributed to higher-quality urban air. To reduce auto emissions in the United States, for example, modifications to automobiles that reduce pollution output have been required since the 1960s. As a result, carbon monoxide emissions have declined by more than three-fourths and nitrogen oxide and hydrocarbon emissions by more than 95 percent. Gains have been offset somewhat, though, by increased numbers of vehicles and other factors. The movement of industrial production from the United States and other developed nations to poorer ones also contributes to pollution in poor countries, and to cleaner air in developed countries.

In many developing countries, urban air pollution is getting worse. Use of motor vehicles is rising rapidly, but infrastructure improvements have not kept up with demand and congestion is a common problem (Figure 14.9.3). In addition, the vehicle fleet contains many older cars and trucks that may not have the pollution control devices found on most modern cars, or the cars may not be adequately maintained. In some areas urban residents burn wood, coal, or dung for cooking and heating. These fuels create serious air pollution problems in poorly ventilated areas, even in the countryside. Such problems can be solved, but the costs are significant. New, technologically advanced vehicles are much more expensive than old ones. Construction of transportation infrastructure such as mass transit that can reduce pollution is extremely expensive and may only serve a portion of the population.

14.10 Forest Resources

- **In general, the tropics are experiencing net deforestation while the midlatitudes are experiencing net forest growth.**
- **Forests are a renewable resource with many uses, ranging from lumber and fuel to recreation and biodiversity.**

Roughly one-third of Earth's land surface (excluding Antarctica) is covered with forest and woodland (Figure 14.1.1). The amount of forest land has decreased substantially as a result of clearing land for agriculture. Forests may originally have covered as much as half of Earth's land surface.

▲ 14.10.1 **BLUE RIDGE MOUNTAINS**
This forest, part of a park, is protected for its recreational value.

A few hundred years ago, many forested regions experienced very low levels of human impact. As human population has grown and technology developed, demands on forests have grown. Today there are few forests that are not being exploited for one purpose or another. Areas that are suitable for agriculture have been or are being cleared and converted to crops or grazing, and elsewhere trees are harvested to produce lumber or paper (Figure 14.10.2). At the same time, many areas that were deforested at some time in the past are regrowing. This is particularly common in the midlatitudes. In much of the eastern United States, for example, land that was cleared for farming or for lumber or fuel in the 1800s is no longer used for those purposes, and forests are expanding. In some areas active reforestation is underway. China's forest area is increasing at over 1 percent per year, and forest area is increasing in most of Europe. The worldwide average at present is a slight decrease in forest area of 0.14 percent per year.

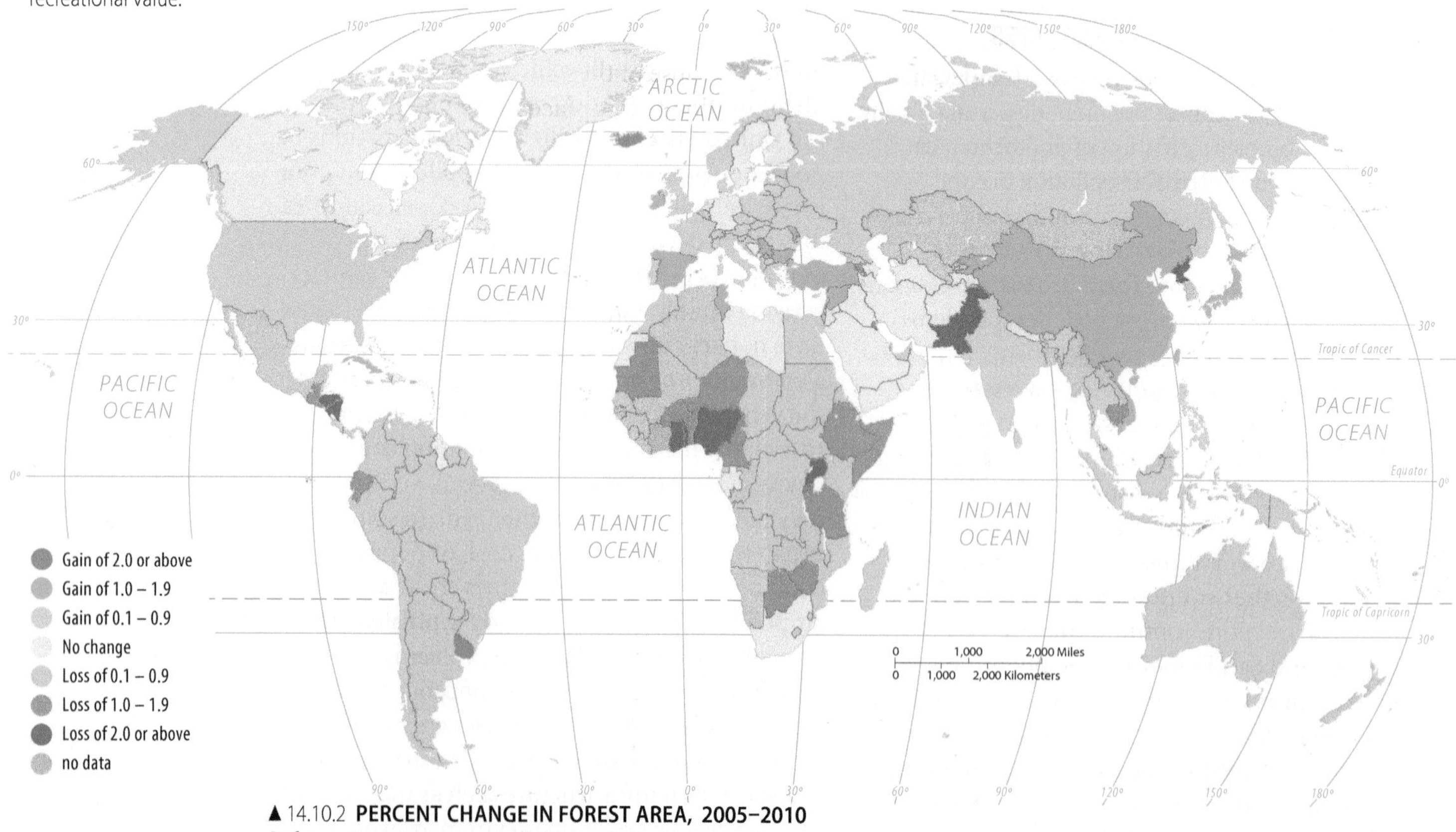

▲ 14.10.2 **PERCENT CHANGE IN FOREST AREA, 2005–2010**
Deforestation rates are generally highest in the tropics; many midlatitude regions are experiencing increases in forest area.

FORESTS AS FIBER RESOURCES

Nearly all forests in the United States, excluding Alaska, have been cut at least once during the past 300 years, so very little original (virgin) forest remains. In Canada, much more original forest remains, especially in the north and west. Original forest is valuable to the timber industry because large, straight trees yield high-quality lumber. Once forests are cut they must be allowed to regrow if future generations are to have them as a source of wood.

The development of wood technology since the early twentieth century has made it possible for us to use much smaller trees than were used in the past. Plywood and various materials that are made from small wood chips or particles glued together are especially important. Because we can manufacture large boards from small trees, it becomes possible to profitably harvest forests after relatively short periods of growth rather than waiting for the trees to reach full size to be sawn into boards. Many areas now have **sustained yield** plantations, in which trees are planted and harvested as crops much as we do in agriculture, although on cycles of a few decades rather than just a single year (Figure 14.10.3).

▲ 14.10.3 **LUMBER PRODUCTION IN THE SOUTHEAST UNITED STATES**
Forests can provide a sustainable wood supply. Use Google Earth to explore a landscape of timber growth and harvest.

Fly to *Chapman, Alabama.*

The main feature is a lumber mill that manufactures plywood from logs supplied from the forest in the region, which are clearly visible around the mill. These forests are fast-growing and are harvested on cycles of 20–40 years.

What evidence can you see that the forests are planted and managed, rather than natural?

OTHER IMPORTANT FOREST USES

In many countries, especially wealthier ones, forests are important recreational resources. They are relatively undeveloped and thus have open space available for hiking, camping, and other outdoor activities. They offer a strong contrast to the noise, commotion, crowding, and pollution of cities. Their shade and water availability provide relief from summer heat. Some forest regions are located in mountainous areas that offer other amenities such as skiing, climbing, and the cooler weather of high elevations. And because they serve as habitats for game animals, they can be preferred areas for hunting and fishing.

Forests are important for their biodiversity, and deforestation has resulted in significant biodiversity losses. Tropical forests are especially important centers of diversity, supporting both larger numbers of tree species and complex vertical habitats. Foremost among these is the Amazon, a vast region of more than 4 million square kilometers (1.5 million square miles), containing literally millions of species. The Amazon has been a focus of attention in part because of the relatively high rates of deforestation taking place there (Figure 14.10.4).

Since the 1980s, the Brazilian Amazon has been cleared at rates of slightly less than 1 percent per year, an area roughly the size of the state of Connecticut. Worldwide, perhaps 40,000 square kilometers (15,000 square miles) of tropical rain forest have been cleared annually in recent years. Concern over this loss is based partly on practical considerations, especially the potential medicinal uses of natural substances, but also on moral grounds, including protecting the rights of indigenous peoples, cultural preservation, and preserving unique environments for future generations.

▼ 14.10.4 **DEFORESTATION FOR FOOD PRODUCTION IN BRAZIL**
Cattle graze on newly cleared rainforest land. Most of the deforestation that is taking place in the Amazon region is for meat production, either through grazing or crops such as soybeans that are fed to meat animals.

14.11 Sustainability

- **Sustainable development utilizes resources at a rate that conserves them for the future.**
- **Biodiversity conservation provides examples of approaches to sustainable development.**

Sustainability is the use of Earth's limited resources by humans in ways that do not constrain resource use by people in the future. Geographers emphasize that each resource has a distinctive capacity for accommodating human activities.

SUSTAINABLE DEVELOPMENT

Sustainable development is "development that meets the needs of the present without compromising the ability of future generations to meet their own needs," according to the United Nations. The UN definition came in the 1987 Brundtland Report, named for the World Commission on Environment and Development's chair, Gro Harlem Brundtland, former prime minister of Norway. Titled *Our Common Future*, the Brundtland Report was a landmark in recognizing sustainable development as a combination of environmental and economic elements.

The report argued that sustainable development had to recognize the importance of economic growth while conserving natural resources. Environmental protection, economic growth, and social equity are linked because economic development aimed at reducing poverty can at the same time threaten the environment.

In general, higher levels of economic development are associated with higher levels of resource consumption, although there is wide variation (Figure 14.11.1). In the early stages of industrialization, pollution-control devices are an unpopular luxury that makes cars and other consumer goods more expensive (Figure 14.11.2), but as levels of income increase more resources can be devoted to pollution control.

Some environmentally oriented critics have argued that it is too late to discuss sustainability. Some would argue that global warming, oxygen-depleted water ("dead zones") in coastal areas, and loss of biodiversity are evidence that we have already exceeded Earth's ability to support the human population. Others criticize sustainability from the opposite perspective: human activities have not exceeded Earth's capacity, they argue, because resource availability has no maximum, and as discussed earlier in this chapter, Earth's resources have no absolute limit because the definition of resources changes drastically and unpredictably over time.

Per capita CO_2 emissions (metric tons): 0, 10, 20, 30, 40, 50, 60, 70

Per capita GNI ($ thousand): $0, $10, $20, $30, $40, $50, $60, $70, $80, $90

Qatar, Kuwait, Trinidad and Tobago, United States, Canada, Estonia, Russia, Singapore, Vietnam, China, Norway, Brazil, Portugal, Switzerland

▲ 14.11.1 **PER CAPITA CARBON EMISSIONS AND NATIONAL INCOME.** In general, richer countries emit more carbon per person than poorer ones, but there are wide variations. For example, Canada emits more than 3 times what Switzerland emits per person, although they have comparable income levels. The main reason for the difference is lifestyle, especially use of private vehicles rather than public transport. Fuel prices are much higher in Switzerland than in Canada, which encourages more sustainable practices.

▼ 14.11.2 **POLLUTION IN CHINA** The rapid economic transformation of China has resulted in rapidly rising levels of pollution. The country has 16 of the 20 most polluted cities, according to the World Bank.

CONSERVATION AND PRESERVATION

Conservation and preservation are two different ways to regard resources:

- **Conservation** is the use and management of natural resources such as wildlife, water, air, and Earth's resources to meet human needs, including food, medicine, and recreation. Renewable resources such as trees are conserved if they are consumed at a rate less than or equal to the rate at which they can be replaced (Figure 14.11.3). Nonrenewable resources such as fossil fuels are conserved if they are used as efficiently as possible, allowing time to develop alternative energy resources for the future. Conservation is important to sustainability.
- **Preservation** is the maintenance of resources in their present condition, with as little human impact as possible. Preservation takes the view that the value of nature does not derive from human needs and interests, but from the fact that every plant and animal living on Earth has a right to exist and should be preserved regardless of the cost.

Preservation does not regard nature as a resource for human use. In contrast, conservation is compatible with development but only if natural resources are utilized in a careful rather than a wasteful manner.

▲ 14.11.3 **LOGGING IN IDAHO.**
Most logging in the United States today is carried out in a sustainable manner.

MANAGING BIODIVERSITY

As described in Chapter 4, biodiversity refers to the variety of species across Earth as a whole or in a specific place. Earth's biodiversity is threatened today by a wide variety of actions such as agriculture, movement of species to new environments where they outcompete native species, and environmental pollution.

To reduce the impact of humans on species diversity, protected areas are being established by national governments and by international agencies. One important international effort is the United Nations Biosphere Reserve Program. Biosphere reserves are typically designed to include a core area in which land use activities are highly restricted, surrounded by a zone in which resources can be used for human purposes but in a way that maintains the environmental integrity of the core. Biosphere reserves exemplify sustainable development because they simultaneously accommodate human needs and nature preservation for the future. Other strategies to protect biodiversity include the Convention on International Trade in Endangered Species of Wild Fauna and Flora, and the curtailing of logging, whaling, and taking of porpoises in tuna seines (nets).

Biodiversity is a barometer of the consequences of human modification of the environment. A commitment to biodiversity means a commitment to a broad range of environmental policies—not just preserving wilderness areas, but maintaining diversity in every biome and ideally every ecosystem (Figure 14.11.4). It is an enormous challenge to which geography and geographers have much to contribute.

▲ 14.11.4 **BIODIVERSITY CONSERVATION**
The welcome sign at a trailhead to the Monteverde Cloud Forest Reserve, Costa Rica.

Natural resources such as energy, water, air quality, and forests are critical to the way we live. Resource availability varies widely and changes over time as our ways of using them change.

Key Questions

What is a natural resource, and what determines the value of resources?

- Resources are defined by their usefulness for humans, which changes over time in relation to technology, economics, and culture.
- Many resources have multiple uses, and many purposes can be served by different resources.

How do we use energy and mineral resources, and what are the supplies of these resources for the future?

- Our use of energy resources has changed over time. Today fossil fuels provide most of our energy needs, and electricity demand is increasing rapidly.
- Nuclear and renewable energy sources offer potential for new electricity production, but nuclear energy has safety concerns and renewable sources currently contribute relatively little to our needs.
- Mineral resource needs change rapidly in response to changing technology and demand.

What is the condition of major renewable resources and how do we manage them?

- Water resources are being stressed as demand grows, and future increases in irrigation needs will be a challenge.
- Considerable progress has been made in reducing water and air pollution, but poorer countries still suffer significant problems.
- Forests are being cleared for agriculture and exploited for fiber in much of the world. Forest area is declining in the tropics but increasing in the midlatitudes.

▼ 14.CR.1 **2008 DROUGHT IN SPAIN**
A tanker delivering water to Barcelona, Spain, during a drought in 2008.

Thinking Geographically

The Middle East is the largest oil-producing region, but oil is also produced in many other areas. North America, Europe, and East Asia are major importing regions, but these regions also include major oil producers.

1. Is it reasonable to think of the world in two parts: oil-exporting and oil-importing countries? Why or why not?

Oil is clearly an important factor in international relations today, and much attention is paid to variations in oil supply, demand, and price. As we look to the future, we can see that water is likely to be increasingly scarce in many areas, and some speak of water as the "oil" of the future (Fig. 14.CR.1).

2. Are oil and water comparable commodities? How are they similar, and how are they different?

Several countries that were very poor only a few decades ago are today experiencing rapid economic growth, and at least portions of those countries are approaching levels of economic development that are comparable to those in rich countries.

3. Is the rapid economic growth taking place in China, India, and other countries good for the environment or bad?

On the Internet

The World Resources Institute (WRI) is a leading organization monitoring and analyzing trends in resources. Visit them at **http://wri.org**.

Comprehensive and up-to-date water resource information, including both water supply and water quality, is provided by the United States Geological Survey: **http://water.usgs.gov**.

Interactive Mapping

GROUNDWATER DEPLETION IN NORTH AMERICA

Groundwater depletion is a symptom of overuse of renewable resources.

Open MapMaster North America in MasteringGEOGRAPHY™

Open *Layered Thematic view* and turn on the *Environmental Issues* layer from the *Physical Environment* menu.

Then select the *Climate* layer from the *Physical Environment* menu.

1. **What regions have the greatest problems with groundwater depletion?**
2. **What are the climates in these regions?**
3. **Where do you find most of the irrigated agriculture?**
4. **How are these things related?**

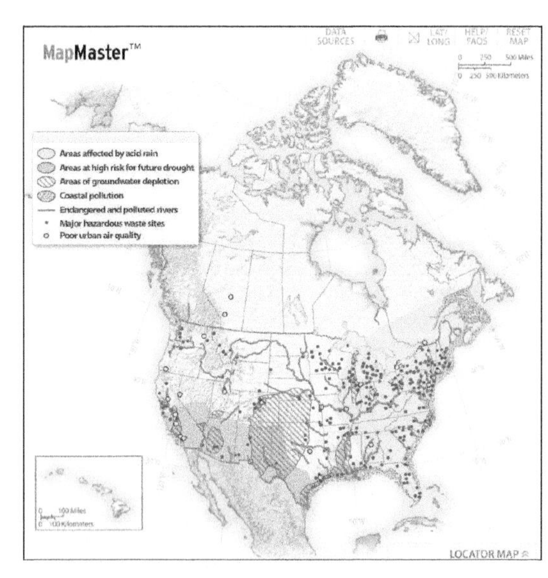

Explore

EXPLORE DEFORESTATION AND MOUNTAINTOP REMOVAL

Deforestation is occuring in many tropical areas.

Fly to *10° S, 63° W*, which is in Rondonia, Brazil.

This area has experienced much deforestation in recent decades.

Can you tell what the land is used for?

Compare this with Figure 10.10.2, which shows a similar area.

Zoom out a little, to an eye altitude of 10 km.

Click on *View* and *Historical imagery.*

Use the time slider to see the landscape change.

When did most of the deforestation take place?

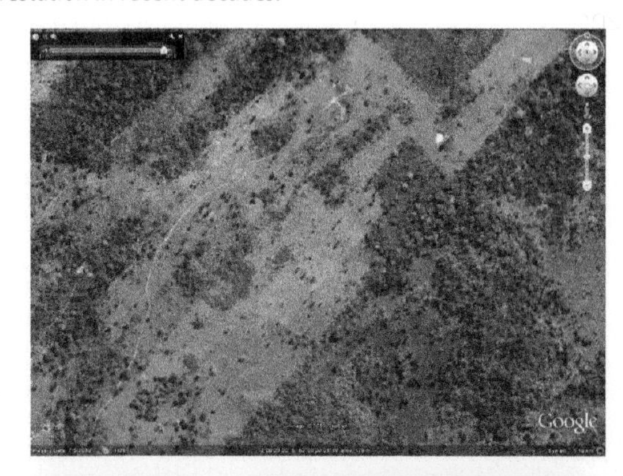

Fly to *38° 20′ N, 81° W*, in West Virginia to see mountaintop removal, an efficient but environmentally destructive method to mine coal for electricity.

Zoom out to an eye altitude of about 5 km.

Click on *View* and *Historical imagery*.

Use the time slider to see the landscape change.

Is the land being revegetated after mining?

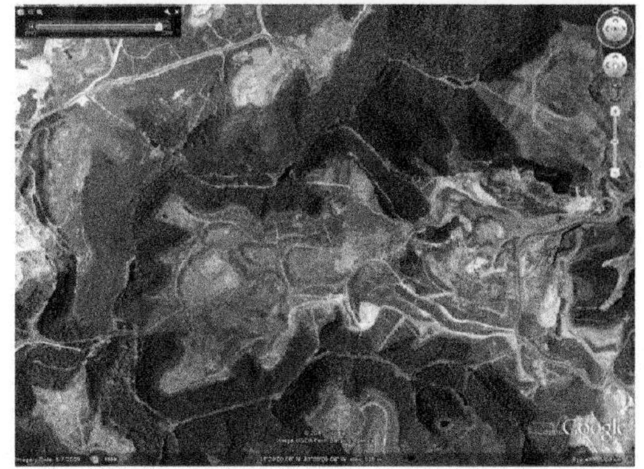

Key Terms

Acid deposition
The deposition of acidic substances on the land/water surface primarily as a result of sulfur and nitrogen oxide pollution of the atmosphere.

Biofuel
Liquid or gaseous fuel produced from biomass.

Cartel
A group of producers that agree to control a particular market by limiting production in order to drive up prices.

Conservation
Efficient use and careful management of resources to attain the maximum possible benefits from them.

Consumptive use
Use of water in which the water is evaporated rather than returned to nature as a liquid.

Fossil fuel
A source of chemical energy stored in formerly living plant and animal tissue. Coal, oil, and natural gas are fossil fuels.

Natural resource
Something that is useful and that exists independent of human activity.

Nonpoint source
A source of pollution that comes from a large, diffuse area.

Nonrenewable resource
A resource that is used at rates far exceeding the rates at which it is replenished in nature.

Particulate
A small solid particle in the air; a component of air pollution.

Photovoltaic electricity
Electric generation using a device that converts light to electricity.

Point source
A source of pollution for which a distinct location can be identified where the pollutant enters the environment.

Pollution
A human-caused increase in the amount of a substance in the environment.

Preservation
The nonuse of resources; limiting resource development for the purpose of saving resources for the future.

Renewable resource
Something that is produced by nature at rates similar to those at which it is consumed by humans.

Reserve
In the context of geologic resources, a deposit that has been identified (located) and is commercially extractable at with present prices and technology.

Sustainability/sustainable development
Resource management and/or economic development that provides for the needs of the present without compromising future opportunities.

Sustained yield
A way of managing a renewable natural resource such that harvest can continue indefinitely.

Temperature inversion
A layer in the atmosphere in which warmer air lies above cooler air, inhibiting vertical circulation.

Wastewater
Water that has been used for some purpose and is returned to the environment as a liquid.

Glossary

Abiotic A system composed of nonliving or inorganic matter.

Acid deposition The deposition of acidic substances on the land/water surface, primarily as a result of sulfur and nitrogen oxide pollution of the atmosphere.

Actual evapotranspiration (actet) The amount of water evaporated and/or transpired in a given environment.

Adiabatic cooling The cooling of air as a result of expansion of rising air; adiabatic means "without heat being involved".

Adolescent fertility rate The number of births per 1,000 women age 15–19.

Advection The horizontal movements of air or substances by wind or ocean currents.

Agribusiness Commercial agriculture characterized by the integration of different steps in the food-processing industry, usually through ownership by large corporations.

Agricultural density The ratio of the number of farmers to the total amount of land suitable for agriculture.

Agriculture The deliberate effort to modify a portion of Earth's surface through the cultivation of crops and the raising of livestock for sustenance or economic gain.

Angle of incidence The angle between the Sun's rays and a horizontal surface at a particular place and time.

Animism Belief that objects, such as plants and stones, or natural events, like thunderstorms and earthquakes, have a discrete spirit and conscious life.

Annexation Legally adding land area to a city in the United States.

Anocracy Type of government with mix of democratic and autocratic features.

Aquaculture (or aquafarming) The cultivation of seafood under controlled conditions.

Arable land Land settled for agriculture.

Arithmetic density The total number of people divided by the total land area.

Arrivals The entry of people into a country.

Atmosphere The thin layer of gases surrounding Earth.

Autocracy Type of government in which leaders are selected from the established elite and have few limits on their powers.

Autumnal equinox In the Northern Hemisphere September 22 or 23, one of two dates when, at noon, the perpendicular rays of the sun strike the equator (meaning that the Sun is directly overhead along the equator).

Balkanization Breakdown of a state because of internal ethnic differences.

Basic industries Industries that sell their products or services primarily to consumers outside the settlement.

Beach A deposit of wave-carried sediment along a shoreline, on which waves break.

Biofuel Liquid or gaseous fuel produced from biomass.

Biogeochemical cycle The environmental recycling process that supplies essential substances such as carbon, nitrogen, and other nutrients to the biosphere.

Biomagnification The tendency for substances that accumulate in body tissues to increase in concentration as they are passed to higher levels in a food chain.

Biomass The dry mass of living or formerly living matter in a given environment.

Biome A large grouping of ecosystems characterized by particular plant and animal types.

Biosphere All living organisms on Earth.

Biotic The system composed of living organisms.

Boreal forest An evergreen needleleaf forest characteristic of cold continental climates.

Brain drain The loss of highly-trained professionals through emigration.

Branch (of a religion) A large and fundamental division within a religion.

Break-of-bulk point A location where transfer is possible from one mode of transportation to another.

Broadleaf deciduous forest A forest with broadleaved trees that lose their leaves in the winter; characteristic of humid midlatitude environments.

Bulk-gaining industry An industry in which the final product weighs more or comprises a greater volume than the inputs.

Bulk-reducing industry An industry in which the final product weighs less or comprises a lower volume than the inputs.

Business services Services that primarily meet the needs of other businesses, including professional, financial, and transportation services.

Carbon cycle The movement of carbon among the atmosphere, hydrosphere, biosphere, and lithosphere as a result of processes such as photosynthesis and respiration, sedimentation, weathering, and fossil-fuel combustion.

Carbon dioxide A trace gas found in the atmosphere with chemical formula CO_2; a major contributor to the greenhouse effect.

Carnivore An animal whose primary food supply is other animals.

Cartel A group of producers that agree to control a particular market by limiting production in order to drive up prices.

Cartography The science of making maps.

Census tract An area delineated by the U.S. Bureau of the Census for which statistics are published; in urbanized areas, census tracts are often delineated to correspond roughly to neighborhoods.

Central business district (CBD) The area of a city where consumer, business, and public services are clustered.

Central place A market center for the exchange of services by people attracted from the surrounding area.

Central place theory A theory that explains the distribution of services, based on the fact that settlements serve as centers of market areas for services; larger settlements are fewer and farther apart than smaller settlements and provide services for a larger number of people who are willing to travel farther.

Centripetal force Political attitude that supports the state.

Cereal grain A grass yielding grain for food.

Chemical weathering The breakdown of rocks or minerals through chemical reactions at Earth's surface.

City An urban settlement that has been legally incorporated into an independent, self-governing unit.

Climate The totality of weather conditions over a period of several decades or more.

Clustered rural settlement An agricultural based community in which a number of families live in close proximity to each other, with fields surrounding the collection of houses and farm buildings.

Cold front The boundary formed when a cold air mass advances against a warmer one.

Combined statistical area (CSA) In the United States, two or more contiguous core based statistical areas tied together by commuting patterns.

Commercial agriculture Agriculture undertaken primarily to generate products for sale off the farm.

Compact state A state whose shape approximates a circle.

Composite cone volcano A volcano formed by a mixture of lava eruptions and more explosive ash eruptions.

Concentration The extent of spread of something over a given area.

Concentric zone model A model of the internal structure of cities in which social groups are spatially arranged in a series of rings.

Condensation Water changing from a gas state (vapor) to a liquid or solid state.

Connection Relationships among people and objects across the barrier of space.

Conservation Efficient use and careful management of resources to attain the maximum possible benefits from them.

Consumer services Businesses that provide services primarily to individual consumers, including retail services and education, health, and leisure services.

Consumptive use Use of water in which the water is evaporated rather than returned to nature as a liquid.

Contagious diffusion The rapid, widespread diffusion of a feature or trend throughout a population.

Contiguous waters Area of the ocean from 12 to 24 nautical miles from a state's shore over which it has limited sovereignty.

Continental glacier A thick glacier hundreds to thousands of kilometers across, large enough to be only partly guided by underlying topography.

Convection Circulation in a fluid caused by temperature induced density differences, such as the rising of warm air in the atmosphere.

Convergent plate boundary A boundary between tectonic plates in which the two plates move toward one another, destroying or thickening the crust.

Core based statistical area (CBSA) In the United States, a term referring to either a metropolitan statistical area or a micropolitan statistical area.

Coriolis effect The tendency of an object moving across Earth's surface to be deflected from its apparent path as a result of Earth's rotation.

Cosmogony A set of religious beliefs concerning the origin of the universe.

Cottage industry Manufacturing based in homes rather than in a factory, commonly found prior to the Industrial Revolution.

Counterurbanization Residential relocation from urban and suburban places to rural ones.

Creole or creolized language A language that results from the mixing of a colonizer's language with the indigenous language of the people being dominated.

Crop Grain or fruit gathered from a field as a harvest during a particular season.

Crop rotation The practice of rotating use of different fields from crop to crop each year, to avoid exhausting the soil.

Crude birth rate (CBR) The total number of live births in a year for every 1,000 people alive in the society.

Crude death rate (CDR) The total number of deaths in a year for every 1,000 people alive in the society.

Cultural boundaries Areal limit of states that follow the distribution of cultural characteristics.

Cultural ecology The geographic study of human-environment relationships.

Cultural landscape Fashioning of a natural landscape by a cultural group.

Cyclone Large low-pressure areas in which winds converge in a counterclockwise swirl in the Northern Hemisphere (or clockwise in the Southern Hemisphere).

Delta A deposit of sediment formed where a river enters a lake or an ocean.

Demic diffusion The movement of people through space over time.

Democracy Type of limited government in which citizens can elect leaders and stand for office.

Demographic transition The process of change in a society's population from a condition of high crude birth and death rates and low rate of natural increase to a condition of low crude birth and death rates, low rate of natural increase, and a higher total population.

Denglish Combination of German and English.

Denomination (of a religion) A division of a branch that unites a number of local congregations in a single legal and administrative body.

Density The frequency with which something exists within a given unit of area.

Dependency ratio The number of people who are considered too young or too old to work (under age 15 or over age 64), compared to the number of people in their productive years.

Desert A vegetation type with sparsely distributed plants, specifically adapted for moisture gathering and moisture retention.

Desert climate A climate with low precipitation and temperatures warm enough to cause potential evapotranspiration to be substantially higher than precipitation.

Desertification The process of a region's soil and vegetation cover becoming more desert like as a result of human land use, usually by overgrazing or cultivation.

Developed country (more developed country or MDC) A country that has progressed relatively far along a continuum of development.

Developing country (less developed country or LDC) A country that is at a relatively early stage in the process of economic development.

Development A process of improvement in the material conditions of people through diffusion of knowledge and technology.

Dialect A regional variety of a language distinguished by vocabulary, spelling, and pronunciation.

Diaspora The widespread diffusion of a people from their region of origin.

Dietary energy consumption The amount of food that an individual consumes.

Diffusion The process of spread of a feature or trend from one place to another over time.

Discharge The quantity of water flowing past a point on a stream per unit time.

Dispersed rural settlement A rural settlement pattern in which farmers live on individual farms isolated from neighbors.

Displacement When people are compelled to move from one place to another.

Distance decay The diminishing in importance and eventual disappearance of a phenomenon with increasing distance from its origin.

Distribution The arrangement of something across Earth's surface.

Divergent plate boundary A boundary between tectonic plates in which the two plates move away from each other, and new crust is created between them.

Doubling time The number of years needed to double a population, assuming a constant rate of natural increase.

Drainage basin The geographic area that contributes runoff to a particular stream, defined with respect to a specific location along that stream—the runoff from the drainage basin passes that point on the stream.

Earthquake A sudden release of energy within Earth, producing a shaking of the crust.

Ecology The scientific study of ecosystems.

Economic base A community's collection of basic industries.

Ecosystem An interrelated collection of plants and animals and the physical environment with which they interact.

Ecotourism Tourism meant to lessen visitors' impact on the environment.

Edge city A large node of office and retail activities on the edge of an urban area.

El Niño A circulation change in the eastern tropical Pacific Ocean, from westward flow to eastward flow, that occurs every few years.

Elderly support ratio The number of working-age people (ages 15–64) divided by the number of persons 65 or older.

Elongated state A state whose shape is long and narrow.

Emigration Out-migration from one area to another area.

Enclave A territory completely surrounded by another state.

Environmental determinism A nineteenth- and early twentieth-century approach to the study of geography that argued that the general laws sought by human geographers could be found in the physical sciences. Geography was therefore the study of how the physical environment caused human activities.

Environmentally displaced persons Individuals compelled to flee natural disasters.

Epicenter The location on Earth's surface immediately above the focus of an earthquake.

Epidemiologic transition Distinctive causes of death in each stage of the demographic transition.

Epidemiology Branch of medical science concerned with the incidence, distribution, and control of diseases that affect large numbers of people.

Ethnic cleansing Process in which a more powerful ethnic group forcibly removes a less powerful one in order to create an ethnically homogeneous region.

Ethnic religion A religion with a relatively concentrated spatial distribution whose principles are likely to be based on the physical characteristics of the particular location in which its adherents are concentrated.

Ethnicity Identity with a group of people who share the cultural traditions of a particular homeland or hearth.

Eutrophication A process in which water bodies receive excess nutrients that stimulate excessive plant growth.

Evapotranspiration The sum of evaporation and transpiration.

Exclave A piece of territory separated from the rest of the country by other countries.

Exclusionary policies Government rules to prevent immigration.

Exclusive Economic Zone (EEZ) The area up to 200 nautical miles from a state's shore in which it has sole authority to exploit natural resources.

Expansion diffusion The spread of a feature or trend among people from one area to another in an additive process.

Extinct language A language that was once used by people in daily activities but is no longer used.

Fair trade Alternative to international trade that emphasizes small businesses and worker-owned and democratically run cooperatives and requires employers to pay workers fair wages, permit union organizing, and comply with minimum environmental and safety standards.

Fault A fracture in Earth's crust along which displacement of rocks has occurred.

Federal state Internal organization of states giving some powers to local government.

Floodplain A low-lying surface adjacent to a stream channel and formed by materials deposited by the stream.

Focus (of an earthquake) The location in Earth where motion originates in an earthquake.

Food chain The sequential consumption of food in an ecosystem, beginning with green plants, followed by herbivores and carnivores, and ending with decomposers.

Food security Physical, social, and economic access at all times to safe and nutritious food sufficient to meet dietary needs and food preferences for an active and healthy life.

Forced migration Any movement undertaken under coercive conditions, such as violence or disaster.

Foreign direct investment Investment made by a foreign company in the economy of another country.

Formal region (or uniform or homogeneous region) An area in which everyone shares in one or more distinctive characteristics.

Fossil fuel A source of chemical energy stored in formerly living plant and animal tissue. Coal, oil, and natural gas are fossil fuels.

Fragmented state A state with disconnected pieces of territory.

Franglais A term used by the French for English words that have entered the French language; a combination of français and anglais, the French words for "French" and "English," respectively.

Front A boundary between warm air and cold air.

Functional region (or nodal region) An area organized around a node or focal point.

Gender Inequality Index (GII) Indicator constructed by the United Nations to measure the extent of each country's gender inequality.

Genocide Murderous campaign to eliminate an ethnic, racial, religious, or national group.

Gentrification A process of converting an urban neighborhood from a predominantly low-income renter-occupied area to a predominantly middle-class owner-occupied area.

Geographic grid A system of imaginary arcs drawn in a grid pattern on Earth's surface.

Geographic Information Science (GIScience) The development and analysis of data about Earth acquired through satellite and other electronic information technologies.

Geographic information system (GIS) A computer system that stores, organizes, analyzes, and displays geographic data.

Gerrymandering Redrawing election districts to favor one party over another.

Glacier A large mass of flowing, perennial ice.

Global Positioning System (GPS) A system that determines the precise position of something on Earth through a series of satellites, tracking stations, and receivers.

Global warming A general increase in temperatures over a period of at least several decades caused primarily by increased levels of carbon dioxide in Earth's atmosphere.

Globalization Actions or processes that involve the entire world and result in making something worldwide in scope.

Grade A condition in which a stream's ability to transport sediment is balanced by the amount of sediment delivered to it.

Grain Seed of a cereal grass.

Gravity model A model that holds that the potential use of a service at a particular location is directly related to the number of people in a location and inversely related to the distance people must travel to reach the service.

Green revolution Rapid diffusion of new agricultural technology, especially new high-yield seeds and fertilizers.

Greenhouse gases Trace substances in the atmosphere that contribute to the greenhouse effect; water vapor, carbon dioxide, ozone, methane, and chlorofluorocarbons are important examples.

Greenwich Mean Time (GMT) The time in that time zone encompassing the prime meridian, or 0° longitude.

Gross domestic product (GDP) The value of the total output of goods and services produced in a country in a year, not accounting for money that leaves and enters the country.

Gross national income (GNI) The value of the output of goods and services produced in a country in a year, including money that leaves and enters the country.

Groundwater The water beneath Earth's surface at a depth where rocks and/or soils are saturated with water.

Guest workers Labor immigrants admitted to meet demand for more workers.

Gyre A circular ocean current beneath a subtropical high pressure cell.

Hearth The region from which innovative ideas originate.

Herbivore An animal whose primary food supply is plants.

Hierarchical diffusion The spread of a feature or trend from one key person or node of authority or power to other persons or places.

Horizon A layer in the soil with distinctive characteristics derived from soil-forming processes.

Human Development Index (HDI) Indicator of level of development for each country, constructed by United Nations, combining income, literacy, education, and life expectancy.

Human origins Where and when modern humans first appeared and how they peopled Earth.

Hurricane An intense tropical cyclone that develops over warm ocean areas in the tropics and subtropics, primarily during the warm season. Hurricanes in the Pacific Ocean are called typhoons; in the Indian Ocean they are called cyclones.

Hydrologic cycle The movement of water from the atmosphere to Earth's surface, across that surface, and back to the atmosphere.

Hydrosphere All of the water on and near Earth's surface.

Igneous rock Rock formed by crystallization of magma.

Immigrant nation A country whose population is primarily composed of immigrants and their descendants.

Immigration In-migration to an area from another area.

Inclusionary policies Government rules meant to accommodate or even encourage immigration.

Industrial Revolution A series of improvements in industrial technology that transformed the process of manufacturing goods.

Inequality-adjusted HDI (IHDI) Indicator of level of development for each country that modifies the HDI to account for inequality.

Infant mortality rate (IMR) The total number of deaths in a year among infants under 1 year old for every 1,000 live births in a society.

Insolation The amount of solar energy intercepted by a particular area of Earth.

Intensive subsistence agriculture A form of subsistence agriculture in which farmers must expend a relatively large amount of effort to produce the maximum feasible yield from a parcel of land.

Internally displaced persons (IDPs) Persons compelled to migrate within their country of origin.

International Date Line A meridian that for the most part follows 180° longitude. When you cross the International Date Line heading east (toward America), the clock moves back 24 hours (one day), and when you go west (toward Asia), the calendar moves ahead one day.

International treaties Agreements between states by which they agree to limit their sovereign rights or to cooperate on shared problems.

Interregional migration Migration between two regions of the same country.

Intertropical Convergence Zone (ITC Z) A low-pressure zone between the Tropic of Cancer and the Tropic of Capricorn where surface winds converge.

Intraregional migration Migration within one region.

Isolated language A language that is unrelated to any other languages and therefore not attached to any language family.

Just-in-time delivery Shipment of parts and materials to arrive at a factory moments before they are needed.

Labor-intensive industry An industry for which labor costs comprise a high percentage of total expenses.

Landform A characteristic shape of the land surface, such as a hill, valley, or floodplain.

Language A system of communication through the use of speech, a collection of sounds understood by a group of people to have the same meaning.

Language branch A collection of languages related through a common ancestor that existed several thousand years ago. Differences are not as extensive or as old as with language families, and archaeological evidence can confirm that the branches derived from the same family.

Language family A collection of languages related to each other through a common ancestor long before recorded history.

Language group A collection of languages within a branch that share a common origin in the relatively recent past and display relatively few diff erences in grammar and vocabulary.

Latent heat Heat stored in water and water vapor, not detectable by people; latent means "hidden".

Latitude The numbering system used to indicate the location of parallels drawn on a globe and measuring distance north and south of the equator (0°).

Lava Magma that reaches Earth's surface.

Life expectancy The average number of years an individual can be expected to live, given current social, economic, and medical conditions. Life expectancy at birth is the average number of years a newborn infant can expect to live.

Lingua franca A language mutually understood and commonly used in trade by people who have different native languages.

Literacy rate The percentage of a country's people who can read and write.

Literary tradition A language that is written as well as spoken.

Lithosphere Earth's crust and a portion of the upper mantle directly below the crust.

Little Ice Age The period between about 1500 to 1750, when climates on Earth were especially cool.

Location The position of anything on Earth's surface.

Longitude The numbering system used to indicate the location of meridians drawn on a globe and measuring distance east and west of the prime meridian (0°).

Longshore current A current in the surf zone along a shoreline, parallel to the shore.

Longshore transport Sediment transport by a longshore current.

Longwave energy Energy reradiated by Earth in wavelengths of about 5.0 to 30.0 microns. Includes infrared radiation, which we sense as heat.

Magma Molten rock beneath Earth's surface.

Mantle The portion of Earth above the core and below the crust.

Map A two-dimensional, or flat, representation of Earth's surface or a portion of it.

Map scale The relationship between the size of an object on a map and the size of the actual feature on Earth's surface.

Marine terrace A nearly level surface along a shoreline, elevated above present sea level, formed by coastal erosion at a time when sea level at the location was higher than at present.

Market area (or hinterland) The area surrounding a central place, from which people are attracted to use the place's goods and services.

Mass movement Downslope movement of rock and soil at Earth's surface, driven mainly by the force of gravity acting on those materials.

Maternal mortality rate The number of women who die giving birth per 100,000 births.

Meandering The tendency of flowing water to follow a sinuous course with alternating right- and left-hand bends.

Mechanical weathering The breakdown of rocks into smaller particles caused by application of physical or mechanical forces.

Meridian An arc drawn on a map between the North and South poles.

Metamorphic rock Rock formed by modification of other rock types, usually by heat and/or pressure.

Methane A trace gas found in the atmosphere with chemical formula CH_4; a major contributor to the greenhouse effect.

Metropolitan statistical area (MSA) In the United States, a central city of at least 50,000 population, the county within which the city is located, and adjacent counties meeting one of several tests indicating a functional connection to the central city.

Micropolitan statistical area An urban area of between 10,000 and 50,000 inhabitants, the county in which it is found, and adjacent counties tied to the city.

Midlatitude cyclone A storm characterized by a center of low pressure in the midlatitudes usually associated with a warm front and a cold front.

Midlatitude low-pressure zones Regions of low pressure and air converging from the subtropical and polar high-pressure zones.

Migration stream A sustained movement of people from the one source area to a common destination area.

Milkshed The ring surrounding a city from which milk can be supplied without spoiling.

Missionary An individual who helps to diffuse a universalizing religion.

Monotheism The doctrine or belief of the existence of only one god.

Monsoon circulation Seasonal reversal of pressure and wind in Asia, in which winter winds from the Asian interior produce dry winters, and summer winds blowing inland from the Indian and Pacific oceans produce wet summers.

Moraine An accumulation of rock and sediment deposited by a glacier, usually in or near the melting area.

Multiple nuclei model A model of the internal structure of cities in which social groups are arranged around a collection of nodes of activities.

Nationality Identity with a group of people who share legal attachment and personal allegiance to a particular country.

Native speakers People for whom a particular language is their first language.

Natural increase rate (NIR) The percentage growth of a population in a year, computed as the crude birth rate minus the crude death rate.

Natural resource Something that is useful and that exists independent of human activity.

Naturalize The process of becoming a citizen of a country other than one's country of origin.

Net migration The numerical difference between immigration and emigration.

Nonbasic industries Industries that sell their products primarily to consumers in the community.

Nonpoint source A source of pollution that comes from a large, diffuse area.

Nonrenewable resource A resource that is used at rates far exceeding the rates at which it is replenished in nature.

Official language The language adopted for use by the government for the conduct of business and publication of documents.

Omnivore An animal that feeds on both plants and other animals.

Orographic precipitation Precipitation caused by air being forced to rise over mountains.

Outwash plain An accumulation of sand and gravel carried by meltwater streams from a glacier, usually deposited immediately beyond the terminal moraine from the glacier.

Overfishing Capturing fish faster than they can reproduce.

Overland flow Water flowing across the soil surface on a hillslope, usually resulting from precipitation falling faster than the ground can absorb it.

Overpopulation The number of people in an area exceeds the capacity of the environment to support life at a decent standard of living.

Ozone A gas composed of molecules with three oxygen atoms; it is a highly corrosive gas at ground level, but in the upper atmosphere essential to protecting life on Earth by absorbing ultraviolet radiation.

Pandemic Disease that occurs over a wide geographic area and affects a very high proportion of the population.

Parallel A circle drawn around the globe parallel to the equator and at right angles to the meridians.

Particulate A small solid particle in the air; a component of air pollution.

Pastoral nomadism A form of subsistence agriculture based on herding domesticated animals.

Pattern The regular arrangement of something in a study area.

Perforated state A country with other state territories enclosed within it.

Peripheral model A model of North American urban areas consisting of an inner city surrounded by large suburban residential and business areas tied together by a beltway or ring road.

Photosynthesis A chemical reaction that occurs in green plants in which carbon dioxide and water are converted to carbohydrates and oxygen.

Photovoltaic electricity Electric generation using a device that converts light to electricity.

Physical boundaries Areal limit of states that coincide with significant features of the natural landscape.

Physiological density The number of people per unit of area of arable land, which is land suitable for agriculture.

Pilgrimage A journey to a place considered sacred for religious purposes.

Place A specific point on Earth distinguished by a particular characteristic.

Plantation A large farm in tropical and subtropical climates that specializes in the production of one or two crops for sale, usually to a more developed country.

Pleistocene Epoch A period of geologic time consisting of the first part of the Quaternary Period beginning about 3 million years ago and ending about 12,000 years ago.

Point source A source of pollution for which a distinct location can be identified where the pollutant enters the environment.

Polar front A boundary between cold polar air and warm subtropical air that circles the globe in the mid latitudes.

Polar high-pressure zones Regions of high pressure and descending air near the North and South poles.

Polder Land created by the Dutch by draining water from an area.

Pollution A human-caused increase in the amount of a substance in the environment.

Polytheism Belief in or worship of more than one god.

Population Pyramid A bar graph that displays the percentage of a place's population for each age and gender.

Possibilism The theory that the physical environment may set limits on human actions, but people have the ability to adjust to the physical environment and choose a course of action from many alternatives.

Potential evapotranspiration (Potet) The amount of evapotranspiration that would occur if water were available.

Preservation The nonuse of resources; limiting resource development for the purpose of saving resources for the future.

Primary census statistical area (PCSA) In the United States, all of the combined statistical areas plus all of the remaining metropolitan statistical areas and micropolitan statistical areas.

Primary sector The portion of the economy concerned with the direct extraction of materials from Earth's surface, generally through agriculture, although sometimes by mining, fishing, and forestry.

Primate city The largest settlement in a country, if it has more than twice as many people as the second-ranking settlement.

Primate city rule A pattern of settlements in a country, such that the largest settlement has more than twice as many people as the second-ranking settlement.

Prime meridian The meridian, designated as 0° longitude, that passes through the Royal Observatory at Greenwich, England.

Productivity The value of a particular product compared to the amount of labor needed to make it.

Projection The system used to transfer locations from Earth's surface to a flat map.

Prorupted state A state whose shape is compact with a protruding extension.

Public services Services offered by the government to provide security and protection for citizens and businesses.

Pull factor A feature of a destination that attracts in-migration.

Push factor A feature of a place that fuels out-migration.

Quaternary Period The period of geologic time encompassing approximately the last 3 million years.

Radiation Energy in the form of electromagnetic waves that radiate in all directions.

Ranching A form of commercial agriculture in which livestock graze over an extensive area.

Range (of a service) The maximum distance people are willing to travel to use a service.

Rank-size rule A pattern of settlements in a country, such that the nth largest settlement is 1/n the population of the largest settlement.

Recognition Formal acknowledgement by existing states of a new state's claim to sovereign independence.

Refugee A person compelled to migrate outside their country of origin as defined by an international convention.

Region An area distinguished by a unique combination of trends or features.

Relative humidity The actual water content of the air compared to how much water the air could potentially hold, expressed as a percentage.

Relocation diffusion The spread of a feature or trend through bodily movement of people from one place to another.

Remittances Money migrants send to family or others in their place of origin.

Remote sensing The acquisition of data about Earth's surface from a satellite orbiting the planet or other long-distance methods.

Renewable resource Something that is produced by nature at rates similar to those at which it is consumed by humans.

Reserve In the context of geologic resources, a deposit that has been identified (located) and is commercially extractable at with present prices and technology.

Residential mobility The movement of households from one place to another.

Respiration A chemical reaction that occurs in plants and animals in which carbohydrates and oxygen are combined, releasing water, carbon dioxide, and heat.

Ridge tillage System of planting crops on ridge tops in order to reduce farm production costs and promote greater soil conservation.

Right-to-work state A U.S. state that has passed a law preventing a union and company from negotiating a contract that requires workers to join a union as a condition of employment.

Runoff Flow of water from the land, either on the soil surface or in streams.

Scale The relationship between the portion of Earth being studied and Earth as a whole.

Sea level The general elevation of the sea surface, averaging out variations caused by waves, storms, and tides.

Seasonal migrants Those who move in response to regular but temporary conditions, for example, to work jobs at harvest time or to escape cold winters.

Secondary sector The portion of the economy concerned with manufacturing useful products through processing, transforming, and assembling raw materials.

Sect (of a religion) A relatively small group that has broken away from an established denomination.

Sector model A model of the internal structure of cities in which social groups are arranged around a series of sectors, or wedges, radiating out from the central business district (CBD).

Sediment transport The movement of rock particles by surface erosional processes.

Sedimentary rock Rock formed through accumulation and fusing of many small rock fragments at Earth's surface.

Selective immigration policies Government rules to include some migrants and exclude others.

Semiarid climate A climate with precipitation slightly less than potential evapotranspiration for most of the year.

Sensible heat Heat detectable by sense of touch, or with a thermometer.

Service Any activity that fulfills a human want or need and returns money to those who provide it.

Settler migration Individuals and households that migrate to new colonies.

Shield The ancient core of a continent.

Shield volcano A volcano with relatively gentle slopes formed by eruption of relatively fluid lavas.

Shifting cultivation A form of subsistence agriculture in which people shift activity from one field to another; each field is used for crops for a relatively few years and left fallow for a relatively long period.

Shortwave energy Radiant energy emitted by the Sun in wavelengths about 0.2 to 5.0 microns.

Site The physical character of a place.

Site factors Location factors related to the costs of factors of production inside the plant, such as land, labor, and capital.

Situation The location of a place relative to other places.

Situation factors Location factors related to the transportation of materials into and from a factory.

Slash-and-burn agriculture Another name for shifting cultivation, so named because fields are cleared by slashing the vegetation and burning the debris.

Social area analysis Statistical analysis used to identify where people of similar living standards, ethnic background, and lifestyle live within an urban area.

Soil A dynamic, porous layer of mineral and organic matter at Earth's surface.

Soil creep The slow downslope movement of soil caused by many individual, near-random particle movements such as those caused by burrowing animals or freeze and thaw.

Solar energy The radiant energy from the Sun.

Sovereignty A state's right to independently rule itself without interference from other states.

Space The physical gap or interval between two objects.

Space-time compression The reduction in the time it takes to diffuse something to a distant place, as a result of improved communications and transportation systems.

Spanglish Combination of Spanish and English, spoken by Hispanic Americans.

Spatial interaction The movement of physical processes, human activities, and ideas within and among regions.

Sprawl Development of new housing sites at relatively low density and at locations that are not contiguous to the existing built-up area.

Squatter settlement An area within a city in a developing country in which people illegally establish residences on land they do not own or rent and erect homemade structures.

State An area organized into a political unit and ruled by an established government with a resident population.

Stimulus diffusion The spread of an underlying principle, even though a specific characteristic is rejected.

Storm surge An area of elevated sea level in the center of a hurricane that may be several meters high, and which does most of the damage when a hurricane comes ashore.

Structural adjustment program Economic policies imposed on less developed countries by international agencies to create conditions encouraging international trade, such as raising taxes, reducing government spending, controlling inflation, selling publicly owned utilities to private corporations, and charging citizens more for services.

Subsistence agriculture Agriculture designed primarily to provide food for direct consumption by the farmer and the farmer's family.

Subtropical high-pressure (STH) zones Regions of high pressure and descending air at about 25° north and south latitudes.

Summer solstice For places in the Northern Hemisphere, June 20 or 21 is the date when at noon the Sun is directly overhead along the parallel of 23.5° north latitude; for places in the Southern Hemisphere, December 21 or 22 is the date when at noon the Sun is directly overhead at places along the parallel of 23.5° south latitude.

Sustainability/sustainable development Resource management and/or economic development that provides for the needs of the present without compromising future opportunities.

Sustainable agriculture Farming methods that preserve long-term productivity of land and minimize pollution, typically by rotating soil-restoring crops with cash crops and reducing inputs of fertilizer and pesticides.

Sustained yield A way of managing a renewable natural resource such that harvest can continue indefinitely.

Swidden A patch of land cleared for planting through slashing and burning.

Taboo A restriction on behavior imposed by social custom.

Tectonic plates Large pieces of Earth's crust that move relative to one another.

Temperature inversion A layer in the atmosphere in which warmer air lies above cooler air, inhibiting vertical circulation.

Temporary labor migrants Migrants looking for work but not permanently migrate.

Territorial waters Area of the ocean within 12 nautical miles of a state's shore over which it has full sovereign authority.

Territory The physical space of a political unit. For states, this includes land, subsoil, waters, and airspace.

Terroir French term for the contribution of a location's distinctive physical features to the way food tastes, similar to the English expressions "grounded" or "sense of place."

Terrorism Systematic use of violence by a group in order to intimidate a population or coerce a government into granting its demands.

Tertiary sector The portion of the economy concerned with transportation, communications, and utilities, sometimes extended to the provision of all goods and services to people in exchange for payment.

Threshold The minimum number of people needed to support the service.

Toponym The name given to a portion of Earth's surface.

Tornado A rapidly rotating column of air usually associated with a thunderstorm, often having winds in excess of 300 kilometers/hour (185 miles/hour).

Total fertility rate (TFR) The average number of children a woman will have throughout her childbearing years.

Trade wind The prevailing wind in subtropical and tropical latitudes that blows toward the Intertropical Convergence Zone, typically from the northeast in the Northern Hemisphere and from the southeast in the Southern Hemisphere.

Transform plate boundary A boundary between tectonic plates in which the two plates pass one another in a direction parallel to the plate boundary.

Transnational corporation A company that conducts research, operates factories, and sells products in many countries, not just where its headquarters or shareholders are located.

Transpiration The use of water by plants, normally drawing it from the soil via their roots, evaporating it in their leaves, and releasing it to the atmosphere.

Trophic level A position in the food chain relative to other organisms, such as producer, herbivore, or carnivore.

Tropic of Cancer The parallel of 23.5° north latitude.

Tropic of Capricorn The parallel of 23.5° south latitude.

Truck farming Commercial gardening and fruit farming, so named because truck was a Middle English word meaning bartering or the exchange of commodities.

Tsunami An extremely long wave created by an underwater earthquake; the wave may travel hundreds of kilometers per hour.

Tundra A low, slow-growing vegetation type found in high-latitude and high-altitude conditions in which snow covers the ground most of the year.

Typhoon The name applied to a hurricane in the Pacific Ocean.

Undernourishment Dietary energy consumption that is continuously below the minimum requirement for maintaining a healthy life and carrying out light physical activity.

Undocumented immigrants Migrants who enter a country without fulfilling a country's legal requirements to do so.

Unitary state Internal organization of states keeping most powers for central government.

Universal jurisdiction A legal principle that gives all states the right to prosecute crimes in non-state spaces, for example piracy on the High Seas.

Universalizing religion A religion that attempts to appeal to all people, not just those living in a particular location.

Urbanized area In the United States, a central city plus its contiguous built-up suburbs.

Value added The gross value of the product minus the costs of raw materials and energy.

Vernacular region (or perceptual region) An area that people believe exists as part of their cultural identity.

Vernal (spring) equinox In the Northern Hemisphere March 20 or 21, one of two dates when at noon the perpendicular rays of the Sun strike the equator (the Sun is directly overhead along the equator).

Visa Permission to enter a country that is granted prior to or during arrival.

Volcano A vent in Earth's surface where magma erupts as lava.

Warm front A boundary formed when a warm air mass advances against a cooler one.

Wastewater Water that has been used for some purpose and is returned to the environment as a liquid.

Water vapor Water in the air in gaseous form.

Wavelength The distance between successive waves of radiant energy, or of successive waves on a water body.

Wet rice Rice planted on dryland in a nursery and then moved to a deliberately flooded field to promote growth.

Winter solstice For places in the Southern Hemisphere, June 20 or 21 is the date when at noon the Sun is directly overhead at places along the parallel of 23.5° north latitude; for places in the Northern Hemisphere, December 21 or 22 is the date when at noon the Sun is directly overhead at places along the parallel of 23.5° south latitude.

CREDITS

FM Half Title Page imagebroker/Alamy Title Page Arcaid Images/Alamy About the Authors Debra Bowles

Chapter 1 1.CO.MAIN Blaine Harrington III/Alamy 1.1.1 Blaine Harrington III/Alamy 1.1.2 Jon Arnold Images Ltd/Alamy 1.1.5 Jenny Matthews/Alamy 1.1.7 Jeremy Sutton-Hibbert/Alamy 1.2.1A Images & Stories/Alamy 1.2.1B Blickwinkel/Alamy 1.2.1C Pearson 1.2.2 North Wind Picture Archives/Alamy 1.2.3 World History Archive/Alamy 1.2.5 INTERFOTO/Alamy 1.2.4 Courtesy of the Library of Congress 1.5.1 Google, Inc. 1.5.2 Dennis MacDonald/Alamy 1.5.4 Google, Inc. 1.6.1 Robert Spencer/The New York Times 1.6.2 Jelle van der Wolf/Alamy 1.6.4 Google, Inc. 1.8.1 Ron Yue/Alamy 1.8.3A Kevin Foy/Alamy 1.8.3B Vario Images GmbH & Co.KG/Alamy 1.8.3C Robert Harding Picture Library Ltd/Alamy 1.8.3D Andrew Woodley/Alamy 1.8.3E Robert Harding Picture Library Ltd/Alamy 1.8.3F Andrew Melbourne/Alamy 1.9.3A TAO Images Limited/Alamy 1.9.3B Jeremy Hoare/Alamy 1.10.4 Vario images GmbH & Co.KG/Alamy 1.11.1 Peter Arnold, Inc./Alamy 1.11.2 Imagebroker/Alamy 1.11.3 WILDLIFE/T.Dressler/Still Pictures/Specialist Stock 1.11.5 Balthasar Thomass/Alamy 1.12.1 Picture Contact BV/Alamy 1.12.2 UPPA/Photoshot 1.CR.1 Worldspec/NASA/Alamy 1.CR.2 Pearson/MapMaster 1.CR.3 Google, Inc. 1.CR.4 Google, Inc. 1.EOC.MAIN Gene Rhoden/Photolibrary.com

Chapter 2 2.CO.MAIN Gene Rhoden/Photolibrary 2.CO.1 Paul Souders/Alamy 2.CO.2 Sborisov/Fotolia 2.CO.3 Photolibrary.com 2.1.4 Sborisov/Fotolia 2.2.3 Itdarbs/Alamy 2.3.1a Jamie Marshall, Dorling Kindersley 2.3.1b Photolibrary.com/Getty Images 2.3.1c Peter M Corr/Alamy 2.4.4.1 Danita Delimont/Alamy 2.4.4.2 Isoft/iStockphoto.com 2.5.2 Jesse Allen, Robert Simmon, and the MODIS science team 2.5.3 Denise Dethlefsen/Alamy 2.6.4 John Henry Claude Wilson/Photolibrary.com 2.7.2 Courtesy NASA/NOAA & OceanRemote Sensing Group, Johns Hopkins Univ. Applied Physics Laboratory 2.7.4 CLAVER CARROLL/Photolibrary.com 2.9.1 Dorling Kindersley 2.9.2 NASA 2.9.3 Ryan McGinnis/Photolibrary.com 2.9.3 Ryan McGinnis/AGE/Photolibrary 2.9.5.1 Mike Hill/Alamy 2.9.5.2 Andrew Fox/Alamy 2.9.5.3 Gino's Premium Images/Alamy 2.10.1 Pavel Filatov/Alamy 2.10.3 Google, Inc. 2.11.1 Photolibrary.com 2.11.3 Sue Wilson/Alamy 2.11.4 F1online digitale Bildagentur GmbH/Alamy 2.11.5 Morales/Photolibrary.com 2.11.6 Martin Zwick/Photolibrary.com 2.12.2 Jeff Schultz/Photolibrary.com 2.13.1.1 Blickwinkel/Alamy 2.13.1.2 Brad Perks Lightscapes/Alamy 2.13.1.3 Angie Sharp/Alamy 2.13.1.4 DBURKE/Alamy 2.13.1.5 David Humphreys/Alamy 2.13.1.6 Dennis Hallinan/Alamy 2.13.1.7 Bailey-Cooper Photography/Alamy 2.13.2 Corbis Premium RF/Alamy 2.CR.1 David Woodfall/Photolibrary.com 2.EOC.1 Pearson 2.EOC.2 Pearson 2.EOC.MAIN Bob Gibbons/Photo Researchers, Inc. 2.12.2a Google, Inc. 2.12.2a Google, Inc.

Chapter 3 3.CO.MAIN Bob Gibbons/Photo Researchers, Inc. 3.CO.2 Jim Wark/Airphoto 3.CO.3 RANDY OLSON/National Geographic Stock 3.CO.4 © Radius Images/Glow Images 3.1.1B Mainichi Newspaper/Aflo/Newscom 3.1.2 AIR PHOTO SERVICE/AFLO/Newscom 3.1.3B Hs2/Newscom 3.2.3A Aerialarchives.com/Alamy 3.2.3B NASA/Alamy 3.3.1 Imago stock&people/Newscom 3.3.2 Imago stock&people/Newscom 3.4.ChartA Harry Taylor, Dorling Kindersley 3.4.ChartB Dave King, Dorling Kindersley 3.4.ChartC Gary Ombler, Dorling Kindersley 3.4.2B Skip Brown/National Geographic Stock 3.4.3B Jim Wark/Airphoto 3.4.4B Jim Wark/Airphoto 3.4.5 Google, Inc. 3.5.1 Imagebroker.net/SuperStock 3.5.2 Raymond Klass Danita Delimont Photography/Newscom 3.5.4 Jason Baxter/Alamy 3.6.1 Life File Photo Library Ltd/Alamy 3.6.2 Tom Till/AGEfotostock.com 3.6.3 Tom Meyers/Photo Researchers 3.6.4 Ricardo Funari/Specialist Stock 3.7.1A Sajith Sivasankaran/Alamy 3.7.1B Mike Goldwater/Alamy 3.7.1C Canadabrian/Alamy 3.7.2 USDA NRCS 3.8.2 Thomas & Pat Leeson/Photo Researchers, Inc. 3.8.4 Colin Underhill/Alamy 3.8.5 RANDY OLSON/National Geographic Stock 3.9.3 Julius Fekete/Shutterstock 3.9.4 Google.com 3.10.2 Radius Images/Glow Images 3.10.3 Google, Inc. 3.10.4 NASA 3.10.5 Norman Owen Tomalin/Alamy 3.11.2 Ashley Cooper pics/Alamy 3.11.3A USGS 3.11.3B USGS 3.11.3C BRUCE MOLNIA USGS/Still Pictures 3.12.1 Jim Wark/Airphoto 3.12.3A David Wall/DanitaDelimont.com "Danita Delimont Photography"/Newscom 3.12.3B Kathy Merrifield/Photo Researchers, Inc. 3.CR.1 z03/ZUMA Press/Newscom 3.EOC.1 Pearson 3.EOC.2 Google, Inc. 3.EOC.MAIN Tomas Sereda/Fotolia.com

Chapter 4 4.CO.MAIN Tomas Sereda /Fotolia.com 4.CO.2 AirPhotoNA 4.CO.3 USDA/Natural Resources Conservation Service 4.CO.4 Martin Ruegner/AGEfotostock.com 4.1.2 AirPhotoNA 4.2.3a NHPA/SuperStock 4.2.4b Gary Whitton/AGEfotostock.com 4.2.4a Curt Teich Postcard A/AGEfotostock.com 4.3.3 Ron Elmy/Alamy 4.5.3 Sinopictures/Still Pictures 4.6.1.1 Barrie Watts, Dorling Kindersley 4.6.1.2 Gaston Piccinetti/AGEfotostock.com 4.6.1.3 Christian Heinrich/AGEfotostock.com 4.6.1.4 Drake Fleege/Alamy 4.6.1.5 Phillip Dowell, Dorling Kindersley 4.6.1.6 Phillip Dowell, Dorling Kindersley 4.7.1 Matthew Ward, Dorling Kindersley 4.7.2a USDA/Natural Resources Conservation Service 4.7.2b USDA/Natural Resources Conservation Service 4.7.2c Peter Andersen, Dorling Kindersley 4.7.2d USDA/Natural Resources Conservation Service 4.7.2e USDA/Natural Resources Conservation Service 4.9.2 M Lohmann/AGEfotostock.com 4.9.3 John E Marriott/AGEfotostock.com 4.9.4 Art Wolfe/Photo Researchers, Inc. 4.9.5 Martin Ruegner/AGEfotostock.com 4.9.6 FLPA/Robin Chittenden/AGEfotostock.com 4.9.7 TED MEAD/Photolibrary.com 4.9.8 Boyd E. Norton/Photo Researchers, Inc. 4.9.9 Robert Harding Picture Library/AGEfotostock.com 4.10.1 Google, Inc. 4.10.2a USDA/NRCS 4.10.2b Preble County Auditor 4.10.3 Tim McCabe /USDA/NRCS 4.CR.1 AirPhotoNA 4.EOC.1 Pearson 4.EOC.2 Pearson 4.EOC.MAIN ERProductions Ltd/Photolibrary.com

Chapter 5 5.CO.MAIN ERProductions Ltd/Photolibrary.com 5.CO.A Bertrand Rieger/Hemis/Photoshot 5.CO.B David R. Frazier Photolibrary, Inc./Alamy 5.CO.C Jeremy sutton-hibbert/Alamy 5.1 Sue Cunningham Photographic/Alamy 5.1 Frans Lemmens/Alamy 5.2.2 Bertrand Rieger/Hemis/Photoshot 5.2.4 Pearson 5.3.3 Penny Tweedie/Alamy 5.5.1 Jake Lyell/Alamy 5.5.2 Renato Bordoni/Alamy 5.5.3 David R. Frazier Photolibrary, Inc./Alamy 5.5.4 LOOK die Bildagentur der Fotografen GmbH/Alamy 5.6.4 Jeremy sutton-hibbert/Alamy 5.7.5 Art Directors & TRIP/Alamy 5.7.6 Alain Le Garsmeur/Photolibrary.com 5.8.2 Neil Emmerson/Photolibrary.com 5.8.3 Yvan TRAVERT/Photolibrary.com 5.9.1 Wissam Al-Okaili/AFP/Getty Images 5.9.2 Eye Ubiquitous/Alamy 5.9.3b Pearson 5.CR.1 Bertrand Rieger/Hemis/Photoshot 5.CR.2 Pearson 5.CR.3 Pearson 5.EOC.MAIN AP Photo/Arturo Rodriguez/FILE

Chapter 6 6.CO.MAIN AP Photo/Arturo Rodriguez/FILE 6.CO.3 Blickwinkel/Alamy 6.CO.4 BrandX Pictures/Photolibrary.com 6.CO.5 REUTERS/Ammar Awad 6.1.2 Renata Caland/Alamy 6.1.3a HO/AFP/Getty Images/Newscom 6.1.3b Pablo Fonseca Q./La Nacion de Costa Rica/Newscom 6.1.4 Pascal Goetgheluck/Photo Researchers, Inc. 6.2.2 Library of Congress 6.3.1 Images of Africa Photobank/Alamy 6.3.3 Janine Wiedel Photolibrary/Alamy 6.4.2 Alan Gignoux/Alamy 6.4.3 Pearson 6.4.4 Alain Le Bot/Photononstop/Photolibrary.com 6.5.2a Boris Heger/Photolibrary.com 6.5.2b Ton Koene/Photolibrary.com 6.5.2c Blickwinkel/Alamy 6.5.2d Ton Koene/Photolibrary.com 6.5.2e Ton Koene/Photolibrary.com 6.5.2f Sean Sprague/Photolibrary.com 6.6.2 Leonid Serebrennikov/Photolibrary.com 6.6.3 Thomas Hoepker/Magnum Photos 6.6.4 K Wothe/Agefotostock.com 6.7.2 Pearson 6.7.4 BrandX Pictures/Photolibrary.com 6.8.2 World History Archive/Alamy 6.8.3 Library of Congress 6.8.4a Mark Edwards/Photolibrary.com 6.8.4b David Grossman/Alamy 6.9.2 Pearson 6.9.3 Aerial Archives/Alamy 6.9.4 Google, Inc. 6.9.5 Robert Harding Picture Library Ltd/Alamy 6.9.6 Jim Wark/Photolibrary.com 6.9.7 Sean Sprague/Photolibrary.com 6.10.1 REUTERS/Ammar Awad 6.10.3 REUTERS/Danny Moloshok 6.CR.1 Jim West/ZUMA Press/Newscom 6.EOC.1 Pearson 6.EOC.2 Pearson 6.EOC.MAIN Pietro Scozzari/Agefotostock.com

Chapter 7 7.CO.MAIN Pietro Scozzari/AGEfotostock.com 7.CO.C Peter M. Wilson/Alamy 7.CO.B Erwin Gavic, autosnelwegen.net 7.CO.D Israel images/Alamy 7.4.3 CandyAppleRed Images/Alamy 7.4.4 Picture Contact BV/Alamy 7.4.5 Jeff Morgan 01/Alamy 7.5.1 Andre Jenny/Alamy 7.5.3 David R. Frazier Photolibrary, Inc./Alamy 7.5.4 CulturalEyes-AusSoc/Alamy 7.6.1 Erwin Gavic, autosnelwegen.net 7.6.4 swissworld.org 7.6.6 Stephen Rees/IStockphoto 7.7.2 Ray Roberts/Alamy 7.7.3 Michael Ventura/Alamy 7.8.1a FALKENSTEINFOTO/Alamy 7.8.1b David Lyons/Alamy 7.8.1c Peter Barritt/Alamy 7.8.2a Luminous/Alamy 7.8.2b Peter M. Wilson/Alamy 7.8.3a Alan Novelli/Alamy 7.8.3b Imagebroker/Alamy 7.8.4 Jim Zuckerman/Alamy 7.9.1 Aurora Photos/Alamy 7.9.2 VojtechVlk/Shutterstock 7.9.3 JTB Photo/Photolibrary 7.9.4 Google, Inc. Earth 7.10.3 Eye Ubiquitous/Photolibrary.com 7.10.4 Iain Lowson/Alamy 7.11.1 Courtesy of Library of Congress 7.11.2 ArkReligion.com/Alamy 7.11.4 Dallas and John Heaton/Photolibrary.com 7.11.3B Linda Whitwam (c) Linda Whitwam, Dorling Kindersley 7.11.5 Dinodia Images/Alamy 7.12.1 Dan Porges/Peter Arnold Inc 7.12.2 Zou Yanju/Photolibrary.com 7.12.3 Robert Estall photo agency/Alamy 7.12.4 Raghu Rai/Magnum Photos, Inc. 7.13.2 Israel images/Alamy 7.Internet 1 Ethnologue.com 7.Internet 2 adherents.com 7.Internet 3 glenmary.org 7.CR.1 REUTERS/Mathieu Belanger 7.CR.2 Pearson/MapMaster 7.CR.3 Pearson 7.EOC.MAIN Gary Moon/Photolibrary.com 7.EOC.MAIN Mark Henley/agefotostock.com

Chapter 8 8.CO.MAIN Mark Henley/agefotostock.com 8.CO.MAIN Gary Moon/Photolibrary.com 8.EOC.MAIN Neil Cooper/Alamy 8.CO.TOP LEFT Bill Bachmann/Alamy 8.CO.TOP CENTER Jeff Greenberg/Alamy 8.CO.TOP RIGHT Trinity Mirror/Mirrorpix/Alamy 8.1.2 Bill Bachmann/Alamy 8.2.3 Google, Inc. 8.4.1a Wim van Cappellen/Photolibrary/Getty Images 8.4.1b David R. Frazier Photolibrary, Inc./Alamy 8.4.1c Wolfgang Rattay/Reuters Pictures 8.4.3 Dbimages/Alamy 8.5.3 Philippe de Poulpiquet/PHOTOPQR/LE PARISIEN/Newscom 8.5.5 Chromorange/Alamy 8.5.6 Jeff Greenberg/Alamy 8.7.1 Robert Francis/Photolibrary/Getty Images 8.7.2 Garry Black/Alamy 8.7.3 Mike Goldwater/Alamy 8.7.4 Gavin Hellier/Photolibrary/Getty Images 8.7.5 REUTERS/Ranko Cukovic 8.8.1 Imagestate Media Partners Limited - Impact Photos/Alamy 8.8.3 Movementway Movementway/Photolibrary.com 8.8.6 Picture Contact BV/Alamy 8.9.2b PCL/Alamy 8.9.4 United States Department of Defense 8.10.1 Michael Runkel/AGEfotostock.com 8.10.2 RichardBakerSudan/Alamy 8.10.3 Wim van Cappellen/Photolibrary 8.11.1 Laperruque/Alamy 8.CR.1 AP Photo/Roshan Mughal 8.EOC.2 Pearson 8.EOC.3 Pearson

Chapter 9 9.CO.MAIN Neil Cooper/Alamy 9.CO.A JORGEN SCHYTTE/Photolibrary.com 9.CO.B Agencja FREE/Alamy 9.CO.C AP Photo/Beth A. Keiser 9.CO.D Steven May/Alamy 9.2.2 APA APA/ Photolibrary.com 9.2.4 H. Mark Weidman Photography/Alamy 9.3.3 MapMaster/Pearson 9.3.5 Agencja FREE/Alamy 9.4.4 Design Pics Inc. - RM Content/Alamy 9.4.7 Picture Contact BV/Alamy 9.5.2 Jeff Morgan 07/Alamy 9.5.4 Imagebroker/Alamy 9.5.5 Kablonk!/Photolibrary.com 9.6.1 REUTERS/Ajay Verma 9.6.2 Robert Harding Picture Library Ltd/Alamy 9.7.1 MapMaster/Pearson 9.7.2a PeerPoint/Alamy 9.7.2b AP Photo/Beth A. Keiser 9.7.3 Eye Ubiquitous/Photolibrary.com 9.7.4 Wim van Cappellen/Photolibrary.com 9.7.6 Jim Holmes/Photolibrary.com 9.8.3 Yobidaba/Shutterstock.com 9.8.5 Google, Inc. 9.9.2 JOERG BOETHLING/Photolibrary.com 9.9.3 GIL MOTI/Photolibrary.com 9.10.6 Robert Harding World Imagery/Alamy 9.10.7 Visions LLC/Photolibrary.com 9.CR.1 JOHN G. MABANGLO/AFP/Newscom 9.CR.2 MapMaster/Pearson 9.CR.3 Google, Inc. 9.EOC.MAIN JOERG BOETHLING/Specialist Stock

Chapter 10 10.CO.MAIN Neil Emmerson/Photolibrary.com 10.CO.2 MAISANT Ludovic/Photolibrary.com 10.CO.3 JOERG BOETHLING/Photolibrary.com 10.1.1 Nigel Pavitt/Photolibrary.com 10.1.4 Craig Lovell/Eagle Visions Photography/Alamy 10.2.2 MAISANT Ludovic/Photolibrary.com 10.2.3 Guenter Fischer/Imagebroker.net/Photolibrary.com 10.2.4 Nigel Pavitt/John Warburton-Lee Photography/Photolibrary.com 10.2.5 Eye Ubiquitous/Photolibrary.com 10.3.1 Fischer B/Photolibrary.com 10.3.4 Philippe Giraud/Photolibrary.com 10.4.2 Jeff Greenberg/Alamy 10.4.7 Steve Morgan/Photolibrary.com 10.5.2 Lite Productions/Glow Images RF/Photolibrary.com 10.5.5 JAUBERT IMAGES/Alamy 10.6.3 Lite Productions/Glow Images RF/Photolibrary.com 10.7.1 Jacques Jangoux/Alamy 10.7.2 Dmytro Korolov/

IStockphoto.com 10.7.4 Google, Inc. 10.7.5 Sopose/Shutterstock.com 10.8.2 Imagebroker/Alamy 10.8.7 Inga spence/Alamy 10.9.2 Mark Edwards/Still Pictures/Photolibrary.com 10.9.6 David R. Frazier Photolibrary, Inc./Alamy 10.9.8 MIXA Co. Ltd./Photolibrary.com 10.10.2 JOERG BOETHLING/Still Pictures/Photolibrary.com 10.10.3 Lana Sundman/Alamy 10.11.2 Inga Spence/Alamy 10.11.4A Malcolm Case-Green/Alamy 10.11.4B Justin Kase zsixz/Alamy 10.11.4C Nigel Cattlin/Alamy 10.11.4D DWP Imaging/Alamy 10.11.4E Stock Connection Blue/Alamy 10.12.3 Mar Photographics/Alamy 10.CR.1 Steve McCurry/Magnum Photos 10.EOC.1 Pearson 10.EOC.2 Google, Inc. 10.EOC.MAIN David Lyons/Alamy

Chapter 11 11.CO.MAIN Eye Ubiquitous/Photolibrary.com 11.CO.2 LHB Photo/Alamy 11.CO.3 Melba/Photolibrary.com 11.1.1A The Print Collector/Alamy 11.1.1B Image Works/Mary Evans Picture Library Ltd 11.1.2 Dave King, Dorling Kindersley, Courtesy of The Science Museum, London 11.1.3 The Art Gallery Collection/Alamy 11.2.2 blickwinkel/Alamy 11.3.1 Google, Inc. 11.3.2 Pearson 11.3.3 AP Photo/Mark Duncan 11.3.4 Huw Jones/Alamy 11.4.2 XenLights/Alamy 11.4.3 CNS/Photolibrary.com 11.5.2 Jim West/Alamy 11.5.4a Courtesy Ford/ZUMA Press/Newscom 11.5.4b Toyota/ZUMA Press/Newscom 11.5.4c Courtesy of Honda/ZUMA Press/Newscom 11.5.4d s06/ZUMA Press/Newscom 11.5.4e eVox Productions LLC/Newscom 11.6.1 ACE STOCK LIMITED/Alamy 11.6.3 LHB Photo/Alamy 11.6.4 Aurora Photos/Alamy 11.6.5 Images-USA/Alamy 11.7.1 Eye Ubiquitous/Glow Images 11.7.3A D.Hurst/Alamy 11.7.3C REUTERS/Ho New 11.7.4 Melba Melba/Photolibrary.com 11.7.5 Aerial Archives/Alamy 11.8.1 Joerg Boethling/Alamy 11.8.4 Wang Dingchang/Photoshot 11.9.5 Marcelo Rudini/Alamy 11.CR.1 Google, Inc. 11.CR.2 Pearson 11.CR.3 Google, Inc. 11.EOC.MAIN Kay Maeritz/AGEfotostock.com

Chapter 12 12.CO.MAIN Kay Maeritz/AGEfotostock.com 12.CO.1 Alex Segre/Alamy 12.CO.2 REUTERS/Sherwin Crasto 12.CO.3 AfriPics.com/Alamy 12.1.6 Jim West/Photolibrary.com 12.2.1A Mechika/Alamy 12.2.1B JG Photography/Alamy 12.2.1C Alex Segre/Alamy 12.2.1D Myrleen Pearson/Alamy 12.2.1E Spencer Grant/Alamy 12.3.4A Radius Images/Alamy 12.3.4B Walter Bibikow/Photolibrary.com 12.4a Michael Neelon(misc)/Alamy 12.4B Bob Pardue/Alamy 12.4C Tom Carter/Alamy 12.5.3A Chad Ehlers/Alamy 12.5.3B PCL/Alamy 12.5.3C noel moore/Alamy 12.6.2 Jan Greune/Photolibrary.com 12.6.3 REUTERS/Sherwin Crasto 12.6.4 Katherine S Miles/Alamy 12.7.1 Pearson 12.7.3 Hendrik Holler/Photolibrary.com 12.8.1 National Geographic Image Collection/Alamy 12.8.2 Pearson 12.8.3 Pearson 12.8.4 AfriPics.com/Alamy 12.9.2 Travelshots.com/Alamy 12.9.4 Ingolf Pompe 55/Alamy 12.9.5 ClassicStock/Alamy 12.10.4 Jeremy Woodhouse/Photolibrary.com 12.CR.1 Peter Adams Photography Ltd/Alamy 12.EOC.1 Pearson 12.EOC.2 Google, Inc. 12.EOC.MAIN BrandX Pictures/Photolibrary.com

Chapter 13 13.CO.MAIN BrandX Pictures/Photolibrary.com 13.CO.2 Mark Edwards/Photolibrary.com 13.CO.4 Tom Uhlman/Alamy 13.1.2 Andre Jenny/Alamy 13.1.3 Andre Jenny/Alamy 13.1.4 Andre Jenny/Alamy 13.1.5 Vespasian/Alamy 13.2.2 SUNNYPhotography.com/Alamy 13.2.3 David R. Frazier Photolibrary, Inc./Alamy 13.2.5 Chad Ehlers/Alamy 13.2.7 Steve Dunwell/Photolibrary.com 13.2.8 Megapress/Alamy 13.3.1 Richard Stockton/Photolibrary.com 13.3.2b David Hodges/Alamy 13.3.2c Typhoonski/Dreamstime.com 13.3.3b Amar and Isabelle Guillen - Guillen Photography/Alamy 13.3.3c Corbis/Photolibrary.com 13.3.4b Dionne McGill/Alamy 13.3.4c Lorie Leigh Lawrence/Alamy 13.4.1 Hemis/Alamy 13.4.2 Universal History Archive/Photolibrary.com 13.4.3 JTB Photo/Photolibrary.com 13.4.4 GARDEL Bertrand/Photolibrary.com 13.4.4 Directphoto.org/Alamy 13.4.5c Directphoto.bz/Alamy 13.4.6b Paris Street/Alamy 13.4.6c Bernard Rouffignac/Photolibrary.com 13.4.7 Directphoto.org/Alamy 13.5.1a DEA/G DAGLI ORTI/Photolibrary.com 13.5.1b DEA/G DAGLI ORTI/Agefotostock.com 13.5.1c Ken Welsh/Photolibrary.com 13.5.5 aerialarchives.com/Alamy 13.5.5 Danita Delimont/Alamy 13.6.3 Pearson 13.7.3 Marvin Dembinsky Photo Associates/Alamy 13.7.4 Lee Madden/Splash News/Splash News/Newscom 13.8.2 Tom Uhlman/Alamy 13.8.3 David R. Gee/Alamy 13.8.3a D Hale-Sutton/Alamy 13.8.4 Google, Inc. 13.9.2 B.A.E. Inc./Alamy 13.9.2 Chad Ehlers/Photolibrary.com 13.9.4 SG cityscapes/Alamy 13.10.1 Aerial Archives/Alamy 13.10.2 SHOUT/Alamy 13.10.3 Jose Carlos Fajardo/ZUMA Press/Newscom 13.10.4 Robert Clay/Photolibrary.com 13.10.5 David Muscroft/Photolibrary.com 13.CR.1 Kordcom/Photolibrary.com 13.EOC.1 Pearson 13.EOC.2 Google, Inc. 13.EOC.MAIN Imagebroker/Alamy

Chapter 14 14.CO.MAIN Imagebroker/Alamy 14.CO.2 All Canada Photos/Alamy 14.CO.3 REUTERS/Stringer Shanghai 14.CO.4 Robert Harding Picture Library Ltd/Alamy 14.1.1a Ernst Klinker/Alamy 14.1.1b Jordan Siemens/Alamy 14.1.2 Jim Wark/Photolibrary.com 14.2.1 Chad Ehlers/Alamy 14.2.2 Third Millenium Alliance 14.2.3a All Canada Photos/Alamy 14.2.3b Stuart Forster/Alamy 14.3.2a Mackertich/Dowse/Alamy 14.3.2b Prisma Bildagentur AG/Alamy 14.5.2 AFP/Getty Images/Newscom 14.5.4 Lite Productions/Photolibrary.com 14.6.3 REUTERS/Stringer Shanghai 14.7.1 Pearson 14.7.2 REUTERS/Stringer Shanghai 14.7.3 Google, Inc. 14.8.1 Ian Lamond/Alamy 14.8.2 REUTERS/Stringer Shanghai 14.9.3 Janusz Wrobel/Alamy 14.10.3 Google, Inc. 14.10.4 Worldwide Picture Library/Alamy 14.11.5 Robert Harding Picture Library Ltd/Alamy 14.CR.1 REUTERS/Albert Gea 14.EOC.1 Pearson 14.EOC.2 Google, Inc. 14.EOC.3 Google, Inc.

Cover COV Caro/Alamy

Index

D

V

W

Y

Z